D0548674

Annual Review of Biophysics

Editorial Committee (2011)

Ken A. Dill, University of California, San Francisco
Stuart G. McLaughlin, State University of New York, Stony Brook
Rob Phillips, California Institute of Technology
Douglas C. Rees, California Institute of Technology
Viola Vogel, ETH Hönggerberg
James R. Williamson, The Scripps Research Institute
Cynthia Wolberger, Johns Hopkins University School of Medicine
Xiaowei Zhuang, Harvard University

**Responsible for the Organization of Volume 40
(Editorial Committee, 2009)**

Ken A. Dill
Stuart G. McLaughlin
Rob Phillips
Douglas C. Rees
Michael P. Sheetz
Viola Vogel
James R. Williamson
Cynthia Wolberger
Xiaowei Zhuang
Frederick Lanni (Guest)

Production Editor: Cleo X. Ray
Bibliographic Quality Control: Mary A. Glass
Electronic Content Coordinator: Suzanne K. Moses
Illustration Editor: Douglas Beckner

Annual Review of Biophysics

Volume 40, 2011

Douglas C. Rees, *Editor*
California Institute of Technology

Ken A. Dill, *Associate Editor*
University of California, San Francisco

James R. Williamson, *Associate Editor*
The Scripps Research Institute

www.annualreviews.org • science@annualreviews.org • 650-493-4400

Annual Reviews
4139 El Camino Way • P.O. Box 10139 • Palo Alto, California 94303-0139

Annual Reviews
Palo Alto, California, USA

COPYRIGHT © 2011 BY ANNUAL REVIEWS, PALO ALTO, CALIFORNIA, USA. ALL RIGHTS RESERVED. The appearance of the code at the bottom of the first page of an article in this serial indicates the copyright owner's consent that copies of the article may be made for personal or internal use, or for the personal or internal use of specific clients. This consent is given on the condition that the copier pay the stated per-copy fee of $20.00 per article through the Copyright Clearance Center, Inc. (222 Rosewood Drive, Danvers, MA 01923) for copying beyond that permitted by Section 107 or 108 of the U.S. Copyright Law. The per-copy fee of $20.00 per article also applies to the copying, under the stated conditions, of articles published in any *Annual Review* serial before January 1, 1978. Individual readers, and nonprofit libraries acting for them, are permitted to make a single copy of an article without charge for use in research or teaching. This consent does not extend to other kinds of copying, such as copying for general distribution, for advertising or promotional purposes, for creating new collective works, or for resale. For such uses, written permission is required. Write to Permissions Dept., Annual Reviews, 4139 El Camino Way, P.O. Box 10139, Palo Alto, CA 94303-0139 USA.

International Standard Serial Number: 1936-122X
International Standard Book Number: 978-0-8243-1840-6
Library of Congress Catalog Card Number: 79-188446

All Annual Reviews and publication titles are registered trademarks of Annual Reviews.

⊗ The paper used in this publication meets the minimum requirements of American National Standards for Information Sciences—Permanence of Paper for Printed Library Materials, ANSI Z39.48-1992.

Annual Reviews and the Editors of its publications assume no responsibility for the statements expressed by the contributors to this *Annual Review*.

TYPESET BY APTARA
PRINTED AND BOUND BY FRIESENS CORPORATION, ALTONA, MANITOBA, CANADA

Contents

Annual Review of
Biophysics

Volume 40, 2011

Index

Errata

An online log of corrections to *Annual Review of Biophysics* articles may be found at
http://biophys.annualreviews.org/errata.shtml

Related Articles

Annual Reviews is a nonprofit scientific publisher established to promote the advancement of the sciences. Beginning in 1932 with the *Annual Review of Biochemistry*, the Company has pursued as its principal function the publication of high-quality, reasonably priced *Annual Review* volumes. The volumes are organized by Editors and Editorial Committees who invite qualified authors to contribute critical articles reviewing significant developments within each major discipline. The Editor-in-Chief invites those interested in serving as future Editorial Committee members to communicate directly with him. Annual Reviews is administered by a Board of Directors, whose members serve without compensation.

2011 Board of Directors, Annual Reviews

Richard N. Zare, *Chairperson of Annual Reviews, Marguerite Blake Wilbur Professor of Natural Science, Department of Chemistry, Stanford University*

Karen S. Cook, *Vice-Chairperson of Annual Reviews, Director of the Institute for Research in the Social Sciences, Stanford University*

Sandra M. Faber, *Vice-Chairperson of Annual Reviews, Professor of Astronomy and Astronomer at Lick Observatory, University of California at Santa Cruz*

John I. Brauman, *J.G. Jackson-C.J. Wood Professor of Chemistry, Stanford University*

Peter F. Carpenter, *Founder, Mission and Values Institute, Atherton, California*

Susan T. Fiske, *Eugene Higgins Professor of Psychology, Princeton University*

Eugene Garfield, *Emeritus Publisher,* The Scientist

Samuel Gubins, *President and Editor-in-Chief, Annual Reviews*

Steven E. Hyman, *Provost, Harvard University*

Roger D. Kornberg, *Professor of Structural Biology, Stanford University School of Medicine*

Sharon R. Long, *Wm. Steere-Pfizer Professor of Biological Sciences, Stanford University*

J. Boyce Nute, *Palo Alto, California*

Michael E. Peskin, *Professor of Particle Physics and Astrophysics, SLAC, Stanford University*

Harriet A. Zuckerman, *Senior Vice President, The Andrew W. Mellon Foundation*

Management of Annual Reviews

Samuel Gubins, President and Editor-in-Chief
Paul J. Calvi Jr., Director of Technology
Steven J. Castro, Chief Financial Officer and Director of Marketing & Sales
Jennifer L. Jongsma, Director of Production
Jeanne M. Kunz, Corporate Secretary
Jada Pimentel, Human Resources Manager

Annual Reviews of

Analytical Chemistry
Anthropology
Astronomy and Astrophysics
Biochemistry
Biomedical Engineering
Biophysics
Cell and Developmental
  Biology
Chemical and Biomolecular
  Engineering
Clinical Psychology
Condensed Matter Physics
Earth and Planetary Sciences
Ecology, Evolution, and
  Systematics
Economics

Entomology
Environment and Resources
Financial Economics
Fluid Mechanics
Food Science and Technology
Genetics
Genomics and Human Genetics
Immunology
Law and Social Science
Marine Science
Materials Research
Medicine
Microbiology
Neuroscience
Nuclear and Particle Science
Nutrition

Pathology: Mechanisms
  of Disease
Pharmacology and Toxicology
Physical Chemistry
Physiology
Phytopathology
Plant Biology
Political Science
Psychology
Public Health
Resource Economics
Sociology

SPECIAL PUBLICATIONS
Excitement and Fascination of
  Science, Vols. 1, 2, 3, and 4

Harold A. Scheraga

Respice, Adspice, and Prospice

Harold A. Scheraga

Baker Laboratory of Chemistry and Chemical Biology, Cornell University, Ithaca, New York 14853-1301; email: has5@cornell.edu

Annu. Rev. Biophys. 2011. 40:1–39

The *Annual Review of Biophysics* is online at biophys.annualreviews.org

This article's doi:
10.1146/annurev-biophys-042910-155334

Copyright © 2011 by Annual Reviews.
All rights reserved

1936-122X/11/0609-0001$20.00

Keywords

blood clotting, hydrodynamics, hydrogen bonding, hydrophobic interactions, protein folding theory, ribonuclease folding, molecular dynamics

Abstract

The title, "Look to the past, Look to the present, and Look to the future," the motto of City College of New York, expresses how my family life and education led me to an academic career in physical chemistry and ultimately to a study of proteins. The economic depression of the 1930s left a lasting impression on my outlook and career aspirations. With fortunate experiences at several stages in my life, I was able to participate in the great adventure of the last half of the twentieth century: the revolution in biology that advanced the field of protein chemistry to so great an extent. The future is bright and limitless, with greater understanding of biology yet to come.

Contents

MONTICELLO

I was born in Brooklyn, New York, on October 18, 1921. My father, Samuel, came to the United States from Romania with his family when he was three and grew up in the Brownsville section of Brooklyn. My mother, Etta, was a first-generation American whose immigrant parents had come from the Ukraine. The early part of her life was spent on the Lower East Side of Manhattan but then later also in Brownsville. My parents married in Brooklyn and shortly after that they moved to Monticello in upstate New York. I was born in my maternal grandparents' home in Brooklyn because my mother was visiting her parents there at that time. Within a month, I was brought to Monticello, where I lived from 1921 to 1930, after which we moved from Monticello to Brooklyn.

My father was a machinist and opened a store on Broadway in Monticello, selling radios and musical instruments. But, in the 1929 economic crash, he lost the business, and we moved back to Brooklyn, where my father was frequently unemployed during the depression of the 1930s. He finally obtained a position as a machinist in the late 1930s before the start of World War II and retired at age 65. I had wanted to drop out of school and help support the family during the depression, but my father strongly insisted against it. I'm forever grateful to him because my career would surely have been derailed if I had left school. My parents instilled a set of values, including a love of education, in my brothers and me. Both parents lived until the age of 94 and died within 10 days of each other.

I entered public school at the age of five and also attended a Hebrew school on weekday afternoons and on Sunday mornings. I had almost four years of schooling in Monticello and continued my education in elementary and high school (and in Hebrew school until my bar mitzvah) in Brooklyn.

BROOKLYN: ELEMENTARY SCHOOL AND HIGH SCHOOL

My elementary school in Brooklyn was P.S. 114, in the Canarsie area. I was there for two years, through the sixth grade. I went to John Marshall Junior High, P.S. 210, in another section of Brooklyn and then to Boys High School.

In junior high, I was very much interested in Latin and was awarded the Latin medal when I graduated in 1934. In high school, I began to develop an interest not only in classics, but also in mathematics. My Latin and mathematics teachers at Boys High were excellent. Also, in high school, I was coeditor-in-chief, with the late Isidore Diamond, of a semiannual Latin periodical, *Forum Latinum*.

In the 1930s, the *New York Herald Tribune* sponsored a weekly essay contest for high school students to write about important current events. In 1936, Franklin Roosevelt won a landslide re-election victory, carrying every state in the union except Maine and Vermont. I wrote an essay about the significance of this victory and its implications for the New Deal and won seventh prize that week.

My interest in mathematics continued to grow at Boys High, where Isidore Diamond and I were also on a math team that competed with teams from other high schools under the auspices of the Interscholastic Algebra League (IAL). During the summer after I graduated from high school in 1937, I studied math on my own. I taught myself analytical geometry and differential calculus. After that summer, I enrolled in City College of New York (CCNY).

Two years of science were required for college entrance. I had taken introductory biology in junior high and a very dull chemistry course in high school. Unfortunately, I was never exposed to physics in high school. I was clearly headed for a career in classics, but with a strong interest in mathematics.

Both of my younger brothers went to college; Morton went to CCNY and became an electrical engineer, and David worked his way through Cornell University to become a mechanical engineer. Neither my mother nor her brothers or sisters attended college; my father had a few years of college at Cooper Union. His twin brother, Morris, went to the College of Veterinary Medicine at Cornell and became a microbiologist. When he left Cornell, he joined the faculty at the University of Kentucky and for many years was head of the Department of Bacteriology, which later became microbiology. Morris was the only college-educated person in the immediate family, and he clearly was a role model for me when I was growing up. Inspired by him, I wanted to go to Cornell but that was impossible because it was too expensive, even with a partial scholarship. Later, when I was on the Cornell faculty, I used to meet my uncle almost every year at the FASEB meetings.

During my high school days, I was very much interested in sports. With my friends in the neighborhood, I played punch ball in the street and handball, baseball, and touch football in the schoolyard, whatever the season. I had aspirations for trying out for the baseball team at Boys High, but unfortunately I was too short.

I was (and still am) a voracious reader of fiction and nonfiction, primarily biographies. I also played chess, especially during the summer with Ruby Schaeffer, who later retired from the Bureau of Standards, and Cy Sroog, who later retired from Du Pont. Through a cultural club to which I belonged in college, I played tennis, which I kept up for many years. One of my mother's brothers was a golfer and gave me golf lessons. I couldn't afford the greens' fees until I joined the Cornell faculty, when I resumed golf lessons in order to play golf with my son, and later with his son.

As for any thought about career plans as I was leaving high school, the world was in the depths of the Depression and the only professional models, in addition to my uncle, were my teachers. Becoming a high school teacher was the kind of career to which I aspired, and I thought of becoming a Latin teacher, even though my Latin teachers had told me that New York State hadn't given an exam for teacher certification in many years. But I persisted.

I received the Latin medal at my high school graduation and planned to study classics. Both Isidore Diamond and I competed for a Pulitzer scholarship, which covered all expenses at Columbia University and required us to take the college entrance board exams. Unfortunately, neither of us obtained the scholarship. Nevertheless, Diamond went to Columbia, wrote its Varsity Shows, and went on to a career in Hollywood as the script writer known as IAL Diamond, working with Billy Wilder. When Wilder received an Academy Award, shortly after Diamond had passed away, he held the Oscar statue high and said, "This is for you, IAL." I, instead, enrolled in City College. I found it to be a very stimulating place, and I benefited from an extraordinary education there.

CITY COLLEGE OF NEW YORK

I was at City College from 1937 to 1941. When I entered, I had to indicate whether I would pursue an arts or a science career. Being interested in classics, I opted for arts, but I also wanted to study mathematics. I was informed that this was impossible, and that the only way that I could study mathematics would be if I enrolled in science; consequently, I chose the science option. But I wasn't completely committed to mathematics. I didn't feel that I wanted to embark on a career in mathematics and enrolled in freshman chemistry, selecting chemistry as my major essentially at random.

Unfortunately, I had not taken physics in high school. My only experience in science, aside from that early biology class, was with mathematics and one chemistry course. As soon as I started taking chemistry, I became excited about it. I had excellent chemistry teachers. In fact, in my freshman course, I had a lab and recitation instructor, Edgar Leifer, who had graduated from City College a year or two before my freshman year. He was carrying out his PhD research work at Columbia under Harold Urey while teaching at City College. Ed lived near me in Brooklyn. After labs, we rode home together at night on the subway, and I would hear about his graduate research work. By coincidence, his uncle had been one of my teachers in Hebrew school. Hearing about the kinds of activities in which Ed was involved as a graduate student, I soon realized that chemistry was going to be my career.

During my undergraduate studies, I took mathematics courses and, happily, I enrolled in physics courses. Ultimately, I had enough math

and physics credits to also major in both of these areas, in addition to my chemistry major. Actually, I found the math teachers at City College rather dull, even though I enjoyed the subject. I remember a course in complex variables in which we used a German text, *Funktionentheorie*, by Konrad Knopp. The instructor spent half the lecture translating the German for us.

One of my outstanding physics professors was Mark Zemansky, the author of the textbook *Heat and Thermodynamics*. I had several courses with Zemansky, including physical optics. However, it was chemistry that excited me the most. I did undergraduate senior research and honors work with Alexander Lehrman on the phase rule and with Morris U. Cohen on X-ray diffraction. I was elected to Phi Beta Kappa and was awarded a BS degree with honors in 1941. That was both an exciting and sad time because, in my senior year in 1940–1941, a committee of the New York State legislature started a communist witch hunt in the City Colleges. The committee was headed by two legislators, Herbert A. Rapp and Frederic René Coudert, Jr. Many professors were subpoenaed to testify before the investigating committee and ultimately lost their jobs. Morris Cohen was one of them. Later, after the war, an amnesty rehabilitated the City College professors who had lost their jobs in that witch hunt. It had taken all that time to rectify the situation.

I would estimate that about twenty-five professors from City College and Brooklyn College lost their jobs. It was much like the post–World War II era of McCarthyism, but this investigation was a product of the New York State legislature. During my youth, the atmosphere was dominated by the Rapp-Coudert Committee, the Dies Committee of the House of Representatives, and the McCarthy Committee of the Senate. City College was a target for these investigations because the students were experimenting with various political philosophies. Although there was always a great deal of political activity on campus, I suspect that the number of communists was very small. Some of the students volunteered to fight in the Spanish Civil War in support of the Spanish government forces. The rebels, under General Francisco Franco, were supported by the German Nazis and the Italian Fascists. I had strong sympathy for those students who put their lives on the line for their beliefs.

During my senior year, in the midst of all this turmoil, the New York Academy of Sciences organized a symposium on X-ray crystallography. Because of his involvement with the Rapp-Coudert Committee, Morris Cohen couldn't attend and suggested that it would be a good experience for me to attend in his place. This was the first scientific meeting that I attended, and I was impressed. The discussions of every lecture were dominated by Isidor Fankuchen of Brooklyn Polytechnic Institute and Peter Debye. Debye had just left Germany to give the Baker Lectures and assume the chairmanship of chemistry at Cornell.

I was definitely heading toward physical chemistry. That's why I also took many math and physics courses. Incidentally, I was not interested in biochemistry as an undergraduate. My uncle in Kentucky had suggested I take a biochemistry course, but I didn't think it was a rigorous scientific discipline. I've since done a 180° phase shift on that one.

Quite a few of my classmates have since become well-known chemists. Ernie Wilson and I worked together with Morris Cohen. Ernie subsequently was on the staff of Columbia Medical School and later moved to Colorado. Saul Roseman at Johns Hopkins, Oscar Touster at Vanderbilt, Seymour Lewin at New York University, and Henry Freiser at the University of Arizona were all classmates of mine. It was quite a class. Ernie Wilson and Seymour Lewin ranked first and second, respectively. I ranked third among the chemistry majors in the class. The whole class of 1941 might have included about 500 or 600 people, with perhaps fifty chemistry majors.

At that time, CCNY produced one of the largest numbers of bachelor's degrees in chemistry. A significant fraction of those who went on for PhDs came from City College. At that time, CCNY was an undergraduate all-male school;

there was no graduate school. When I was there, a woman was admitted to the School of Engineering. City College soon became coed and then established a graduate school.

During my college years, I spent one summer taking a required course in economics but spent the other summers working at camp, first as a waiter and then as a counselor. After graduation from CCNY in 1941, I tried to find a summer job as a shipping clerk at Macy's department store. I waited in line with other applicants who were screened by a personnel executive. She asked only one question, "From which college did you receive your bachelor's degree?" Only Ivy League college graduates survived that interview.

CLUB RAIM

I had many friends at school, but my closest ones with whom I had much in common were those who lived closer to my home. My social life revolved around a Jewish cultural club in Brooklyn, Club Raim. Its members were students from City College and Brooklyn College. Most of us became lifelong friends. In fact, I met my wife, Miriam Kurnow, there. She was a student at Brooklyn College. There were a number of marriages from that club, and the couples continued to be friends for many years, and we still see them quite frequently.

GRADUATE SCHOOL AND WAR WORK AT DUKE UNIVERSITY

At CCNY, I took it for granted that I was going to graduate school. At that time, most students at City College were Jewish. CCNY did not have the religious and racial quotas prevalent in many colleges. City College students were also regarded as politically leftist. Consequently, it was very hard for a City College student to be accepted into graduate or medical school. I graduated in June of 1941 before we entered the war. I sent out twenty-five applications for graduate school but received no favorable replies. I was third in my class and I had very good letters of recommendation. Later in the spring of 1941, I received a telegram from

Duke University, offering me a $550 teaching assistantship out of which I would have to pay $250 for graduate school tuition. The chairman of the Department of Chemistry at Duke was Paul M. Gross, who had been a classmate of Alex Lehrman at City College in the class of 1917 or 1918. He went from CCNY to Trinity College via a PhD at Columbia. Trinity College became Duke University, with Gross as the head of the chemistry department. In the late 1930s, he began to take one City College student every six months (at City College one could graduate at midyear). Because Duke had a young graduate school, it was difficult to attract graduate students. Some of the best undergraduates from City College went to Duke for that reason, including David Breslow, who went on to the Hercules company; Philip Skell, who retired from Penn State; and Henry Linschitz, at Brandeis, all of whom were my predecessors there. Henry Freiser preceded me by six months; I roomed with Henry when I went to Duke in September 1941.

Incidentally, in the summer of 1941, I received an offer from the University of Chicago, but I had already accepted the Duke offer. I was concerned about going to Duke, but went there because Fritz London was at Duke. I wanted to work with London, but he wouldn't take on any graduate students because he thought they weren't experienced enough. Gross assigned me to work for him on the Kerr effect to study polarizability. This was shortly after London had developed the quantum mechanical theory of van der Waals forces, which involved the polarizability of molecules. Under London's encouragement, Gross wanted to start an experimental program to determine anisotropic polarizabilities; all the students who worked for Gross were mentored by Marcus Hobbs, an excellent young faculty member.

London had come from France in 1939. Gross was very astute in attracting him to Duke. London had left Germany as soon as Hitler became Chancellor, and Paul Langevin accommodated him in Paris. I remember London telling me that he saw the handwriting on the wall in 1939 and was going to leave

France because he could see that Hitler was going to precipitate World War II. As he told it, Langevin tried to prevail upon him to remain in France, saying, "On les resistera"; London's reply was, "Non, non. On capitulara." London sailed on the last boat that left France in 1939. He was at Duke for about two years by the time that I arrived there in September 1941.

London was one of the best teachers I had at Duke. He gave a year-long course in quantum mechanics, which I took my first year. The next year, I took his year-long course in statistical mechanics. Because I felt I had a lot more to gain from it, I audited the quantum mechanics course when he taught it again my third year. London taught a personalized and historical view of quantum mechanics and had participated in its development. The group taking his courses consisted of only five or six students. We felt we had more to learn from him and prevailed upon him to give us an extra course at night. He was working on superconductivity at the time and gave us a series of lectures on that subject.

I began experimental work on the Kerr effect on small molecules when I arrived at Duke. On Sunday, December 7, I was working in my lab on a problem set for one of my physics courses, with my radio providing background classical music. The program was interrupted by an announcement of the Japanese attack on Pearl Harbor. The next day, before a joint session of Congress, Roosevelt asked for and obtained a Declaration of War, and academic life changed. Everything was in a state of flux, but I managed to continue for the remainder of that academic year studying electrical birefringence, for which I received a master's degree for work I had done. This work increased my interest in molecular structure as related to optical and electrical properties. Ultimately, I returned to that subject in my postdoctoral work. To help in the war effort, we all engaged in war work. I was assigned to two war projects, but I was still able to continue my graduate work. I worked on the war projects during the day and on my graduate research during the night, related to another war project to which I was not assigned.

The first project was sponsored by the Navy to surmount problems with fires onboard ships, which used to generate a lot of smoke. The smoke particles would penetrate the filters of the gas mask and destroy the impermeability to poison gases. The project involved a study of the interaction of smoke particles with cellulosic matter, that is, paper. This project was carried out in collaboration with Columbia University under the supervision of Victor LaMer, a colloid chemist. We would measure the sizes of smoke particles in electric fields, essentially a Millikan oil drop experiment. However, I was never made aware of the application of our results.

The second project I participated in was sponsored at Duke by the U.S. Air Force and run by Paul Gross and was called the frangible bullet project. By that time, the U.S. Air Force was making bombing raids over Europe with B-17s, but there was no way to train the gunners in the nose, tail, and waist who were defending the B-17s against German fighter planes. In this project, a 30-caliber plastic bullet containing a bit of lead to give it some additional mass was developed. The idea was to provide the gunner with a bullet that he could use during training in a simulated attack on a B-17 by a fighter plane flown by a friendly pilot. In that way, the gunner in the B-17 could shoot at the fighter plane as if in combat without causing any damage to the fighter plane because these plastic bullets would break up upon impact. However, in actual combat the B-17s were armored with 50-caliber machine guns, whose projectiles were too large even for a frangible plastic bullet. Therefore, we had to modify a 30-caliber machine gun to simulate a 50-caliber one so that the gunner would think he was firing a 50-caliber gun. But, even the 30-caliber plastic/lead projectile at the normal velocity could still penetrate the fuselage.

Therefore, the fuselage of the training fighter plane was armored with thicker metal plating with a microphone attached to every plate, with an electrical connection to the nose that would cause it to light up upon impact in order to enable visualization of a hit. It was nicknamed the Pinball Machine. But even with the extra armor, at the velocity at which the plastic 30-caliber projectile was fired, the bullet would still penetrate upon impact. Therefore, we had to reduce the velocity of the projectile but then the machine gun wouldn't recoil. My part of the project was to perform the calculations and supporting experiments on the interior and exterior ballistics of these projectiles. To obtain enough force for the recoil, we put a flange on the nose of the gun barrel with a cylinder around it so that, as the gases emerged from the muzzle, there was enough back pressure against this large flange to develop sufficient force to make the gun recoil. Then, we altered the gun sight so that the gunner would think he was firing a 50-caliber gun. In addition to all this theoretical work that I was doing, we had to test it in a firing range in the attic of the chemistry building, where we would fire the machine gun and measure the velocity of the bullet. We had to decide on how much gun powder to put into the shells to obtain the desired velocity of the projectile. One couldn't calculate the required amount of gun powder. Therefore, we had to determine it experimentally because there was a friendly fighter pilot in that plane who was going to be shot at with live ammunition. With this kind of training, the number of hits that a gunner in the B-17 scored on his first combat mission in Europe increased considerably, making it a very worthwhile research project. When the war was coming to an end in Europe, we were using B-29s for the bombing raids in Japan. The velocity of the B-29 was sufficiently greater than that of the B-17, such that, when added to the muzzle velocity of the bullet, it was already too high even for this scheme to work. We never solved the problem, but the atomic bomb ended the war before we were faced with the need to improve our method.

My thesis research centered on a problem involving kinetics of gas-phase chlorination of an aromatic hydrocarbon. My lab partner at that time was Milton Manes, who has since been a lifelong friend. He later joined the faculty at Kent State University and is now retired. When several students were killed at Kent State during the Vietnam War, and classes were disrupted, he continued to run his physical chemistry lab with his students at one of the government bureaus in Pittsburgh.

At Duke, Manes and I shared a two-man research lab. This gas-phase kinetics research required high-vacuum work, involving much glassblowing and putting up and taking down high-vacuum systems. We needed liquid air in order to operate the diffusion pump. We obtained liquid air from nearby University of North Carolina in Chapel Hill, which operated a machine to produce liquid air. We would pick up 15 liters of liquid air at UNC and bring it back to Duke. But when their liquefaction machine broke down, they couldn't get parts during the war to replace it. Fortunately, we located a source in Washington, DC. A 15-liter Dewar was placed on the Southern Railroad, but by the time it reached Durham, we never received 15 liters from a 15-liter Dewar. And we could never make a run until we could start up the diffusion pump. These were the conditions under which we were trying to do research at Duke.

As for my course work at Duke, besides London's courses, I had to take a year-long course in organic chemistry; one term was taught by Charles Hauser and the other by Lucius Bigelow. I took a course in chemical thermodynamics, which Gross and Hobbs taught from Lewis and Randall. I didn't take any mathematics courses because I had taken considerable math at CCNY, but I did take several physics courses as a graduate student. In addition to London, two other refugees from Germany were at Duke, on the physics faculty. I took three semesters of electrodynamics and relativity with Lothar Wolfgang Nordheim, having already taken electricity and magnetism as an undergraduate. Then I took a year-long course in atomic and molecular spectroscopy with Hertha Sponer, who had been a student of James Franck.

In the midst of my graduate work at Duke, Miriam and I married. During that time, I was working full-time during the day on the frangible bullet project and wasn't allowed any vacation time. However, I was given a few days in 1943 to go home to Brooklyn to be married. Our honeymoon consisted of only a day's stopover in Washington, DC, on the way back to Durham. Because all the hotels in Washington were completely occupied by people whose activity was related to the war effort, a friend of my wife's provided us with a room in a private house as a wedding present.

Miriam had to work to help support us. She was trained as a sociologist, and took some graduate courses in that field, while working for Nordheim. Nordheim also worked on a war project and Miriam operated a Marchant calculator for him. He later left Duke for some unknown place that I subsequently learned was Los Alamos. Thus, she lost her job with Nordheim but then obtained work in the chemistry department on a tobacco project. As I understand it, American cigarettes are a blend of Virginia-type bright-leaf tobacco and Turkish tobacco and, during the war, the supply of Turkish tobacco was cut off. The rumor was that Hitler was trying to corner the market on Turkish tobacco and use it to trade tobacco for munitions. The North Carolinians decided that they were going to try to grow Turkish tobacco in North Carolina, and this required much chemical analysis. As a matter of fact, after the war ended in 1945, my war project ended and I still had a year to finish my PhD research in 1946. To earn some money during our last year at Duke, Manes and I worked on that tobacco project, collecting juice from tobacco leaves and separating the components chromatographically. Miriam was doing the analytical work for which she was trained by Dr. Darkis, with whom she worked. Then she became pregnant and we had our first child about four months before I left Duke to go to Harvard Medical School as a postdoc.

At Duke, Hans Neurath, who later became head of biochemistry at the University of Washington, Seattle, was a young biochemistry faculty member. He used to join a group of graduate students and postdocs sitting under a tree after lunch for lots of chitchat. Being a young faculty member, he mixed more with graduate students and postdocs than with his older colleagues.

POSTDOCTORAL FELLOW: HARVARD MEDICAL SCHOOL

On a particular visit to the Duke Chemistry Library, I found a new book that changed my whole career, *Peptides, Amino Acids, and Proteins*, a multi-authored volume edited by Edwin Cohn and John Edsall. Most of the chapters were written by Cohn or Edsall, but the book also contained chapters by George Scatchard and John Kirkwood. These chapters looked like interesting physical chemistry applied to interesting biological systems. So, I wrote to Cohn, who was head of the Department of Physical Chemistry at Harvard Medical School, and asked if he would consider me for a postdoc position. He never answered the letter; instead, he turned it over to Edsall, who answered it. Edsall said he would be glad to take me on if I could come up with my own money.

Toward the end of the war, the American Chemical Society (ACS) seemed to be concerned that not enough people would return from their war jobs to academia. Consequently, they set up a postdoctoral fellowship program and awarded ten fellowships; I received one of them. This was before there were NSF and NIH postdoctoral fellowships. My ACS postdoctoral fellowship included a stipend of $2,500 for the year. The terms were that the school at which one attended had to supplement it with $1,000. Therefore, I was made the equivalent of a teaching assistant (TA) at Harvard as service for the other $1,000. I was a TA for Edsall and Jeffries Wyman, both of whom gave a course in biophysical chemistry, the lectures for which they later published as a textbook. As a TA, I graded the problem sets for this course.

I had never had any contact with biochemistry and had a lot to learn when I arrived at Harvard. During the 1930s, Cohn, Edsall, and coworkers had been carrying out physical chemical studies on amino acids, peptides, and proteins. Their book was a summary of all that work. George Scatchard and John Kirkwood,

who was a faculty member at Cornell at the time, were collaborating with Cohn and Edsall on protein research. When I later came to Cornell in 1947, I was very disappointed to learn that Kirkwood had left Cornell to go to Caltech. With the advent of World War II, Cohn started a blood plasma fractionation program to satisfy the needs for plasma components, specifically serum albumin for shock treatment. After the war when I arrived at Harvard as a postdoc in 1946, they were still operating a pilot plant to fractionate blood plasma on a scale of 200 liters of plasma at a time. All new postdocs were introduced to blood protein work by spending two or three weeks in the pilot plant. Never having handled any of this, I immediately had to do a lot of reading and learn how to fractionate blood plasma on that scale.

Larry Oncley, who was on the faculty, taught a course in methods of biophysical chemistry. This was supplemented by weekly lunch meetings in Cohn's large office, where I saw Scatchard, who was a regular visitor there from MIT; Kirkwood would come occasionally from Cornell. There were also visitors from other universities. That's where I first met Irving Klotz from Northwestern. Alex Rich, who is now at MIT, was a medical student at Harvard and halfway through his medical studies. While at Harvard, Alex and I did some work together on nucleic acids. Harry Saroff, with whom I collaborated on research on serum albumin, was also there, and he subsequently retired from NIH. I actually worked very closely with Geoffrey A. Gilbert under Edsall's direction, but not with Cohn. Gilbert and I worked on an outgrowth of the plasma fractionation program on a protein called cold-insoluble globulin, which appeared in a plasma fraction with fibrinogen. It has since been renamed fibronectin and is now an important topic, but in those days, all we were able to do was determine its size and shape. I had written my proposal for the ACS postdoctoral fellowship, trying to resurrect some of my old interest in the Kerr effect, electric birefringence. I knew Oncley was working on dielectric dispersion of proteins and Edsall was involved in flow birefringence of proteins. I tried to make a combination out of these two disciplines to determine the size and shape of proteins. When I arrived at Harvard, Oncley didn't seem too interested at the time, so I worked with Edsall on flow birefringence; Geoffrey Gilbert and I collaborated on preparing cold-insoluble globulin and then carrying out flow birefringence measurements. That's how I obtained my first experience with proteins, aside from the pilot plant training, the Edsall-Wyman biophysical chemistry course, and the lectures by Oncley on methodologies to study proteins.

During that time, we noticed an anomaly in which the ascending and descending limbs of the Tiselius electrophoresis apparatus were giving different behaviors. It was Gilbert who figured out what was going on, and he went on to utilize the phenomenon and make some very fundamental contributions to the transport properties of proteins. Gilbert was from the University of Birmingham in England and later returned to Birmingham.

CORNELL UNIVERSITY

I went to Harvard in September 1946 and left in August 1947. My fellowship was for only one year. It wasn't renewable, and I started looking for a job by going to the spring 1947 ACS meeting in Atlantic City, New Jersey, where I met Paul Gross on the boardwalk. I told him that I was looking for a job, and he told me that Peter Debye was looking for a faculty candidate. Gross had previously spent a sabbatical leave with Debye in Leipzig, Germany, and was a friend of Debye's, and told him about me. Debye asked the late Professor A.W. Laubengayer to contact me through the ACS employment clearinghouse for an interview. On the basis of that interview, I was invited to Cornell to present a seminar and go through the usual interviewing process. I spent a full day talking with various faculty members, and at the end of the day I gave my seminar. Nowadays, the candidate goes home and hears the outcome weeks later, but that was not Debye's way of doing things. Professor Simon Bauer of our

department was seminar chairman and told me to wait in the Chemistry Library while Debye called the faculty together for a meeting. In about a half hour, Bauer came in and told me to see Debye, who said, "We are going to offer you a job," and he "wanted my answer right away." Of course, I was ready to accept it, but I felt that I at least ought to let Miriam know. I promised to let Debye know soon. This was on a Thursday; I called him on Monday to accept the job and started at Cornell in September 1947.

I arrived at Cornell in September 1947 with Miriam and our daughter Judy. Many war veterans were returning to Ithaca, and it was difficult to find housing. We ended up having to live in Watkins Glen, New York. The university took over a resort hotel in Watkins Glen and converted it to apartments and transported us by bus back and forth to the campus, a thirty-mile commute each way, every day. We finally found an apartment in Ithaca in May 1948, and in the fall of 1955, Miriam and I were able to build a house, where we have lived ever since. Living in Ithaca, we were able to take advantage of the cultural activities at Cornell University, at Ithaca College, and in the City of Ithaca. This provided access to lectures, concerts, theater, and movies throughout the year, and excellent schools for our three children, Judy, Debby, and Danny. All three later graduated from Cornell and have gone on to independent careers. Our five grandchildren, Jonathan, Alexandra, Rebecca, Dana, and Jeffrey, have completed college and are embarking on their own careers.

After our children went off to college, Miriam enrolled in the Master of Library Science program at Syracuse University. This involved taking night courses over a period of three years, after which she completed the degree program. Many of those nighttime trips from Ithaca to Syracuse involved hazardous driving conditions, especially during the winter. Since receiving a Master of Library Science degree in 1973, Miriam has worked at the Cornell Library.

I was hired at Cornell, at the now-nonexistent rank of instructor, to teach quantitative analysis, but I wasn't expected to do my research in that area. I became an assistant professor in 1950, and in 1952, I was asked to teach undergraduate physical chemistry. I served my whole teaching career in undergraduate physical chemistry after that. I also taught a graduate course entitled Physical Chemistry of Proteins.

In 1947, very little was known about proteins that were regarded as rigid colloidal particles, and interest was directed toward determining their size and shape. I began my independent research at Cornell to apply experimental physical chemistry to protein science. Years later, this experimental work led me to the formulation of a theoretical approach to protein structure and function. I started my experimental work by focusing on asymmetrical molecules and built an apparatus to use flow birefringence to determine the size and shape of fibrinogen. At the same time, I began an investigation of the mechanism of action of thrombin on fibrinogen to form the fibrin clot.

In those days, research funds were almost nonexistent. Nowadays, a newly appointed faculty member is given a sizable amount of start-up research money. However, I had free access to the stockroom and to excellent machinists in the machine shop. I also managed to obtain a $200 grant from the Rumford Fund of the American Academy of Arts and Sciences to pay for the precision machining by a Boston firm of the concentric cylinders of the flow birefringence apparatus that I was assembling.

To interpret the data from flow birefringence experiments, it was necessary to solve a partial differential equation to compute the rotational diffusion constant of dissolved asymmetrical molecules. For this purpose, I used the Mark 1 computer as a guest at the Harvard Computation Lab to carry out the computations. Having worked out the solutions of the equation with Legendre polynomials to be programmed for the computer, I asked Debye if he would check them, which he did without having to consult a textbook or table of integrals. As a busy department chairman, he was nevertheless always approachable to discuss a scientific question. I later extended the theory

(110) to treat non-Newtonian viscosity of solutions of asymmetrical molecules (101) as an alternative technique to obtain rotational diffusion constants.

In 1950, I was awarded my first grant, by the Office of Naval Research (ONR), and was able to provide a research assistantship to support my first graduate student, the late Michael Laskowski, Jr. He later became a professor at Purdue. The late John Backus also joined my research group in 1952. At that time, Debye was applying his newly developed light-scattering method to study the properties of detergent micelles, and the question of their sizes and shapes arose. I could answer that question with my flow birefringence apparatus and assigned Backus to investigate that problem (108). Debye was always interested in hearing about these results.

BLOOD CLOTTING

With Laskowski, we investigated the mechanism of the thrombin-induced conversion of fibrinogen to fibrin, a part of the blood clotting process, a subject I continued to work on for many years. Initially, thrombin was used simply as a reagent to activate fibrinogen to fibrin monomer, and we investigated the polymerization of the resulting fibrin monomer by flow birefringence (7) and light scattering (11). In a later review article (107), based on a lecture I gave at the University of Wisconsin as a memorial to the late John Ferry, I provided a summary of the nature of the staggered-overlapped polymers of fibrin monomer, deduced from these measurements and confirmed by electron microscopy. John Ferry was carrying out similar experiments on the polymerization of fibrin monomer (31) at that time. We also provided information on which functional groups were involved in the polymerization, based on the pH dependency of the interactions (122), and on the regions of the fibrinogen molecule where the proteolytic action of thrombin releases short peptides to expose a polymerization site on the resulting fibrin monomer. This action of thrombin is very specific; i.e., whereas trypsin hydrolyzes peptide bonds after every arginine and lysine residue in a protein, thrombin hydrolyzes particular arginine-glycine bonds in fibrinogen.

In subsequent investigations of the mechanism of action of thrombin, we pursued NMR-transferred nuclear Overhauser effect (NOE) studies of complexes of thrombin with appropriate fibrinogen peptide sequences (80) and demonstrated a specific configuration of the complex that accounts for this specificity of thrombin at a particular sequence of fibrinogen (107). The NMR-transferred NOE technique was also used (80) to determine the structure of a nonclottable mutant fibrinogen in which a glycine (not the one at the active site) was replaced by a valine. This mutation altered the structure at the critical arginine-glycine site so that the relevant position of fibrinogen wasn't properly oriented in the active site of thrombin to enable the enzyme to catalyze the hydrolysis required for ultimate clotting, leading to a bleeding disorder (81).

DISTANCE CONSTRAINTS

With Laskowski, we also investigated the thermodynamics of protein reactions. Linus Pauling and Robert Corey had just published their famous papers on the α helix and β sheets in 1951, with the emphasis on backbone hydrogen bonding, and we focused on the role of side chain hydrogen bonding. We were able to account quantitatively for the effect of internal hydrogen bonding in proteins on the pK values of ionizable groups (40), on the reactivity of covalent bonds, including peptide and disulfide bonds (41), and on protein stability (42). We also showed how hydrogen bonds could stabilize peptide bonds (41) so that proteolytic enzymes could be used to catalyze not only the hydrolysis but also the synthesis of peptide bonds.

We followed these studies by trying to locate internal hydrogen bonds to identify internal interactions in proteins in order to determine their three-dimensional structure from their amino acid sequences. Such sequences were just being published; insulin was the first one, and we started to work with that protein. This was

long before X-ray structures of proteins were known. Our idea was that if one knew the amino acid sequence and if one could find some specific noncovalent interactions, one could determine the three-dimensional structure with the aid of such distance constraints. Therefore, we carried out thermodynamic and spectroscopic studies to identify local interactions and published some papers on insulin. However, we were always thwarted by the fact that insulin is insoluble in the neutral pH range, between 6 and 7. For most of the experiments that we wanted to do, insulin precipitates, but it was the only protein that was sequenced in the early 1950s. We had heard that the amino acid sequences of lysozyme and ribonuclease A (RNase A) would soon be available; therefore, I started applying this approach to both of these proteins. The lysozyme work was started with Laskowski, and Jack Donovan also participated in it, but this protein also presented problems. It has several tryptophan residues, and at that time it was difficult to analyze peptide fragments for tryptophan. RNase A, on the other hand, is soluble over the whole pH range and contains no tryptophan. Many years later, we incorporated tryptophan into RNase A (116) with recombinant DNA methodology for a variety of fluorescence experiments. However, we were finally able to use wild-type RNase A to pursue our original objective of acquiring pairwise interactions to determine three-dimensional structures. I started the RNase A work in 1956 when I was on sabbatical leave at the Carlsberg Laboratory in Copenhagen.

When I started my protein work at Cornell, Paul Flory came as a Baker Lecturer in 1948 and then joined the faculty. He was supposed to write a book based on his Baker Lectures, but he didn't write it until several years later and published it in 1953. While he was writing it, he asked one of his postdocs, the late Leo Mandelkern, and me to read the drafts of his chapters and tell him what was and wasn't clear.

HYDRODYNAMIC PROPERTIES

With Debye and Flory on the faculty, Cornell was a very stimulating place. We used to have weekly polymer seminars at night and, of course, Debye and Flory would dominate these. There were three competing theories of the hydrodynamic properties of polymer solutions, all developed at Cornell: Kirkwood-Riseman, Debye-Bueche, and Flory-Fox. Each pair thought their theory was the correct one, which in turn made for some lively discussions, but it never interfered with their friendship. Debye and Flory could argue and then go off as pals and drink beer together.

I had considerable interaction with Flory and collaborated in research with him and his postdocs, the late Leo Mandelkern and the late Bill Krigbaum. We worked on nitrocellulose and on the hydrodynamic properties of polyisobutylene to test the three Cornell polymer theories cited above. Flory was just starting to work on proteins, and we did some collaborative work on collagen. He was interested in the phase transition properties and the regeneration of collagen.

As one of the tests of the three polymer theories, we carried out hydrodynamic studies of solutions of polyisobutylene. Krigbaum fractionated the polymers, Flory made viscosity measurements, and I made sedimentation velocity measurements (with a Model E ultracentrifuge at the Brookhaven National Laboratory) on solutions of these fractions, and Flory and Mandelkern analyzed the data to test the theories (63). The Flory-Fox theory fit the data better than the other two theories did. Incidentally, all four coauthors of this paper (63), each in later, separate years for other work, received the ACS Award in Polymer Science.

I met Leo Mandelkern as soon as I came to Cornell. He was a graduate student of Frank Long's then and was married to Birdie. Birdie and Miriam soon became good friends as did Leo and I; having young children of the same ages made our friendships even closer. When he decided to stay on as a postdoc with Flory, the mutuality of interests cemented our professional and social lives. While I was working with detergent micelles and on the blood clotting system, I was hearing from Leo and Flory

about what was going on at the time in the synthetic polymer field.

I tried to see how Flory's treatment, tested on polyisobutylene, would apply to proteins thought of as rigid particles. With proteins, we had to take into account a shape factor that wasn't necessary with a flexible chain polymer because, being a statistical coil, the polymer molecule was basically a spherical object. I started to work with Flory's equation, got excited about what I was finding, and phoned Leo one morning. We decided that we had better look into this. With the motivation by Flory's theory, ours introduced the concept that a native protein is a flexible, not a rigid, molecule. Of course, Flory was very sympathetic; he wasn't a part of it but he encouraged us to keep going with it, and we finally worked it out and sent it to the *Journal of the American Chemical Society* (115). Our paper contradicted established ideas at the time and we had a difficult time trying to get it published. Albert Noyes was the editor, and he was sending us horrible referee reports. Flory kept encouraging us not to give up and to write rebuttals. During the course of writing the rebuttals we did some more calculations, which only strengthened the paper, and finally the referees relented and Noyes published it. According to *Current Contents*®, it became a citation classic, although some people refer to it as controversial work.

During that time, we had a running battle with Charles Tanford that continued for many years. It was mostly over the work I had done with Mandelkern (115). We had challenged a view of Oncley's, and Tanford tended to support that view. I think it died out as a controversy, although some people still use the old procedures, even in papers published today. Most people make use of our method. However, some use it in a way that we never envisioned. We were using our theoretical treatment to determine the size and shape of a flexible molecule, but some people turned it around and used it to obtain molecular weights. I don't recommend it. We had assumed the molecular weight to be known in order to determine the size and shape. There are more direct ways to obtain a molecular weight, for example, by sedimentation equilibrium or light scattering.

I was an instructor from 1947 through 1950 and assistant professor from 1950 through 1953. In 1953, I was promoted to associate professor and received tenure. Then, Frank Long nominated me for the Eli Lilly Award in Biochemistry, which I received in 1957. I'm sure that the hydrodynamic work that I did with Leo Mandelkern, and the blood clotting work with Mike Laskowski, Jr., were both significant contributing factors that led to that award.

CARLSBERG LABORATORY

With the aid of Fulbright and Guggenheim fellowships, I spent the 1956–1957 academic year on sabbatical leave at the Carlsberg Laboratory in Copenhagen with Kai Linderstrøm-Lang. While there, I used UV absorption spectroscopy on bovine pancreatic RNase A to demonstrate that one or more tyrosyl groups were close to one or more carboxyl groups (102). This was the start of our effort to obtain distance constraints in RNase A. I also learned from Linderstrøm-Lang some of the techniques he had developed for studying deuterium-hydrogen exchange. I applied the technique to insulin there, at a time when I was phasing out my Cornell work on insulin.

Walter Kauzmann was also on leave with Linderstrøm-Lang during the second half of my year at the Carlsberg lab. I had many interesting discussions with Kauzmann and Linderstrøm-Lang about what was becoming known as the hydrophobic interaction and about the idea that the solvent water must play a dominant role in this interaction.

During my year at Carlsberg, Linderstrøm-Lang celebrated his sixtieth birthday. It was a great birthday party; Niels Bjerrum and Niels Bohr were there. It was a banquet at the laboratory with plenty of wine, beer, aquavit, and cognac. The way the Danes celebrate is that everybody makes a brief speech and offers a toast after each speech. Some of the people at the lab thought I ought to make a speech in Danish on behalf of the foreign visitors. So I memorized

a speech in Danish, which my children (who learned Danish in elementary school that year) still enjoy reciting to me to this day. I remember Niels Bohr telling me that he appreciated my speech in Danish. I had heard Bohr lecture on several occasions. One had to sit in the first row and still couldn't hear him because he whispered. He always gave the impression of being a gentle sort of person. Of course, for a young scientist, just to be in his presence was exciting. Old Niels Bjerrum, who was also there, used to come to the weekly seminars at the Carlsberg Laboratory. He would make very sarcastic remarks; one couldn't slip anything past him. Lang was more of a diplomat and a gentleman when he criticized. For me scientifically, this period of the hydrophobic interaction was very important.

Stimulated by the musical atmosphere at the Carlsberg lab, I began violin lessons. On several occasions with Linderstrøm-Lang, Kauzmann, and others, we did our best to try to play chamber music. I continued with my violin lessons at Cornell for three years until I assumed the chairmanship of the chemistry department in 1960. I was, however, able to keep up my research activity during the period of my chairmanship.

Shortly after I returned to Cornell, the Soviet Union launched Sputnik. President Eisenhower appointed a science advisor, with great emphasis placed on science education in the public schools to try to catch up with the Soviets. Consequently, I agreed to stand for election to the Ithaca School Board and served a one-year term in that capacity.

At the same time, following up on my observation of possible tyrosyl-carboxyl interactions, I started a whole series of proteolytic cleavage and spectroscopic experiments on RNase A, involving many graduate students and postdocs. As a result, before the X-ray structure was known, we had identified three specific tyrosyl . . . aspartyl interactions, that is, between groups that were near each other (45, 103). The identification of these interactions was based on the observations that 3 out of 6 tyrosines had abnormal pK values, and 3 out of 11 carboxyls also had abnormal pK values. There are over 19,000 ways to pair 3 out of 11 carboxyls with 3 out of 6 tyrosines. In 1966, we proposed a specific pairing (45, 103) based on the series of proteolytic cleavage and spectroscopic experiments. When the X-ray structure was published in 1988 (138), we were right about the locations of these three tyrosyl-aspartyl interactions from those experiments. I think that was a triumph of protein physical chemistry. All this work using physical chemistry to obtain distance constraints had its origins in the work with Michael Laskowski.

HYDROPHOBIC INTERACTIONS

While this work on RNase A was being carried out after I returned to Cornell in 1957, I also acquired a new graduate student, the late George Némethy. George had come to Cornell to work with Flory, but when I returned, Flory had left for the Mellon Institute. So I inherited George, and he did a beautiful thesis on the structure of liquid water (73), on the solubility of hydrocarbons (74), and on hydrophobic bonding (75). We realized that water was playing an important role in determining hydrophobic bond strengths, so we felt we had to understand aqueous hydrocarbon solutions as a model for the hydrophobic bond. Van der Waals interactions play only a minor role in forming hydrophobic bonds; most of the free energy of the interaction arises from changes in liquid water structure as two nonpolar molecules approach each other (74, 75). If one wants to understand hydrocarbon solutions, one had better first understand the solvent and what hydrocarbons are doing in it. The model that we proposed, which was based on some earlier qualitative discussion by Frank & Evans (19), was one in which a partial clathrate forms around the hydrocarbon. We then carried out a statistical mechanical treatment of that model. Fifteen years later, that model now arises naturally as a result of the intermolecular potentials (84). We now have good potentials, and we—and many other people—have done Monte Carlo or molecular dynamics calculations on a methane molecule in water, and the clathrate can be seen to

develop. That's the model that we had assumed, but now it arises naturally as a result of the potential function.

The theory of hydrophobic bonding (75) provided numerical values for the thermodynamic parameters for the pairwise interactions between all nonpolar side chains of proteins. These were then verified by a variety of experiments on model compounds in aqueous solution, such as association of a homologous series of aliphatic carboxylic acids, fluorescence quenching of phenolic compounds by carboxylic acids, and effect of side chain hydrophobic bonding on the stability of α helices, among others, as summarized in Reference 106. With Izchak Steinberg, a postdoc from the Weizmann Institute, we applied the hydrophobic bonding theory to compute entropy changes in protein association (121). At about the same time, Kauzmann returned to Princeton and wrote his famous chapter on hydrophobic interactions in *Advances in Protein Chemistry* (32).

A question that puzzled protein chemists at that time was, how can hydrogen bonds between polar side chains provide any stabilization free energy if they must also form hydrogen bonds with water? We answered this question with George Némethy and Izchak Steinberg (79) by pointing out that so-called polar side chains have a nonpolar neck, as illustrated in **Figure 1**. As shown in **Figure 1c**, a lysine–glutamic acid hydrogen bond is stabilized by hydrophobic interactions involving isoleucine and leucine with each other and with the corresponding nonpolar parts of the polar residues, respectively. These hydrophobic interactions restrict the internal rotational freedom of the polar residues and provide a nonpolar environment around the polar head groups.

This subject was revisited recently with Jane Dyson and Peter Wright (14), with whom I have interacted for many years during annual visits to the Scripps Research Institute. Their experiments on the folding of mutant apomyoglobins provided corroboration for models based on the hypothesis that folding initiation sites arise from hydrophobic interactions even

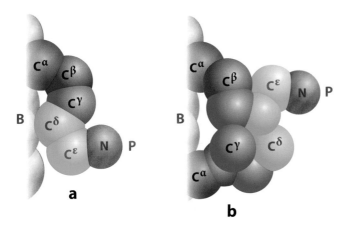

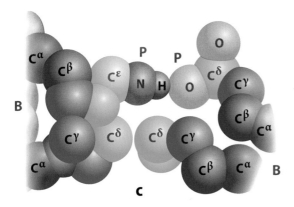

Figure 1

Schematic representation of various hydrophobic interactions of a polar side chain with its surroundings. (*a*) Interaction of a lysine side chain with the backbone; (*b*) interaction of a lysine side chain with a nearby isoleucine side chain; (*c*) interaction of two polar side chains (lysine and glutamic acid), engaged in hydrogen bonding near two nonpolar side chains (isoleucine and leucine, respectively). A hydrophobic interaction is also formed between the two nonpolar side chains. Abbreviations: B, backbone; P, polar head; α, α-carbon. Adapted from figure 1 of *Biopolymers*, 1:43–69 (1963) with permission.

Némethy thought he would like to go into industry, and went to General Electric. But, as soon as he arrived there, he realized that it wasn't a good fit for him. So he took a postdoc appointment with Dan Koshland at Rockefeller University. Koshland was at Brookhaven at that time but with a joint appointment at Rockefeller. When he went to Berkeley, Némethy remained at Rockefeller. Némethy and Koshland developed one of the theories for allosterism. Némethy is a coauthor of three Citation Classics, his two papers with me on both the structure of water and hydrophobic bonding, and his paper with Koshland on allosterism. While Némethy was at Rockefeller, we still maintained contact. I thought there was something to be learned about how the zero-point energy of the hydrogen bond affected the thermodynamic properties of liquid D_2O and the effect of deuterium substitution on hydrophobic interactions; so we continued to collaborate and we published a paper on that subject (76).

HELIX-COIL TRANSITIONS

Shortly after that, Douglas Poland joined my group as both a graduate student and a postdoc. With Poland, we did considerable work, mostly on the equilibrium (88) and kinetic (89) theories of the helix-coil transition in homopolymers (88, 89) and copolymers (93), and ultimately we wrote a textbook on the helix-coil transition (94). We also treated the theoretical aspects of phase transitions in polyamino acids (90) and nucleic acids (91). In addition, we carried out work on oligonucleotides, with experiments on single-stranded polymers, followed by application of helix-coil transition theory to obtain estimates of the stacking energies between bases in these polynucleotides (16, 134, 135). Simultaneously, a theory for equilibrium unwinding in finite chains of DNA was developed (92). Our research on phase transitions in nucleic acids (91) has been followed up recently by physicists such as by Monthus & Garel (69), Everaers et al. (17), and Einert et al. (15). Our recent work clarified the nature of the cooperative two-state transition

among side chains containing charged or polar residues.

Joel Hildebrand questioned our use of the term hydrophobic bonding, and Némethy and I, together with Kauzmann (78), published a reply to Hildebrand to point out what the physics was, irrespective of the nomenclature. We have since then adopted the term hydrophobic interaction.

in proteins (26, 58, 142). Poland was subsequently chairman of the chemistry department at Johns Hopkins. Later, with Nobuhiro Gō, we provided a statistical mechanical treatment of the equilibrium conformations of polymers (24, 25).

THEORETICAL APPROACHES TO PROTEIN STRUCTURE

While we were carrying out experiments to detect noncovalent interactions to determine the structure of RNase A (103), it became apparent that, if we had enough of such constraints, we could determine the structure theoretically from the amino acid sequence. That was the start of our theoretical approach to determine protein structure. Just before Némethy left Cornell in 1962, I had a discussion with him and with a postdoc, John Rupley, who was working with me on ribonuclease. We began to think that we ought to be able to determine protein structure theoretically, by making use of experimentally determined distance constraints such as those three tyrosyl... aspartyl interactions (102, 103). Because George was waiting to take his PhD thesis exam, we decided to examine a loop of RNase A and write out the analytical geometry, generate a structure, and identify steric overlaps. This was the start of our theoretical approach to structure (77), which involved a long series of activities with Syd Leach, Roy Scott, and Doug Poland, among others. Némethy collaborated from a distance and then, when he later returned to Cornell, he participated in this research full time.

From then onward, my main research interests evolved into a combination of experimental and theoretical studies to gain an understanding of how interresidue interactions determine the structure, folding pathway, thermodynamics, and biological activity of a protein. Besides that, I continued with an ongoing interest in the mechanism of the thrombin-fibrinogen reaction in blood clotting and the structure and mode of action of various growth factors such as epidermal growth factor (68).

DEPARTMENT CHAIRMAN

In 1960, I accepted a five-year appointment as chair of the Department of Chemistry. I was reelected in 1965 for a second five-year period. I attribute the collegial atmosphere of our department to my predecessor as chair, Frank Long, and I attempted to emulate his expert leadership.

As department chair, I was able to maintain an active research program. In 1967, I politely refused an invitation to serve as dean of the College of Arts and Sciences and resigned the department chairmanship because I wanted to continue what was becoming an even more active and exciting research program.

During my tenure as chair, I led the department into two new research areas, molecular biology and materials science, and recruited new faculty in these and the more traditional areas. Our new recruits included James Burlitch, Vincent du Vigneaud, Elliott Elson, Robert Fay, Michael Fisher, Jack Freed, Gordon Hammes, Roald Hoffman, Robert Hughes, George Morrison, Hans Muxfeldt, Martin Semmelhack, and David Usher. I also established the position of a professional executive officer, which was filled admirably, in later years, by Earl Peters.

To provide lab space for this increase in faculty, I convinced the university administration to build the Olin Chemistry Research Wing on Baker Laboratory, and I participated in raising funds from industry and government for this building project. With the help of my colleague Robert Plane, who succeeded me as chairman in 1967, we submitted proposals to NIH and NSF for building funds for the Olin wing, and our proposals were funded. I moved my own lab facilities from Baker Laboratory to the new Olin wing when I finished my chairmanship in 1967.

WEIZMANN INSTITUTE

I took a second sabbatical leave in 1963 and was lucky to obtain another Fulbright Grant and Guggenheim Fellowship. I had previously met the Katchalsky brothers, but Linderstrøm-Lang's lab was the Mecca for protein chemistry in the 1950s, and that's why I went there. I had met Aharon Katchalsky at a macromolecule meeting that Flory and I attended in Stockholm in 1952. Then I met his brother, Ephraim, a few years later at a Gordon Conference and was equally impressed with both of them. Having spent a whole year on my previous sabbatical, I felt it was too long to be away from my students, so I took only a half-year as I did on subsequent sabbaticals. At the Weizmann Institute, I carried out experimental and theoretical work with both Katchalsky brothers (29, 111) and with Shneior Lifson (8). With Lifson and one of his students, we developed a quantitative treatment of the helix-coil theory, taking the interactions involving side chains into account to determine how they contributed to helix stability. Because we had already proposed (75) that a weak hydrophobic interaction must always be present between the β carbon of the ith residue and the α carbon of the $(i + 3)$ residue in an α helix (a $\beta_1-\alpha_4$ interaction, with the subscript numbers increasing toward the N terminus), we extended Lifson's helix-coil theory with this interaction and showed that the melting point of a poly-L-alanine helix would be about $100°$ higher than that of a polyglycine helix, if a polyglycine helix existed (8). That started a whole series of experimental and theoretical studies at Cornell on the effects of side chain interactions on helix stability that I then continued with Doug Poland. That also led us into random copolymers because most homopolymer helices aren't soluble in water and we wanted information about the interactions in water. The only way to surmount the solubility problem was to make random copolymers with a water-soluble host, and the residue of interest as a guest. We were helped with this random copolymer work about that time, in the 1960s, by one of the postdocs from the Weizmann Institute, Noah Lotan, who had developed a good model system that we used as a host. With this host-guest technique, we obtained experimental parameters to characterize the helix-forming tendencies of all 20 naturally occurring amino acids; the

parameters for the twentieth, and a summary for all 20, were published in Reference 139.

The sabbatical leave in 1963 served to foster strong ties with the Weizmann Institute and with Israel. As a result, I have continued to make annual visits there. One of them occurred in January 1967 on my way back from Madras, India, where I had attended a conference organized by Ramachandran. The mood in Israel was quite unsettling at that time because Gamal Abdel Nasser, the president of Egypt, was broadcasting threats to destroy Israel. Later in the spring, when I was back in Ithaca, the threats became ominous. In order to try to do something to calm the situation, I phoned several friends and colleagues from U.S. universities. As a result, we organized the American Professors for Peace in the Middle East (APME) to encourage dialogue to facilitate a peaceful solution to the standoff between the Arab states and Israel. Unfortunately, the threats from Cairo escalated to such an extent that Israel was forced to strike first in June 1967. Although they were successful in the 1967 war in warding off the threatened destruction, Israel was threatened again in 1973 when the Egyptians invaded. Throughout all these years, and since, APME tries to add its voice to hopefully encourage a peaceful resolution to this conflict.

In 1970, with the aid of an NIH Senior Fellowship, I spent another sabbatical period at the Weizmann Institute. Aside from interactions with Weizmann scientists, the atmosphere was enhanced by a simultaneous sabbatical period with several other scientists, Chris Anfinsen, Bill Harrington, and John and Charlotte Schellman. The discussions with this group of scientists concerned much of our future outlook and scientific research. At that time, Anfinsen and I were appointed to the Board of Governors of the Weizmann Institute and served for many years on its scientific advisory committee.

In 1970, I was offered a professorship at the Weizmann Institute; however, I was reluctant to leave Cornell. As a compromise, I spent every other spring semester at the Weizmann Institute as a visiting professor from 1972 to 1980.

The original plan was for me to establish a research group there and to be supervised by Ephraim Katchalski (Katzir) and Arieh Berger during the intervening times that I would be in residence at Cornell. This plan never materialized because Ephraim became President of Israel in 1973 and Arieh died that year. Instead, I always brought two or three graduate students or postdocs with me to the Weizmann Institute for close collaboration. I also taught a course there on statistical mechanics of the helix-coil transition.

FRET MEASUREMENTS

When Ephraim finished his term as President of Israel, he was considering returning to an active research career. In a meeting with Ephraim and Izchak Steinberg (his former PhD student who also did postdoctoral research with me and became a faculty member at the Weizmann Institute), I suggested that we start a collaborative project using fluorescence resonance energy transfer (FRET) to study RNase A folding. However, at that time, Ephraim was interested in starting a biotechnology program that would contribute to the economy of Israel. Izchak had just returned to the Weizmann Institute from a second sabbatical leave and was interested in pursuing neuroscience. At Izchak's suggestion, we invited Elisha Haas, one of his recent PhD students as a collaborator.

Elisha was receptive to this suggestion, and we started a fruitful collaboration with him and his graduate student Ami Navon, and with Chuck McWherter, David Buckler, and John Laity from my lab. Interesting results emerged from this collaboration using FRET measurements to obtain several interresidue distances in various states of RNase A. With reduced RNase A in refolding buffer, in particular, we found a wide distribution of interresidue distances in the C-terminal β-like portion of the molecule, much larger than in native RNase A, but with mean values close to those in the native protein. In addition, the N-terminal α-like portion of the chain was separated from the C-terminal core by very large intramolec-

ular distances, larger than those in the crystal structure. It seemed that the backbone fold in the C-terminal β-like portion appears to adopt a native-like overall fold. Thus, under folding conditions, reduced RNase A is poised for the final folding steps with a native-like trace of the chain fold but with a large separation between the two subdomains that is then decreased to the native value upon introduction of three of the four native disulfide cross-links (72).

FOLDING PATHWAYS OF RNase A

Upon returning to Cornell after my 1963 sabbatical at the Weizmann Institute, I began to continue actively with our experimental and theoretical approach to the folding of polypeptides and proteins. An experimental effort was initiated with a postdoc, Yasuo Konishi, to elucidate the relatively slow oxidative folding pathways of RNase A from the reduced protein to the native structure by using a redox mixture of oxidized and reduced glutathione, GSSG and GSH, respectively (37, 38, 113). With a wide range of GSSG and GSH concentrations, 7,191 theoretically possible intermediates, in addition to the fully reduced and native forms, exhibiting a combination of intramolecular disulfide bonds, intermolecular disulfide bonds between a cysteine residue of the protein and GSH, and free sulfhydryl groups could appear. With chromatographic analyses of mixtures of these intermediate forms at any time during the oxidation, it was possible to determine the concentrations of all these forms, arranged in nine groups, and the unreacted fully reduced form, and the fully formed native species. By analysis of all possible elementary reactions between all species, six different pathways involved in the folding process were identified as dominant. The relative apparent standard-state conformational chemical potentials of the intermediates were estimated by using data for the apparent equilibrium constants (among the species in preequilibrium) and for the redox potentials of cysteine/cystine and GSSG/GSH. Two models, designated growth-type and rearrangement-type, respectively, were deduced from the equilibrium and kinetic data. In the growth-type model, nucleation of the native-like structure occurs in the folding process, in the rate-determining steps, and subsequent folding around the nucleation sites proceeds smoothly to form the native disulfide bonds and conformation. In the rearrangement-type model, proper nucleation does not occur in the folding process; instead, nonnative interactions play a significant role in the folding pathways and lead to metastable intermediate species. Such nonnative interactions must be disrupted or rearranged to nucleate the native interactions (in the rate-determining steps) for the protein to fold.

To simplify the kinetic study of RNase A folding, with David Rothwarf, and later with Mahesh Narayan, we subsequently used a mixture of oxidized and reduced dithiothreitol, DTT^{ox} and DTT^{red}, respectively, instead of GSSG and GSH. In using the DTT mixture, formation of mixed-disulfide bonds does not occur because of the rapid recyclization of the dithiothreitol involved in the mixed disulfide. With DTT, the number of possible species is, therefore, reduced from 7,193 to 764. Using similar procedures as those used with the GSSS/GSH mixture, the model of **Figure 2** with two parallel pathways was deduced (99, 100), where R_U is the completely reduced form, nS_U are ensembles of unfolded (U) one-, two-, three-, and four-disulfide species, des [65–72] and des [40–95] are native-like intermediates formed from $3S_U$ by SH/S-S interchange in

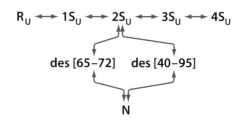

Figure 3

Regeneration of native RNase A along two different folding pathways by oxidation of $2S_U$. Adapted from figure 6 of *Biochemistry*, 37:4490–501 (1998) with permission.

rate-determining steps and containing three native disulfide bonds but lacking the 65–72 and 40–95 disulfide bonds, respectively, at a ratio of about 1:4. With the aid of mutants, each lacking one pair of cysteines, 65, 72 or 40, 95, respectively, another folding mechanism was found (30, 141), namely, the one in **Figure 3**; i.e., the rate-determining step involved oxidation of the unfolded $2S_U$ ensemble to form the native-like species des [65–72] and des [40–95], respectively, where in this case N is a native three-disulfide mutant. In the scheme of **Figure 3**, the rate and equilibrium constants obtained for the mutants imply that the combination of pathways leading to des [65–72] and des [40–95] in wild-type RNase A in the scheme of **Figure 2** accounts for only about 6% of the regeneration of native RNase A in **Figure 2**. Therefore, the pathways leading to des [65–72] and des [40–95] in **Figure 2** are still the dominant ones, accounting for 17% and 77%, respectively, of the folding process. These studies identified four alternative pathways in the oxidative folding of RNase A. There are undoubtedly other pathways for folding this protein, consistent with the concept that there is an ensemble of pathways, rather than a unique one, to fold a protein (113). The native-like character of des [65–72] and des [40–95] was established by 2D NMR investigations of the three-dimensional structures of mutants that could not form the 65–72 and 40–95 disulfide bonds, respectively (39, 118).

In a related investigation, the distribution of disulfide bonds in the 1S ensemble of wild-type

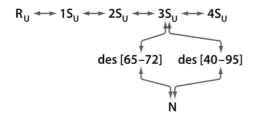

Figure 2

Regeneration of native RNase A along two different folding pathways by S-S/SH reshuffling of $3S_U$. The subscript U indicates that the species are unfolded. Adapted from figure 3 of *Biochemistry*, 37:3767–76 (1998) with permission.

RNase A was examined (140). Of the initially formed 28 theoretically possible one-disulfide species found in this ensemble, 40% had the native 65–72 disulfide bond, 10% had the non-native 58–65 disulfide bond, in a 4:1 ratio, and the remaining 50% were distributed among the 26 remaining species. The same 4:1 ratio of the native 65–72 to the nonnative 58–65 disulfide bonds was found in a small oligopeptide containing Cys residues 58, 65, and 72, demonstrating the importance of local interactions in determining disulfide-bond formation in both the oligopeptide and the parent protein (2).

Onconase is a sequence and structural protein homolog of RNase A. Three of its four disulfide bonds are in positions corresponding to those of RNase A, but the fourth one (analogue of the dominant 65–72 in RNase A, leading to des [40–95]) is in a different position. Therefore, it was expected that onconase would adopt a different folding mechanism. Consequently, the same type of oxidative folding of onconase with dithiothretol was carried out and, indeed, a different folding mechanism was found (20), as shown in **Figure 4**. Unlike the behavior of RNase A, the formation of the structured species of I_1 from 1S is a relatively fast critical step compared to the oxidation of I_1 to I_2 and I_3, or to the oxidation of I_2 and I_3 to N. The mechanisms in **Figures 2**, **3**, and **4** illustrate the sensitivity of folding pathways to subtle differences in amino acid sequence.

By contrast to oxidative folding, the folding of disulfide-intact RNase A from the unfolded to the native form is a very rapid reaction, requiring a stopped-flow technique to investigate the folding. RNase A is first unfolded with guanidine hydrochloride (Gdn · HCl) and then refolded by diluting out the Gdn · HCl. In the native protein, X-Pro 93 and 114 are *cis* and 42 and 117 are *trans*. The resulting pathway, which involves *cis-trans* isomerization of X-Pro peptide bonds, is represented as the box model (27) (**Figure 5**), where U_{vf}, in which the *cis-trans* isomerization has not taken place, under a particular set of unfolding conditions, is consequently a very-fast folding species. Then, three independent *cis-trans* and *trans-cis*

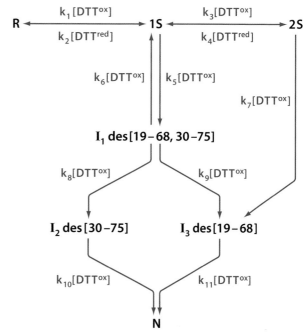

Figure 4

Oxidative folding mechanism for onconase. No 3S or 4S ensembles were detected under all folding conditions. The only 3S species are I_2 and I_3, and the only 4S species is N. Adapted from figure 4 of *Biochemistry*, 48:2740–51 (2009) with permission.

isomerization events take place at peptide bonds X-Pro 93, 114, and 117 in the unfolded state of the protein, with c and t referring to the peptide bonds in that order (X-Pro 42 does not change its native *trans* state during unfolding under this set of conditions). Such rapid *cis-trans* isomerization also occurs in slower oxidative folding.

More recently, folding and structural studies were carried out with peptide fragments of the B3 domain of the immunoglobulin binding protein G from *Streptococcus* (44) in order to provide information about the folding pathway of the wild-type protein as an aid in parameterization of a united-residue potential energy function for theoretical studies of protein folding.

In the late 1990s, I felt that I had succeeded in providing considerable information about the thrombin-fibrinogen interaction (107). With the aid of various physicochemical techniques, we had demonstrated how clotting

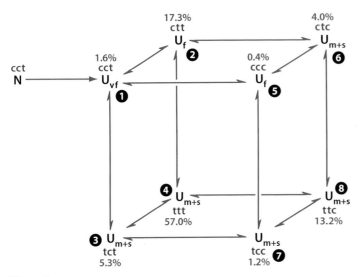

Figure 5

Distributions of different unfolded species (U_{vf}, U_f, and U_{m+s}) under a particular set of folding conditions. The subscripts represent the different refolding phases observed. c and t refer to the *cis* or *trans* conformation about the X-Pro 93, 114, and 117 peptide bonds, in that order. The unfolded species are defined according to the conformations about these X-Pro peptide bonds. The percentage next to each unfolded species refers to the relative concentration of that species in the equilibrium unfolded state of the protein at 4.2 M Gdn·HCl, pH 2.0, and 15°C. Adapted from figure 5A of *Biochemistry*, 35:11719–33 (1996) with permission.

was initiated in a reversible (114) proteolytic reaction to liberate fibrinopeptides A and B and expose sites for polymerization of the resulting fibrin monomer. With a second polymerization site on the original fibrinogen molecule, the now two polymerization sites fostered a reversible (11) polymerization of fibrin monomer into elongated, staggered-overlapping, rod-like clottable intermediates that then aggregated reversibly (114) to form a cross-linked gel whose structure John Ferry and Peter Morrison had characterized in 1947 (18). As evidence of the reversibility of this final reaction, it is possible to dissolve fibrin clots in buffer at neutral pH, but only if the reaction is carried out with pure fibrinogen (114). In blood plasma, there is an additional enzyme that, as shown by Laszlo Lorand (57), introduces covalent cross-links into the fibrin clot to stabilize it. Further subsequent work with transferred NOE provided information about the mechanism of action of thrombin on fibrinogen (80). With this

information now available, I decided to close my experimental research lab on blood clotting and confine my future research efforts to experimental and theoretical aspects of protein folding.

Having satisfactorily carried out experiments on oxidative folding and reductive unfolding of RNase A, subsequently with Dave Rothwarf, and similar folding/unfolding of disulfide-intact RNase A with Walid Houry, I decided to stop all experimental work on protein folding in December 2008 and devote my remaining research effort to theoretical work on both protein folding and biological processes involving protein-protein and protein–nucleic acid interactions. I waited until Robert Gahl, my last graduate student in my experimental group, finished his PhD thesis research, and until I could find an alternative opportunity in the laboratory of my colleague, George Hess, for Lovy Pradeep, my last experimental postdoc. I then moved my office from the Olin Research Wing to Baker Laboratory to be close to my postdocs, who are now carrying out theoretical research on the protein-folding and related problems.

ECEPP

Turning to our theoretical activity (77, 104), initiated in the 1960s to compute structure, folding pathways, thermodynamics, and kinetics from amino acid sequences, we started with a simplified model with which we treated atoms as billiard balls, with the hard-sphere approximation. Atoms weren't allowed to overlap; therefore, many stereochemical conformations were eliminated. Also the Ramachandran map was used as a constraint. I remember Peter Debye commenting after my lecture at a Welch Foundation meeting, "But atoms aren't hard billiards balls." Of course we knew that, and we were already working on developing more realistic interatomic potentials (67), namely ECEPP (Empirical Conformational Energy Program for Peptides).

As originally developed with my long-time associate Frank Momany, ECEPP was based on structures from the Protein Data Bank and also on experimental data. The effect of water was treated as a continuum in terms of several exposed-surface or volume-overlap protocols added to ECEPP.

It is of interest that, as shown by the superposition of the hard-sphere potential map for terminally blocked alanine on that for the rigid-geometry ECEPP potential in figure 15 of Reference 104, the hard-sphere map captures most of the character of the map for the rigid-geometry potential. This is simply a demonstration that atoms occupy space and have to avoid steric overlaps; i.e., the hard-sphere potential is a dominant factor determining polypeptide stereochemistry, with minor modifications from the electrostatic, hydrogen-bonding, and torsional contributions.

In addition to having a good potential energy function (force field), one must have an efficient procedure to search conformational space to locate the global minimum of the free energy of the system, i.e., the protein plus the solvent. For this purpose, it is necessary to globally optimize a function of hundreds of independent variables in a space with many local minima especially if, as in our original work, we wanted to minimize the potential energy. Several local-minimization algorithms were evaluated with Ken Gibson (22), and we found the most efficient one to be that of Gay (21) for an all-atom force field.

Many global optimization algorithms have been reported in the literature, but they deal largely with problems involving a small number of independent variables. Because of the large number of such variables in an all-atom force field, we have spent many years developing and evaluating a large menu of global, rather than local, optimization procedures (105).

The particularly useful procedure Electrostatically Driven Monte Carlo (EDMC) (96) was developed by Daniel Ripoll as a combination of an electrostatically optimized procedure formulated by Lucjan Piela (86) and Monte Carlo-plus-minimization developed by Zhenqin Li (46). Using EDMC and ECEPP, Jorge Vila and Daniel Ripoll computed the native structure of the 46-residue α-helical protein A (128), the largest protein simulated at that time with an all-atom force field. Piela, a visiting scientist from the University of Warsaw, also developed a diffusion-equation procedure (85) with Jaroslaw Kostrowicki to smooth the potential energy surface to locate the global minimum.

ECEPP was developed for a rigid polypeptide, i.e., one with fixed bond lengths and bond angles, for which the variables to change conformation are the torsional angles for rotation about the bonds of the backbone and side chain, and parameterized originally on experimental data. In principle, bond lengths and bond angles should be allowed to vary, as in subsequently developed force fields such as CHARMM (9) and AMBER (137), but, being initially constricted by limited computational facilities, we wanted to keep the number of independent variables as low as possible. When the bond lengths and bond angles are allowed to vary, a potential energy function must be included in the force field for this purpose. In most force fields that use such flexible geometry, a harmonic potential is introduced to describe the variation of bond lengths and bond angles. However, with a postdoc, Irena Roterman (98), we showed that, because a harmonic potential allows only for small variations around the minimum, the use of such a potential leads to small ($\sim 3°$) variations of bond angles, which raises a question about how chemical bonding treats a $3°$ variation in the difference of only $10.5°$ between an sp^2 and an sp^3 carbon. This error can be avoided by adopting an anharmonic potential.

Being limited by the computer power available at the time of our initial theoretical work on conformational energy calculations, we focused on relatively simple problems such as the physical factors leading to the preference of particular homopolymer α helices to adopt either a right- or a left-handed twist (82). Conformational energy calculations were also carried out to elucidate the molecular basis for the helix-coil transition in homopolymers (23), for the cis-trans isomerism in the transition between forms I and II of poly-L-proline (123), for the formation of the triple helix of collagen-like polytripeptides such as poly-Gly-Pro-Pro (65), and for the metastable form of native cellulose (119).

A natural peptide whose structure we computed with Mary Dygert and Nobuhiro Gō, based on earlier calculations of symmetrical and nonsymmetrical cyclic peptides with Gō, was the cyclic decapeptide gramicidin S (13). Because there were no experimental data with which to compare our calculated structure, we decided to publish the Cartesian coordinates (13). Sometime later, Mirau & Bovey (66) used these coordinates to compute a 2D NMR spectrum in comparison with their experimental one; they obtained good agreement for the α-carbon coordinates.

To examine the physical features of the common regular structures of proteins, with Kuo-Chen Chou and George Némethy, I've identified the interactions that led to the twist of β sheets and the interactions between pairs of β sheets, a β sheet with an α helix, and pairs of α helices (109). A visiting scientist, Wayne Mattice, developed a theory for the transition between a statistical coil and an antiparallel β sheet (64). Another attempt to examine the physical features appearing in proteins was the application of differential geometry by my postdoc Shalom Rackovsky (95). He was able to use this technique to characterize the protein landscape on a larger length scale than the single residue one of Ramachandran.

Venturing into large globular proteins with Paul Warme, a postdoc, we tried to use homology modeling to compute the structure of α-lactalbumin from that of lysozyme (136). When the X-ray structure of α-lactalbumin was later determined (1), there was agreement between it and the computed structure, except in the flexible C-terminal region of the molecule.

ENZYME-SUBSTRATE COMPLEXES

I had a long-term collaboration with Matthew Pincus that started when he joined my lab as a postdoc. One of his important contributions was the development of methodology to compute the structures of enzyme-substrate complexes. An example of this work is the complex of hen egg-white lysozyme with a hexasaccharide substrate. Lysozyme cleaves this hexasaccharide between the fourth and fifth glucose units, and David Phillips' crystal structure of lysozyme with a trisaccharide showed the position of this substrate in the active site and placed the remaining three glucose units by model building (28). Our calculations (87) placed the last few glucose units in a different position than that reported by Imoto et al. (28). In a subsequent experimental study with Sandy Smith-Gill et al. (120), our calculated position and orientation of the whole hexasaccharide in the active site of lysozyme were verified.

UNITED RESIDUE MODEL

In 1989, I had the opportunity to attend a polymer symposium in Warsaw and, among other things, to reconnect with Piela, who arranged for Miriam and me to visit several sites in Poland. Among them was a trip to the University of Gdansk, where my host was Professor Zbigniew Grzonka, who recommended that I accommodate one of his PhD students, Adam Liwo, in my lab as a postdoc. This was the start of a very fruitful, and thus far, 20-year collaboration with Adam and other Polish colleagues (Stanislaw Ołdziej, Cezary Czaplewski, Jaroslaw Pillardy, Mariusz Makowski, and others).

Adam has played a leading role in developing and applying a coarse-grained united-residue (UNRES) hierarchical approach to the protein-folding problem. As originally developed in 1997–1998 (50, 55, 56), it was based on structures from the Protein Data Bank but gradually all its terms, except for side chain–side chain interactions, were replaced by the results of quantum mechanical calculations on model compounds. More recently, side chain interactions are beginning to be expressed as potentials of mean force of the interaction of amino acid side chains in water (M. Makowski, A. Liwo, E. Sobolewski & H.A. Scheraga, unpublished observations; M. Makowski, A. Liwo & H.A.

Scheraga, unpublished observations). UNRES is therefore based primarily on physics rather than on knowledge-based information, in order to provide an understanding about the interactions leading to the structure and properties of proteins. The development of this coarse-grained model was necessary because it was clear that, with an all-atom force field, we would not be able to compute the structures of proteins containing 100 or 200 amino acid residues. The only all-atom globular protein structure that we had previously been able to compute was the 46-residue protein A (128). Our coarse-grained hierarchical approach involved a search of the UNRES space with a conformational space annealing (CSA) method developed by a postdoc, Jooyoung Lee (43), in order to locate only the region of the global minimum. Then, the low-energy UNRES conformations were converted to all-atom conformations by a procedure developed with a postdoc, Rajmund Kazmierkiewicz (33, 34), and the search was continued with the all-atom ECEPP force field (67) and the EDMC search (96) to locate the global minimum of the ECEPP energy. Frequently, it is not necessary to implement the final step, i.e., the EDMC search. In 2001, the UNRES model was implemented by Adam Liwo to include cumulant-based expressions for multibody terms for the correlation between local and electrostatic interactions in the UNRES force field (49). In 2002, Adam developed a method to optimize potential-energy functions by a hierarchical design of the folding pathway, and applied it to the UNRES landscape (47). To judge the validity of our approach with UNRES, we participated in the CASP3 exercise. One of our computed structures (53), shown in **Figure 6**, provided excellent results for the N-terminal portion and the larger C-terminal portion. In the evaluation of the results of **Figure 6**, the judges (83) reported that "for protein HDEA... the most impressive prediction was that of Scheraga's group using... ab initio methods.... Their method uses no information from sequence alignments, secondary structure prediction, or threading."

CRYSTAL STRUCTURES

With long-time collaborator Yelena Arnautova, we improved the all-atom ECEPP potential function (3, 4, 6). We have also participated in a CASP-like blind test, run by the Cambridge Crystallographic Data Center, to compute the crystal structures of small organic molecules, given their covalent structure. As with our UNRES approach in CASP, we computed crystal structures using only a physics-based empirical force field without knowledge-based information such as space groups, unit cell dimensions, or experimental powder diffraction patterns. Thus far, because of problems such as polymorphism and variable kinetics of crystallization, significant success has not yet been achieved, although the observed structure frequently is found among a group of low-energy structures, but not the lowest-energy one (12).

MOLECULAR DYNAMICS WITH UNRES

Initially, UNRES was developed to compute structure. Subsequently, because the degrees of freedom, which limit the folding simulation time, were averaged out in UNRES, it was possible to extend the timescale of molecular dynamics to follow the kinetics of protein folding (35, 36, 52). Some examples are shown in **Figure 7** (52) for the folding of single-chain proteins. All simulations were started from the fully extended chain, and the timescale depends on the force field. The trajectory at the top pertains to an all-α-helix bundle (the largest folded structure obtained thus far by molecular dynamics); and the trajectory at the bottom pertains to an $\alpha + \beta$ protein.

The molecular dynamics treatment with UNRES was also extended to include ab initio folding simulations of multichain proteins (97). **Figure 8** illustrates the folding trajectory for an α-helical homodimer with 48 residues per chain. This methodology was applied to simulate the molecular mechanism of fiber assembly in aggregation of the Aβ peptide (96a) (**Figure 9**).

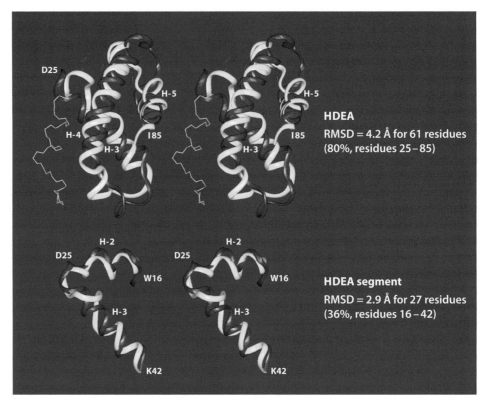

Figure 6

(*Top*) Stereo view of superposition of the C^α atoms of the fragment included between residues D25 and I85 of the crystal (*red*) and predicted (*yellow*) structures of HDEA. Helices 3, 4, and 5 are indicated as H-3, H-4, and H-5, respectively. (*Bottom*) Stereo view of superposition of the C^α atoms of the 27-residue segment (W16 to K42) of HDEA. Helices 2 and 3 are indicated as H-2 and H-3, respectively. Adapted from figures 1 and 2 of *Proc. Natl. Acad. Sci. USA*, 96:5482–85 (1999) with permission.

IMPROVEMENT OF UNRES

To improve the efficiency of covering the UNRES free energy surface, replica exchange molecular dynamics (REMD) (70, 71) and multiplexed replica exchange molecular dynamics (MREMD) (10) were implemented. Further, to introduce entropic effects, i.e., to compute free energies instead of potential energies, and thereby evaluate thermodynamic properties, Liwo et al. (51) included temperature dependency in the UNRES force field. With this enhancement, the peak of the calculated heat capacity curve, and the folding temperature, matched experimental results.

Attention is continually being paid to improvement of the UNRES force field.

Recognizing the importance of side chain–side chain interactions, Adam Liwo and Mariusz Makowski are carrying out calculations of potentials of mean force for the interaction of amino acid side chains in water (M. Makowski, A. Liwo, E. Sobolewski & H.A. Scheraga, unpublished observations; M. Makowski, A. Liwo & H.A. Scheraga, unpublished observations). In order to extend the applicability of UNRES to very large proteins, Liwo et al. (54) implemented a massively parallel architecture into the UNRES code. The parallelized code provides an effective 1,000-fold speed-up compared to the experimental timescale and therefore enables us to effectively carry out millisecond-scale molecular dynamics

1CLB (α, 75 residues)

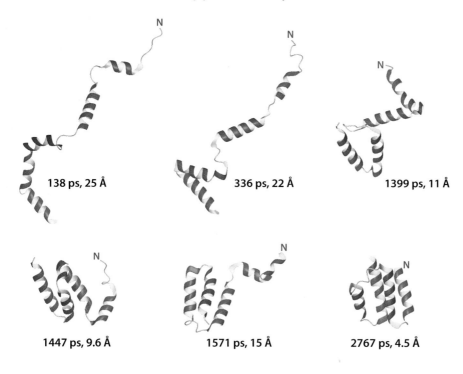

138 ps, 25 Å   336 ps, 22 Å   1399 ps, 11 Å

1447 ps, 9.6 Å   1571 ps, 15 Å   2767 ps, 4.5 Å

1E0G (α+β, 48 residues)

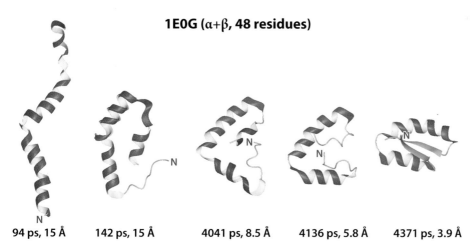

94 ps, 15 Å   142 ps, 15 Å   4041 ps, 8.5 Å   4136 ps, 5.8 Å   4371 ps, 3.9 Å

Figure 7

(*Top*) Example of a fast-folding pathway of 1CLB obtained in Langevin dynamics simulations, starting from an extended structure. (*Bottom*) Example of a fast-folding pathway of 1E0G. Adapted from figures 4 and 5 of *Proc. Natl. Acad. Sci. USA*, 102:2362–67 (2005) with permission.

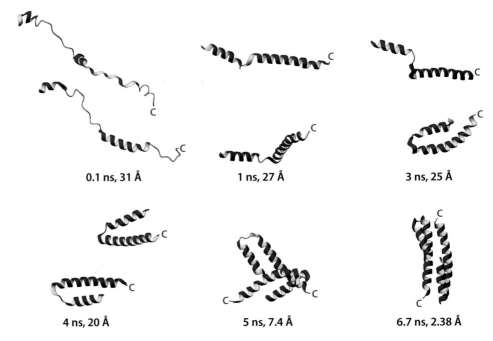

0.1 ns, 31 Å   **1 ns, 27 Å**   **3 ns, 25 Å**

4 ns, 20 Å   **5 ns, 7.4 Å**   **6.7 ns, 2.38 Å**

Figure 8

Example of a successful trajectory of 1 G6U obtained with Langevin dynamics, starting from an extended structure. Adapted from figure 5 of *J. Phys. Chem. B*, 111:293–309 (2007) with permission.

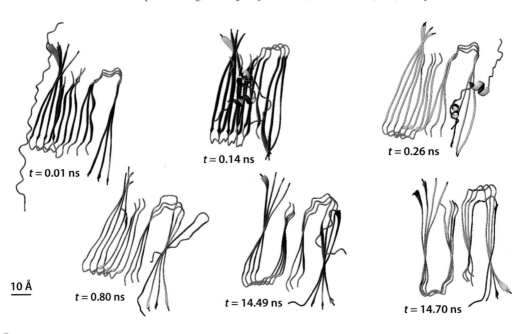

$t = 0.01$ ns $t = 0.14$ ns $t = 0.26$ ns

10 Å

$t = 0.80$ ns $t = 14.49$ ns $t = 14.70$ ns

Figure 9

Selected snapshots along a representative trajectory of an initially extended monomer binding to an Aβ fibril. Adapted from figure 7 of *J. Mol. Biol.*, 404:537–52 (2010) with permission.

simulations of proteins with 500 and more amino acid residues in days of wall-clock time.

The simulations of molecular dynamics trajectories are now being subjected to principal component analysis (PCA) to evaluate free energy landscapes and dynamics (59–62). This work is being carried out by Gia Maisuradze, a research associate, and Adam Liwo. Dynamics of fluctuations in the native state of a protein are being examined by Patrick Senet (117), a visiting scientist, and by Cote et al. (143). Much of our recent work on dynamics has been summarized in two review chapters (48, 112).

BIOLOGICAL PROBLEMS

Recently, we have been trying to apply UNRES molecular dynamics to treat biological problems such as the aggregation of the Aβ peptide (**Figure 9**). In collaboration with Harel Weinstein of Cornell Medical School, Yi He, a postdoc, and Adam Liwo used this approach to elucidate the binding of the PDZ domain to the BAR domain of PICK1 (144). PICK1 is a key regulator of receptor traffic and interacts with over 40 other proteins, including receptors, transporters, and ionic channels. Adam and others have also deciphered the mechanism of opening of an Hsp70 chaperone with coarse-grained canonical and replica-exchange molecular dynamics (A. Liwo, G.G. Maisuradze, P. Senet, S. Ołdziej, C. Czaplewski & H.A. Scheraga, unpublished observations). This involved simulation of the transition from the closed (substrate bound) to the open (substrate nonbound) conformation of the Hsp70 chaperone from *Escherichia coli* (PDB code: 2KHO).

CHEMICAL SHIFTS

Most recently, we have begun to exploit the use of NMR $^{13}C^{\alpha}$ chemical shifts to validate and determine protein structures. This approach was formulated and developed by Jorge Vila, who was a postdoc in my lab and is now professor and researcher at the National University of San Luis, in Argentina. For several years, he has been working in my group as a senior research associate, and he initiated a $^{13}C^{\alpha}$ chemical shift-based validation and determination protocol to exploit the following properties: (*a*) The assignment of $^{13}C^{\alpha}$ chemical shifts is a fundamental step in protein structure determination by NMR spectroscopy, with no extra experimental work needed. (*b*) In addition to the impact of the covalent structure, $^{13}C^{\alpha}$ chemical shifts, but no other nuclei, are determined mainly by the intraresidue backbone and side chain dihedral angles. (*c*) $^{13}C^{\alpha}$ is ubiquitous in proteins. (*d*) $^{13}C^{\alpha}$ chemical shifts, for any residue **Xxx** in a protein, can be computed with high accuracy at the quantum chemistry level of theory by using an ECEPP-rigid-geometry model peptide: Ac-Gly-**Xxx**-Gly-NMe. Consequently, the discrete Fourier transform (DFT)-based methodology has been used recently to (*a*) establish the factors affecting the computation of $^{13}C^{\alpha}$ chemical shifts (127, 130, 132); (*b*) accurately validate complete protein structures solved by X-ray crystallography and NMR spectroscopy (5, 131–133); (*c*) unambiguously identify flaws in the sequence at the residue level (131); (*d*) determine α-helical (124, 129) and β-sheet protein structures (126); and (*e*) develop a purely physics-based server, *Che*Shift (125), available to the scientific community that enables one to predict $^{13}C^{\alpha}$ chemical shifts with reasonable accuracy in seconds and, hence, provides a standard with which to evaluate the quality of any reported protein structure solved by either X-ray crystallography or NMR-spectroscopy, provided that the experimentally observed $^{13}C^{\alpha}$ chemical shifts are available.

FUTURE OUTLOOK

Looking ahead to the next five years, we can only hope that the UNRES model will be sufficiently refined so that one can feel assured that the physics of the protein-folding problem is understood. If so, we should be able to predict protein structure and protein-folding pathways, and thermodynamics and kinetics of folding, from amino acid sequences. We can also envision extensions to predict protein structures and folding pathways in a membrane

Table 1   Members of the Scheraga research group (1950–2010)

Adler, Harvey	Cohly, Mauj	Gō, Mitiko	Kim, J.S.
Adler, Mark	Crippen, Gordon	Gō, Nobuhiro	Kincaid, Robert
Ali, Akhbar	Culver, Diane	Gonzelez-Prolongo, M.	Klein, Terry
Alter, John	Czaplewski, Cezary	Grant, Andrew	Kleinerman, Dana
Altmann, Karl-Heinz	Czurylo, E.	Greenwall, Ann	Kobayashi, Yuji
Amas, Amando	DaCosta, H.	Griffith, Joe	Konishi, Yasuo
Amir, Dan	Dalzell, Bonnie	Gruen, Lise	Kostrowicki, Jaroslaw
An, Alex	Davenport, Gary	Hageman, Tom	Kotelchuck, Dave
Ananthanarayanan, V.	Denton, James	Hages, Nick	Kozlowska, Urszula
Anderson, Herbert R.	Denton, Mary	Hagler, Arnold	Kresheck, Gordon
Anderson, Janet	Derbenwick, Jean	Hamori, Eugene	Kretch, Laura
Andreatta, Rudolph	DiBella, Elsie	Han, Shaw-Lin	Krieger, Neil
Arnautova, Yelena	Dodge, Robert	Hanna, Lewis	Kunimitsu, Donald
Ashton, Robert	Donnelly, Thomas H.	Hao, Ming-Hong	Kusko, Stephanie
Auer, Henry	Donovan, John	Hatcher, Paul	Kwon, Oh Young
Augspurger, Joseph	Dudek, Mike	Hathaway, Laura	Laity, John
Backus, John K.	Dunfield, Lawrence	He, Siqian	Lambert, Millard
Ball, Robert	Dygert, Mary	He, Yi	Larsen, Paul
Beals, John	Dygert, Steve	Hecht, Michael	Laskowski, Michael
Bhat, Rajiv	Dzyabchenko, Alex	Heckel, A.	Leach, Syd
Bond, Jeff	Eastman, Margaret	Hermans, Jan	Lee, Eok
Bosenberg, Max	Eckstein, Bernard H.	Hesselink, Theo	Lee, Jooyoung
Bradbury, Howard	Ehrenpreis, Seymour	Hill, David	Lee, S.
Bratley, Margaret	Endres, Gerry	Hocking, Colin	Leed, Andrew
Broomfield, Clarence	English, Brian	Hodes, Zachary	Lentz, Barry
Browman, Jean	Epand, Raquel	Houry, Walid	LeRoy, Lise
Bryan, William	Epand, Richard	Horwitz, Bruce	Lester, Cathy
Buckler, David	Erenrich, Eric	Howard, James	Leung, Howard
Bunville, Lyle	Falcomer, Caterina	Hughes, Lou	Lewandowska, A.
Burgess, Antony	Feinberg, Tom	Incledon, Bev	Lewis, Aaron
Bush, Allen	Fitzwater, Susan	Ink, Elaine (Asst.)	Lewis, Peter
Cao, Aoneng	Forcada, M.	Ingwall, Joanne	Li, Lu-Ku
Carbone, Frank	Fossey, Steve	Ingwall, Richard	Li, Yue-Jin
Cardinaux, Francis	Frederickson, Robert	Isogai, Yoshinori	Li, Zhenqin
Carruthers, Lucy	Friedman, Michael	Iwaoka, Michio	Lieberman, Paul
Cerovsky, Vaclav	Fu, Janet	Jagielska, Ana	Liebman, Joe
Cha, Chul-Yung	Fu, Frank	Joseph, Marie-Pierre	Liem, Ron
Chan, H.	Fujioka, Hideko	Joubert, Franz	Lin, S.
Chang, Mary	Gabel, Detlev	Jourdan, Natalie	Liwo, Adam
Chase, Lynn	Gahl, Robert	Juminaga, Darmawi	Loeb, George
Chavez, Lloyd	Galuska, Barbara	Kang, Young Kee	Lotan, Noah
Cheng, Betty	Garduno, R.	Kappel, Joan	Low, L.
Cherayil, Binney	Gerritsen, Mirjam	Karpinski, M.	Lynn, R.
Chinchio, Maurizio	Gerry, Janice	Kashuba, Karen	Maciejczyk, Maciej
Chipot, Chris	Ghosh, Avijit	Kazmierkiewicz, R.	Maigret, Bernard
Cho, Kwang Hwi	Gibson, Kenneth	Khalili, Mei	Maisuradze, Gia
Chou, Kuo-Chen	Giglio, Jose	Kidera, Akinori	Makowska, Joanna
Chou, Peter	Glasser, Leslie	Kikuchi, Takeshi	Makowski, Mariusz

(Continued)

Table 1   (*Continued*)

Mark, James E.	Owicki, Jack	Saunders, Jeff	Tijn Van, B.
Matheson, Robert	Paine, Greg	Schafroth, Heather	Timms, David
Marsh, Henry	Paiva, Anna	Schapiro, Bernard L.	Toupin, Barbara
Martin, Osvaldo	Paiva, Tony	Scheule, Ron	Trosset, Jean-Yves
Martinelli, Richard	Palmer, Kate	Schildkraut, Carl	Troxell, Terry
Mattice, Wayne	Parr, Gwen (Secy.)	Schneider, Henry	Tuttle, Robert
Maurer, Muriel	Paterson, Yvonne	Schrier, Eugene	Ueda, Yuzo
Maxfield, Fred	Payne, Phil	Schussler, Helga	Vali, Zsofi
McDonald, M.	Peer, William	Scott, Roy	Vanderkooi, Gary
McFaddin, Mary	Perlman, Sidney	Seidman, Steve	Van Nispen, Jan
McGuire, Robert	Perrot, George	Sendak, Rebecca	Van Wart, Harold
McHale, Janet	Piela, Lucjan	Senet, Patrick	Varadi, Andras
McQuie, Jerome	Pillardy, Jaroslaw	Shames, Peter	Vásquez, Max
McWherter, Charles	Pincus, Matthew	Shanbhag, Niraj	Vila, Jorge
Mei, Dorothy	Platzer, Karen	Shen, Hujun	Volles, Michael
Meinwald, Yvonne	Poland, Douglas	Shi, P.T.	Von Dreele, Patricia
Meirovitch, Hagai	Ponnuswamy, P.K.	Shieh, Huey	Vorobjev, Yury
Mernan, John P.	Pottle, Marcia	Shimotakahara, S.	Vournakis, John
Milburn, Peter	Powers, Susan	Shin, Hang-Cheol	Wagner, Myron
Miller, Martha	Poyer, Joyce (Secy.)	Shipman, Les	Wako, Hiroshi
Minteer, James	Pradeep, Lovy	Siegel, Maxine	Wang, Z.X.
Momany, Frank	Prusinowski, Longine	Silverman, Dave	Wanjalla, Celestine
Montelione, Gaetano	Purisima, Enrico	Simon, Istvan	Warme, Paul
Moon, A. Young	Rabow, Al	Simon, Martha	Wawak, Ryszard
Mooney, Richard W.	Rackovsky, Shalom	Sippl, Manfred	Wedemeyer, William
Moxley, Nancy	Rae, Ian	Skwierawska, A.	Weetall, Marsha
Moy, Franklin	Rakowitz, David	Slutsky, Leon	Weiner, Michael
Mui, Phil	Rapaport, David	Snir, Joseph	Weinstein, Larry
Murarka, Rajesh	Rasse, Daniel	Solomon, Leonard	Welker, Ervin
Nacuta, Dan	Regan, Mark	Song, Myeong-Cheol	Wertz, David
Nagy, Janice	Rehr, Sherry	Spasic, Alex	Widom, Joanne
Nakajima, Akio	Riddiford, Lynn	Steinberg, Izchak	Williams, Roger
Nanias, Marian	Riehm, John	Stimson, Evelyn	Windsor, Don
Narayan, Mahesh	Ripoll, Daniel	Stokrova, Stepanka	Wojcik, Jacek
Nayeem, Achbar	Rodziewicz-Motowidlo,	Stone, Nancy	Wolf, R.M.
Nemenoff, Raphael	Sylwia	Streuli, Carl	Woody, Robert
Némethy, George	Rojas, Ana	Sueki, M.	Wu, Victor
Ni, Feng	Rosen, Susan	Sullivan, Mary	Xu, John
Nishikawa, Ken	Roterman, Irena	Sun, Hongzhi	Xu, X.
Niu, Gregory	Rothwarf, David	Swadesh, Joel	Yan, Johnson
No, Kyoung Tai	Rudolph, Reiner	Swegler, Ernest	Yang, P.P.
Oberlin, Dan	Rumball, Silvia	Swenson, Mary	Yoon, Byung-Jip
Oka, Masahito	Rumsey, Shirley	Takahashi, Sho	Yoon, Chang-No
Ołdziej, Stanislaw	Rupley, John	Talluri, Sekhar	Zagari, Adriana
Oliveira, Theresa	Ruterjans, Heinrich	Tanaka, Seiji	Zheng, Z.
Olpin, Jay	Saito, K.	Taylor, Gerry	Zimmerman, Scott
Olszewski, Kris	Saluja, P.P.S.	Taylor, Miriam	Zweifel, Bernard
Ooi, Tatsuo	Sanger, Betsy (Asst.)	Thannhauser, Ted	
Ostroy, Sanford	Saunders, Jean	Thomas, Mary	

environment and those leading to protein-protein and protein–nucleic acid complexes. Our ultimate goal is to gain a fundamental understanding of how interatomic interactions determine these structures and pathways, and the formation of intermolecular complexes, for large systems on a timescale approaching the experimental one.

To achieve this goal, it will be necessary to finish the development of UNRES by replacing the last-remaining knowledge-based term, corresponding to side chain–side chain interactions, by physics-based terms, i.e., potentials of mean force. One can expect to see an extension of UNRES to treat proteins in the lipid-membrane environment and to treat proteins at various pH values. It will also be necessary to complete our current efforts to develop an UNRES model of nucleic acids, and to merge it with UNRES into a viable package to treat protein–nucleic acid interactions. In addition to developments with the coarse-grained model, we can envision improvement in the treatment of proteins at the all-atom level with enhanced use of quantum mechanical calculations of $^{13}C^{\alpha}$ chemical shifts in protein structure simulations.

All this methodology must serve the purpose of clarifying the mechanism of biological processes such as signal transduction and the action of chaperones. Finally, the excitement involved in trying to understand biology through chemistry and physics should continue to attract young scientists into this area of research.

DISCLOSURE STATEMENT

The author is not aware of any affiliations, memberships, funding, or financial holdings that might be perceived as affecting the objectivity of this review.

ACKNOWLEDGMENTS

With many collaborators, as well as an excellent group of graduate students, postdocs, visiting senior scientists, and even undergraduate students, I have been able to carry out the research discussed here. Many of them went on to academic positions in the United States and elsewhere. The domestic and foreign coworkers were helpful in fostering a steady stream of experienced scientists to work in my lab. Initially, they came from Australia, then Japan, England, Israel, Hungary, Argentina, Taiwan, Korea, Russia, China, France, Italy, India, Germany, and, more recently, Poland. They constitute a large family of scientific offspring (**Table 1**) working on protein science.

I want to express my indebtedness to Paul Gross, who helped advance my career on three separate occasions. First, he offered me the opportunity to do graduate work at Duke. Second, he supported my ACS fellowship application and recommended me to John Edsall as a postdoctoral fellow. Third, he recommended me to Debye for a faculty position at Cornell. I appreciate the outstanding research carried out in my lab by my many coworkers of every age, and I am indebted to the Office of Naval Research for my initial grant and to the National Institutes of Health and the National Science Foundation for continued financial support throughout my career. Above all, without Miriam's support over 67 years together, it would have been very difficult to maintain an active research program.

My curriculum vitae (including a list of honors) and my complete publication list can be found on my website at **http://www.chem.cornell.edu/has5/**.

LITERATURE CITED

1. Acharya KR, Stuart DI, Phillips DC, Scheraga HA. 1990. A critical evaluation of the predicted and X-ray structures of α-lactalbumin. *J. Protein Chem.* 9:549–63

2. Altmann KH, Scheraga HA. 1990. Local structure in ribonuclease A. Effect of amino acid substitutions on the preferential formation of the native disulfide loop in synthetic peptides corresponding to residues Cys^{58}-Cys^{72} of bovine pancreatic ribonuclease A. *J. Am. Chem. Soc.* 112:4926–31

3. Arnautova YA, Jagielska A, Scheraga HA. 2006. A new force field (ECEPP-05) for peptides, proteins and organic molecules. *J. Phys. Chem. B* 110:5025–44

4. Arnautova YA, Scheraga HA. 2008. Use of decoys to optimize an all-atom force field including hydration. *Biophys. J.* 95:2434–49

5. Arnautova YA, Vila JA, Martin OA, Scheraga HA. 2009. What can we learn by computing ^{13}C$^{\alpha}$ chemical shifts for X-ray protein models? *Acta Crystallogr. D* 65:697–703

6. Arnautova YA, Vorobjev YN, Vila JA, Scheraga HA. 2009. Identifying native-like protein structures with scoring functions based on all-atom ECEPP force fields, implicit solvent models and structure relaxation. *Proteins Struct. Funct. Bioinform.* 77:38–51

7. Backus JK, Laskowski M Jr, Scheraga HA, Nims LF. 1952. Distribution of intermediate polymers in the fibrinogen-fibrin conversion. *Arch. Biochem. Biophys.* 41:354–66

8. Bixon M, Scheraga HA, Lifson S. 1963. Effect of hydrophobic bonding on the stability of poly-L-alanine helices in water. *Biopolymers* 1:419–29

9. Brooks BR, Bruccoleri RE, Olafson BD, States DJ, Swaminathan S, Karplus M. 1983. CHARMM: a program for macromolecular energy minimization, and dynamics calculations. *J. Comp. Chem.* 4:187–217

10. Czaplewski C, Kalinowski S, Liwo A, Scheraga HA. 2009. Application of multiplexed replica exchange molecular dynamics to the UNRES force field: tests with α and $\alpha + \beta$ proteins. *J. Chem. Theory Comput.* 5:627–40

11. Donnelly TH, Laskowski M Jr, Notley N, Scheraga HA. 1955. Equilibria in the fibrinogen-fibrin conversion. II. Reversibility of the polymerization steps. *Arch. Biochem. Biophys.* 56:369–87

12. Dunitz D, Scheraga HA. 2004. Exercises in prognostication: crystal structures and protein folding. *Proc. Natl. Acad. Sci. USA* 101:14309–11

13. Dygert M, Gō N, Scheraga HA. 1975. Use of a symmetry condition to compute the conformation of gramicidin S. *Macromolecules* 8:750–61

14. Dyson HJ, Wright PE, Scheraga HA. 2006. The role of hydrophobic interactions in initiation and propagation of protein folding. *Proc. Natl. Acad. Sci. USA* 103:13057–61

15. Einert TR, Staple DB, Kreuzer H-J, Netz BR. 2010. A three-state model with loop entropy for the overstretching transition of DNA. *Biophys. J.* 99:578–87

16. Epand RM, Scheraga HA. 1967. Enthalpy of stacking in single-stranded polyriboadenylic acid. *J. Am. Chem. Soc.* 89:3888–92

17. Everaers R, Kumar S, Simm C. 2007. Unified distribution of poly- and oligonucleotide DNA melting: nearest-neighbor, Poland-Scheraga, and lattice models. *Phys. Rev. E* 75:041918

18. Ferry JD, Morrison PR. 1947. Preparation and properties of serum and plasma proteins. VIII. The conversion of human fibrinogen to fibrin under various conditions. *J. Am. Chem. Soc.* 69:388–400

19. Frank HS, Evans MJ. 1945. Free volume and entropy in condensed systems. III. Entropy in binary liquid mixtures; partial molal entropy in dilute solutions; structure and thermodynamics in aqueous electrolytes. *J. Chem. Phys.* 13:507–32

20. Gahl RF, Scheraga HA. 2009. Oxidative folding pathway of onconase, a ribonuclease homologue: insight into oxidative folding mechanisms from a study of two homologues. *Biochemistry* 48:2740–51

21. Gay DM. 1983. Algorithm 611: subroutine for unconstrained minimization using a model/trust-region approach. *ACM Trans. Math. Soft.* 9:503–4

22. Gibson KD, Scheraga HA. 1967. Minimization of polypeptide energy. I. Preliminary structures of bovine pancreatic ribonuclease S-peptide. *Proc. Natl. Acad. Sci. USA* 58:420–27

23. Gō M, Hesselink FT, Gō N, Scheraga HA. 1974. Molecular theory of the helix-coil transition in poly(amino acids). IV. Evaluation and analysis of s for poly(L-valine) in the absence and presence of water. *Macromolecules* 7:459–67

24. Gō N, Scheraga HA. 1969. Analysis of the contribution of internal vibrations to the statistical weights of equilibrium conformations of macromolecules. *J. Chem. Phys.* 51:4751–67

25. Gō N, Scheraga HA. 1976. On the use of classical statistical mechanics in the treatment of polymer chain conformation. *Macromolecules* 9:535–42

26. Hao M-H, Scheraga HA. 1998. Theory of two-state cooperative folding of proteins. *Acc. Chem. Res.* 31:433–40

27. Houry WA, Scheraga HA. 1996. The nature of the unfolded state of ribonuclease A: effect of *cis-trans* X-Pro peptide bond isomerization. *Biochemistry* 35:11719–33

28. Imoto T, Johnson LN, North ACT, Phillips DC, Rupley JA. 1972. Vertebrate lysozymes. In *The Enzymes, Vol. 7*, ed. PD Boyer, 21:665–868. New York/London: Academic

29. Ingwall RT, Scheraga HA, Lotan N, Berger A, Katchalski E. 1968. Conformational studies of poly-L-alanine in water. *Biopolymers* 6:331–68

30. Iwaoka M, Juminaga D, Scheraga HA. 1998. Regeneration of three-disulfide mutants of bovine pancreatic ribonuclease A missing the 65–72 disulfide bond: characterization of a minor folding pathway of ribonuclease A and kinetic roles of Cys65 and Cys72. *Biochemistry* 37:4490–501

31. Katz S, Gutfreund K, Shulman S, Ferry JD. 1952. The conversion of fibrinogen to fibrin: X. Light scattering studies of bovine fibrinogen. *J. Am. Chem. Soc.* 74:5706–9

32. Kauzmann W. 1959. Some factors in the interpretation of protein denaturation. *Adv. Protein Chem.* 14:1–63

33. Kazmierkiewicz R, Liwo A, Scheraga HA. 2002. Energy-based reconstruction of a protein backbone from its α-carbon trace by a Monte-Carlo method. *J. Comput. Chem.* 23:715–23

34. Kazmierkiewicz R, Liwo A, Scheraga HA. 2003. Addition of side chains to a known backbone with defined side-chain centroids. *Biophys. Chem.* 100:261–80. Erratum. 2003. *Biophys. Chem.* 106:91

35. Khalili M, Liwo A, Jagielska A, Scheraga HA. 2005. Molecular dynamics with the united-residue model of polypeptide chains. II. Langevin and Berendsen-bath dynamics and tests on model α-helical systems. *J. Phys. Chem. B* 109:13798–810

36. Khalili M, Liwo A, Scheraga HA. 2006. Kinetic studies of folding of the B-domain of staphylococcal protein A with molecular dynamics and a united-residue (UNRES) model of polypeptide chains. *J. Mol. Biol.* 355:536–47

37. Konishi Y, Ooi T, Scheraga HA. 1982. Regeneration of ribonuclease A from the reduced protein. Energetic analysis. *Biochemistry* 21:4741–48

38. Konishi Y, Ooi T, Scheraga HA. 1982. Regeneration of RNase A from the reduced protein: models of regeneration pathways. *Proc. Natl. Acad. Sci. USA* 79:5734–38

39. Laity JH, Lester CC, Shimotakahara S, Zimmerman DE, Montelione GT, Scheraga HA. 1997. Structural characterization of an analog of the major rate-determining disulfide folding intermediate of bovine pancreatic ribonuclease A. *Biochemistry* 36:12683–99

40. Laskowski M Jr, Scheraga HA. 1954. Thermodynamic considerations of protein reactions. I. Modified reactivity of polar groups. *J. Am. Chem. Soc.* 76:6305–19

41. Laskowski M Jr, Scheraga HA. 1956. Thermodynamic considerations of protein reactions. II. Modified reactivity of primary valence bonds. *J. Am. Chem. Soc.* 78:5793–98

42. Laskowski M Jr, Scheraga HA. 1961. Thermodynamic considerations of protein reactions. III. Kinetics of protein denaturation. *J. Am. Chem. Soc.* 83:266–74

43. Lee J, Scheraga HA, Rackovsky S. 1997. New optimization method for conformational energy calculations on polypeptides: conformational space annealing. *J. Comput. Chem.* 18:1222–32

44. Lewandowska A, Ołdziej S, Liwo A, Scheraga HA. 2010. β-hairpin-forming peptides; models of early stages of protein folding. *Biophys. Chem.* 151:1–9

45. Li L-K, Riehm JP, Scheraga HA. 1966. Structural studies of ribonuclease. XXIII. Pairing of the tyrosyl and carboxyl groups. *Biochemistry* 5:2043–48

46. Li Z, Scheraga HA. 1987. Monte Carlo–minimization approach to the multiple-minima problem in protein folding. *Proc. Natl. Acad. Sci. USA* 84:6611–15

47. Liwo A, Arlukowicz P, Czaplewski C, Ołdziej S, Pillardy P, Scheraga HA. 2002. A method for optimizing potential-energy functions by a hierarchical design of the potential-energy landscape: application to the UNRES force field. *Proc. Natl. Acad. Sci. USA* 99:1937–42

48. Liwo A, Czaplewski C, Ołdziej S, Rojas AV, Kazmierkiewicz R, et al. 2008. Simulation of protein structure and dynamics with the coarse-grained UNRES force field. In *Coarse-Graining of Condensed Phase and Biomolecular Systems*, ed. GA Voth, pp. 107–22. Boca Raton, FL: CRC Press

49. Liwo A, Czaplewski C, Pillardy J, Scheraga HA. 2001. Cumulant-based expressions for the multibody terms for the correlation between local and electrostatic interactions in the united-residue force field. *J. Chem. Phys.* 115:2323–47

50. Liwo A, Kazmierkiewicz R, Czaplewski C, Groth M, Oldziej S, et al. 1998. A united-residue force field for off-lattice protein-structure simulations. III. Origin of backbone hydrogen-bonding cooperativity in united-residue potentials. *J. Comput. Chem.* 19:259–76

51. Liwo A, Khalili M, Czaplewski C, Kalinowski S, Ołdziej S, et al. 2007. Modification and optimization of the united-residue (UNRES) potential-energy function for canonical simulations. I. Temperature dependence of the effective energy function and tests of the optimization method with single training proteins. *J. Phys. Chem. B* 111:260–85

52. Liwo A, Khalili M, Scheraga HA. 2005. Ab initio simulations of protein-folding pathways by molecular dynamics with the united-residue model of polypeptide chains. *Proc. Natl. Acad. Sci. USA* 102:2362–67

53. Liwo A, Lee J, Ripoll DR, Pillardy J, Scheraga HA. 1999. Protein structure prediction by global optimization of a potential energy function. *Proc. Natl. Acad. Sci. USA* 96:5482–85

54. Liwo A, Ołdziej S, Czaplewski C, Kleinerman DS, Blood P, Scheraga HA. 2010. Implementation of molecular dynamics and its extensions with the coarse-grained UNRES force field on massively parallel systems; towards millisecond-scale simulations of protein structure, dynamics, and thermodynamics. *J. Chem. Theory Comput.* 6:890–909

55. Liwo A, Oldziej S, Pincus MR, Wawak RJ, Rackovsky S, Scheraga HA. 1997. A united-residue force field for off-lattice protein-structure simulations. I. Functional forms and parameters of long-range side-chain interaction potentials from protein crystal data. *J. Comput. Chem.* 18:849–73

56. Liwo A, Pincus MR, Wawak RJ, Rackovsky S, Oldziej S, Scheraga HA. 1997. A united-residue force field for off-lattice protein-structure simulations. II. Parameterization of short-range interactions and determination of weights of energy terms by Z-score optimization. *J. Comput. Chem.* 18:874–87

57. Lorand L. 1950. Fibrin clots. *Nature* 166:694–95

58. Maisuradze GG, Liwo A, Ołdziej S, Scheraga HA. 2010. Evidence, from simulations, of a single state with residual native structure at the thermal denaturation midpoint of a small globular protein. *J. Am. Chem. Soc.* 132:9444–52

59. Maisuradze GG, Liwo A, Scheraga HA. 2009. Principal component analysis for protein folding dynamics. *J. Mol. Biol.* 385:312–29

60. Maisuradze GG, Liwo A, Scheraga HA. 2009. How adequate are one- and two-dimensional free-energy landscapes for protein-folding dynamics? *Phys. Rev. Lett.* 102:238102-1-4

61. Maisuradze GG, Liwo A, Scheraga HA. 2010. Relation between free energy landscapes of proteins and dynamics. *J. Chem. Theory Comput.* 6:583–95

62. Maisuradze GG, Senet P, Czaplewski C, Liwo A, Scheraga HA. 2010. Investigation of protein folding by coarse-grained molecular dynamics with the UNRES force field. *J. Phys. Chem. A* 114:4471–85

63. Mandelkern L, Krigbaum WR, Scheraga HA, Flory PJ. 1952. Sedimentation behavior of flexible chain molecules: polyisobutylene. *J. Chem. Phys.* 20:1392–97

64. Mattice WL, Scheraga HA. 1984. Matrix formulation of the transition from a statistical coil to an intramolecular antiparallel β sheet. *Biopolymers* 23:1701–24

65. Miller MH, Scheraga HA. 1976. Calculation of the structures of collagen models. Role of interchain interactions in determining the triple-helical coiled-coil conformation. I. Poly(glycyl-prolyl-prolyl). *J. Polymer Sci. Polymer Symp.* 54:171–200

66. Mirau PA, Bovey FA. 1990. 2D and 3D NMR studies of polypeptide structure and function. *Polymer Preprints, Division of Polymer Chemistry, POLY58, 199th A.C.S. August Meeting, Boston, MA,* 31:206

67. Momany FA, McGuire RF, Burgess AW, Scheraga HA. 1975. Energy parameters in polypeptides. VII. Geometric parameters, partial atomic charges, nonbonded interactions, hydrogen bond interactions, and intrinsic torsional potentials for the naturally occurring amino acids. *J. Phys. Chem.* 79:2361–81

68. Montelione GT, Wüthrich K, Burgess AW, Nice EC, Wagner G, et al. 1992. Solution structure of murine epidermal growth factor determined by NMR spectroscopy and refined by energy minimization with restraints. *Biochemistry* 31:236–49. Erratum. 1992. *Biochemistry* 31:10138

69. Monthus C, Garel T. 2005. Distribution of pseudo-critical temperatures and lack of self-averaging in disordered Poland-Scheraga models with different loop exponents. *Eur. Phys. J. B* 48:393–403

70. Nanias M, Chinchio M, Ołdziej S, Czaplewski C, Scheraga HA. 2005. Protein structure prediction with the UNRES force-field using Replica-Exchange Monte Carlo-with-Minimization; comparison with MCM, CSA and CFMC. *J. Comput. Chem.* 26:1472–86

71. Nanias M, Czaplewski C, Scheraga HA. 2006. Replica exchange and multicanonical algorithms with the coarse-grained united-residue (UNRES) force field. *J. Chem. Theor. Comput.* 2:513–28

72. Navon A, Ittah V, Landsman P, Scheraga HA, Haas E. 2001. Distributions of intramolecular distances in the reduced and denatured states of bovine pancreatic ribonuclease A. Folding initiation structures in the C-terminal portions of the reduced protein. *Biochemistry* 40:105–18

73. Némethy G, Scheraga HA. 1962. The structure of water and hydrophobic bonding in proteins. I. A model for the thermodynamic properties of liquid water. *J. Chem. Phys.* 36:3382–400

74. Némethy G, Scheraga HA. 1962. The structure of water and hydrophobic bonding in proteins. II. A model for the thermodynamic properties of aqueous solutions of hydrocarbons. *J. Chem. Phys.* 36:3401–17

75. Némethy G, Scheraga HA. 1962. The structure of water and hydrophobic bonding in proteins. III. The thermodynamic properties of hydrophobic bonds in proteins. *J. Phys. Chem.* 66:1773–89. Erratum. 1963. *J. Phys. Chem.* 67:2888

76. Némethy G, Scheraga HA. 1964. Structure of water and hydrophobic bonding in proteins. IV. The thermodynamic properties of liquid deuterium oxide. *J. Chem. Phys.* 41:680–89

77. Némethy G, Scheraga HA. 1965. Theoretical determination of sterically allowed conformations of a polypeptide chain by a computer method. *Biopolymers* 3:155–84

78. Némethy G, Scheraga HA, Kauzmann W. 1968. Comments on the communication "a criticism of the term 'hydrophobic bond'" by Joel H. Hildebrand. *J. Phys. Chem.* 72:1842

79. Némethy G, Steinberg IZ, Scheraga HA. 1963. The influence of water structure and of hydrophobic interactions on the strength of side-chain hydrogen bonds in proteins. *Biopolymers* 1:43–69

80. Ni F, Gibson KD, Scheraga HA. 1992. Nuclear magnetic resonance studies of thrombin-fibrinopeptide and thrombin-hirudin complexes. In *Thrombin: Structure and Function*, ed. LJ Berliner, pp. 63–85. New York: Plenum

81. Ni F, Konishi Y, Bullock LD, Rivetna MN, Scheraga HA. 1989. High-resolution NMR studies of fibrinogen-like peptides in solution: structural basis for the bleeding disorder caused by a single mutation of Gly(12) to Val(12) in the Aα chain of human fibrinogen Rouen. *Biochemistry* 28:3106–19

82. Ooi T, Scott RA, Vanderkooi G, Scheraga HA. 1967. Conformational analysis of macromolecules. IV. Helical structures of poly-L-alanine, poly-L-valine, poly-β-methyl-L-aspartate, poly-γ-methyl-L-glutamate, and poly-L-tyrosine. *J. Chem. Phys.* 46:4410–26

83. Orengo CA, Bray JE, Hubbard T, LoConte L, Sillitoe I. 1999. Analyses and assessment of ab initio three-dimensional prediction, secondary structure, and contacts prediction. *Proteins Struct. Funct. Genet.* 37(S3):149–70

84. Owicki JC, Scheraga HA. 1977. Monte Carlo calculations in the isothermal-isobaric ensemble. 2. Dilute aqueous solution of methane. *J. Am. Chem. Soc.* 99:7413–18

85. Piela L, Kostrowicki J, Scheraga HA. 1989. The multiple-minima problem in the conformational analysis of molecules. Deformation of the potential energy hypersurface by the diffusion equation method. *J. Phys. Chem.* 93:3339–46

86. Piela L, Scheraga HA. 1987. On the multiple-minima problem in the conformational analysis of polypeptides. I. Backbone degrees of freedom for a perturbed α-helix. *Biopolymers* 26:S33–58

87. Pincus MR, Scheraga HA. 1981. Theoretical calculations on enzyme-substrate complexes: the basis of molecular recognition and catalysis. *Acc. Chem. Res.* 14:299–306

88. Poland D, Scheraga HA. 1965. Comparison of theories of the helix-coil transition in polypeptides. *J. Chem. Phys.* 43:2071–74. Erratum. 1965. *J. Chem. Phys.* 43:3774

89. Poland D, Scheraga HA. 1966. Kinetics of the helix-coil transition in polyamino acids. *J. Chem. Phys.* 45:2071–90

90. Poland D, Scheraga HA. 1966. Phase transitions in one dimension, and the helix-coil transition in polyamino acids. *J. Chem. Phys.* 45:1456–63

91. Poland D, Scheraga HA. 1966. Occurrence of a phase transition in nucleic acid models. *J. Chem. Phys.* 45:1464–69

92. Poland D, Scheraga HA. 1969. The equilibrium unwinding in finite chains of DNA. *Physiol. Chem. Phys.* 1:389–446

93. Poland D, Scheraga HA. 1969. The Lifson-Allegra theories of the helix-coil transition for random copolymers: comparison with exact results and extension. *Biopolymers* 7:887–908

94. Poland D, Scheraga HA. 1970. *Theory of Helix-Coil Transitions in Biopolymers*. New York: Academic

95. Rackovsky S, Scheraga HA. 1984. Differential geometry and protein folding. *Acc. Chem. Res.* 17:209–14

96. Ripoll DR, Scheraga HA. 1988. On the multiple-minima problem in the conformational analysis of polypeptides. II. An electrostatically driven Monte Carlo method-tests on poly(L-alanine). *Biopolymers* 27:1283–303

96a. Rojas A, Liwo A, Browne D, Scheraga HA. 2010. Mechanism of fiber assembly: treatment of Aβ-peptide aggregation with a coarse-grained united-residue force field. *J. Mol. Biol.* 404:537–52

97. Rojas A, Liwo A, Scheraga HA. 2007. Molecular dynamics with the united-residue (UNRES) force field. Ab initio folding simulations of multi-chain proteins. *J. Phys. Chem. B* 111:293–309

98. Roterman IK, Lambert MH, Gibson KD, Scheraga HA. 1989. A comparison of the CHARMM, AMBER and ECEPP potentials for peptides. II. φ-ψ maps for *N*-acetyl alanine *N'*-methyl amide: comparisons, contrasts and simple experimental tests. *J. Biomol. Struct. Dyn.* 7:421–53

99. Rothwarf DM, Li Y-J, Scheraga HA. 1998. Regeneration of bovine pancreatic ribonuclease A. Identification of two nativelike three-disulfide intermediates involved in separate pathways. *Biochemistry* 37:3760–66

100. Rothwarf DM, Li Y-J, Scheraga HA. 1998. Regeneration of bovine pancreatic ribonuclease A. Detailed kinetic analysis of two independent folding pathways. *Biochemistry* 37:3767–76

101. Scheraga HA. 1955. Non-Newtonian viscosity of solutions of ellipsoidal particles. *J. Chem. Phys.* 23:1526–32

102. Scheraga HA. 1957. Tyrosyl-carboxylate ion hydrogen bonding in ribonuclease. *Biochim. Biophys. Acta* 23:196–97

103. Scheraga HA. 1967. Structural studies of pancreatic ribonuclease. *Fed. Proc.* 26:1380–87

104. Scheraga HA. 1968. Calculations of conformations of polypeptides. *Adv. Phys. Org. Chem.* 6:103–84

105. Scheraga HA. 1996. Recent developments in the theory of protein folding: searching for the global energy minimum. *Biophys. Chem.* 59:329–39

106. Scheraga HA. 1998. Theory of hydrophobic interactions. *J. Biomol. Struct. Dyn.* 16:447–60

107. Scheraga HA. 2004. The thrombin-fibrinogen interaction. *Biophys. Chem.* 112:117–30

108. Scheraga HA, Backus JK. 1951. Flow birefringence in solutions of *N*-hexadecyltrimethylammonium bromide. *J. Am. Chem. Soc.* 73:5108–12

109. Scheraga HA, Chou KC, Némethy G. 1983. Interactions between the fundamental structures of polypeptide chains. In *Conformation in Biology*, ed. R Srinivasan, RH Sarma, pp. 1–10. Schenectady, NY: Adenine

110. Scheraga HA, Edsall JT, Gadd JO Jr. 1951. Double refraction of flow: numerical evaluation of extinction angle and birefringence as a function of velocity gradient. *J. Chem. Phys.* 19:1101–8

111. Scheraga HA, Katchalsky A, Alterman Z. 1969. Electrochemical properties of permeable multichain polyamino acids. *J. Am. Chem. Soc.* 91:7242–49

112. Scheraga HA, Khalili M, Liwo A. 2007. Protein folding dynamics: overview of molecular simulation techniques. *Annu. Rev. Phys. Chem.* 58:57–83

113. Scheraga HA, Konishi Y, Ooi T. 1984. Multiple pathways for regenerating ribonuclease A. *Adv. Biophys.* 18:21–41

114. Scheraga HA, Laskowski M Jr. 1957. The fibrinogen-fibrin conversion. *Adv. Protein Chem.* 12:1–131

115. Scheraga HA, Mandelkern L. 1953. Consideration of the hydrodynamic properties of proteins. *J. Am. Chem. Soc.* 75:179–84

116. Sendak RA, Rothwarf DM, Wedemeyer WJ, Houry WA, Scheraga HA. 1996. Kinetic and thermodynamic studies of the folding/unfolding of a tryptophan-containing mutant of ribonuclease A. *Biochemistry* 35:12978–92

117. Senet P, Maisuradze GG, Foulie C, Delarue P, Scheraga HA. 2008. How main-chains of proteins explore the free-energy landscape in native states. *Proc. Natl. Acad. Sci. USA* 105:19708–13

118. Shimotakahara S, Rios CB, Laity JH, Zimmerman DE, Scheraga HA, Montelione GT. 1997. NMR structural analysis of an analog of an intermediate formed in the rate-determining step of one pathway in the oxidative folding of bovine pancreatic ribonuclease A: automated analysis of ^{1}H, ^{13}C, and ^{15}N resonance assignments for wild-type and [C65S, C72S] mutant forms. *Biochemistry* 36:6915–29

119. Simon I, Glasser L, Scheraga HA, Manley RSJ. 1988. Structure of cellulose. 2. Low-energy crystalline arrangements. *Macromolecules* 21:990–98

120. Smith-Gill SJ, Rupley JA, Pincus MR, Carty RP, Scheraga HA. 1984. Experimental identification of a theoretically predicted "left-sided" binding mode for (GlcNAc) in the active site of lysozyme. *Biochemistry* 23:993–97

121. Steinberg IZ, Scheraga HA. 1963. Entropy changes accompanying association reactions of proteins. *J. Biol. Chem.* 238:172–81

122. Sturtevant JM, Laskowski M Jr, Donnelly TH, Scheraga HA. 1955. Equilibria in the fibrinogen-fibrin conversion. III. Heats of polymerization and clotting of fibrin monomer. *J. Am. Chem. Soc.* 77:6168–72

123. Tanaka S, Scheraga HA. 1975. Theory of the cooperative transition between two ordered conformations of poly(L-proline). III. Molecular theory in the presence of solvent. *Macromolecules* 8:516–21

124. Vila JA, Aramini JM, Rossi P, Kuzin A, Su M, et al. 2008. Quantum chemical ^{13}C$^\alpha$ chemical shift calculations for protein NMR structure determination, refinement, and validation. *Proc. Natl. Acad. Sci. USA* 105:14389–94

125. Vila JA, Arnautova YA, Martin OA, Scheraga HA. 2009. Quantum-mechanics-derived ^{13}C$^\alpha$ chemical shift server (*Che*Shift) for protein structure validation. *Proc. Natl. Acad. Sci. USA* 106:16972–77

126. Vila JA, Arnautova YA, Scheraga HA. 2008. Use of ^{13}C$^\alpha$ chemical shifts for accurate determination of β-sheet structures in solution. *Proc. Natl. Acad. Sci. USA* 105:1891–96

127. Vila JA, Baldoni HA, Scheraga HA. 2009. Performance of density functional models to reproduce observed ^{13}C$^\alpha$ chemical shifts of proteins in solution. *J. Comput. Chem.* 30:884–92

128. Vila JA, Ripoll DR, Scheraga HA. 2003. Atomically detailed folding simulation of the B domain of staphylococcal protein A from random structures. *Proc. Natl. Acad. Sci. USA* 100:14812–16

129. Vila JA, Ripoll DR, Scheraga HA. 2007. Use of ^{13}C$^\alpha$ chemical shifts in protein-structure determination. *J. Phys. Chem. B* 111:6577–85

130. Vila JA, Scheraga HA. 2008. Factors affecting the use of ^{13}C$^\alpha$ chemical shifts to determine, refine and validate protein structures. *Proteins Struct. Funct. Bioinform.* 71:641–54

131. Vila JA, Scheraga HA. 2009. Assessing the accuracy of protein structures by quantum mechanical computations of ^{13}C$^\alpha$ chemical shifts. *Acc. Chem. Res.* 42:1545–53

132. Vila JA, Serrano P, Wüthrich K, Scheraga HA. 2010. Sequential nearest-neighbor effects on computed ^{13}C$^\alpha$ chemical shifts. *J. Biomol. NMR* 48:23–30

133. Vila JA, Villegas ME, Baldoni HA, Scheraga HA. 2007. Predicting ^{13}C$^\alpha$ chemical shifts for validation of protein structures. *J. Biomol. NMR* 38:221–35

134. Vournakis JN, Poland D, Scheraga HA. 1967. Anti-cooperative interactions in single-strand oligomers of deoxyriboadenylic acid. *Biopolymers* 5:403–22

135. Vournakis JN, Scheraga HA, Rushizky GW, Sober HA. 1966. Sober-neighbor-neighbor interactions in single-strand polynucleotides; optical rotatory dispersion studies of the ribonucleotide ApApCp. *Biopolymers* 4:33–41

136. Warme PK, Momany FA, Rumball SV, Tuttle RW, Scheraga HA. 1974. Computation of structures of homologous proteins; α-lactalbumin from lysozyme. *Biochemistry* 13:768–82

137. Weiner SI, Kollman PA, Case DA, Singh UC, Ghio C, et al. 1984. A new force field for molecular mechanical simulation of nucleic acids and proteins. *J. Am. Chem. Soc.* 106:765–84

138. Wlodawer A, Svensson LA, Sjölin L, Gilliland GL. 1988. Structure of phosphate-free ribonuclease A refined at 1.26 Å. *Biochemistry* 27:2705–17

139. Wojcik J, Altmann KH, Scheraga HA. 1990. Helix-coil stability constants for the naturally occurring amino acids in water. XXIV. Half-cystine parameters from random poly(hydroxybutylglutamine-co-S-methylthio-L-cysteine). *Biopolymers* 30:121–34

140. Xu X, Rothwarf DM, Scheraga HA. 1996. Nonrandom distribution of the one-disulfide intermediates in the regeneration of ribonuclease A. *Biochemistry* 35:6406–17

141. Xu X, Scheraga HA. 1998. Kinetic folding pathway of a three-disulfide mutant of bovine pancreatic ribonuclease A missing the [40–95] disulfide bond. *Biochemistry* 37:7561–71

142. Ye Y-J, Ripoll DR, Scheraga HA. 1999. Kinetics of cooperative protein folding involving two separate conformational families. *Comput. Theor. Polym. Sci.* 9:359–70

143. Cote P, Senet P, Delarue P, Maisuradze GG, Scheraga HA. 2010. Nonexponential decay of internal rotational correlation functions of native proteins and self-similar structural fluctuations. *Proc. Natl. Acad. Sci. USA* 107:19844–49

144. He Y, Liwo A, Weinstein H, Scheraga HA. 2011. PDZ binding to the BAR domain of PICK1 is elucidated by coarse-grained molecular dynamics. *J. Mol. Biol.* 405:298–314

Equilibrium Sampling in Biomolecular Simulations

Daniel M. Zuckerman

Department of Computational and Systems Biology, University of Pittsburgh, Pittsburgh, Pennsylvania 15213; email: ddmmzz@pitt.edu

Annu. Rev. Biophys. 2011. 40:41–62

First published online as a Review in Advance on March 1, 2011

The *Annual Review of Biophysics* is online at biophys.annualreviews.org

This article's doi: 10.1146/annurev-biophys-042910-155255

Copyright © 2011 by Annual Reviews. All rights reserved

1936-122X/11/0609-0041$20.00

Keywords

effective sample size, algorithms, efficiency, hardware, timescales

Abstract

Equilibrium sampling of biomolecules remains an unmet challenge after more than 30 years of atomistic simulation. Efforts to enhance sampling capability, which are reviewed here, range from the development of new algorithms to parallelization to novel uses of hardware. Special focus is placed on classifying algorithms—most of which are underpinned by a few key ideas—in order to understand their fundamental strengths and limitations. Although algorithms have proliferated, progress resulting from novel hardware use appears to be more clear-cut than from algorithms alone, due partly to the lack of widely used sampling measures.

Contents

INTRODUCTION

Why Sample?

Biomolecular behavior can be substantially characterized by the states of the system of interest—that is, by the configurational energy basins reflecting the coordinates of all constituent molecules in a system. These states might be largely internal to a single macromolecule, such as a protein, or more generally involve binding partners and their relative coordinates. But regardless of the complexity of a system, the states represent key functional configurations along with potential intermediates for transitioning among the major states. The primary goal of equilibrium computer simulation is to specify configurational states and their populations. A more complete, mechanistic description would also include kinetic properties and dynamical pathways (147). Nevertheless, even the equilibrium description provides a basic view of the range of structural motions of biomolecules and can serve more immediately practical purposes such as for ensemble docking (56, 65).

The basic algorithm of biomolecular simulation, molecular dynamics (MD) simulation, has not changed substantially since the first MD study of a protein more than 30 years ago (82). Although routine explicit-solvent MD simulations are now four or five orders of magnitude longer (i.e., $100–10^3$ ns currently), modern MD studies still appear to fall significantly short of what is needed for statistically valid equilibrium simulation (36, 38). Roughly speaking, one would like to run a simulation at least 10 times longer than the slowest important timescale in a system. Unfortunately, many biomolecular timescales exceed 1 ms, and in some cases by orders of magnitude (44).

Despite the outpouring of algorithmic ideas over the past decades, MD largely remains the tool of choice for biomolecular simulations. While this staying power partly reflects the ready availability of software packages, and perhaps some psychological inertia, it also is indicative of a simple fact: No other method can routinely and reliably outperform MD by a significant amount.

This review employs several points of view in considering efforts to improve sampling. First, it attempts to define the equilibrium sampling problem(s) as precisely as possible, which necessarily includes discussing the quantification of sampling. Significant and substantiated progress is not possible without clear yardsticks. Second, a brief discussion of how sampling happens—i.e., how simulations generate ensembles and averages—provides a basis for understanding numerous methods. Third, the review attempts to examine a fairly wide array of modern algorithms—most of which are underpinned by a surprisingly small number of key ideas. The algorithms and studies are too numerous to be reviewed on a case-by-case basis, but a bird's-eye view is informative; more focused attention is paid to apparently conflicting statements about the replica exchange approach, however. Fourth, special emphasis is be placed on novel uses of hardware—graphics processing units (GPUs), random-access memory (RAM), and special central processing units (CPUs); this new front in sampling efforts has yielded rather clean results in some cases. Using hardware in new ways necessarily involves substantial algorithmic efforts.

This review is limited. At least a full volume would be required to comprehensively describe sampling methods and results for biomolecular systems. This review therefore aims to primarily catalog key ideas and principles of sampling, along with enough references for the reader to delve more deeply into areas of interest. The author apologizes for the numerous studies that have not been mentioned because of space limitations or because he was not aware of them. Other review articles (e.g., 4, 17) and books on simulation methods (2, 32, 62, 105) are available.

WHAT IS THE SAMPLING PROBLEM?

Although scientists actively working in the field of biomolecular simulation may assume the essential meaning of the sampling problem is universally accepted, a survey of the literature indicates that several somewhat different interpretations are implicitly assumed. An individual's understanding of the sampling problem and, accordingly, success or failure in addressing the problem surely will dictate the choice of methodology.

A Simple View of the Equilibrium Sampling Problem

Equilibrium sampling at constant temperature and fixed volume can be concisely identified with the generation of full-system configurations, $\mathbf{x}$, distributed according to the Boltzmann-factor distribution:

$$\rho(\mathbf{x}) \propto \exp[-U(\mathbf{x})/k_B T], \qquad 1.$$

where ρ is the probability density function, U is the potential energy function, k_B is Boltzmann's constant, and T is the absolute temperature in Kelvin units. The coordinates $\mathbf{x}$ refer (in the classical picture assumed here) to the coordinates of every atom in the system, including solute and solvent. Note that other thermodynamic conditions, such as constant pressure or constant chemical potential, require additional energy-like terms in the Boltzmann factor (147), but here we shall consider only the distribution of Equation 1 for clarity and simplicity.

Performing equilibrium sampling requires access to all regions of configuration space, or at least to those regions with significant populations, and that configurations have the correct relative probabilities. This point is further discussed below.

Ideal Sampling as a Reference Point

In practice, it is nearly impossible to generate completely independent and identically distributed (i.i.d.) configurations—obeying Equation 1—for biomolecules. Both dynamical and nondynamical methods tend to produce correlated samples (38, 41, 73, 86, 141). Nevertheless, such ideal sampling is a highly useful

MD: molecular dynamics
GPU: graphics processing unit
CPU: central processing unit

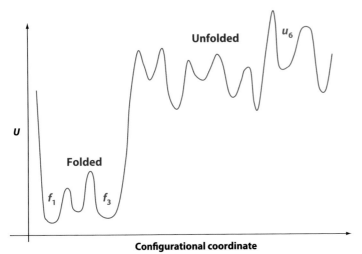

Figure 1

A schematic energy landscape of a protein. Both the folded and unfolded states generally can be expected to consist of multiple substates, f_i and u_i, respectively. Transitions among substates themselves could be slow, compared with simulation timescales. The sampling problem is sometimes construed to involve generating configurations from both folded and unfolded states; the text discusses when this view is justified.

reference point for disentangling complications arising in practical simulation methods that generate correlated configurations.

Consider a hypothetical example in which an equilibrium ensemble of N i.i.d. configurations has been generated according to Equation 1 for a biomolecule with a complex landscape. Regions of configuration space with probability $p > 1/N$ are likely to be represented in the ensemble, and this is true regardless of kinetic barriers among states. Conversely, regions with $p < 1/N$ are not likely to be represented, whether or not the region is interesting. For instance, the unfolded state might be an appropriate part of a size N ensemble, but this will depend on the system (i.e., on the free energy difference between folded and unfolded states) and on N (see **Figure 1** and the discussion below).

Goals and Ambiguities in Equilibrium Sampling

The question, "What should be achieved in equilibrium sampling?" can be divided into at least three subquestions. First, for a given problem (defined by the specific system and a specified initial condition), there is the question of defining success—When is sampling sufficient? Usually, the goal of equilibrium simulation is to calculate observables of interest to a level of precision sufficient to draw physical conclusions. Yet the variables of interest vary from study to study. It therefore seems more reasonable to set a goal of generating an ensemble of configurations from which arbitrary observables can be calculated. In rough terms, let us therefore assume that the goal is to calculate N^{eff} effectively independent configurations, where presumably it is desirable to achieve $N^{eff} > 10$. Larger values may often be necessary, however, because fractional uncertainty will vary as $1/\sqrt{N^{eff}}$ for slower-to-converge properties such as state populations. This issue is addressed further below.

The other questions are probably the most pressing, starting with, "What type of system is to be sampled?" A basic distinction that seems to appear implicitly in the literature is whether the unfolded state is important in the system. In this context, importance would usually be defined by whether a system exhibits a significant equilibrium unfolded fraction—say, roughly equal probabilities $p(\text{folded}) \sim p(\text{unfolded})$. Indeed, some evaluations of sampling methods have specifically targeted such a balance (48, 102), e.g., by adjusting the temperature. A third question concerns initial conditions, which sometimes correspond to the unfolded state. When will the initial conditions be important? This issue is addressed separately below.

These questions arise in the first place because all available practical sampling methods for biomolecules are imperfect in the sense of producing correlated configurations. In ideal equilibrium sampling, by contrast, the sufficiency of sampling is fully quantified in terms of the number N of independent configurations and can be assessed objectively in terms of observables of interest. The notion of an initial condition is irrelevant in ideal sampling because there are no correlations. Also, the values of $p(\text{folded})$ and $p(\text{unfolded})$ have no effect on sampling quality, which is fully embodied in N;

instead, it is N that determines which states are represented in the ensemble and the appropriate frequencies of occurrence.

Issues surrounding initial conditions and the unfolded state.

A typical simulation of a protein can be described as one intended to sample the folded (presumably functional) state, which may consist of a number of substates as in **Figure 1**. One or more of the folded substates may be partially disordered. So long as the simulation is initiated in or near one of the nominally folded substates, the resources required to sample it will not depend significantly on the initial conditions. Rather, timescales for transitioning among all folded substates will dominate the computing cost. In a dynamical simulation, after all, it is the transitions among substates that yield estimates for substate populations and hence the equilibrium ensemble.

Sensitivity to initial conditions can become important when the unfolded state is involved. Even when the goal is to sample the equilibrium ensemble of an overwhelmingly folded protein, the initial condition can come into play if it introduces an anomalously long timescale. Consider beginning an equilibrium simulation of the landscape of **Figure 1** in the unfolded substate labeled u_6. Depending on details of the system, the time required to find the folded state could easily exceed the time for sampling folded substates. Thus, such a simulation must first solve a search problem before sampling can begin.

Aside from the issue of initial conditions, the relative importance of the unfolded state can be considered in the context of the folding free energy, which is defined on the basis of the ratio of probabilities for unfolded and folded states:

$$p(\text{folded})/p(\text{unfolded}) = \exp(-\Delta G_{\text{fold}}/k_B T).$$
2.

Consider the value $\Delta G_{\text{fold}} \sim -3\,\text{kcal/mol} \sim -5 k_B T$, which implies the unfolded state is occupied 1% of the time or less. In ensembles with fewer than 100 independent configurations ($N^{\text{eff}} < 100$), such an unfolded state typically should not be represented. In practical

terms, because key folded-state timescales often exceed microseconds or even milliseconds, it is unlikely that a simulation of a protein in folding conditions in the current era will contain a sufficient number of independent configurations for the unfolded state to be represented.

Note that we have assumed it is straightforward to distinguish folded and unfolded configurations. In reality, there may be a spectrum of disordered states in many systems and hence ambiguity in defining folded and unfolded states.

SAMPLING BASICS: MECHANISM, TIMESCALE, AND COST

Transitions Provide the Key Information in Sampling

The problem of equilibrium sampling can be summarized as determining the metastable states and their relative populations, but how are populations actually determined by the simulation? In almost every practical algorithm, the information comes from transitions among the states. By switching back and forth between pairs of states, with dwell times between transitions acting as proxies for rate constants k, a dynamical simulation gathers information about relative populations based on the equilibrium balance relation $p_i k_{ij} = p_j k_{ji}$ (147). With more transitions, the error in the population estimates p_i decreases (102). Without transitions, most algorithms have no information about relative populations. Consider the case of two independent simulations started from different states that exhibit no transitions: Determining the populations of the states then requires more advanced analysis (136) that often may not be practical.

Most advanced algorithms—such as the varieties of exchange simulation—also require ordinary transitions to gather population information. That is, transitions among states within individual continuous trajectories are required to obtain sampling, as has been discussed previously (38, 102, 143). Any algorithm that uses dynamical trajectories as a component can be

Correlation time:
the time that must
elapse, on average,
between evaluations of
a specific observable
for the values to be
statistically
independent

MC: Monte Carlo

expected to require transitions among metastable states in those trajectories.

It must also be pointed out that transitions themselves do not necessarily equate to good sampling. An example is when a high temperature is used to aid sampling at a lower temperature. Although transitions are necessary, they are not sufficient because of the overlap issue: The sampled configurations ultimately must be important in the targeted (e.g., lower temperature) ensemble after they are reweighted (72, 109).

Timescales of Sampling

Much of the foregoing discussion can be summed up based on two key timescales. (Even if a sampling method is not fully dynamical, analogous quantities can be defined in terms of computing times.) The timescale that typically is limiting for simulations can be denoted t^*_{corr}, which denotes the longest important equilibrium correlation time. Roughly, this is the time required to explore all the important parts of configuration space once (73)—starting from any well-populated (sub)state. For sampling the landscape of **Figure 1** under folding conditions, t^*_{corr} would be the time to visit all folded substates starting from any of the folded basins. However, as also discussed above, initial conditions may play an important role, and thus we can define t_{init} as the equilibration time (or burn-in time in Monte Carlo lingo). This is not an equilibrium timescale, but the time necessary to relax from a particular nonequilibrium initial condition to equilibrium. Thus, t_{init} is specific to the initial configuration of each simulation. The landscape of **Figure 1** under folding conditions suggests $t_{init} \ll t^*_{corr}$ for a simulation started from a folded substate, but it is possible to have significant values—even $t_{init} > t^*_{corr}$—when starting from an unfolded state.

Factors Contributing to Sampling Cost

A little thought about the ingredients of sampling can help one understand successes and failures of various efforts and, more importantly, aid in planning future research. The most important ingredient to understand is single-trajectory sampling, i.e., the class of dynamical methods including MD, single-Markov-chain Monte Carlo (MC), or any other method that generates an ensemble as a sequentially correlated list of configurations. Single-trajectory methods tend to be the engine underlying traditional and advanced methods.

The sampling cost of a trajectory method can be divided into two intuitive factors:

$$\text{Trajectory sampling cost}$$
$$= (\text{Cost per trajectory step})$$
$$\times (\text{Number of steps needed for sampling})$$
$$\sim (\text{Cost of energy call})$$
$$\times (\text{Roughness of energy landscape}). \qquad 3.$$

In words, a given algorithm requires a certain number of steps to achieve good sampling (e.g., $N^{eff} \gg 1$), and the total cost of generating such a trajectory is simply the total cost per step multiplied by the required number of steps. Although this decomposition is trivial, it is informative. For example, if one wants to use a single-trajectory algorithm (e.g., MD or similar) where a step corresponds roughly to 10^{-15} s, then 10^9–10^{12} steps are required to reach the microsecond to millisecond range. Therefore, unless the cost of a step can be reduced by several orders of magnitude, MD and similar methods likely cannot achieve sampling with typical current resources. In a modified landscape (e.g., different temperature or model), the roughness may be reduced, but again that reduction should be several orders of magnitude or else be accompanied by a compensating reduction in the cost per step.

Among noteworthy examples, increasing temperature decreases roughness but not the cost per step, whereas changing a model (i.e., modifying the potential energy function U) can affect both the roughness and the step cost. Both strategies sample a different distribution, which may be nontrivial to convert to the targeted ensemble. This type of understanding, building on Equation 3, can aid in the analysis of multilevel (e.g., exchange) simulations, which require good sampling at some level (143, 148) by single continuous trajectories.

QUANTITATIVE ASSESSMENT OF SAMPLING

Although the challenge of sampling biomolecular systems has been widely recognized for years, and although new algorithms are regularly proposed, systematic quantification of sampling has often not received sufficient attention. Most importantly, there does not seem to be a widely accepted (and widely used) yardstick for quantifying the effectiveness of sampling procedures. Given the lack of a broadly accepted measure, it is tempting for individual investigators to report measures that cast the most favorable light on their results.

Two key benefits would flow from a universal measure of sampling quality. As an example, assume we are able to assign an effective sample size, N^{eff}, to any ensemble generated in a simulation; N^{eff} would characterize the number of statistically independent configurations generated. Sampling efficiency could then be quantified by the CPU time (or number of cycles) required per independent sample. The first benefit is that we would no longer be able to fool ourselves as to the efficiency of a method based on qualitative evidence or perhaps the elegance of a method. Thus, the field would be pushed harder to focus on a bottom-line measure. The second benefit is that demonstrating the importance of a new algorithm would be more straightforward. It would no longer be necessary to directly compare a new method with, perhaps, a competitor's approach requiring subtle optimization. Instead, each method could be compared with a standard (e.g., MD), removing ambiguities in the outcome. It would still be important to test a method on a number of systems (e.g., small and large, stiff and flexible, implicitly and explicitly solvated), but the results would be quantitative and readily verified by other groups. On a related note, there has been a proposal to organize a sampling contest or challenge event to allow head-to-head comparison of different methodologies (B.R. Brooks, unpublished communication).

Numerous ideas for assessing sampling have been proposed over the years (36, 38, 41, 73, 86, 88, 114, 116, 141), but it is important to divide such proposals into absolute and relative measures. Absolute measures attempt to give a binary indication of whether convergence has been achieved, whereas relative measures estimate how much sampling has been achieved, e.g., by an effective sample size N^{eff}. The perspective of the present review is that absolute measures fail to account for the fundamental statistical picture underlying the sampling problem. To understand why, note that any measured observable has an uncertainty associated with it; roughly speaking, sampling quality is reflected in typical sizes of error bars, which decrease with better sampling. There does not seem to an unambiguous point where sampling can be considered absolutely converged (although a simulation effectively does not begin to sample equilibrium averages until t_{init} has been surpassed).

It is certainly valuable, nevertheless, to be able to gauge in an absolute sense when sampling is wholly inadequate. Absolute methods that may be useful at detecting extremely poor sampling include casual use of the ergodic measure (86) (i.e., whether or not it approaches zero) and cluster counting (114).

Assessing Dynamical Sampling

The assessment of dynamical methods illustrates the key ideas behind the relative measures of sampling. For this purpose, dynamical is defined as when trajectories consist of configurations with purely sequential correlations—i.e., where a given configuration was produced based solely on the immediately preceding configuration(s). Thus, MD, Langevin dynamics, and simple Markov-chain MC are dynamical methods, but exchange algorithms are not (147). Correlation times or their analogs for MC can then readily be associated with any trajectory that was dynamically generated. Assuming, for the moment, that there is a fundamental

Effective sample size: the number of statistically independent configurations to which a given ensemble of correlated configurations is equivalent

RE: replica exchange

correlation time for overall sampling of the system, t^*_{corr}, then an effective sample size can be calculated from the simple relation

$$N^{eff} \simeq t_{sim}/t^*_{corr}, \qquad 4.$$

where t_{sim} is the total CPU time (or number of cycles) used to generate the trajectory. Thus, sampling quality is relative in that it should increase linearly with t_{sim} (86). In rough terms, the absolute lack of sampling would correspond to $N^{eff} \lesssim 1$; physically, this would indicate that important parts of configuration have been visited only once—e.g., a simulation shorter than $t_{init} + t^*_{corr}$. (It would seem impossible to detect parts of the space never visited.)

Work by Lyman & Zuckerman (73) has suggested that a reasonable overall correlation time t^*_{corr} can be calculated from a dynamical trajectory. Those authors derived their correlation time from the overall distribution in configuration space, estimating the time that must elapse between trajectory frames so that they behave as if statistically independent. The approach has the twin advantages of being based on the full configuration-space distribution (as opposed to isolated observables) and of being blindly and objectively applicable to any dynamical trajectory. Other measures meeting these criteria would also be valuable; note, for instance, the work by Hess (41) using principal components.

Assessing Nondynamical Sampling

If sampling is performed in a nondynamical way, one cannot rely on sequential correlations to assess sampling as in Equation 4. Many of the modern algorithms that attempt to enhance sampling, such as those reviewed below, are not dynamical. Replica exchange (RE) (34, 51, 118) is a typical example: If one is solely interested in the ensemble at a single temperature, a given configuration may be strongly correlated with other configurations distant in the sequence at that temperature and/or be uncorrelated with sequential neighbors. At the same time, RE does generate trajectories that are continuous in configuration space, if not temperature, and it may be possible to analyze these in a dynamical

sense (73) but care will be required (15). Other algorithms, for instance, based on polymer growth procedures (29, 35, 126, 138–140), are explicitly nondynamical.

Two recent papers (38, 141) have argued that nondynamical simulations are best assessed by multiple independent runs. The lack of sequential correlations—but the presence of more complex correlations—in nondynamical ensembles means that the list of configurations cannot be divided into nearly independent segments for blocking-based analysis. Zhang et al. (141) suggest that multiple runs be used to assess variance in the populations of physical states for two related reasons: Such states are defined to be separated from one another by the slowest timescales in a system, and relative populations of states cannot be estimated properly without good sampling within each state. State-population variances, in turn, can be used to estimate N^{eff} based on simple statistical arguments (141); nevertheless, it is important that states be approximated in an automated fashion (see References 14, 91, and 141) to eliminate the possibility for bias.

PURELY ALGORITHMIC EFFORTS TO IMPROVE SAMPLING

How can we beat MD? Despite 30 years of effort, there is no algorithm that is significantly more efficient than MD for the full range of systems of interest. Further, although dozens of different detailed procedures have been suggested, there are a limited number of qualitatively distinct ideas. Some of the strategies that have been proposed are described below.

Replica Exchange and Multiple Temperatures

The most common strategy is to employ elevated temperature. Many variations on this strategy have been proposed; one of the earliest suggestions was for spin systems (27). The approach that has been applied to biomolecules most often is RE (also called parallel

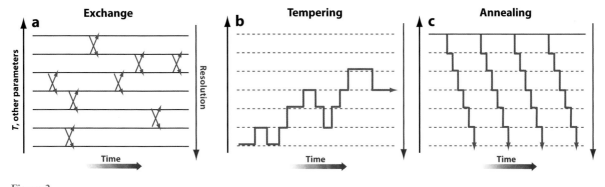

Figure 2

Many ways to use many levels. Sampling algorithms are schematized in the panels: Different horizontal levels represent different simulation conditions such as varying temperature, model parameters, and/or resolution. A given ladder or set of conditions can be used in different formulations: (*a*) exchange, (*b*) tempering, or (*c*) annealing.

tempering), in which parallel simulations wander among a set of fixed temperature values with swaps governed by a Metropolis criterion (34, 51, 118). Many variations and optimizations have been proposed for RE (e.g., 8, 11, 26, 57, 64, 67, 93–95, 99–101, 103, 112, 121). Closely related to RE is simulated tempering, in which a single trajectory wanders among a set of temperatures (75); once again, optimizations have been proposed (e.g., 137) (**Figure 2**).

Less well known, but formally closely related to exchange and tempering, is annealed importance sampling (AIS), in which a high-temperature ensemble is annealed to lower temperatures in a weighted way that preserves canonical sampling (49, 87) (**Figure 2**). AIS has been applied to biomolecules and optimizations have been suggested (72, 74). AIS is nominally a nonequilibrium approach, but in precise analogy to the Jarzynski equality (52), it yields equilibrium ensembles. The J-walking approach also starts from high temperature (28); a good discussion of several related methods is given in Reference 9.

How effective is replica exchange? RE simulation is both popular and controversial. Some authors have noted weaknesses (19, 92, 148), others have described successes (e.g., 48, 104, 102), and some conclusions have been more ambiguous (3, 76, 96, 143). What is the real story—how much better is RE than ordinary MD?

Examination of the various claims and studies reveals that, in fact, there is little disagreement so long as the particular sampling problem is explicitly accounted for as discussed above. In brief, when observables of interest depend significantly on states that are rapidly accessed at high temperature, then RE and related methods can be efficient (48, 102). A prime example would be estimating the folded fraction near the folding transition. On the other hand, when the goal is to sample an overwhelmingly folded ensemble, there does not appear to be significant evidence of RE's efficiency—so long as the initial condition is a folded configuration (i.e., in terms of the timescales above, if $t_{\text{init}} < t_{\text{corr}}^*$). This last qualification is important in understanding the data presented in Reference 48.

An intuitive picture underpinning the preceding conclusions is readily gained by revisiting the idea that state populations are estimated by transitions between states in continuous trajectories. Indeed, this same picture is the basis for discrete-state analyses (102, 143), which we can consider along with the schematic landscape of **Figure 1**. Specifically, the average of any observable A can be written as $\langle A \rangle = \sum_i p_i \bar{A}_i$, where i indexes all folded and unfolded substates and $\bar{A}_i$ is the average of A within substate i. If p_i values are significant in both folded and unfolded substates, then

Canonical distribution: the distribution of configurations in which the probability of each is proportional to the Boltzmann factor of the energy at constant temperature, volume, and number of atoms

Force field: the classical potential energy function that approximates the quantum mechanical Hamiltonian and governs a simulation

Reweighting: assigning probabilistic values to configurations based on the ratio of the targeted distribution and the distribution actually sampled

the elevated temperatures employed in RE simulation can usefully promote transitions to sample unfolded states (102). On the other hand, if p(unfolded) $\ll$ p(folded), where p(folded) $= \sum_i^{\text{folded}} p_i$ is the sum of folded probabilities and p(unfolded) $= 1 - p$(folded), then computer resources devoted to unfolded substates can detract from estimating folded-state observables; this line of analysis echoes earlier studies (92, 143).

Whether or not RE is superior to MD, another key issue is whether RE can provide sufficient sampling ($N^{\text{eff}} > 10$) for protein-sized systems. It is far from clear that this is the case. RE cannot provide full sampling unless some level of the ladder can be fully sampled (148; see above), and the ability for a simple dynamical trajectory to sample the large configuration space of a fully or partly unfolded state of a protein has not been demonstrated to date.

Energy-based sampling methods. Energy-based schemes must be seen as closely related to temperature-based strategies because the two variables are conjugate to each other in the Boltzmann factor (Equation 1). Yet because any fixed-temperature ensemble can include arbitrary energy values, energy-based schemes must account for this. Multicanonical schemes based on sampling energy values (typically uniformly) have been implemented for biomolecules (40, 85), including with the Wang-Landau device (55, 132). Nevertheless, it should be remembered that energy is not expected to be a good proxy for the configurational coordinates of primary interest (**Figure 1**).

Hamiltonian Exchange and Multiple Models

As schematized in **Figure 2**, the different temperature-based methods sketched above can readily be generalized to variations in force field parameters (117), and even resolution as discussed below. This is because the methods are based on the Boltzmann factor, which is a general form that can apply to different force fields U_λ, each characterized by the set of parameters $\lambda = \{\lambda_1, \lambda_2, \ldots\}$. Thus, the fundamental distribution Equation 1 can be written more explicitly as

$$\rho(\mathbf{x}; T, \lambda) \propto \exp[-U_\lambda(\mathbf{x})/k_B T]. \qquad 5.$$

The different schemes for using ladders of different types—based on temperature, force field parameters, or resolution—are depicted in **Figure 2**. One of the first proposals for using multiple models was given in Reference 69.

The simplest way to reduce the roughness of a landscape (cf. Equation 3) is to maintain the functional form of U but change parameters—e.g., coefficients of dihedral terms. Changing parameters in parameter files of common software packages makes this route fairly straightforward. Note that unless force field terms are explicitly removed, the step cost in Equation 3 remains unchanged.

Hamiltonian exchange has been applied in a number of cases, for instance, using models with softened van der Waals' interactions and hence decreased roughness (45). The approach called accelerated MD employs ordinary MD on a smoothed potential energy landscape and requires reweighting to obtain canonical averages (39, 109); it can therefore be considered a two-level Hamiltonian-changing algorithm.

Resolution Exchange and Multiple-Resolution Approaches

Extending the multiple-model ideas a step further leads to the consideration of multiple resolutions. To account for varying resolution in the formulation of Equation 5, some of the λ_i parameters may be considered prefactors that can eliminate detailed interactions when set to zero. Different, formally exact approaches to using multiple resolutions have been proposed (16, 68, 70, 71, 145), primarily echoing ideas in RE and annealing.

Although multiple-resolution methods have not produced dramatic results for canonical sampling of large atomistic systems, they appear to hold unique potential in the context of the sampling costs embodied in Equation 3. In particular, low-resolution models reduce cost and roughness to such an extent that some coarse models permit good sampling for proteins ($N^{eff} > 10$) with typical resources (77, 135). It is not clear that a similar claim can be made for high-temperature or modified-potential simulations of stably folded proteins.

Multiple Trajectories Without Multiple Levels

There is a qualitatively different group of algorithms that also uses multiple trajectories, but all under the same conditions—i.e., unicanonically (6, 47, 89)—in contrast to RE and related methods. The basic idea of these methods is that when a trajectory reaches a new region of configuration space, it can spawn multiple daughter trajectories, all on the unaltered landscape. The daughter trajectories enable better sampling of regions that otherwise would receive less computer resources, and resources can be directed away from regions that already are well sampled. In the context of Markov-state models, this strategy can be optimized to increase precision in the estimate of any observable (42). Exact canonical sampling can also use a steady-state formulation of the weighted ensemble path sampling method (6; for related approaches see References 98, 125, and 133).

Single-Trajectory Approaches: Dynamics, Monte Carlo, and Variants

It is useful to combine a large group of approaches—including MD, simple MC, and Langevin dynamics (2, 32, 105)—and consider them basic dynamics (147). These methods generate a single trajectory (or Markov chain) in which configurations are strongly sequentially correlated and have no nonsequential correlations. In other words, all these methods should be expected to behave roughly like MD simulation because the motion is intrinsically constrained by the high density and landscape complexity of biomolecular systems. On the one hand, basic dynamical methods generate physically realizable trajectories or nearly so; on the

other hand, they are severely limited by physical timescales as described above. In this section, we explore some basic ideas of dynamical approaches and also some interesting variants.

Are there real dynamics? Unlike many other sampling algorithms, MD also simulates the fundamental classical dynamics of a system. That is, the trajectories produced by MD are also intended to model the time dependence of physical processes—over and above their value for sampling.

Should MD trajectories provide much better depictions of the underlying processes than other basic simulation methods, e.g., Langevin or MC simulation? The answer is not so clear. First, like other methods, MD necessarily uses an approximate force field (31). Second, on the long timescales ultimately of interest, roundoff errors can be expected to lead to significant deviations from the exact trajectories for the given force field.

Somewhat more fundamentally, there appears to be a physical inconsistency in running finite-temperature MD simulations for finite-sized systems. Thermostats of various types are used (32), some stochastic and some deterministic: Is there a correct method? By definition, a finite-temperature system is not isolated but is coupled to a bath by some physical process, presumably molecular collisions. Because the internal degrees of freedom of a bath are not explicitly simulated, again by definition, a bath should be intrinsically stochastic. The author is unaware of a first-principles prescription for modeling the coupling to a thermal bath, although sophisticated thermostats have been proposed (32, 63). For MD simulations with periodic boundary conditions, it is particularly difficult to imagine truly physical coupling to a bath.

A finite-temperature process in a finite system is intrinsically stochastic, rendering questionable the notion that standard MD protocols produce correct trajectories. To consider this point another way, should we really expect inertial effects to be important in dense aqueous and macromolecular media?

Markov chain: a sequence of configurations in which each is generated by a stochastic rule depending on only the immediately prior configuration

The foregoing discussion suggests, certainly, that Langevin simulations should also be considered physical, but what about Monte Carlo? Certainly, one can readily imagine unphysical MC protocols (e.g., attempting trial moves for atoms in the C-terminal domain of a protein ten times as often as for those in the N terminus). Nevertheless, so long as trial moves are small and performed in a spatially uniform way, one can expect MC dynamics to be an approximation to overdamped (noninertial) Langevin simulation (90, 110, 146).

Modifying dynamics to improve sampling. Although changing temperature or model parameters automatically changes the dynamics observed in MD or LD simulation, the dynamics can be modified and maintain canonical sampling on the original landscape, $U(\mathbf{x})$ (84, 144). This possibility defies the simple formulation of the second line of Equation 3, because the landscape roughness remains unchanged while the number of required steps is reduced. The method has been applied to alkane and small-peptide systems, but its generalization to arbitrary systems has not been presented to the author's knowledge. A better known and related approach is the use of multiple time steps, with shorter steps for faster degrees of freedom (119, 122), which can save computer time up to a limit set by resonance phenomena (106).

Another approach to modifying dynamics is to avoid revisiting regions already sampled by means of a memory potential that repels the trajectory from prior configurations (50, 60). This strategy is called metadynamics, or the local-elevation method, and typically employs preselection of important coordinates to be assigned repulsive terms. Although trajectories generated in a history-dependent manner do not satisfy detailed balance, it is possible to correct for the bias introduced and recover canonical sampling (10). This approach could also be considered a potential-of-mean-force method.

The potential surface and dynamics can also be modified by using a strategy of raising basins while keeping barriers intact (39, 97, 129, 149). The accelerated MD approach is a well-known

example, but achieving canonical sampling requires a reweighting procedure (109). Such reweighting is limited by the overlap issues confronting many methods: The sampled distribution must be sufficiently similar to the targeted distribution in Equation 1 so that the data are not overwhelmed by statistical noise (72, 109).

Qualitatively, how can we think about the methods just described? As dynamical methods, they all gather information by transitions among states : More transitions suggest greater statistical quality of the data. However, methods that modify the potential surface may increase the number of transitions, but the overlap between the sampled and targeted distributions is generally expected to decline with more substantial changes to the potential. In this context, the modification of dynamics without perturbing the potential (84, 144) seems particularly intriguing, even though technical challenges in implementation may remain.

Monte Carlo approaches. The term Monte Carlo can be used in many ways, so it useful to first delimit our discussion. Here, we want to focus on single-Markov-chain Monte Carlo, in which a sequence of configurations is generated, each one based on a trial move away from the previous configuration. Typically, such MC simulations of biomolecules have a strong dynamical character because trial moves tend to be small and physically realizable in a small time increment (90, 110, 146), whereas large trial moves would tend to be rejected in any reasonably detailed model. As we shall see, however, less physical moves sometimes can be used— including move sets that do not strictly obey detailed balance but instead conform to the weaker balance condition (79). The various exchange simulations discussed above can be considered MC simulations because exchange is a special kind of trial move and indeed governed by a Metropolis acceptance criterion (147).

A key advantage of single-chain MC simulation is that it can be used with any energy function, whether continuous or not, because forces do not need to be calculated (32). MC simulation is available for use with standard force fields in some simulation packages (46, 53), but it seems to be most commonly used in connection with simplified models (110, 146). The choice to use MC is usually for convenience: It readily enables the use of simple or discontinuous energy functions; furthermore, MC naturally handles constraints such as fixed bond lengths and/or angles. Substantial effort has gone into developing trial moves useful for biomolecules, with a focus on moves that act fully locally (5, 113, 123, 127, 134). Two studies have reported that MC employing quasi-physical trial moves can be more efficient than MD. (53, 124). MC has also been applied to large implicitly solvated peptides (80). Further, it is widely used in nonstatistical approaches such as docking (62), and it also is used in the Rosetta folding software (18).

For implicitly solvated all-atom peptides, a novel class of trial moves has proven extremely efficient. Ding et al. (20) showed that when libraries of configurations for each amino acid were computed in advance, swap moves with library configurations provided extremely rapid motion in configuration space. Efficiency gains based on measuring N^{eff} suggested the simulations were 100–1,000 times faster than Langevin dynamics. Trial moves involving entire residues apparently were successful because subtle correlations among the atoms were accounted for in the precomputed libraries; by comparison, internal-coordinate trial moves of a similar magnitude (e.g., twisting ϕ and ψ dihedrals) were rarely accepted (20). More detailed discussion of Monte Carlo methods can be found in recent reviews (23, 128) and textbooks (2, 32).

Potential-of-Mean-Force Methods

Many methods are designed to calculate a potential of mean force (PMF) for a specified set of coordinates χ, and such approaches implicitly are sampling methods. After all, the Boltzmann factor of the PMF is defined to be proportional to the probability distribution for the specified coordinates: $\rho(\chi) \propto \exp[-\text{PMF}(\chi)/k_B T]$ (147). If the PMF has been calculated well (with sufficient sampling along the χ coordinates and orthogonal to them), then the configurations can be suitably reweighted into a canonical ensemble.

Numerous approaches are geared toward calculating a PMF. Some, such as lambda dynamics, explore the full configuration space in a single trajectory (58, 120), and can be categorized with modified dynamics methods, although this is just a semantic issue. Many PMF calculations employ the weighted histogram analysis method (WHAM) (59), but alternatives have been developed (1, 25, 81, 111, 142).

The main advantage that PMF methods hope to attain is faster sampling along the χ coordinates than is possible using brute-force sampling. This perspective allows several observations. First, to aid sampling, the investigator must be able to select all important slow coordinates in advance—otherwise, sampling in directions orthogonal to χ will be impractical. Second, the maximum advantage that can be gained over a brute-force simulation depends on how substantial the barriers are along the χ coordinates; if sampling along a coordinate is slow because many states are separated by only small barriers, then the advantage may be modest. Last, for a PMF calculation to be successful, different values of the χ coordinates must be quantitatively related to each other: Depending on the details of the method, this can occur via numerous transitions (e.g., 25, 58, 120, 142) as given above, or by requiring well-sampled subregions of overlap (59).

Nondynamical Methods

Methods that do not rely on dynamics for sampling use a variety of distinct strategies, but as there are relatively few such efforts, they are grouped together for convenience. Recently, for example, there have been a number of applications to biomolecules of old polymer growth ideas (83, 131), sometimes termed sequential importance sampling (66). Here, monomers (e.g., amino acid) are added one at a time to an ensemble of partially grown configurations while (*a*) keeping track of appropriate statistical

PMF: potential of mean force

weights and/or (b) resampling (66). These approaches have been applied to simplified models of proteins and nucleic acids (29, 138–140), to all-atom peptides at high temperature (126), and more recently to all-atom peptides by using presampled libraries of amino acid configurations (78). The intrinsic challenge in polymer growth methods is that configurations important in the full molecule may have low statistical probability in early stages of growth; thus, application to large, detailed systems likely will require biasing toward structural information known for the full molecule (78, 138).

Some methods attempt a semienumerative description of energy basins. For instance, the mining minima method uses a quasi-harmonic procedure to estimate the partition function of each basin, which in turn determines the relative probability of the basin (7, 12). Although such approaches have the disadvantage of requiring an exponentially large number of basins (130), they largely avoid timescale issues associated with dynamical methods because the basins are located with faster search methods.

Another class of approaches has attempted to treat the problem of estimating the free energy of a previously generated canonical ensemble (13, 43, 54, 136). In analogy to PMF methods, knowledge of free energies for independent simulations in separate states (for which no transitions have been observed) enables the states to be combined in a full ensemble: The free energies directly imply the relative probabilities of the states. The ideas behind such methods are fairly technical and beyond the scope of this review.

SPECIAL HARDWARE USE FOR SAMPLING

In several cases, rather impressive improvements in sampling speed have been reported based on using novel hardware—and based on novel uses of ordinary hardware. It is almost always the case that using new hardware also requires algorithmic development, and that component should not be minimized. Little is plug and play in this business.

Parallelization, Special-Purpose CPUs, and Distributed Computing

Perhaps the best known way to exploit hardware is by parallelization. On the one hand, parallelization typically makes sampling less efficient than single-core simulation when measured by sampling per core (or per dollar) due to overhead costs. On the other hand, parallelization allows by far the fastest sampling for a given amount of wall-clock time (for a single system) including record-setting runs (22, 30, 108). Recent examples of parallelization of single-trajectory MD include a 10-μs simulation of a small domain (30); microsecond and longer simulations of membrane proteins on the BlueGene (37) and Desmond (21) platforms; and the longest reported to date, a millisecond simulation of a small globular protein on the Anton machine (24, 108). The Anton simulation reflects parallelization and the use of special-purpose chips, and the chips alone can be inferred to provide a speedup of at least 10 times compared to standard chips (107, 108).

Long simulations have significant value. They allow the community to study selected systems in great detail and to appreciate phenomena that could not otherwise be observed. Of equal importance, long simulations can alert the community to limitations of MD and force fields (31, 36).

There are other parallel strategies, such as exchange simulation (see above) and distributed computing. Distributed computing employs many simultaneous independent simulations or—via repeated rounds of simulation—quasi-parallel computing with minimal communication among simulations. Although distributed computing has been applied primarily to the folding problem (115), recent work has shown the value of multiple short simulations for producing Markov-state models (47, 89). Such models can be used to deduce both nonequilibrium and equilibrium information—thus canonical ensembles if desired. A distributed computing framework can also be used for multilevel simulations such as RE (100) and in principle for other quasi-parallel methods (6).

Use of GPUs and RAM for Sampling

Means other than parallelization are used to exploit hardware. Implicit-solvent simulations of all-atom proteins can be performed hundreds of times faster by combining a GPU and a CPU than with a CPU alone (33). Other work has exploited the RAM available on modern computers: When libraries of amino acid configurations were precalculated, library-based Monte Carlo sampled implicitly solvated peptides hundreds and sometimes more than 1,000 times faster than standard simulation (20). In a similar spirit, tabulation of a scaled form of the generalized Born implicit-solvent model led to significant speed gain for the tabulated part (61).

CONCLUDING PERSPECTIVE

The main goal of this review has been to array various methodologies into qualitative groupings to aid the critical analysis necessary to make progress in equilibrium sampling of biomolecules. Where appropriate, an effort has been made to offer a point of view on essential strengths and weaknesses of various methods. As an example, when considering popular multilevel schemes (see above), users should be confident that some level of the ladder can be well sampled.

Interesting conclusions result from surveying the sampling literature. Except in small systems, purely algorithmic improvements have yet to demonstrably accelerate equilibrium sampling of biomolecules by a significant amount. Hardware-based advances have been more dramatic, however. In fairness, demonstrating the effectiveness of new hardware for MD is much more straightforward than assessing an algorithm.

To aid future progress, developing and using sampling yardsticks should be a key priority for the field. Such measures should probe the configuration-space distribution in an objective, automatic way to measure the effective sample size. Once a small number of standard measures of sampling quality are accepted and used, efforts naturally will focus on approaches that make a significant difference. Currently, there is a proliferation of nuanced modifications of a small number of central ideas, without a good basis for distinguishing among them. Unlike in the history of theoretical physics, where elegance has sometimes served as a guide for truth, the sampling problem cries out for a pragmatic focus on efficiency. After all, it is efficiency that ultimately permits us to address biophysical and biochemical questions with confidence.

In summary, there seems little choice but to express pessimism on the current state of equilibrium sampling of important biomolecular systems. Keys to moving forward would seem to be (*a*) exploiting hardware, (*b*) quantifying sampling to determine which algorithms are more efficient than MD, and (*c*) employing large-resource simulation data to provide benchmarks and guide future efforts.

SUMMARY POINTS

1. The subtleties in defining the equilibrium sampling problem have to do with initial conditions and the targeted number of independent configurations.

2. MD simulations on typical hardware remain several orders of magnitude shorter than known biological timescales.

3. Dozens of algorithmic variants are available for biomolecular simulation, but none has been documented to yield an order of magnitude improvement over standard MD in the range of systems and conditions of primary interest.

4. Algorithms typically have not been assessed by a standard measure, but some such measures are now available.

5. Isolated instances of high algorithmic efficiency (compared to MD) have been reported in small systems and/or special conditions.

6. Novel uses of hardware (CPU, GPU, and RAM) have yielded some of the most dramatic and demonstrable advances.

FUTURE ISSUES

1. Better sampling should reveal much more of the biophysics and biochemistry that motivates biomolecular simulations. We still do not know the scales of typical equilibrium fluctuations in large biomolecules.

2. The degree of sampling quality achieved by simulations must be assessed objectively and quantitatively.

3. Progress in sampling can be expected to come from the combination of novel algorithms and novel hardware use.

4. Although current simulations typically use all-atom models at the expense of poor sampling, it is possible that reduced or hybrid all-atom/coarse-grained models will yield a better overall picture of equilibrium fluctuations.

DISCLOSURE STATEMENT

The author is not aware of any affiliations, memberships, funding, or financial holdings that might be perceived as affecting the objectivity of this review.

ACKNOWLEDGMENTS

The author has benefitted from numerous stimulating discussions of sampling with colleagues and group members over the years. Financial support has been provided by grants from the National Science Foundation (grant no. MCB-0643456) and the National Institutes of Health (grant no. GM076569).

LITERATURE CITED

1. Abrams JB, Tuckerman ME. 2008. Efficient and direct generation of multidimensional free energy surfaces via adiabatic dynamics without coordinate transformations. *J. Phys. Chem. B* 112(49):15742–57

2. Allen MP, Tildesley DJ. 1987. *Computer Simulation of Liquids*. Oxford: Oxford Univ. Press

3. Beck DAC, White GWN, Daggett V. 2007. Exploring the energy landscape of protein folding using replica-exchange and conventional molecular dynamics simulations. *J. Struct. Biol.* 157(3):514–23

4. Berne BJ, Straub JE. 1997. Novel methods of sampling phase space in the simulation of biological systems. *Curr. Opin. Struct. Biol.* 7:181–89

5. Betancourt MR. 2005. Efficient Monte Carlo trial moves for polypeptide simulations. *J. Chem. Phys.* 123(17):174905

6. Bhatt D, Zhang BW, Zuckerman DM. 2010. Steady state via weighted ensemble path sampling. *J. Chem. Phys.* 133:014110

7. Bogdan TV, Wales DJ, Calvo F. 2006. Equilibrium thermodynamics from basin-sampling. *J. Chem. Phys.* 124(4):044102

8. Brenner P, Sweet CR, VonHandorf D, Izaguirre JA. 2007. Accelerating the replica exchange method through an efficient all-pairs exchange. *J. Chem. Phys.* 126(7):074103

9. Brown S, Head-Gordon T. 2002. Cool walking: a new Markov chain Monte Carlo sampling method. *J. Comp. Chem.* 24:68–76

10. Bussi G, Laio A, Parrinello M. 2006. Equilibrium free energies from nonequilibrium metadynamics. *Phys. Rev. Lett.* 96(9):090601

11. Calvo F. 2005. All-exchanges parallel tempering. *J. Chem. Phys.* 123(12):124106

12. Chang CE, Chen W, Gilson MK. 2007. Ligand configurational entropy and protein binding. *Proc. Natl. Acad. Sci. USA* 104(5):1534–39

13. Cheluvaraja S, Meirovitch H. 2004. Simulation method for calculating the entropy and free energy peptides and proteins. *Proc. Natl. Acad. Sci. USA* 101:9241–46

14. Chodera JD, Singhal N, Pande VS, Dill KA, Swope WC. 2007. Automatic discovery of metastable states for the construction of Markov models of macromolecular conformational dynamics. *J. Chem. Phys.* 126(15):155101–17

15. Chodera JD, Swope WC, Pitera JW, Seok C, Dill KA. 2007. Use of the weighted histogram analysis method for the analysis of simulated and parallel tempering simulations. *J. Chem. Theory Comput.* 3(1):26–41

16. Christen M, van Gunsteren WF. 2006. Multigraining: an algorithm for simultaneous fine-grained and coarse-grained simulation of molecular systems. *J. Chem. Phys.* 124(15):154106

17. Christen M, van Gunsteren WF. 2008. On searching in, sampling of, and dynamically moving through conformational space of biomolecular systems: a review. *J. Comput. Chem.* 29(2):157–66

18. Das R, Baker D. 2008. Macromolecular modeling with Rosetta. *Annu. Rev. Biochem.* 77:363–82

19. Denschlag R, Lingenheil M, Tavan P. 2008. Efficiency reduction and pseudo-convergence in replica exchange sampling of peptide folding-unfolding equilibria. *Chem. Phys. Lett.* 458:244–48

20. Ding Y, Mamonov AB, Zuckerman DM. 2010. Efficient equilibrium sampling of all-atom peptides using library-based Monte Carlo. *J. Phys. Chem. B* 114(17):5870–77

21. Dror RO, Arlow DH, Borhani DW, Jensen MÃ, Piana S, Shaw DE. 2009. Identification of two distinct inactive conformations of the beta2-adrenergic receptor reconciles structural and biochemical observations. *Proc. Natl. Acad. Sci. USA* 106(12):4689–94

22. Duan Y, Kollman PA. 1998. Pathways to a protein folding intermediate observed in a 1-microsecond simulation in aqueous solution. *Science* 282(5389):740–44

23. Earl DJ, Deem MW. 2008. Monte Carlo simulations. *Methods Mol. Biol.* 443:25–36

24. Ensign DL, Kasson PM, Pande VS. 2007. Heterogeneity even at the speed limit of folding: large-scale molecular dynamics study of a fast-folding variant of the villin headpiece. *J. Mol. Biol.* 374(3):806–16

25. Fasnacht M, Swendsen RH, Rosenberg JM. 2004. Adaptive integration method for Monte Carlo simulations. *Phys. Rev. E* 69:056704

26. Fenwick MK, Escobedo FA. 2003. Expanded ensemble and replica exchange methods for simulation of protein-like systems. *J. Chem. Phys.* 119(22):11998–2010

27. Ferrenberg AM, Swendsen RH. 1988. New Monte Carlo technique for studying phase transitions. *Phys. Rev. Lett.* 61:2635–38

28. Frantz DD, Freeman DL, Doll JD. 1990. Reducing quasi-ergodic behavior in Monte Carlo simulation by J-walking: applications to atomic clusters. *J. Chem. Phys.* 93:2769–84

29. Frauenkron H, Bastolla U, Gerstner E, Grassberger P, Nadler W. 1998. New Monte Carlo algorithm for protein folding. *Phys. Rev. Lett.* 80(14):3149–52

30. Freddolino PL, Liu F, Gruebele M, Schulten K. 2008. Ten-microsecond molecular dynamics simulation of a fast-folding WW domain. *Biophys. J.* 94(10):L75–77

31. Freddolino PL, Park S, Roux B, Schulten K. 2009. Force field bias in protein folding simulations. *Biophys. J.* 96(9):3772–80

32. Frenkel D, Smit B. 1996. *Understanding Molecular Simulation*. San Diego, CA: Academic

33. Friedrichs MS, Eastman P, Vaidyanathan V, Houston M, Legrand S, et al. 2009. Accelerating molecular dynamic simulation on graphics processing units. *J. Comput. Chem.* 30:864–72

34. Geyer CJ. 1991. Markov chain Monte Carlo maximum likelihood. In *Comput. Sci. Stat.: Proc. 23rd Symp. Interface*, ed. EM Keramidas, pp. 156–63. Fairfax Station: Interface Found.

33. Describes GPU-based simulation of implicitly solvated proteins which are hundreds of times faster than CPU simulation.

35. Grassberger P. 1997. Pruned-enriched Rosenbluth method: simulations of theta polymers of chain length up to 1,000,000. *Phys. Rev. E* 56(3):3682–93

36. Grossfield A, Feller SE, Pitman MC. 2007. Convergence of molecular dynamics simulations of membrane proteins. *Proteins* 67(1):31–40

37. Grossfield A, Pitman MC, Feller SE, Soubias O, Gawrisch K. 2008. Internal hydration increases during activation of the G-protein-coupled receptor rhodopsin. *J. Mol. Biol.* 381(2):478–86

38. Grossfield A, Zuckerman DM. 2009. Quantifying uncertainty and sampling quality in biomolecular simulations. *Annu. Rep. Comput. Chem.* 5:23–48

39. Hamelberg D, Mongan J, McCammon JA. 2004. Accelerated molecular dynamics: a promising and efficient simulation method for biomolecules. *J. Chem. Phys.* 120(24):11919–29

40. Hansmann UH, Okamoto Y. 1996. Monte Carlo simulations in generalized ensemble: multicanonical algorithm versus simulated tempering. *Phys. Rev. E* 54(5):5863–65

41. Hess B. 2002. Convergence of sampling in protein simulations. *Phys. Rev. E* 65(3):031910

42. Hinrichs NS, Pande VS. 2007. Calculation of the distribution of eigenvalues and eigenvectors in Markovian state models for molecular dynamics. *J. Chem. Phys.* 126(24):244101

43. Hnizdo V, Tan J, Killian BJ, Gilson MK. 2008. Efficient calculation of configurational entropy from molecular simulations by combining the mutual-information expansion and nearest-neighbor methods. *J. Comput. Chem.* 29(10):1605–14

44. Howard J. 2001. *Mechanics of Motor Proteins and the Cytoskeleton.* Sunderland, MA: Sinauer

45. Hritz J, Oostenbrink C. 2008. Hamiltonian replica exchange molecular dynamics using soft-core interactions. *J. Chem. Phys.* 128(14):144121

46. Hu J, Ma A, Dinner AR. 2006. Monte Carlo simulations of biomolecules: the MC module in CHARMM. *J. Comput. Chem.* 27(2):203–16

47. Huang X, Bowman GR, Bacallado S, Pande VS. 2009. Rapid equilibrium sampling initiated from nonequilibrium data. *Proc. Natl. Acad. Sci. USA* 106(47):19765–69

48. Huang X, Bowman GR, Pande VS. 2008. Convergence of folding free energy landscapes via application of enhanced sampling methods in a distributed computing environment. *J. Chem. Phys.* 128(20):205106

49. Huber GA, McCammon JA. 1997. Weighted-ensemble simulated annealing: faster optimization on hierarchical energy surfaces. *Phys. Rev. E* 55(4):4822

50. Huber T, Torda AE, van Gunsteren WF. 1994. Local elevation: a method for improving the searching properties of molecular dynamics simulation. *J. Comput. Aided Mol. Des.* 8(6):695–708

51. Hukushima K, Nemoto K. 1996. Exchange Monte Carlo method and application to spin glass simulation. *J. Phys. Soc. Jpn.* 65:1604–8

52. Jarzynski C. 1997. Nonequilibrium equality for free energy differences. *Phys. Rev. Lett.* 78:2690–93

53. Jorgensen WL, Tirado-Rives J. 1996. Monte Carlo vs molecular dynamics for conformational sampling. *J. Phys. Chem.* 100:14508–13

54. Killian BJ, Kravitz JY, Gilson MK. 2007. Extraction of configurational entropy from molecular simulations via an expansion approximation. *J. Chem. Phys.* 127(2):024107

55. Kim J, Straub JE, Keyes T. 2006. Statistical-temperature Monte Carlo and molecular dynamics algorithms. *Phys. Rev. Lett.* 97(5):050601

56. Knegtel RMA, Kuntz ID, Oshiro CM. 1997. Molecular docking to ensembles of protein structures. *J. Mol. Biol.* 266:424–40

57. Kofke DA. 2002. On the acceptance probability of replica-exchange Monte Carlo trials. *J. Chem. Phys.* 117(15):6911–14

58. Kong X, Brooks CM. 1996. Lambda-dynamics: a new approach to free energy calculations. *J. Chem. Phys.* 105:2414–23

59. Kumar S, Bouzida D, Swendsen RH, Kollman PA, Rosenberg JM. 1992. The weighted histogram analysis method for free-energy calculations on biomolecules. I. The method. *J. Comp. Chem.* 13:1011–21

60. Laio A, Parrinello M. 2002. Escaping free-energy minima. *Proc. Natl. Acad. Sci. USA* 99(20):12562–66

61. Larsson P, Lindahl E. 2010. A high-performance parallel-generalized born implementation enabled by tabulated interaction rescaling. *J. Comput. Chem.* 31(14):2593–600

62. Leach AR. 2001. *Molecular Modelling: Principles and Applications.* Upper Saddle River, NJ: Prentice Hall

63. Leimkuhler BJ, Sweet CR. 2004. The canonical ensemble via symplectic integrators using nose and nose-poincare chains. *J. Chem. Phys.* 121(1):108–16

64. Li H, Li G, Berg BA, Yang W. 2006. Finite reservoir replica exchange to enhance canonical sampling in rugged energy surfaces. *J. Chem. Phys.* 125(14):144902

65. Lin J-H, Perryman AL, Schames JR, McCammon JA. 2002. Computational drug design accommodating receptor flexibility: the relaxed complex scheme. *J. Am. Chem. Soc.* 124(20):5632–33

66. Liu JS. 2002. *Monte Carlo Strategies in Scientific Computing.* New York: Springer

67. Liu P, Kim B, Friesner RA, Berne BJ. 2005. Replica exchange with solute tempering: a method for sampling biological systems in explicit water. *Proc. Natl. Acad. Sci. USA* 102:13749–54

68. Liu P, Shi Q, Lyman E, Voth GA. 2008. Reconstructing atomistic detail for coarse-grained models with resolution exchange. *J. Chem. Phys.* 129(11):114103

69. Liu Z, Berne BJ. 1993. Method for accelerating chain folding and mixing. *J. Chem. Phys.* 99(8):6071–77

70. Lyman E, Ytreberg FM, Zuckerman DM. 2006. Resolution exchange simulation. *Phys. Rev. Lett.* 96:028105

71. Lyman E, Zuckerman DM. 2006. Resolution exchange simulation with incremental coarsening. *J. Chem. Theory Comput.* 2:656–66

72. Lyman E, Zuckerman DM. 2007. Annealed importance sampling of peptides. *J. Chem. Phys.* 127(6):065101–6

73. Lyman E, Zuckerman DM. 2007. On the structural convergence of biomolecular simulations by determination of the effective sample size. *J. Phys. Chem. B* 111(44):12876–82

74. Lyman E, Zuckerman DM. 2009. Resampling improves the efficiency of a "fast-switch" equilibrium sampling protocol. *J. Chem. Phys.* 130(8):081102

75. Lyubartsev AP, Martsinovski AA, Shevkunov SV, Vorontsov-Velyaminov PN. 1992. New approach to Monte Carlo calculation of the free energy: method of expanded ensembles. *J. Chem. Phys.* 96(3):1776–83

76. Machta J. 2009. Strengths and weaknesses of parallel tempering. *Phys. Rev. E Stat. Nonlin Soft Matter Phys.* 80(5 Pt. 2):056706

77. Mamonov AB, Bhatt D, Cashman DJ, Ding Y, Zuckerman DM. 2009. General library-based Monte Carlo technique enables equilibrium sampling of semi-atomistic protein models. *J. Phys. Chem. B* 113(31):10891–904

78. Mamonov AB, Zhang X, Zuckerman D. 2011. Rapid sampling of all-atom peptides using a library-based polymer-growth approach. *J. Comput. Chem.* 32:396–405

79. Manousiouthakis VI, Deem MW. 1999. Strict detailed balance is unnecessary in Monte Carlo simulation. *J. Chem. Phys.* 110(6):2753–56

80. Mao AH, Crick SL, Vitalis A, Chicoine CL, Pappu RV. 2010. Net charge per residue modulates conformational ensembles of intrinsically disordered proteins. *Proc. Natl. Acad. Sci. USA* 107(18):8183–88

81. Maragliano L, Vanden-Eijnden E. 2008. Single-sweep methods for free energy calculations. *J. Chem. Phys.* 128(18):184110

82. McCammon JA, Gelin BR, Karplus M. 1977. Dynamics of folded proteins. *Nature* 267(5612):585–90

83. Meirovitch H. 1982. A new method for simulation of real chains. Scanning future steps. *J. Phys. A* 15:L735–40

84. Minary P, Tuckerman ME, Martyna GJ. 2004. Long time molecular dynamics for enhanced conformational sampling in biomolecular systems. *Phys. Rev. Lett.* 93(15):150201

85. Mitsutake A, Sugita Y, Okamoto Y. 2001. Generalized-ensemble algorithms for molecular simulations of biopolymers. *Biopolymers* 60(2):96–123

86. Mountain RD, Thirumalai D. 1989. Measures of effective ergodic convergence in liquids. *J. Phys. Chem.* 93(19):6975–79

87. Neal RM. 2001. Annealed importance sampling. *Stat. Comput.* 11:125–39

88. Neirotti JP, Freeman DL, Doll JD. 2000. Approach to ergodicity in Monte Carlo simulations. *Phys. Rev. E* 62(5):7445–61

89. Noe F, Schütte C, Vanden-Eijnden E, Reich L, Weikl TR. 2009. Constructing the equilibrium ensemble of folding pathways from short off-equilibrium simulations. *Proc. Natl. Acad. Sci. USA* 106(45):19011–16

90. Nowak U, Chantrell RW, Kennedy EC. 2000. Monte Carlo simulation with time step quantification in terms of Langevin dynamics. *Phys. Rev. Lett.* 84(1):163–66

77. Introduces a memory-intensive approach to simulation of all-atom, coarse-grained, and hybrid models.

91. Noé F, Horenko I, Schütte C, Smith JC. 2007. Hierarchical analysis of conformational dynamics in biomolecules: transition networks of metastable states. *J. Chem. Phys.* 126(15):155102

92. Nymeyer H. 2008. How efficient is replica exchange molecular dynamics? An analytic approach. *J. Chem. Theory Comput.* 4(4):626–36

93. Okur A, Roe DR, Cui G, Hornak V, Simmerling C. 2007. Improving convergence of replica-exchange simulations through coupling to a high-temperature structure reservoir. *J. Chem. Theory Comput.* 3(2):557–68

94. Opps SB, Schofield J. 2001. Extended state-space Monte Carlo methods. *Phys. Rev. E* 63(5 Pt. 2):056701

95. Paschek D, Garcia AE. 2004. Reversible temperature and pressure denaturation of a protein fragment: a replica-exchange molecular dynamics simulation study. *Phys. Rev. Lett.* 93:238105

96. Periole X, Mark AE. 2007. Convergence and sampling efficiency in replica exchange simulations of peptide folding in explicit solvent. *J. Chem. Phys.* 126(1):014903

97. Rahman JA, Tully JC. 2002. Puddle-skimming: an efficient sampling of multidimensional configuration space. *J. Chem. Phys.* 116(20):8750–60

98. Raiteri P, Laio A, Gervasio FL, Micheletti C, Parrinello M. 2006. Efficient reconstruction of complex free energy landscapes by multiple walkers metadynamics. *J. Phys. Chem. B* 110(8):3533–39

99. Rathore N, Chopra M, de Pablo JJ. 2005. Optimal allocation of replicas in parallel tempering simulations. *J. Chem. Phys.* 122(2):024111

100. Rhee YM, Pande VS. 2003. Multiplexed-replica exchange molecular dynamics method for protein folding simulation. *Biophys. J.* 84(2 Pt. 1):775–86

101. Rick SW. 2007. Replica exchange with dynamical scaling. *J. Chem. Phys.* 126(5):054102

102. Rosta E, Hummer G. 2009. Error and efficiency of replica exchange molecular dynamics simulations. *J. Chem. Phys.* 131(16):165102

103. Ruscio JZ, Fawzi NL, Head-Gordon T. 2010. How hot? Systematic convergence of the replica exchange method using multiple reservoirs. *J. Comput. Chem.* 31(3):620–27

104. Sanbonmatsu KY, García AE. 2002. Structure of met-enkephalin in explicit aqueous solution using replica exchange molecular dynamics. *Proteins* 46(2):225–34

105. Schlick T. 2002. *Molecular Modeling and Simulation.* New York: Springer

106. Schlick T, Mandziuk M, Skeel RD, Srinivas K. 1998. Nonlinear resonance artifacts in molecular dynamics simulations. *J. Comput. Phys.* 140(1):1–29

107. Shaw DE, Deneroff MM, Dror RO, Kuskin JS, Larson RH, et al. 2007. Anton: a special-purpose machine for molecular dynamics simulation. In *Proc. 34th Annu. Int. Symp. Comput. Archit., San Diego, CA, (ISCA '07)*, pp. 1–12. New York: Assoc. Comput. Mach.

108. Shaw DE, Dror RO, Salmon JK, Grossman JP, Mackenzie KM, et al. 2009. Millisecond-scale molecular dynamics simulations on Anton. In *Proc. Conf. High Perform. Comput. Netw., Storage Anal., (SC '09)*, pp. 1–11. New York: Assoc. Comput. Mach.

109. Shen T, Hamelberg D. 2008. A statistical analysis of the precision of reweighting-based simulations. *J. Chem. Phys.* 129(3):034103

110. Shimada J, Kussell EL, Shakhnovich EI. 2001. The folding thermodynamics and kinetics of crambin using an all-atom Monte Carlo simulation. *J. Mol. Biol.* 308(1):79–95

111. Shirts MR, Chodera JD. 2008. Statistically optimal analysis of samples from multiple equilibrium states. *J. Chem. Phys.* 129(12):124105

112. Sindhikara D, Meng Y, Roitberg AE. 2008. Exchange frequency in replica exchange molecular dynamics. *J. Chem. Phys.* 128(2):024103

113. Smith CA, Kortemme T. 2008. Backrub-like backbone simulation recapitulates natural protein conformational variability and improves mutant side-chain prediction. *J. Mol. Biol.* 380(4):742–56

114. Smith LJ, Daura X, van Gunsteren WF. 2002. Assessing equilibration and convergence in biomolecular simulations. *Proteins* 48:487–96

115. Snow CD, Nguyen H, Pande VS, Gruebele M. 2002. Absolute comparison of simulated and experimental protein-folding dynamics. *Nature* 420(6911):102–6

116. Straub JE, Thirumalai D. 1993. Exploring the energy landscape in proteins. *Proc. Natl. Acad. Sci. USA* 90(3):809–13

108. A millisecond-scale simulation of an explicitly solvated protein, performed with special hardware and parallelization algorithms.

117. Sugita Y, Kitao A, Okamoto Y. 2000. Multidimensional replica-exchange method for free-energy calculations. *J. Chem. Phys.* 113:6042–51

118. Swendsen RH, Wang J-S. 1986. Replica Monte Carlo simulation of spin-glasses. *Phys. Rev. Lett.* 57:2607–9

119. Teleman O, Jonsson B. 1986. Vectorizing a general purpose molecular dynamics simulation program. *J. Comput. Chem.* 7(1):58–66

120. Tidor B. 1993. Simulated annealing on free energy surfaces by a combined molecular dynamics and Monte Carlo approach. *J. Phys. Chem.* 97:1069–73

121. Trebst S, Troyer M, Hansmann UHE. 2006. Optimized parallel tempering simulations of proteins. *J. Chem. Phys.* 124(17):174903

122. Tuckerman M, Berne BJ, Martyna GJ. 1992. Reversible multiple time scale molecular dynamics. *J. Chem. Phys.* 97(3):1990–2001

123. Ulmschneider JP, Jorgensen WL. 2003. Monte Carlo backbone sampling for polypeptides with variable bond angles and dihedral angles using concerted rotations and a Gaussian bias. *J. Chem. Phys.* 118(9):4261–71

124. Ulmschneider JP, Ulmschneider MB, Nola AD. 2006. Monte Carlo vs molecular dynamics for all-atom polypeptide folding simulations. *J. Phys. Chem. B* 110(33):16733–42

125. Valeriani C, Allen RJ, Morelli MJ, Frenkel D, ten Wolde PR. 2007. Computing stationary distributions in equilibrium and nonequilibrium systems with forward flux sampling. *J. Chem. Phys.* 127(11):114109

126. Velikson B, Garel T, Niel JC, Orland H, Smith JC. 1992. Conformational distribution of heptaalanine: analysis using a new Monte Carlo chain growth method. *J. Comput. Chem.* 13(10):1216–33

127. Vendruscolo M. 1997. Modified configurational bias Monte Carlo method for simulation of polymer systems. *J. Chem. Phys.* 106(7):2970–76

128. Vitalis A, Pappu RV. 2009. Methods for Monte Carlo simulations of biomacromolecules. *Annu. Rep. Comput. Chem.* 5:49–76

129. Voter AF. 1997. Hyperdynamics: accelerated molecular dynamics of infrequent events. *Phys. Rev. Lett.* 78(20):3908–11

130. Wales D. 2004. *Energy Landscapes: Applications to Clusters, Biomolecules and Glasses.* Cambridge, UK: Cambridge Univ. Press

131. Wall FT, Erpenbeck JJ. 1959. New method for the statistical computation of polymer dimensions. *J. Chem. Phys.* 30(3):634–37

132. Wang F, Landau DP. 2001. Efficient, multiple-range random walk algorithm to calculate the density of states. *Phys. Rev. Lett.* 86(10):2050–53

133. Warmflash A, Bhimalapuram P, Dinner AR. 2007. Umbrella sampling for nonequilibrium processes. *J. Chem. Phys.* 127(15):154112–18

134. Wu MG, Deem MW. 1999. Analytical rebridging Monte Carlo: application to cis/trans isomerization in proline-containing, cyclic peptides. *J. Chem. Phys.* 111(14):6625–32

135. Ytreberg FM, Aroutiounian SK, Zuckerman DM. 2007. Demonstrated convergence of the equilibrium ensemble for a fast united-residue protein model. *J. Chem. Theory Comput.* 3(5):1860–66

136. Ytreberg FM, Zuckerman DM. 2008. A black-box re-weighting analysis can correct flawed simulation data. *Proc. Natl. Acad. Sci. USA* 105(23):7982–87

137. Zhang C, Ma J. 2010. Enhanced sampling and applications in protein folding in explicit solvent. *J. Chem. Phys.* 132(24):244101

138. Zhang J, Lin M, Chen R, Liang J, Liu JS. 2007. Monte Carlo sampling of near-native structures of proteins with applications. *Proteins* 66(1):61–68

139. Zhang J, Lin M, Chen R, Wang W, Liang J. 2008. Discrete state model and accurate estimation of loop entropy of RNA secondary structures. *J. Chem. Phys.* 128(12):125107

140. Zhang JL, Liu JS. 2002. A new sequential importance sampling method and its application to the two-dimensional hydrophobic–hydrophilic model. *J. Chem. Phys.* 117(7):3492–98

141. Zhang X, Bhatt D, Zuckerman DM. 2010. Automated sampling assessment for molecular simulations using the effective sample size. *J. Chem. Theory Comput.* 6:3048–57

142. Zheng L, Chen M, Yang W. 2008. Random walk in orthogonal space to achieve efficient free-energy simulation of complex systems. *Proc. Natl. Acad. Sci. USA* 105(51):20227–32

118. Introduces replica exchange simulation.

133. Introduces a method for sampling steady states (of which equilibrium is one).

141. Provides an automated assessment of sampling quality for the effective sample size based on the configuration-space distribution.

143. Zheng W, Andrec M, Gallicchio E, Levy RM. 2007. Simulating replica exchange simulations of protein folding with a kinetic network model. *Proc. Natl. Acad. Sci. USA* 104(39):15340–45

144. Zhu Z, Tuckerman ME, Samuelson SO, Martyna GJ. 2002. Using novel variable transformations to enhance conformational sampling in molecular dynamics. *Phys. Rev. Lett.* 88(10):100201

145. Zuckerman D. 2009. Principles and practicalities of canonical mixed-resolution sampling of biomolecules. In *Coarse-Graining of Condensed Phase and Biomolecular Systems*, ed. GA Voth, pp. 171–84. Boca Raton, FL: Taylor & Francis

146. Zuckerman DM. 2004. Simulation of an ensemble of conformational transitions in a united-residue model of calmodulin. *J. Phys. Chem. B* 108:5127–37

147. Zuckerman DM. 2010. *Statistical Physics of Biomolecules: An Introduction*. Boca Raton, FL: CRC Press

148. Zuckerman DM, Lyman E. 2006. A second look at canonical sampling of biomolecules using replica exchange simulation. *J. Chem. Theory Comput.* 2(4):1200–2

149. Zuckerman DM, Woolf TB. 1999. Dynamic reaction paths and rates through importance-sampled stochastic dynamics. *J. Chem. Phys.* 111:9475–84

143. Describes kinetic analysis of replica exchange explaining strengths and weaknesses.

144. Describes a method for accelerating dynamics in the canonical ensemble.

Decision Making in Living Cells: Lessons from a Simple System

Ido Golding

Verna and Marrs McLean Department of Biochemistry and Molecular Biology, Baylor College of Medicine, Houston, Texas 77030

Department of Physics and Center for the Physics of Living Cells, University of Illinois, Urbana, Illinois 61801; email: igolding@illinois.edu

Annu. Rev. Biophys. 2011. 40:63–80

The *Annual Review of Biophysics* is online at biophys.annualreviews.org

This article's doi: 10.1146/annurev-biophys-042910-155227

Copyright © 2011 by Annual Reviews. All rights reserved

1936-122X/11/0609-0063$20.00

Keywords

Escherichia coli, bacteriophage lambda, lysis/lysogeny, gene expression, physics of living cells, systems biology

Abstract

The life cycle of bacteriophage lambda serves as a simplified paradigm for cell-fate decisions. The ongoing quantitative, high-resolution experimental investigation of this life cycle has produced some important insights in recent years. These insights have to do with the way cells choose among alternative fates, how they maintain long-term memory of their gene-expression state, and how they switch from one stable state to another. The recent studies have highlighted the role of spatiotemporal effects in cellular processes and the importance of distinguishing chemical stochasticity from possible hidden variables in cellular decision making.

Contents

INTRODUCTION

Bacteriophage:
a virus that infects
bacterial cells, also
called phage

Prophage: a dormant
intracellular phage

One of the most striking features exhibited by living cells is their ability to process information and make decisions (3, 15, 69). In response to signals from the environment and from other cells, a cell may modulate its behavior either continuously (e.g., changing the expression level of a gene) or in a digital manner, choosing between a discrete set of predefined behaviors. Once a cell has chosen a specific option, it can remember that decision and maintain its state even as the environment keeps changing and the original stimulus is gone. However, this stability in face of fluctuations does not come at the expense of the ability to switch to an alternative state when the proper signal is given (43, 69).

To a person trained in statistical physics, these properties are strongly reminiscent of the emergent states of multi-particle systems (36, 89). One is then tempted to emulate for living cells what has been done in azoic systems: create a narrative for the behavior of the system, which is quantitative, simple, and universal all at the same time. This narrative would arise not from a molecular understanding of every detail, but from a coarse-grained (systems-level) approach. As in the physics of nonliving systems, the way to construct this narrative is through the formulation of basic laws characterizing the process of cellular decision making. Over the last decade or so, such an endeavor has been pursued by a growing number of researchers. It is arguably one of the greatest challenges to physics in this century.

So where should one begin to study the process of cellular decision making? The answer is clear. The bacterium *Escherichia coli* has long served as the "hydrogen atom" of biology. The way *E. coli* modulates its swimming behavior (14, 15) and the expression of its genes (3, 68) has proven a fertile ground for studying cellular information processing. And when it comes to discrete cell-fate decisions and the long-term memory of the cellular state, the system composed of *E. coli* and one of its viruses, bacteriophage lambda, is a perfect place to start. The life cycle of phage lambda is depicted in **Figure 1**. Following infection of the bacterial cell, a decision is made between two alternative pathways: The invading phage can either replicate and lyse (kill) the host cell (lytic pathway), or it can integrate into the host chromosome—becoming a prophage—where it replicates as part of the bacterial genome while all lytic functions are repressed (lysogenic pathway). In response to cell damage, a switch back to the lytic

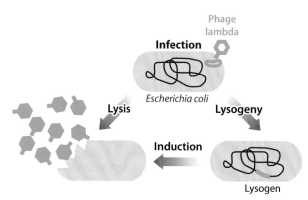

Figure 1

The life cycle of bacteriophage lambda. Following infection, the bacteriophage can either replicate and kill the host cell (lysis), or it can integrate into the bacterial chromosome, where it replicates as part of the host genome (lysogeny). The lysogenic state is extremely stable, but a lysogen can be switched (induced) back to the lytic pathway either spontaneously or in a directed manner. Adapted from Reference 21.

pathway can be induced: The prophage exits the host chromosome, replicates, and lyses the bacterial cell (67, 69).

Despite its relative simplicity, phage lambda already displays some of the intriguing features of cellular decision making seen in higher systems; this is what makes it an ideal starting point in the endeavor to unveil universal principles in cellular decision making.

- Noisy decision. When examined at the single-cell level, cellular decision making often appears imprecise, or noisy, in the sense that individual cells in a clonal population undergo different fates even when subject to identical conditions (6, 19, 49, 57, 58, 79, 81, 84, 88). This phenotype is commonly attributed to biochemical stochasticity (6, 19, 49, 57, 58, 79, 84).

- A self-regulating fate determining gene. Once a decision has been made towards lysogeny, the lysogenic state of an *E. coli* cell harboring a dormant bacteriophage serves as one of the simplest examples for a differentiated cellular state (67, 69, 70). Lysogenic stability is maintained by the activity of a single protein species, the lambda repressor (CI), that acts as a transcription factor to repress all lytic functions from the prophage, as well as to regulate its own production (69). This feature of autoregulation by the fate-determining protein is also observed in higher systems displaying long-term cellular memory (26, 43, 54).

- Stability and switchability. The lambda lysogeny system exhibits extremely high stability: Under the proper conditions, spontaneous switching events occur less than once per 10^8 cell generations (56). Yet, fast and efficient switching can be observed in response to the appropriate stimulus, e.g., damage to the bacterial genome (67). Thus, the lysogenic state, like the differentiated state in higher systems, combines long-term stability with efficient state switching (reprogramming) in response to the proper stimulus (43).

A verbal, nonquantitative narrative already exists for the lambda life cycle, based on half a century of genetics and biochemistry (46, 47, 69). In addition, considerable theoretical effort has been invested in forming a dynamical systems picture for the lysis/lysogeny decision and the maintenance of lysogeny, based on the known biochemical players (see e.g., 1, 6, 10, 11, 60, 78, 87). Earlier studies are phrased in deterministic terms (1, 60, 78); later ones often incorporate the effects of chemical stochasticity (6, 10, 11).

Although constituting important first steps toward the formation of a quantitative narrative, these theoretical works were limited by the fact that the kinetics characterizing the relevant cellular processes are largely unknown. Despite a small number of pioneering efforts to quantitatively phenotype the lambda system (4, 48, 52, 74), the kinetic assumptions and parameter values used in modeling typically arise from either (*a*) ad hoc choices made to reproduce the observed phenomena; (*b*) measurements made on whole populations, which average out much of the relevant dynamics in both space and time; or (*c*) in vitro measurements, far removed from actual physiological conditions (31). The bottom line is that a wide quantitative gap still exists between the vast genetic and biochemical knowledge on the one hand and the observed phenotype on the other hand. In this regard, phage lambda serves as an example of the inadequacy of our current comprehension of living systems—but it is also a system in which this gap can arguably be bridged for the first time.

To begin this process, we have been endeavoring in my lab over the last few years to characterize the life cycle of phage lambda in a quantitative manner, in real time, at the resolution of individual phages and cells. By doing so, we are hoping to obtain a quantitative understanding of the lysis/lysogeny switch: from the initial cell-fate decision following infection, through the long-term maintenance of the lysogenic state, to the kinetics of switching from lysogeny to lysis during induction. Below, I describe some of the insights that have emerged

Lysogeny: the condition of a bacterial cell following phage infection, in which the phage does not kill the cell but instead lies dormant

Lysis: the death of a bacterial cell following phage infection

Induction: a switch from lysogeny to lysis; can occur spontaneously or as a result of an external signal

LamB: a receptor
protein on the surface
of *E. coli* that triggers
the injection of phage
lambda DNA into the
cell

THE LIFE CYCLE OF BACTERIOPHAGE LAMBDA

Target Finding by the Infecting Phage: Rethinking Spatiotemporal Dynamics

The bacterial cell is traditionally considered a well-mixed "sack of soup" where interactions are largely governed by diffusion kinetics. In recent years, however, this simple picture is being challenged by a wealth of new information regarding the complex organization and intricate spatiotemporal dynamics inside bacterial cells, which are reminiscent of those observed in eukaryotes. Specific cellular functions are performed at designated locations in the cell (66); active transport mechanisms are used to shuttle macromolecules to their destination (85). During our own studies of transcription in *E. coli*, we found that mRNA molecules accumulate at the site of transcription to a larger degree than was previously believed (37). Furthermore, mRNA molecules moving randomly in the cytoplasm exhibit anomalous rather than normal (Fickian) diffusion, pointing to a unique interaction between the moving macromolecules and the medium surrounding them (38). All these deviations from the ideal homogenous, diffusion-dominated picture may have a critical effect on the kinetics of intracellular processes, such as gene expression, by bringing closer together molecules that have to interact with each other while keeping apart those that should not. This intricacy should also remind us why biochemical parameters obtained in vitro should always be taken with a grain of salt.

The lambda life cycle offers an opportunity to characterize in a quantitative manner the spatiotemporal aspects of basic cellular processes and examine the effects of spatial organization on the observed kinetics when compared to the naïve picture of diffusion-limited reactions. Here I discuss one such case, that of target finding by the infecting phage. The initial step in viral infection is the attachment of a virus to the surface of the host cell, followed by delivery of the viral genome into the cell (59, 80). This process involves a highly specific interaction between the virus and a receptor on the cell surface (59, 80). In the case of lambda, the target receptor (LamB) is originally used by the cell for uptaking maltose sugar but has been hijacked by the phage (as is often done by viruses) for the purpose of injecting its DNA into the host. The number of LamB receptors on the cell surface varies depending on environmental conditions, typically by a few hundred under laboratory conditions (20, 64).

How does the phage find its target receptor? In other words, what are the spatiotemporal dynamics by which a phage, initially diffusing in bulk, arrives at a specific site on the cell surface and there inject its DNA? The classical biophysical picture of this process goes back to Adam & Delbrück (2), who introduced the concept of "reduction of dimensionality" as a means of minimizing target-finding time. According to this picture, the phage freely diffuses in three dimensions (3D) until it encounters the surface of the cell. The 3D motion is then replaced by a two-dimensional (2D) diffusion on the cell surface, until the target—a receptor—is found; the phage injects its DNA and the infection cycle ensues. The reduction in spatial dimension, from three to two, is seen as a way to accelerate the process of target finding (2). Bulk infection experiments exhibit kinetics that are consistent with this two-step picture (64, 76). However, in recent years a number of intriguing observations have been made that suggested that our view of the target-finding process should be re-examined: First, a study of infection kinetics in bulk suggested that the presence of LamB receptors on the host cell is required not only for the final injection step, but also for the initial virus-host association (64). Second, imaging of fluorescently labeled phages on individual cells indicated that infecting viruses bind preferentially to the bacterial poles rather than cover the cell surface uniformly (29). Last, a time-varying, nonuniform spatial organization of the LamB receptors in live *E. coli* cells was revealed using fluorescently labeled tails of phage lambda (35).

To examine the process by which the phage arrives at its final attachment site, we recently performed single-particle tracking of individual fluorescent phages on live bacterial cells during the early stages of infection and quantified viral trajectories with nanometer accuracy and ~50-ms resolution (E. Rothenberg, L.A. Sepúlveda, S.O. Skinner, L. Zeng, P.R. Selvin & I. Golding, manuscript in preparation) (**Figure 2a**). The inspection of hundreds of phage infection events at first suggested that the classical picture was appropriate: Phages diffused in the bulk until randomly encountering the surface of a bacterial cell. A phage would then move randomly on the cell surface. The surface motion ended either with an attachment to a specific site on the cell or—with a similar probability—with the phage falling off the cell and diffusing away (**Figure 2b**). The different modes of motion exhibited markedly different diffusion coefficients (**Figure 2c**).

We next examined closely the portion of viral motion that occurs on the cell surface prior to attachment. To our surprise, we found that phages exhibited an anisotropic motion pattern, with a tendency to move along the short axis of the cell (**Figure 2d,e**). Their motion was also inhomogeneous. Specifically, the viruses exhibited what we termed spatial focusing along the cell (**Figure 2f**): The initial virus-cell points of encounter were uniformly distributed along the cell, but subsequent virus trajectories showed an affinity to reside in specific regions along the cell, one of which is the cellular pole; eventually, bound viruses were spatially focused, showing a distinct preference for the cell poles.

These unique features of viral motion led us to ask whether they reflect an interaction of the moving viruses with an ordered pattern present on the surface of the cell, specifically with the viral receptors, LamB. Fluorescently labeling the LamB receptors on the cell revealed striped patterns of receptors on the surface, reminiscent of rings and helices (**Figure 2g**). The spatial organization of the receptors—position along the cell and angles of receptor bands—suggests that the observed viral motion is largely influenced by the receptor network. Next, dual-labeling experiments confirmed that phages moving on the surface spend the majority of their time in the vicinity of receptors. The presence of receptors also determines both the diffusion coefficient and the dwell time of phages on the cell surface. A control experiment demonstrated that when phages land on cells lacking LamB, they fall off almost instantaneously rather than spend time moving on the cell surface.

The emerging picture is that the interaction of phage lambda with the cell surface occurs exclusively through the LamB receptor. In other words, the receptor is not only the final target for the phage, the place where DNA injection occurs, it is also the element that guides the phage through the search process, on its way to the final attachment site. Although we have no clear molecular picture of this search process, it likely involves a weak, reversible interaction between the lambda tail and LamB (64). To examine the plausibility of this scenario, we have formulated a simple coarse-grained model of phage motion on the cell surface (E. Rothenberg, L.A. Sepúlveda, S.O. Skinner, L. Zeng, P.R. Selvin & I. Golding, manuscript in preparation). In this model, the cell surface is divided into receptor-rich and receptor-free regions. The presence or absence of receptors in turn affects phage dynamics by determining the diffusion coefficient as well as the probabilities of attachment and of falling off the surface. By numerical simulations, we found that this simple model reproduces much of the observed dynamics, thus suggesting that it captures at least some of the essential features of the target-finding process.

Finally, it is interesting to speculate whether what we observed represents a reduction-of-dimensionality scheme, albeit different from the one originally envisioned: The interaction of phages with the network of LamB receptors on the cell surface limits their motion to a fraction of the surface, in effect rendering the motion quasi 1D, or possibly of a fractal dimension between one and two. Thus, the reduction in dimensionality in the transition

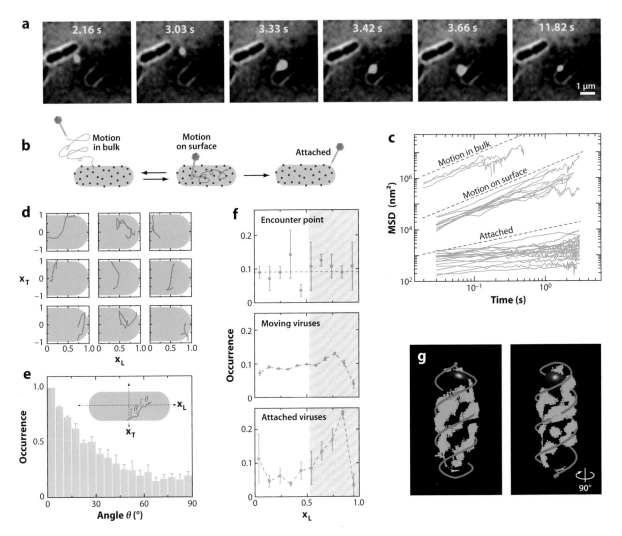

Figure 2

Target finding by the infecting phage. (*a*) Time-lapse images of a lambda phage (*green spot*) moving near and on the surface of an *Escherichia coli* cell. (*b*) Illustration of the different stages observed in the phage target-finding process. The phage first freely diffuses in bulk until it encounters a bacterial cell. The phage then moves on the cell surface, a motion that ends in either falling off the cell (back to bulk motion) or attaching to a target receptor and ceasing movement. (*c*) The calculated mean squared displacement (MSD) as a function of lag time for individual viral trajectories during free diffusion in bulk (*top cluster*), motion on the cell surface (*middle cluster*), and attachment (*bottom cluster*). The different motion types form distinct clusters with more than an order of magnitude separation in MSD values. (*d*) Typical phage trajectories on the cell surface plotted in normalized coordinates along (X_L) and across (X_T) the bacterial cell. (*e*) The distribution of instantaneous angles relative to the cell short axis, from >100 viral trajectories. The tendency to move along the short axis is evident. (*f*) The distribution of phage positions along the cell length during the target-finding process. The initial points of encounter are uniformly distributed (*top*). Phages moving on the cell surface show an affinity toward the cell pole (*middle*). The final attachment sites show a pronounced polar localization (*bottom*). Gray shading highlights the area to which phages converge during the search process. (*g*) Spatial organization of LamB receptors on the cell surface. One typical cell is shown. Quantum dots (*green*) were used to label the receptors, and a 3D reconstruction of the spatial structure was obtained by imaging multiple z-positions. The observed pattern is well described by two helices out of phase (*blue*). The cell outline is highlighted in red.

from bulk (3D) to cell surface is larger than previously assumed. Whether this scenario optimizes the search process—for example, by minimizing the time it takes the phage to arrive at the cell pole—is a promising avenue for further theoretical investigation.

The Postinfection Decision: Stochasticity Versus Hidden Variables

Following infection by the bacteriophage, a decision is made between cell death (lysis) and viral dormancy (lysogeny) (69). During the decision process, the regulatory circuit encoded by viral genes (primarily *cI*, *cII*, and *cro*) integrates multiple physiological and environmental signals, including the number of infecting viruses and the metabolic state of the cell, in order to reach a decision (52, 67, 87). This postinfection choice serves as a simple paradigm for decision making between alternative cell fates during development (25, 70, 83). It also serves as a test case for the role of stochasticity in cellular processes.

Our limited understanding of cellular decision making is aptly demonstrated by the contradictory views of researchers toward the role of stochasticity in the lambda system. On the one hand, the genetic circuitry of lambda serves as a paradigm for the intricacy and precision of gene regulation, as exemplified by the function of the lysis/lysogeny switch (25, 69) and the timing of cell lysis by the phage (42, 86). On the other hand, one of the first demonstrations of heterogeneity in a cell phenotype within a clonal population was Ellis & Delbrück's (30) study of the wide distribution of phage burst sizes following lysis (we have obtained similar results in the lambda system; data not shown). The cell-to-cell variability in event timing following induction was also recently quantified (4). Almost sixty years after Ellis & Delbrück's work, Arkin et al. (6) used a numerical study of the lambda lysis/lysogeny decision following infection to emphasize the role of chemical stochasticity in genetic circuits. Their work led to the emergence of the widely accepted picture of cell variability driven by spontaneous biochemical stochasticity, not only in lambda but in other systems as well (19, 57, 58, 79, 84). More recently, however, the pendulum has shown signs of swinging back to the deterministic direction. St-Pierre and Endy have shown that, at the single-cell level, cell size is correlated with cell fate following infection. They thus demonstrated how previously undetected hidden variables can explain away some of the observed cell fate heterogeneity and reduce (though not eliminate) the expected role of biochemical stochasticity in the decision (83).

We recently developed an assay for following the postinfection decision under the microscope in real time, at the level of individual phages and cells (**Figure 3a,b**) (91). Examination of thousands of infection events showed that the probability of lysogenization f increases with the number of infecting phages (multiplicity of infection, m) and decreases with the cell length l. Both of these observations were in agreement with earlier studies (53, 83). A simple theoretical model of the lysis/lysogeny decision circuit predicts that the probability of lysogenization should scale like (m/l), approximating the concentration of viral genomes in the cell, because this concentration determines the dosage of the fate-determining genes (87). However, we found that plotting $f(m/l)$ failed to collapse the data from different multiplicities of infection (91). In trying to understand the failure of this model, we asked: What if each individual phage makes an independent decision? We hypothesized that the decision by each infecting phage depends only on the viral concentration (m/l) as predicted by the simple model. However, as an added feature, we assumed that the individual decisions by all infecting phages have to be considered when deciding the fate of the cell. Based on the known genetic circuitry, we assumed that lysis is the default route (25, 67), and thus only if all phages independently vote for lysogeny, that fate will be chosen and the cell will survive. The new model predicted a different scaling for the probability of lysogeny: $f(m,l) = [f_1(m/l)]^m$, where f_1 is the probability of choosing lysogeny by an individual phage when m phages infect a

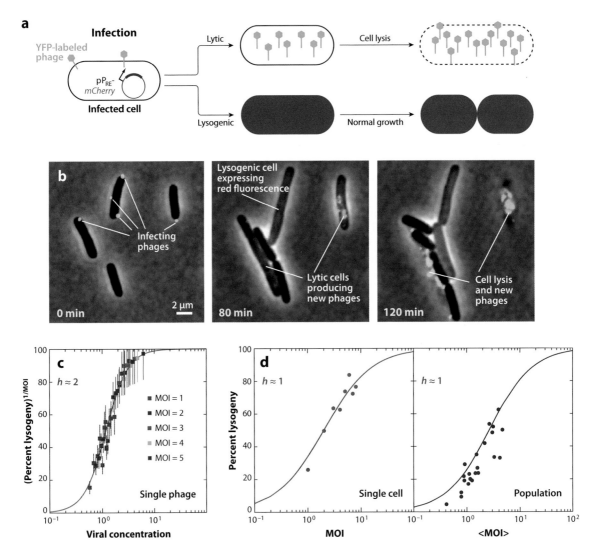

Figure 3

Cell-fate decision following infection. (*a*) A schematic description of our cell-fate assay. Multiple fluorescently labeled phages (*green*) simultaneously infect individual cells of *Escherichia coli*. The postinfection fate can be detected in each infected cell. Choice of the lytic pathway is indicated by the intracellular production of new fluorescent phages, followed by cell lysis. Choice of the lysogenic pathway is indicated by the production of red fluorescence from the P_{RE} promoter, followed by resumed growth and cell division. The three stages of the process correspond to the three images seen in panel *b*. (*b*) Frames from a time-lapse movie depicting infection events. At time $t = 0$ (*left*), two cells are each infected by a single phage (*green spots*), and one cell is infected by three phages. At $t = 80$ min (*middle*), the two cells infected by single phages have each gone into the lytic pathway, as indicated by the intracellular production of new phages (*green*). The cell infected by three phages has gone into the lysogenic pathway, as indicated by the production of red fluorescence from P_{RE} (red). At $t = 2$ h (*right*), the lytic pathway has resulted in cell lysis, whereas the lysogenic cell has divided. (*c*) Scaled probability of lysogeny $[f(m,l)]^{1/m}$ as a function of viral concentration (m/l). Data from different multiplicities of infection (MOIs) collapse into a single curve, representing the probability of lysogeny for each individual infecting phage (f_1) in a cell of length l infected by a total of m phages. f_1 can be fitted to a Hill function, $f_1(m/l) = (m/l)^b/(K^b + (m/l)^b)$, with $b \approx 2$. (*d*) The probability of lysogeny as a function of the relevant input parameter, at the single-cell (input is MOI of the individual cell) and population-average (input is the average MOI over all cells) levels. Circles represent experimental data. Solid lines represent theoretical prediction, fitted to a Hill function. The decision becomes more "noisy" (lower Hill coefficient, $b \approx 1$) when moving from the single-phage (panel *c*) to the single-cell level. Moving from the single cell to the population average does not decrease the Hill coefficient further. Reprinted from Reference 91, with permission from Elsevier.

cell of length *l*. The exponent *m* arises from the requirement of unanimous vote by all phages for producing lysogeny. As seen in **Figure 3c**, this scaling successfully collapsed the data from all experiments. Additional experiments using fluorescent reporters for lysogenic and lytic genes supported the hypothesis of individual decisions by infecting phages and the requirement of a unanimous vote for obtaining lysogeny (91).

Having quantified the "decision curve" of the individual phage, $f_1(m/l)$, we next reversed the process and reconstructed the observed decision-making phenotype at the whole-cell and whole-population levels (**Figure 3d**). This was done by integrating over the different degrees of freedom that remained hidden in the lower-resolution (coarse-grained) experiments. Thus, when going from individual phages to the whole cell, we integrated over the spatial positions of phage infections and their effect on infection efficiency, as well as the length distribution of cells in the population, obtaining the predicted single-cell response curve $f(m)$. We then integrated further over the random phage-bacterium collision probabilities (64) to obtain the predicted population-averaged response, $f(M)$. When comparing the decision curves at the different resolution levels (**Figure 3c,d**), we found that most of the apparent noise in the decision arises at the transition from the single-phage to the single-cell level, when integrating over individual phage decisions and the distribution of cell ages in the population. Moving further from individual cells to the population average did not add significantly to the observed imprecision of the decision. In other words, measurements at the single-cell level mask as much of the critical degrees of freedom as do measurements made in bulk—counter to the widely accepted view (6, 79).

The concept of decision making at the subcellular level may at first appear counterintuitive: Presumably, all the relevant regulatory proteins produced from the individual viral genomes achieve perfect mixing in the bacterial cytoplasm within seconds of their production due to diffusion (32). How viral individuality is maintained inside the cell is an open question. The answer may lie in the discreteness of viral genomes and of the gene expression events underlying the decision-making process (18, 39, 52, 90). Another possibility is that subcellular decision making is enabled by spatial separation of key players in the process (66). It is intriguing to contemplate the possibility of subcellular decision making by individual genomes at the other end of the complexity spectrum, in higher eukaryotic systems. In those systems, multiple copies of a gene circuit often exist, and copy number variations play a critical role in health and disease (24). Subcellular decision making may thus have a profound effect if present.

Maintenance of Lysogeny: A Simple Model for Cell State Stability

The ability of cells to maintain an inheritable memory of their gene expression state is key to cellular differentiation (65). A differentiated cellular state may be maintained for a long time while at the same time allowing efficient state switching (reprogramming) in response to the proper stimulus (43). However, even in the absence of external perturbation, the cell state may not be infinitely stable (irreversible) (54). Stochastic fluctuations may switch a cell from one state to another. A natural question then arises: How stable is a cell's gene expression state in the absence of an external perturbation? In other words, how long will a differentiated cell stay in the same state before spontaneously switching to an alternative one? What features of the underlying gene regulatory network determine this stability?

The lambda lysogen serves as one of the simplest examples for a stable cellular state (67, 69, 70). Lysogenic stability is maintained by the activity of the lambda repressor (CI), which acts as a transcription factor to repress all lytic functions from the prophage in the bacterial cell and to regulate its own production (**Figure 4a**) (69). As mentioned above, autoregulation by the fate-determining protein is also observed in higher systems displaying long-term cellular memory (26, 43, 54). The resulting lysogenic

Differentiation: the process by which a cell acquires a unique gene expression state and maintains it for a long time

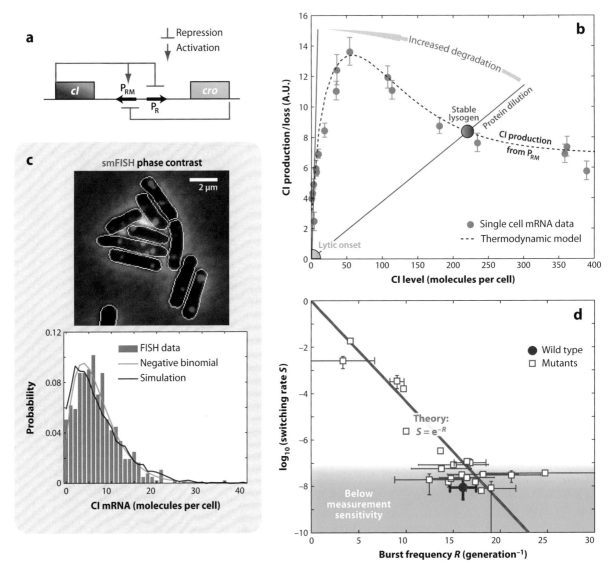

Figure 4

Maintenance of the lysogenic state. (*a*) The lysis/lysogeny switch is governed (to a first approximation) by a mutual repression loop between two genes: the lambda repressor, *cI*, and the antirepressor, *cro*. *cI* also regulates its own production. Adapted from Reference 17 with permission from Elsevier. (*b*) According to the standard theoretical picture, the stable lysogenic state is defined by the balance between CI production (from the P_{RM} promoter) and elimination. P_{RM} activity was measured in individual cells (*dark yellow circles*) and can also be reproduced using a thermodynamic model for promoter occupancy (*dashed dark brown line*) (92). In wild-type lysogens, CI elimination is dominated by cell growth and division (*solid blue line*). During forced induction, CI degradation is increased (*blue arrow*), shifting the steady state from [CI] ≫ 0 (lysogeny) to [CI] ≈ 0 (lytic onset). (*c*) Top: *cI* mRNA in lysogens, labeled using single-molecule fluorescence in situ hybridization (smFISH). Fluorescent foci (*red*) indicate the presence of *cI* mRNA molecules. The photon count from these foci was used to estimate the number of mRNA molecules in each cell. Bottom: The resulting *cI* mRNA copy number distribution. The experimental histogram (*gray bars*) was fitted to a negative binomial distribution (*blue curve*) whose parameters were used to calculate the transcriptional burst frequency and burst size. The results of a stochastic simulation of the gene circuit (*red curve*) are also shown for comparison. Adapted from Reference 92. (*d*) The relation between lysogen stability and P_{RM} activity. The measured switching rate (*S*) for the wild-type lysogen (*red circle*) and the mutants in *cI* and P_{RM} (*blue squares*) is plotted as a function of the number of activity bursts from P_{RM} in one cell generation (*R*). The points fall close to the theoretical prediction given by $S = \exp(-R)$ (*solid gray line*). Adapted from Reference 92.

state is extremely stable: Spontaneous switching events occur less than once per 10^8 cell generations in the absence of cellular RecA activity (56).

To understand what determines cell state stability, one first needs to define the steady state of the system—the gene expression level averaged over time and over cells in the population. The lambda system has been well characterized in terms of the regulatory circuitry that creates the stable lysogenic state. The regulation of the two key promoters, P_{RM} (producing CI) and P_R (producing the antirepressor, or Cro, which initiates the lytic cascade at low repressor levels), has been mapped as a function of CI and Cro concentrations (28, 69, 92) (**Figure 4b**). A thermodynamic model using a grand canonical ensemble has been used to describe the occupancy of the six operator sites controlling promoter activities (O_{R1-3}, O_{L1-3}) and the resulting protein levels (5, 27, 78).

In the resulting theoretical picture, the lysogenic state is seen as the stable attractor of a dynamical system, or the point of minimum potential. Taking this analogy further, the event of spontaneous induction, when prophage escapes lysogenic repression and initiates lytic development, is (somewhat loosely) mapped to the thermally driven escape of a particle from a potential well—a so-called Kramers problem (11, 16, 44). A significant degree of complexity is added by the properties of the effective temperature, given in this case by the stochasticity of gene expression (16). Thus, characterization of the steady state has to be accompanied by quantification of the stochastic dynamics of gene activity, which shift individual cells away from the average state, and may switch a cell from lysogeny to lysis. It is now known that both transcription (39) and translation (18, 90) exhibit intermittent, non-Poissonian kinetics. Such bursty gene activity has been previously suggested to affect cell state switching (23, 40, 50, 61).

To predict the stability of the lysogenic state, we recently formulated and then tested experimentally a simple model for the maintenance of lysogeny (92). We assumed that CI molecules are produced in discrete bursts and that the occurrence of these bursty activity events obeys Poissonian statistics (33, 39). We denote that average frequency of activity bursts as r. For a switch from lysogeny to lysis to occur, a cell needs to lose most of its repressor molecules. This happens if no P_{RM} activity occurs for one generation time τ, as this is the typical protein lifetime (due to cell growth and dilution). Under the assumption that transcription burst occurrence follows Poisson statistics, the probability of a cell not producing cI mRNA (and therefore repressor proteins) for duration τ is $P_0 = \exp(-r\tau)$. Thus, the probability of switching from lysogeny to lysis during one cell generation is approximately equal to the exponent of the number of activity bursts from P_{RM} during the same time (a number we denote $R = r\tau$).

Note that the model above is extremely simple. We have neglected some of the features considered essential in the maintenance of lysogeny. First, the fact that CI regulates its own production—see the $P_{RM}([CI])$ curve in **Figure 4b**, whose properties supposedly determine the stable lysogenic state at $[CI] > 0$. In our case, we approximated P_{RM} as a constitutive promoter whose activity does not vary as CI levels change. Second, the attractor at $[CI] \approx 0$ (the lytic onset) was described simply using an absorbing wall at $[CI] = 0$, instead of explicitly modeling the gradual derepression of the P_R promoter (producing Cro) as CI numbers decrease (75).

Our model is thus simple. The important question is, of course, is it too simple? To test the prediction of lysogen stability, we first had to measure the frequency of transcription bursts from P_{RM} (92). We used single-molecule fluorescence in situ hybridization (smFISH) (73) to quantify cI mRNA copy number statistics in lysogenic cells (**Figure 4c**). The measured histograms followed a negative binomial distribution, consistent with the notion of transcription bursts following Poissonian statistics (71, 77). The parameters of this distribution allowed us to estimate the burst frequency r. We then measured the spontaneous switching rate from the lysogenic state S (92). This was done by measuring the rate of appearance of free phages in a culture of exponentially growing lysogenic cells (56).

We thus had the ability to relate promoter activity and lysogen stability. Such a measurement, however, only gave us a single data point. To create a stability curve, we needed to tune system parameters—promoter activity and the resulting lysogen stability. This was done by using a series of phage mutants: the temperature-sensitive allele $cI857$, in which the efficiency of the repressor decreases as temperature is increased (45, 46, 48); as well as 18 additional mutants modified in either the P_{RM} or the cI sequence (62, 92). When examining the relation between lysogen stability S and the P_{RM} burst frequency R in the set of lysogens, we found that the data were consistent with the theoretical prediction, $S \approx \exp(-R)$ (**Figure 4d**).

At first glance, it may seem surprising that such an oversimplified model captured the behavior of a real-life, naturally evolved system in which stability is believed to be an important phenotype (56). However, in line with this observation that many system parameters are coarse-grained to produce the stability phenotype, there is a body of work from the past decade, mainly from the Little lab (7–9, 56, 62, 63), pointing to the robust performance of the lambda lysogeny switch even when system parameters are genetically perturbed. Although only semiquantitative in nature, these studies are consistent with the idea that an intricate genetic circuitry with fine-tuned parameters is not needed for obtaining a stable lysogenic state.

It is tempting to contemplate the possible relevance of our results concerning the stability of the cellular state to higher systems, where the ability to maintain an inheritable memory of the gene expression state is key to cellular differentiation (43, 65). Admittedly, the maintenance of bacterial lysogeny does not exhibit the complexity of cell differentiation in higher eukaryotic systems, where a range of additional mechanisms play a role in cellular memory (17). Nevertheless, the fundamental feature of autoregulation by the fate-determining protein

appears central there too (26, 43, 54). Thus, investigating the stability of differentiated states in higher, and in particular multicellular, systems is a promising direction for experimental investigation.

Forced Induction: Switching Dynamics

In contrast to the case of spontaneous induction discussed above, switching of the cellular state from lysogeny to lysis can also be achieved in a directed manner. Experimentally, forced induction is commonly achieved by one of two methods: (*a*) using DNA-damaging agents, for example, irradiating the cells with UV light. When cellular DNA is damaged, the protease activity of RecA is turned on as part of the cell's SOS response (55). This leads to degradation of CI proteins. (*b*) The second method uses a temperature-sensitive allele of the lambda repressor ($cI857$) that becomes inactivated at high temperature (45, 46, 48). In both cases, the dynamical systems picture is that the increased elimination rate of CI leads to a loss of the [CI] > 0 (lysogenic) steady state (see **Figure 4b**).

Traditional biochemical assays demonstrated the expected decrease in cellular CI levels preceding cell lysis (12). However, population-averaged characterization of induction obscures the histories of individual cells— for example, the presence of temporal oscillations in the SOS response (34). The quantitative features of induction kinetics remain largely unexplored. One example of such a feature is depicted in **Figure 5a** (22, 51). A culture of lysogens was irradiated with UV, and the fraction of cells induced (switched to lysis) was measured as a function of UV amount. It was found that the UV dose response can be closely approximated by a Hill function $y(x) = x^h/(k^h + x^h)$ with a Hill coefficient of $h = 4$. Moreover, the same coefficient holds for phage strains with mutated cI genes (which otherwise exhibit a different phenotype in terms of their spontaneous induction levels and the amount of UV required to induce the cells) as well as for different numbers of phage genomes in the lysogen. Thus,

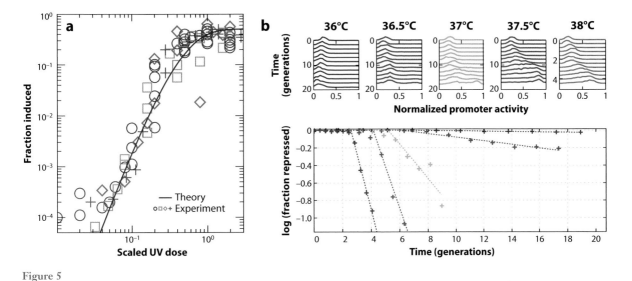

Figure 5

Cell state switching. (*a*) The fraction of lysogens in a population that were induced, as a function of the amount of UV radiation. Experimental data (markers) is from wild-type and different lambda mutants. The solid line is a fit to the theoretical expression using reliability theory (22), yielding a power law with an exponent of 4. Reprinted with permission from Reference 22, Copyright (2009) by the American Physical Society. (*b*) Single-cell measurements. The activity of P_{RM} and P_R was measured in individual cells using a two-color fluorescence reporter. Temperature was used to tune the state of the lysis/lysogeny circuit, and the population statistics of promoter activities was measured at different times after the temperature shift (*top*). When examining the fraction of switched cells as a function of time (*bottom*), one observes a delay period, in which the population remains fully lysogenic, followed by a temperature-dependent rate of switching to the Cro-dominated state.

the power of 4 appears to be a universal feature of the switch.

A possible way to think about this result is to consider the lysogeny maintenance circuit as consisting of four redundant elements—the CI dimers bound at four operator sites (O_{R1}, O_{R2}, O_{L1}, O_{L2})—and to assume that only the failure of all elements will lead to lysis. Using the formalism of reliability theory, one can write down the expected fraction of failure events as a function of the UV dose and arrive at the observed power law (**Figure 5***a*). For more details see Reference 22.

To go beyond this simple phenomenology and elucidate the kinetics of cell state switching, the activity of P_{RM} has to be followed in individual cells during the induction process. If the standard picture is correct, then as the level of CI monotonically diminishes [a process that takes ∼30–45 min in the case of UV induction (12)], P_{RM} is expected to react by scanning through a large range of its response curve in

an attempt to counter the decreasing CI levels (**Figure 4***b*). In most cases this attempt will fail. Occasionally, however, CI production succeeds in overcoming elimination, leading to the reestablishment of lysogeny (abortive induction; 74). In any case, forced induction thus offers a possible window into the shape of the P_{RM} autoregulatory curve, which lies at the heart of the lysogeny maintenance system.

To examine the process of forced induction at single-cell resolution, we recently characterized the induction kinetics using a two-color fluorescent reporter, which allows us to examine simultaneously the activity of both P_{RM} (maintaining lysogeny) and P_R (initiating lysis) in individual cells (M.W. Bednarz, J.A. Halliday, C. Herman, I. Golding, manuscript in preparation). Promoter activities were followed over time, under a range of induction strengths (again using the temperature-sensitive allele *cI857*) (**Figure 5***b*). Our results indicate that the fraction of induced cells over time exhibits a biphasic behavior: First, there is a delay (of up to a few generations), in which all cells remain in the lysogenic state. This is followed by a constant flux of cells into the derepressed (Cro-dominated) state. During the delay period there are already changes in gene activity; however, these changes are still insufficient to tilt the individual cells from lysogeny to lysis. It will be interesting to examine whether this behavior can be explained using a simple stochastic model of the underlying gene circuitry.

SUMMARY

Despite heroic efforts to illuminate cellular decision-making in higher systems (see e.g., 13, 41, 72, 82), important insights can be gained by attacking the same processes in their simplest instantiation. In this review I used the system comprising *E. coli* and bacteriophage lambda to make that case. We are obviously still far from achieving the desired quantitative narrative for the viral life cycle. However, as described above, the quantitative experimental study of the system has already yielded new insights regarding the way randomly moving viruses find their target on the cell surface, the way cell fate decisions are made at a level finer than that of the whole cell, the way bursty gene activity affects the stability of the cellular state, and more.

At the same time, before applying the principles learned here to higher organisms, we must beware: As physicists studying living systems, we are always in jeopardy of overassuming universality. Against that ever-present temptation, one has to keep in mind the distinction between Occam's razor and Occam's rug[1]: The former, of course, is the guiding rule for physicists, who will always choose the simplest, most universal explanation for an observed phenomenon. The principle of Occam's rug, on the other hand, states the following: When studying a living system, a simple elegant narrative often implies that too much of the data was swept under the rug. In other words, always be wary of claims of simplicity and universality in biology.

_____

[1] I am indebted to Ted Cox for this insight.

SUMMARY POINTS

1. The life cycle of bacteriophage lambda serves as a paradigm for decision making in living systems.

2. Quantitative, high-resolution experimental investigation of the lambda system has already yielded new insights.

3. Receptors on the cell surface guide the infecting phage through its target-finding process.

4. Following infection, the decision between alternative fates is first made at the level of individual phages.

5. Stability of the lysogenic (dormant) cellular state is determined by the frequency of activity bursts from the fate-determining gene.

DISCLOSURE STATEMENT

The author is not aware of any affiliations, memberships, funding, or financial holdings that might be perceived as affecting the objectivity of this review.

ACKNOWLEDGMENTS

I thank all my lab members for their help in preparing this manuscript, especially S. Skinner for creating the figures and proofreading the text. I am grateful to many members of the lambda community for their assistance and encouragement to a newcomer in the field. Work in my lab is supported by NIH grant R01GM082837, HFSP grant RGY 70/2008, and NSF Grant 082265 (PFC: Center for the Physics of Living Cells).

I dedicate this paper with gratitude to Ted Cox, who worked hard to make a molecular biologist out of me.

LITERATURE CITED

1. Ackers GK, Johnson AD, Shea MA. 1982. Quantitative model for gene regulation by lambda phage repressor. *Proc. Natl. Acad. Sci. USA* 79:1129–33

2. Adam G, Delbrück M. 1968. Reduction of dimensionality in biological diffusion processes. In *Structural Chemistry and Molecular Biology*, ed. A Rich, N Davidson, pp. 198–215. San Francisco: W.H. Freeman & Company

3. Alon U. 2007. *An Introduction to Systems Biology: Design Principles of Biological Circuits*. Boca Raton, FL: Chapman & Hall/CRC. 301 pp.

4. Amir A, Kobiler O, Rokney A, Oppenheim AB, Stavans J. 2007. Noise in timing and precision of gene activities in a genetic cascade. *Mol. Syst. Biol.* 3:71

5. Anderson LM, Yang H. 2008. DNA looping can enhance lysogenic CI transcription in phage lambda. *Proc. Natl. Acad. Sci. USA* 105:5827–32

6. Arkin A, Ross J, McAdams HH. 1998. Stochastic kinetic analysis of developmental pathway bifurcation in phage lambda-infected *Escherichia coli* cells. *Genetics* 149:1633–48

7. Atsumi S, Little JW. 2004. Regulatory circuit design and evolution using phage lambda. *Genes Dev.* 18:2086–94

8. Atsumi S, Little JW. 2006. A synthetic phage lambda regulatory circuit. *Proc. Natl. Acad. Sci. USA* 103:19045–50

9. Atsumi S, Little JW. 2006. Role of the lytic repressor in prophage induction of phage lambda as analyzed by a module-replacement approach. *Proc. Natl. Acad. Sci. USA* 103:4558–63

10. Aurell E, Brown S, Johanson J, Sneppen K. 2002. Stability puzzles in phage lambda. *Phys. Rev. E Stat. Nonlin. Soft. Matter. Phys.* 65:051914

11. Aurell E, Sneppen K. 2002. Epigenetics as a first exit problem. *Phys. Rev. Lett.* 88:048101

12. Bailone A, Levine A, Devoret R. 1979. Inactivation of prophage lambda repressor in vivo. *J. Mol. Biol.* 131:553–72

13. Ben-Zvi D, Shilo BZ, Fainsod A, Barkai N. 2008. Scaling of the BMP activation gradient in *Xenopus* embryos. *Nature* 453:1205–11

14. Berg HC. 1993. *Random Walks in Biology*. Princeton, NJ: Princeton Univ. Press. 152 pp.

15. Berg HC. 2004. E. coli *in Motion*. New York: Springer. 133 pp.

16. Bialek W. 2001. Stability and noise in biochemical switches. In *Advances in Neural Information Processing Systems 13*, ed. TK Leen, TG Dietterich, V Tresp, pp. 103–9. Cambridge, MA: MIT Press

17. Burrill DR, Silver PA. 2010. Making cellular memories. *Cell* 140:13–18

18. Cai L, Friedman N, Xie XS. 2006. Stochastic protein expression in individual cells at the single molecule level. *Nature* 440:358–62

19. Chang HH, Hemberg M, Barahona M, Ingber DE, Huang S. 2008. Transcriptome-wide noise controls lineage choice in mammalian progenitor cells. *Nature* 453:544–47

20. Chapman-McQuiston E, Wu XL. 2008. Stochastic receptor expression allows sensitive bacteria to evade phage attack. Part I: experiments. *Biophys. J.* 94:4525–36

21. Chen Y, Golding I, Sawai S, Guo L, Cox EC. 2005. Population fitness and the regulation of *Escherichia coli* genes by bacterial viruses. *PLoS Biol.* 3:e229

22. Chia N, Golding I, Goldenfeld N. 2009. Lambda-prophage induction modeled as a cooperative failure mode of lytic repression. *Phys. Rev. E Stat. Nonlin. Soft. Matter. Phys.* 80:030901

23. Choi PJ, Cai L, Frieda K, Xie XS. 2008. A stochastic single-molecule event triggers phenotype switching of a bacterial cell. *Science* 322:442–46

24. Cohen J. 2007. Genomics. DNA duplications and deletions help determine health. *Science* 317:1315–17

25. Court DL, Oppenheim AB, Adhya S. 2006. A new look at bacteriophage lambda genetic networks. *J. Bacteriol.* 189:298–304

26. Crews ST, Pearson JC. 2009. Transcriptional autoregulation in development. *Curr. Biol.* 19:R241–46

27. Darling PJ, Holt JM, Ackers GK. 2000. Coupled energetics of lambda cro repressor self-assembly and site-specific DNA operator binding II: cooperative interactions of cro dimers. *J. Mol. Biol.* 302:625–38

28. Dodd IB, Perkins AJ, Tsemitsidis D, Egan JB. 2001. Octamerization of lambda CI repressor is needed for effective repression of P(RM) and efficient switching from lysogeny. *Genes Dev.* 15:3013–22

29. Edgar R, Rokney A, Feeney M, Semsey S, Kessel M, et al. 2008. Bacteriophage infection is targeted to cellular poles. *Mol. Microbiol.* 68:1107–16

30. Ellis EL, Delbrück M. 1939. The growth of bacteriophage. *J. Gen. Physiol.* 22:365–84

31. Ellis RJ. 2001. Macromolecular crowding: obvious but underappreciated. *Trends Biochem. Sci.* 26:597–604

32. Elowitz MB, Surette MG, Wolf PE, Stock JB, Leibler S. 1999. Protein mobility in the cytoplasm of *Escherichia coli*. *J. Bacteriol.* 181:197–203

33. Friedman N, Cai L, Xie XS. 2006. Linking stochastic dynamics to population distribution: an analytical framework of gene expression. *Phys. Rev. Lett.* 97:168302

34. Friedman N, Vardi S, Ronen M, Alon U, Stavans J. 2005. Precise temporal modulation in the response of the SOS DNA repair network in individual bacteria. *PLoS Biol.* 3:e238

35. Gibbs KA, Isaac DD, Xu J, Hendrix RW, Silhavy TJ, Theriot JA. 2004. Complex spatial distribution and dynamics of an abundant *Escherichia coli* outer membrane protein, LamB. *Mol. Microbiol.* 53:1771–83

36. Goldenfeld N. 1992. *Lectures on Phase Transitions and the Renormalization Group*. Reading, MA: Addison-Wesley

37. Golding I, Cox EC. 2004. RNA dynamics in live *Escherichia coli* cells. *Proc. Natl. Acad. Sci. USA* 101:11310–15

38. Golding I, Cox EC. 2006. Physical nature of bacterial cytoplasm. *Phys. Rev. Lett.* 96:098102

39. Golding I, Paulsson J, Zawilski SM, Cox EC. 2005. Real-time kinetics of gene activity in individual bacteria. *Cell* 123:1025–36

40. Gordon AJ, Halliday JA, Blankschien MD, Burns PA, Yatagai F, Herman C. 2009. Transcriptional infidelity promotes heritable phenotypic change in a bistable gene network. *PLoS Biol.* 7:e44

41. Gregor T, Tank DW, Wieschaus EF, Bialek W. 2007. Probing the limits to positional information. *Cell* 130:153–64

42. Grundling A, Manson MD, Young R. 2001. Holins kill without warning. *Proc. Natl. Acad. Sci. USA* 98:9348–52

43. Gurdon JB, Melton DA. 2008. Nuclear reprogramming in cells. *Science* 322:1811–15

44. Hänggi P, Talkner P, Borkovec M. 1990. Reaction-rate theory—50 years after Kramers. *Rev. Model. Phys.* 62:251–341

45. Hecht MH, Nelson HC, Sauer RT. 1983. Mutations in lambda repressor's amino-terminal domain: implications for protein stability and DNA binding. *Proc. Natl. Acad. Sci. USA* 80:2676–80

46. Hendrix RW, Roberts JW, Stahl FW, Weisberg RA. 1983. *Lambda II (Monograph 13)*. Cold Spring Harbor, NY: Cold Spring Harbor Lab. 694 pp.

47. Hershey AD. 1971. *The Bacteriophage Lambda*. Cold Spring Harbor, NY: Cold Spring Harbor Lab. 792 pp.

48. Isaacs FJ, Hasty J, Cantor CR, Collins JJ. 2003. Prediction and measurement of an autoregulatory genetic module. *Proc. Natl. Acad. Sci. USA* 100:7714–19

49. Kaern M, Elston TC, Blake WJ, Collins JJ. 2005. Stochasticity in gene expression: from theories to phenotypes. *Nat. Rev. Genet.* 6:451–64

50. Kaufmann BB, Yang Q, Mettetal JT, van Oudenaarden A. 2007. Heritable stochastic switching revealed by single-cell genealogy. *PLoS Biol.* 5:e239

51. Kneser H. 1966. Repair of ultraviolet lesions and induction of lambda prophage. *Virology* 28:701–6

52. Provides a quantitative study of gene expression kinetics during the postinfection decision.

52. Kobiler O, Rokney A, Friedman N, Court DL, Stavans J, Oppenheim AB. 2005. Quantitative kinetic analysis of the bacteriophage lambda genetic network. *Proc. Natl. Acad. Sci. USA* 102:4470–75

53. Kourilsky P, Knapp A. 1974. Lysogenization by bacteriophage lambda. III. Multiplicity dependent phenomena occurring upon infection by lambda. *Biochimie* 56:1517–23

54. Lawrence PA. 1992. *The Making of a Fly: The Genetics of Animal Design*. Oxford/Boston: Blackwell Sci. Publ. 228 pp.

55. Little JW, Mount DW. 1982. The SOS regulatory system of *Escherichia coli*. *Cell* 29:11–22

56. Little JW, Shepley DP, Wert DW. 1999. Robustness of a gene regulatory circuit. *EMBO J.* 18:4299–307

57. Losick R, Desplan C. 2008. Stochasticity and cell fate. *Science* 320:65–68

58. Maamar H, Raj A, Dubnau D. 2007. Noise in gene expression determines cell fate in *Bacillus subtilis*. *Science* 317:526–29

59. Marsh M, Helenius A. 2006. Virus entry: open sesame. *Cell* 124:729–40

60. McAdams HH, Shapiro L. 1995. Circuit simulation of genetic networks. *Science* 269:650–56

61. Mehta P, Mukhopadhyay R, Wingreen NS. 2008. Exponential sensitivity of noise-driven switching in genetic networks. *Phys. Biol.* 5:26005

62. Michalowski CB, Little JW. 2005. Positive autoregulation of cI is a dispensable feature of the phage lambda gene regulatory circuitry. *J. Bacteriol.* 187:6430–42

63. Michalowski CB, Short MD, Little JW. 2004. Sequence tolerance of the phage lambda PRM promoter: implications for evolution of gene regulatory circuitry. *J. Bacteriol.* 186:7988–99

64. Moldovan R, Chapman-McQuiston E, Wu XL. 2007. On kinetics of phage adsorption. *Biophys. J.* 93:303–15

65. Monod J, Jacob F. 1961. General conclusions: teleonomic mechanisms in cellular metabolism, growth, and differentiation. *Cold Spring. Harb. Symp. Quant. Biol.* 26:389–401

66. Montero Llopis P, Jackson AF, Sliusarenko O, Surovtsev I, Heinritz J, et al. 2010. Spatial organization of the flow of genetic information in bacteria. *Nature* 466:77–81

67. Oppenheim AB, Kobiler O, Stavans J, Court DL, Adhya S. 2005. Switches in bacteriophage lambda development. *Annu. Rev. Genet.* 39:409–29

68. Phillips R, Kondev J, Theriot J. 2009. *Physical Biology of the Cell*. New York: Garland Sci. 807 pp.

69. Ptashne M. 2004. *A Genetic Switch: Phage Lambda Revisited*. Cold Spring Harbor, NY: Cold Spring Harbor Lab. Press. 154 pp.

70. Ptashne M. 2007. On the use of the word 'epigenetic'. *Curr. Biol.* 17:R233–36

71. Raj A, Peskin CS, Tranchina D, Vargas DY, Tyagi S. 2006. Stochastic mRNA synthesis in mammalian cells. *PLoS Biol.* 4:e309

72. Raj A, Rifkin SA, Andersen E, van Oudenaarden A. 2010. Variability in gene expression underlies incomplete penetrance. *Nature* 463:913–18

73. Raj A, van den Bogaard P, Rifkin SA, van Oudenaarden A, Tyagi S. 2008. Imaging individual mRNA molecules using multiple singly labeled probes. *Nat. Methods* 5:877–79

74. Rokney A, Kobiler O, Amir A, Court DL, Stavans J, et al. 2008. Host responses influence on the induction of lambda prophage. *Mol. Microbiol.* 68:29–36

75. Schubert RA, Dodd IB, Egan JB, Shearwin KE. 2007. Cro's role in the CI Cro bistable switch is critical for lambda's transition from lysogeny to lytic development. *Genes Dev.* 21:2461–72

76. Schwartz M. 1975. Reversible interaction between coliphage lambda and its receptor protein. *J. Mol. Biol.* 99:185–201

77. Shahrezaei V, Swain PS. 2008. Analytical distributions for stochastic gene expression. *Proc. Natl. Acad. Sci. USA* 105:17256–61

78. Shea MA, Ackers GK. 1985. The OR control system of bacteriophage lambda. A physical-chemical model for gene regulation. *J. Mol. Biol.* 181:211–30

79. Singh A, Weinberger LS. 2009. Stochastic gene expression as a molecular switch for viral latency. *Curr. Opin. Microbiol.* 12:460–66

80. Smith AE, Helenius A. 2004. How viruses enter animal cells. *Science* 304:237–42

81. Spencer SL, Gaudet S, Albeck JG, Burke JM, Sorger PK. 2009. Non-genetic origins of cell-to-cell variability in TRAIL-induced apoptosis. *Nature* 459:428–32

82. Sprinzak D, Lakhanpal A, Lebon L, Santat LA, Fontes ME, et al. 2010. Cis-interactions between Notch and Delta generate mutually exclusive signalling states. *Nature* 465:86–90

83. St-Pierre F, Endy D. 2008. Determination of cell fate selection during phage lambda infection. *Proc. Natl. Acad. Sci. USA* 105:20705–10

56. Demonstrates the robustness of the lambda lysogeny system to genetic perturbations.

67. Reviews the genetic circuitry underlying decision making in the lambda system.

69. Provides a brief introduction to bacteriophage lambda. Makes for a great weekend read.

83. Demonstrates that cell length has a deterministic effect on cell fate following lambda infection.

84. Suel GM, Kulkarni RP, Dworkin J, Garcia-Ojalvo J, Elowitz MB. 2007. Tunability and noise dependence in differentiation dynamics. *Science* 315:1716–19

85. Thanbichler M, Shapiro L. 2008. Getting organized—how bacterial cells move proteins and DNA. *Nat. Rev. Microbiol.* 6:28–40

86. Wang IN, Smith DL, Young R. 2000. Holins: the protein clocks of bacteriophage infections. *Annu. Rev. Microbiol.* 54:799–825

87. Weitz JS, Mileyko Y, Joh RI, Voit EO. 2008. Collective decision making in bacterial viruses. *Biophys. J.* 95:2673–80

88. Yamanaka S. 2009. Elite and stochastic models for induced pluripotent stem cell generation. *Nature* 460:49–52

89. Yeomans JM. 1992. *Statistical Mechanics of Phase Transitions.* Oxford/New York: Clarendon/Oxford Univ. Press. 154 pp.

90. Yu J, Xiao J, Ren X, Lao K, Xie XS. 2006. Probing gene expression in live cells, one protein molecule at a time. *Science* 311:1600–3

91. Zeng L, Skinner SO, Zong C, Sippy J, Feiss M, Golding I. 2010. Decision making at a subcellular level determines the outcome of bacteriophage infection. *Cell* 141:682–91

92. Zong C, So L, Sepúlveda LA, Skinner SO, Golding I. 2010. Lysogen stability is determined by the frequency of activity bursts from the fate-determining gene. *Mol. Syst. Biol.* 6:440

87. Demonstrates how multiplicity of infection can serve as a control parameter for the postinfection cell fate.

91. Demonstrates that the decision between lysis and lysogeny is first made at the level of individual phages infecting the cell.

92. Demonstrates that lysogen stability depends in a simple manner on the frequency of activity bursts from the fate-determining gene, *cI*.

High-Pressure Protein Crystallography and NMR to Explore Protein Conformations

Marcus D. Collins,[1] Chae Un Kim,[2] and Sol M. Gruner[2,3]

[1]Department of Physiology and Biophysics, University of Washington, Seattle, Washington 98195-7290

[2]Cornell High Energy Synchrotron Source, [3]Department of Physics, Cornell University, Ithaca, New York 14853; email: smg26@cornell.edu

Annu. Rev. Biophys. 2011. 40:81–98

First published online as a Review in Advance on January 28, 2011

The *Annual Review of Biophysics* is online at biophys.annualreviews.org

This article's doi: 10.1146/annurev-biophys-042910-155304

Copyright © 2011 by Annual Reviews. All rights reserved

1936-122X/11/0609-0081$20.00

Keywords

conformational substates, energy landscape, protein thermodynamics, pressure cryocooling

Abstract

High-pressure methods for solving protein structures by X-ray crystallography and NMR are maturing. These techniques are beginning to impact our understanding of thermodynamic and structural features that define not only the protein's native conformation, but also the higher free energy conformations. The ability of high-pressure methods to visualize these mostly unexplored conformations provides new insight into protein function and dynamics. In this review, we begin with a historical discussion of high-pressure structural studies, with an eye toward early results that paved the way to mapping the multiple conformations of proteins. This is followed by an examination of several recent studies that emphasize different strengths and uses of high-pressure structural studies, ranging from basic thermodynamics to the suggestion of high-pressure structural methods as a tool for protein engineering.

Contents

INTRODUCTION

In 1914, Percy Bridgman (5) reported that egg white coagulates under pressure, thereby launching the field of high-pressure biology. High-pressure biology is not esoteric: Most of Earth's biosphere exists under pressures greater than hundreds of atmospheres (43, 50). Further, pressure effects on biomolecular function are both numerous and often of significant magnitude (7, 8, 14, 17, 23, 40, 41, 49, 50, 54).

Many experimental methods have been developed since Bridgman's time to explore high-pressure biophysics. We review some of the more recent of these, namely, the use of pressure as a tool for detailed studies of proteins via X-ray scattering, crystallography, and NMR. A recent review (16) focused largely on methodological improvements in crystallography; hence, the present review focuses on the structural biology findings in sum. The goal is to convince the reader that this structural data provides new ways to use pressure to explore the conformational space available to proteins, to test hypotheses of protein function, and to engineer proteins.

Consider the variables conjugate in free energy that the experimenter can directly control. For temperature, the conjugate variable is entropy, a vitally important but less intuitive quantity that in a protein energy landscape picture (6, 18, 58) can be thought of as the accessible area near an equilibrium structure—that is, the fluctuations about the equilibrium. Similarly, the presence of salts or denaturing chemicals perturbs the structure of a protein in complex, indirect ways that may be difficult to visualize. Pressure has the advantage that its conjugate is volume, something readily and directly related to structure and to the conformational changes needed for protein function.

Pressure has remarkable effects on proteins, including a tendency to unfold them. At lower pressures at which the protein still maintains a functional form, many proteins undergo dramatic shifts in activity (7, 8, 14, 17, 40, 41, 49, 50, 54). These effects can be related directly to measurable structural deformations of the protein. The perspective adopted in this review is that these deformations guide a mechanistic understanding of the molecule.

We first introduce the reader to thermodynamic versus microstructural viewpoints of protein energy landscapes in the context of high-pressure experiments. Next, we review the historical development of high-pressure structural studies of macromolecules and provide recent examples of insights gained from crystallography and NMR. Finally, we show that pressure allows the exploration of

Energy landscape:
a map of the internal (or sometimes free) energy of a protein as a function of its atomic coordinates

Microstructural:
refers to the detailed atomic coordinates of a protein structure

enzyme functional pathways and a means of modulating those pathways.

THERMODYNAMIC AND MICROSTRUCTURAL PERSPECTIVES OF PROTEIN ENERGY LANDSCAPES

Proteins are now frequently described using the concept of a multidimensional free energy landscape (6, 18). This landscape is a function of all conformational variables of a protein system such as the protein molecule, solvent, ions, and ligands.

The landscape may be viewed from either thermodynamic or statistical mechanical perspectives. In the thermodynamic perspective we think about states of the protein system that have macroscopic properties such as temperature, volume, and compressibility. In general, these states are not tied to any one exact conformation of the protein, but rather include very many such conformations (6). Until recently, most protein pressure studies measured only macroscopic thermodynamic variables, for instance, a determination of protein compressibility by measuring the fractional volume change of a protein system upon application of a unit pressure. Ultimately, one wishes to understand how the system free energy changes as a function of such variables. An extension of this approach is to consider the shape of the multidimensional free energy surface (i.e., the energy landscape) versus a set of conformational coordinates that can represent any of many degrees of freedom of the system. **Figure 1** shows a simplified energy landscape using citrine as an example.

Thermodynamics, by definition, provides averaged information about the behavior of large numbers of molecules. While this information is useful, it is difficult to gain insight from it about detailed changes around specific residues when proteins are pressurized. Today the experimenter has exquisite control of microscopic details of proteins. Crystallography allows visualization of the internal organization of a protein and genetic engineering allows

synthesis of precise mutants. More direct questions are needed about how the energy landscape varies with detailed perturbations. For example, consider an enzyme whose activity is pressure dependent. What is the effect of pressure on the structure of the active site? If the protein is mutated by substituting one amino acid for another at an active site, does the pressure dependency of enzymatic activity increase or decrease? If so, can we devise models of the energy landscape that allow us to explain or, more ideally, to predict the sign of the change? Can we make the same kinds of predictions about distant, allosterically coupled amino acids? This type of information would directly assist engineering protein activity.

High-pressure studies may also help us understand the basis of both allostery and the conformational substates (CS) that represent well-defined minima in the free energy landscape. For example, hemoglobin exists in various distinct CS, each readily identified spectroscopically and with known, distinct structures (27). Spectroscopic pressure studies have shown that the equilibrium between relaxed (R) and tensed (T) hemoglobin in solution shifts toward R at high pressure (52). In a crystallographic study of primarily T hemoglobin, would applied high pressure shift the hemoglobin atomic coordinates to a more R-like structure? That is, is there a progressive decrease in the root-mean-square (RMS) difference in atomic coordinates of the high-pressure T-state hemoglobin and the known R-state structure? Although this experiment has not yet been performed on hemoglobin, it posits well-defined, experimentally accessible, quantitative questions about the relative atomic coordinates in the molecule that directly elucidate the interconversion between CS. A similar experiment has been performed on myoglobin (see below) (53).

HISTORICAL DEVELOPMENT OF HIGH-PRESSURE STRUCTURAL STUDIES OF PROTEINS

This section reviews the use of high-pressure crystallography and NMR to explore protein

Conformational coordinate: an averaged coordinate, such as the center of mass of a protein subdomain, or the angle between two helices

Allostery: the tendency of two or more conformational changes in a protein to favor each other

CS: conformational substate(s)

RMS: root-mean-square (usually in reference to atomic displacements)

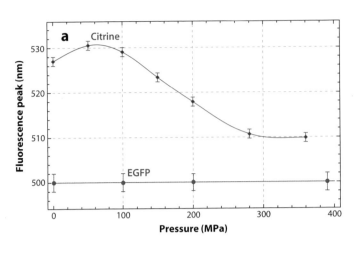

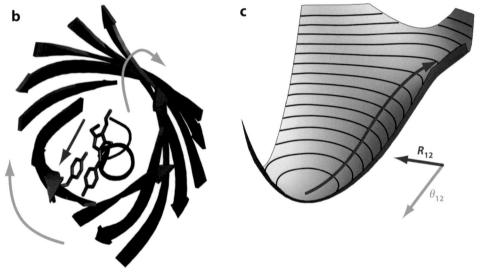

Figure 1

High-pressure deformations reveal the free energy landscape of citrine. (*a*) Peak fluorescence wavelength of citrine (*red diamonds*) and enhanced green fluorescent protein (EGFP) (*blue circles*) as a function of pressure. Curves are meant only to guide the eye. Reproduced with permission from Reference 3. (*b*) The citrine structure. Cluster 1 (see text) is shown in red, Cluster 2 in blue. Light blue arrows indicate rotations of Cluster 1 relative to Cluster 2 as pressure increases, while the purple arrow indicates the pressure-induced displacement of Cluster 1's center of mass relative to that of Cluster 2. Adapted from Reference 4. (*c*) A hypothetical (internal) energy landscape with two conformational coordinates: R_{12}, the distance between the clusters' centers of mass, and θ_{12}, the angle between principal axes of the two clusters. As pressure increases, work is done and the protein has higher internal energy, as indicated by the red arrow, which travels up a shallow valley in the energy landscape.

landscapes. The primary focus is on cases for which detailed atomic structures are known and can be connected to the effects of pressure at the macroscopic scale. Generally, one looks for locations in a given protein that display deformation under pressure. Proteins are globally linked objects. Thus, if one region of a protein compresses more than other regions, relative spatial readjustments to accommodate that heterogeneous compressibility may propagate pressure effects on other parts of the protein.

First Steps

In 1973, Thomanek et al. (51) reported X-ray crystallographic data collected from a crystal of sperm whale myoglobin pressurized to 250 MPa in liquid isopentane and frozen in liquid nitrogen. Isopentane was chosen as a pressurization medium because it is hydrophobic, so it hardly penetrated the crystal but could be readily flaked off the crystal when frozen. The extracted crystal, always kept near liquid nitrogen temperatures, was mounted on an X-ray diffraction stage. The first critical observation was that the crystal survived pressurization and diffracted well. A number of Bragg reflections had intensities that differed from crystals at ambient conditions. The authors speculated that pressure-induced structural changes persisted at low temperature, even though the pressure had been released; i.e., structural effects due to pressurization were somehow locked-in as long as the crystal was kept near liquid nitrogen temperatures. The crystal did not compress uniformly; rather the compressibility was small along one crystallographic axis (b^*) and larger along the other axis (a^*). No structure was solved, as the focus of the paper was on freezing crystals to prevent radiation damage and to improve Mössbauer spectroscopy data collection.

In the mid-1980s, Kundrot & Richards (37, 38) developed a beryllium specimen chamber specifically designed for X-ray crystallography and used it to solve the structure of hen egg white lysozyme at up to 100 MPa. The chamber is a rod into which a crystal could be inserted and then pressurized in mother liquor. The beryllium rod was strong enough to operate at 100 MPa, yet sufficiently small and X-ray transparent that it was readily mounted on a standard X-ray goniometer. The motivation was, in part, to identify the most compressible regions in a protein that might be important to function or solvent accessibility. At the time, acoustic spectroscopy was the best means to measure protein compressibility; Kundrot & Richards set out to make a more direct measurement.

They discussed several important themes. The lysozyme domains had measurably different compressibilities—in fact one of the two is incompressible. These domains are not contiguous in the amino acid sequence but are purely structurally defined—in this case by a hypothesized hinge around an active site. However, no net rotation about the suggested hinge was detected. It has since become possible to analyze pairs or sequences of structures for such motions in a better defined way, as discussed below for the case of citrine.

The Debye-Waller temperature factors (or B-factors) decreased on average, but no coherent trends appeared and many B-factors even increased. This proves to be generally true of pressurized proteins. Kundrot & Richards observed that, despite different compressibilities, both domains moved toward the center of mass of the molecule, perhaps suggesting a different nature for the hinge motion inferred by earlier researchers. Side chain reorientations were the main source of compression, and buried side chains reoriented much less than surface side chains, demonstrating the well-packed nature of the protein.

Most importantly, the high-pressure lysozyme structure suggested new ways that one could identify structurally defined subdomains for further analysis. The ability to reduce the number of structural coordinates needed to describe an energy landscape is fundamental to the power of high pressure in studying protein structure and dynamics. This study strongly suggested the most significant motions were of large structural motifs, not an atomic free-for-all. This leads us to speculate whether the distinctly compressible domains seen in lysozyme are coincident with the structural sectors identified by Halabi et al. (20).

The Kundrot & Richards work had several unintended consequences that discouraged high-pressure crystallography. First, the absolute magnitudes of the deformations were generally sub-angstrom, leading others to draw

BPTI: bovine
pancreatic trypsin
inhibitor protein

HPr: histidine-
containing protein

the erroneous conclusion that they were not biologically important. In fact, sub-angstrom motions around active sites can greatly alter enzyme activity. Difficulties with crystal cracking led others to conclude that crystal pressurization is difficult. Instead, recent experiences with dozens of different proteins have shown that crystals are typically robust against cracking while pressurized.

Crystal cracking during pressurization involves some perhaps nonintuitive subtleties. Protein crystals usually contain more water than protein. Consider a protein 50 Å across as a sphere scaled to 30 cm, about the size of a basketball. When stacked into a simple cubic lattice, the basketballs occupy only 52% of the crystal volume; the remaining volume is filled with water molecules (~1 cm across on this scale). Imagine all the 1-cm marbles that can fill the volume between the spheres in the basketball crystal. Although some of the water is bound to the protein surface, most is effectively bulk water several centimeters removed from the protein. Thus, pressurization of a protein crystal is qualitatively different from pressurization of a typical solid crystal: There is no strain gradient across the faces of the crystal because the interstitial water transmits pressure uniformly throughout the crystal to each protein molecule. So why would a crystal crack under pressurization? It must be because pressure is affecting the shape of each protein molecule or the nature of the soft forces that hold the crystal together. Whereas in most solids crystal cracking is a result of macroscopic strain, with proteins crystal cracking is a molecular manifestation of pressure.

NMR Leads the Way

NMR spectroscopists began to study small proteins such as bovine pancreatic trypsin inhibitor (BPTI; the first high-pressure NMR structure we are aware of) (56), histidine-containing protein (HPr) (28), ubiquitin (35) and the GTPase Rap1 (26), among others (25, 39, 42, 48). It remains challenging to solve full structures by NMR, so although it was clear that pressure was

inducing alternative protein conformations, it was not always clear exactly what those conformations were.

Histidine-Containing Protein

These studies are well characterized by the work on HPr (28), which is involved in a phosphorylation pathway ultimately responsible for carbohydrate transport. From biochemical studies it is known that active HPr is itself phosphorylated at Ser46, which affects HPr's substrate binding surface. To account for the allosteric effect of phosphorylation, the authors presumed that fluctuations must be elevated at the binding surface, including the His15 active site, and that Ser46 phosphorylation would suppress or expand such fluctuations. Alternatively, there could be a small number of alternative conformers, one of which might correspond to the conformation favored by phosphorylation at Ser46. In the energy landscape picture, these two hypotheses are only slightly different, the distinction being whether or not the alternative conformers are true local free energy minima. In either case, one expects associated higher and lower volume conformations, so that the active site should have increased compressibility.

Indeed, pressure-induced changes of the ^{1}H–^{15}N heteronuclear single quantum coherence (HSQC) NMR spectra were largest near Ser46 and His15. Unexpectedly, the pressure-induced backbone amide ^{1}H–^{15}N peak shifts of Ser46 and Ile14 were distinctly different from those of their structural neighbors. Ile14 has one of the lowest shifts in the entire molecule, which the authors speculated may indicate it forms a conformational pivot or nucleation point. They did not specifically consider Ser46, but based on the tabulated data, its amide peak changes are smaller than typical for surrounding residues, and the ^{1}H shift is opposite in sign from its neighbors. The rigidity of Ser46 and Ile14 may help transmit phosphorylation-induced displacements to the active site His15.

Dihydrofolate reductase. Perhaps the best characterization of how high-pressure NMR could be used to characterize the energy landscape and important dynamic modes of a protein was the study by Kitahara et al. (34) on dihydrofolate reductase (DHFR). From ambient pressure X-ray crystallography, DHFR was known to adopt different conformations when binding cofactors and substrate. A model had been developed using those X-ray structures to explain the conformational changes associated with enzymatic action, but solution NMR work had revealed only one conformation in the folate-bound state, whereas crystallography had indicated two.

Prior ambient pressure NMR indicated substantial fluctuations. Thus, high-pressure NMR was an excellent tool to isolate CS involved in the fluctuations and to tie those substates to enzymatic action based on the crystal structures. Those structures revealed a potential motion of the M20 loop that could occlude or open the active site. Kitahara et al. examined both the pressure and temperature dependency of ^{1}H–^{15}N HSQC spectra, showing that on either end of this loop the cross-peaks shifted and split into doublets, indicating two conformational states. They measured the free energy, enthalpy, entropy, and volume differences between the two states. In particular they showed that at ambient pressure, with a folate substrate, and without the NADH cofactor, only 10% of the population would be in the open state, functionally significant, but largely undetectable by ambient pressure NMR. High pressure was necessary to identify this state in solution.

Compared with earlier work on BPTI (56), the authors used the sensitivity of the ^{1}H–^{15}N HSQC spectra to amide NH group orientations to estimate changes in backbone torsion angles. Roughly 4° changes at the ends of the M20 loop contrasted sharply with no observable torsion angle changes in the rest of the loop. The high-pressure NMR work conclusively demonstrated a continuous hinge-like motion, confirming the hypothesis generated from X-ray structures. Clearly, the challenge was to generate high-pressure structures and directly observe such changes

with the complete structural information available from X-ray crystallography.

Methods of Performing High-Pressure Protein Crystallography

It is useful at this point to digress from the historical flow to summarize the major methods that have been developed to perform high-pressure protein crystallography. These include ambient temperature beryllium and diamond anvil cell (DAC) methods and cryogenic liquid and gas charged methods.

Ambient temperature methods. Urayama et al. (53) modified Kundrot & Richards' (37) beryllium cell to reach 200 MPa by repositioning an O-ring between the threaded beryllium rod and a steel seat to act as a Bridgman seal. Beryllium cells have limitations. The required high-strength beryllium material is difficult to obtain and machine. The typical one-quarter-inch rod thickness limits pressures to about 200 MPa; in principle a thicker rod and higher energy X-rays can be used to widen the pressure range while mitigating increased X-ray absorption. It is difficult to align the crystal in the X-ray beam because the rod is opaque. Most seriously, intense beryllium powder diffraction rings are superposed on the protein diffraction spots. The lowest order rings begin at an atomic spacing of ~2.3 Å, limiting the collection of higher resolution protein diffraction data.

Diamond anvil cells are alternative X-ray pressure cells capable of reaching higher pressures and resolutions. Early experiments (29) have evolved considerably, as reviewed recently (16). The diamond optical transparency facilitates aligning the protein crystal in the X-ray beam, but X-ray absorption and limited sample size require a high-energy beam line delivering an intense, small-area beam. The diamond cell mount limits the rotation angles of the crystal; this is partly compensated for by short X-ray wavelength reduction of scattering angles into the forward direction. Wider angle DAC

DHFR: dihydrofolate reductase

Diamond anvil cell (DAC): a kind of high-pressure cell for X-ray and optical studies

designs exist but have not yet been applied to protein crystallography.

Cryogenic methods. The method pioneered by Thomanek et al. (51) uses liquid isopentane as a pressure medium. The tedium of removing the pressure cryocooled crystal from isopentane at liquid nitrogen temperature led us to develop a convenient method using helium gas (30). The crystal is pressurized in a column of helium gas and then dropped, still at high pressure, into a zone at liquid nitrogen temperature. Once the crystal is frozen, gas pressure is released and the crystal is removed from the apparatus under liquid nitrogen and then handled as any normal cryocooled crystals. Normally, no penetrating cryoprotectants are required; the only fluid present is mother liquor. The method requires special equipment, but once in place it is relatively easy to use. At Cornell, several crystals per hour may be processed, and dozens of different types of protein crystals have been successfully pressure cryocooled (30).

Crystals cryocooled in this manner tend to produce superior Bragg diffraction. It is also possible to produce bulk vitreous cryocooled mother liquors: The method has been used to cryoprotect crystals and crystallization solution in thick-walled polycarbonate capillaries, which is not possible at ambient pressure without adding high concentrations of chemical cryoprotectants (31). Not only do cryoprotectants add complexity, they frequently compete with ligands for binding sites, compromising structural experiments. A good example of these advantages is a recent study of ligand bound to the RCK domain of the KtrAB K$^+$ transporter (1), which proved difficult in the presence of cryoprotectants and standard ambient pressure methods.

Kim et al. proposed (30), and subsequently confirmed (32), that the method mitigates cooling damage by inducing formation of high-density amorphous ice in the crystal. The glassy water exhibits unusual liquid-like behavior when warmed from 80 K to 160 K (see figure 4 of Reference 33). Although it is beyond the scope of this review, this behavior may be advantageous in studying protein fluctuations or using temperature dependency to separate entropic contributions to conformational free energies.

Conformational Substates of Myglobin

In 2002, Urayama et al. (53) solved multiple structures of sperm whale myoglobin with two goals in mind. The first goal was technological: At the time, high-pressure crystallography on proteins required beryllium cells that limited both the diffraction resolution and the pressure range available. Urayama et al. quantitatively demonstrated that the structural effects of high pressure at room temperature could be "frozen in" if the crystal was cooled to 77 K while under pressure in isopentane. The observation that the diffraction quality was excellent, even if the crystals were slowly cooled without added cryoprotectants, motivated the development of the high-pressure cryocooling technique (30).

The second goal was to map out conformational substates of myoglobin. In a landscape picture, effects that are trapped upon cooling under high pressure represent free energy minima that are linked to each other in a continuous free energy landscape. The low-energy substates will be populated some definite fraction of the time and can be functionally relevant. To show that the frozen-in effects of pressure had in fact shifted the equilibrium population of two or more CS, Urayama et al. began from the observation that increasingly acidic pH and increasing high pressure had similar effects on the relative populations of two spectroscopically defined CS. The structures of sperm whale myoglobin at representative pH values were available, and the object was to compare structures solved at high pressure against those solved at acidic pH to determine whether the conformational changes were similar, as measured by the displacements of individual peptide backbone atoms. Urayama et al. showed that the changes due to pH and pressure are highly correlated. This was the first clear demonstration that pressure could be used to identify and characterize detailed shifts in CS populations that would be functionally relevant to the study of a given protein.

The RMS deviations of the α-carbon backbone positions upon pressurization of myoglobin were typically sub-angstrom in magnitude. Although crystallographic resolutions are usually poorer than an angstrom, deviations of groups of atoms can be determined to higher accuracy. Urayama et al. (53) determined that the experimental uncertainty in the myoglobin experiments was on the order of one-tenth of an angstrom. A related issue is how best to superimpose two distorted structures so that the RMS deviations may be measured (for more detail see References 3 and 10).

Energy Landscapes by Crystallography

An important caveat is that a crystallographic experiment yields a structure representing the ensemble of states. Any side chain or structural subunit of the protein that normally fluctuates between two CS will likely appear as a region of increased thermal *B*-factors. *B*-factors are difficult to interpret quantitatively because structural contributions are convolved with crystal mosaicity and other imperfections.

Furthermore, the implied large conformational and volume fluctuations can also be explained by a relatively flat free energy landscape. To demonstrate conclusively the existence of true CS requires careful observation of the complete transition and, ideally, a model of the CS that describes changes in volume, free energy, and so on, as was done in the pioneering NMR examples described above. The remaining examples below demonstrate how various kinds of CS can be mapped out in detail using X-ray crystallography, either directly under high pressure or in parallel with other high-pressure experiments.

T4 lysozyme, the hydrophobic core, and buried water in proteins. Collins et al. (9, 10) used a beryllium cell to solve the room-temperature, high-pressure structure of the T4 lysozyme (T4L) mutant L99A. Brian Matthews' laboratory (59) had shown that this mutant contained a buried and empty (i.e., vacuum!) cavity of almost 150 Å^3. The cavity is hydrophobic, lined by aliphatic and aromatic residues, buried in the C-terminal helical bundle of the protein, and is formed by a leucine-to-alanine substitution. Defying expectations, the cavity failed to collapse under pressures as high as 200 MPa and instead steadily filled with water, whereas the structure of the protein remained mostly unchanged. This presented an unusual opportunity to closely follow a clear multistate system as pressure altered CS populations, i.e., the cavity water occupancies.

The issue of water buried in proteins is fascinating and sometimes contentious. Many proteins (including T4L) contain buried pockets of water inaccessible to the bulk solvent; this buried water is frequently coordinated with buried polar or charged amino acid side chains (47, 55). The more interesting cases include interleukin-1β, certain cytochromes, and even membrane ion channels, where water is proposed to occupy more hydrophobic pockets. What effect do such waters have on protein function and stability?

Collins et al. (9) examined T4L and determined the water occupancy by integrating the excess electron density inside the L99A cavity. Increasing electron density with pressure indicated that the cavity water content increased slowly up to pressures of about 100 MPa, and then rose steeply above that, suggesting the midpoint of a sigmoidal transition (**Figure 2**).

Molecular dynamics (MD) modeling yielded additional insight and suggested a transition between two cavity states—one empty and one filled with up to four hydrogen-bonded water molecules (**Figure 2**). The protein itself was extraordinarily rigid with a cavity volume that decreased by less than 3%, suggesting that pressure did not alter the energetics of water interaction with the cavity walls. Assuming this, it was estimated that the free energy difference between the cavity's empty and flooded states was only $\sim 1.4 \, kT$/water molecule (Boltzmann's constant, k, times the absolute temperature, T, equal to $\sim 2.5 \text{ kJ mol}^{-1}$ at room temperature). This difference is sufficiently small that even

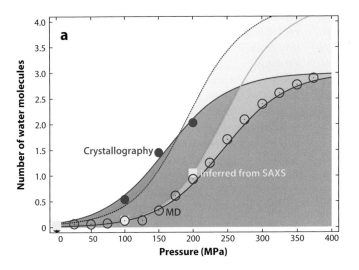

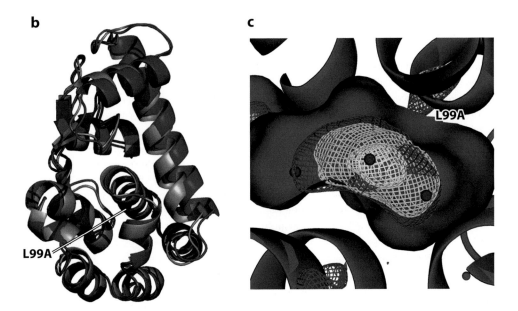

Figure 2

T4 lysozyme (T4L) mutant L99A at high pressure. (*a*) Number of water molecules observed crystallographically (*solid blue circles*), in molecular dynamics simulations (*open purple circles*), and inferred from small-angle X-ray scattering (SAXS) measurements on pressure unfolding (*yellow squares*). For comparison, the SAXS data are shifted horizontally (*blue dotted line*) to have the same midpoint pressure. Adapted from Reference 2 and 9. (*b*) Structural changes in T4L from the ambient (*blue*) to the 200 MPa (*orange*) structure. Differences have been magnified 5X for clarity, and the structures are aligned using the helical bundle that contains the L99A cavity mutation. (*c*) Crystallographic electron density maps show water filling the L99A cavity at 100 (*yellow*), 150 (*blue*), and 200 (*magenta*) MPa. Adapted from Reference 9.

under ambient conditions one would expect this state to be populated around 2% of the time, assuming the cavity had access to the bulk water.

Water diffuses even through very hydrophobic materials such as Teflon™ or lipid bilayers at measurable rates. A protein's hydrophobic core fluctuates constantly and is only angstroms removed from bulk water, so one must expect a slow diffusion of water through the volume of all proteins. In the case of T4L, the effect of pressure is to increase the water occupancy in the cavity by making the free energy sufficiently favorable that the waters remain long enough to be detected crystallographically.

The T4L studies provided crystallographic evidence connecting detailed structural changes in the protein and transitions involved in unfolding at high pressure. It has been long hypothesized that pressure-induced unfolding involves water penetration into the protein core, a subtly different concept than the protein core becoming exposed to bulk water (24). The T4L measurements showed that such penetration by water molecules into the hydrophobic protein core was not only possible, but probably quite common.

Studies of pressure unfolding of staphylococcal nuclease (SNase) had shown that the radius of gyration of pressure-unfolded SNase was smaller than that of thermally or chemically denatured SNase (45). This finding suggested that pressure only partially unfolded SNase. Paliwal et al. (44) used small-angle neutron scattering and MD simulations to infer that pressure unfolding of SNase was a progressive process involving loosening of tertiary contacts, leaving much secondary structure relatively intact. This suggested that water was involved in protein unfolding at high pressures, but direct evidence of water entering at specific sites of the protein was lacking. An important challenge in the T4 lysozyme experiments was to connect the crystallographic result that water penetrated into the protein's hydrophobic core under pressure with partial unfolding.

Ando et al. (2) followed with a careful study of L99A T4L unfolding by high pressure. Using a newly designed small-angle X-ray scattering (SAXS) chamber, Ando et al. determined the volume change in a pressure-induced, unfolding-like transition. This volume change could be substantially accounted for by the empty cavity volume of the ambient pressure structure, supporting an unfolding scenario in which these cavities become hydrated and thereby loosen tertiary contacts. Furthermore, the SAXS data demonstrated the relatively compact structures of the progressively unfolded states with pressure.

Pressure-induced unfolding highlights that there is neither one unfolded state nor one folded state of a protein. High-pressure structural studies, building on mutagenic studies of proteins, help to show this more clearly by allowing the experimenter to slowly tune a perturbation of the structure. There is increasing evidence that the complicated free energy landscapes of proteins, and rarely populated CS, may be involved in diseases such as Alzheimer's and type II diabetes (22). The studies on T4L suggest that high-pressure crystallography may allow such states to be studied in detail.

Ubiquitin and T4L: How rigid is an α helix?
In 2005, Kitahara et al. (36) published the structure of the important small protein ubiquitin, one of the first complete high-pressure structures solved by NMR. Using nuclear Overhauser effect (NOE) spectra, backbone dihedral angle constraints from chemical shifts, and simultaneous fitting of the two structures to the data, the authors generated averaged structures at 3.0 and 300 MPa and independently determined their relative populations. The largest difference between the two structures was a 9° rotation of the main helix (residues 23–40). This helix's internal structure was unaltered by pressure, but it rotated and displaced relative to the rest of the protein. This contrasts with T4L, in which a helix in a similarly exposed position bent by several degrees roughly about its midsection, suggesting that helices might require some stabilization from their surroundings to be truly rigid (10). In ubiquitin, a smaller protein with presumably fewer side chain contacts to stabilize this helix, it appears that 300 MPa

SAXS: small-angle X-ray scattering

was not enough to deform the helix. On the other hand, in T4L, the midpoint of the deformed helix rests against a dense network of contacts that has also been implicated as a nucleation site in folding. It would be especially interesting to examine the issue of side chain contacts in ubiquitin.

Ubiquitin and T4L: Water penetrates the core in high-pressure unfolding. Similar to Collins et al. (9, 10), Kitahara et al. (36) found that the protein core opened up at high pressure, visible both by its three-dimensional structure and by an increasing solvent-accessible surface area even as the molecular volume decreased. While this may lead to unfolding, this core opening is directly related to the shift between the two observed conformational states.

It would be especially interesting to better constrain side chain conformations and any changes in crystallographically observable water in this pressure range and to compare the results with the T4L study (9, 10). Not long after the study by Kitahara et al. on ubiquitin

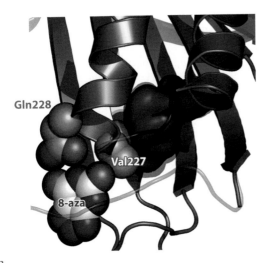

Figure 3

The urate oxidase active site and accessory cavity. Nearby residues Val227 and Gln228 are shown with their carbon atoms in orange. The active site blocker 8-azaxanthine (8-aza) is shown with its carbons in white. The accessory cavity of the 150 MPa structure (PDB code 3F2M; Reference 19) is shown in blue, with the ambient pressure cavity (PDB code 2IBA; Reference 12) shown in transparent blue around it.

structure was published, MD simulations (15) confirmed an increase of water hydrating small, hydrophobic cavities that opened up in ubiquitin, largely validating hypotheses of pressure-induced denaturation (2, 9, 24). It remains unclear whether these cavities become hydrated before unfolding or if they become completely open to the surrounding water as a result of pressure (57).

A VIEW INTO THE FUTURE OF PROTEIN ENGINEERING?

Thus far, we have discussed how NMR and crystallography at high pressure have been used to map out regions of the protein free energy landscape to understand protein action. Now we turn to examples more directly related to practical protein engineering.

Urate Oxidase

The enzyme urate oxidase (Uox) is unusual in that it requires no cofactors or metal ions to break down its target substrate, uric acid's purine ring. It can be used medically to manage hyperuricemia, for instance, due to renal failure or chemotherapy, but is avoided because it requires intravenous administration and can be anaphylactic. Reengineering such enzymes has obvious merit.

Colloc'h and colleagues (11) reported the Uox structure at 140 MPa using DAC X-ray crystallography. Consistent with other work, the Uox β-sheet structure is remarkably rigid, with the all-atom RMS deviation only 0.14 Å, which is barely resolvable. The ambient pressure structure contains a 136 Å³ hydrophobic cavity devoid of water (**Figure 3**). Unlike the T4L cavity, its volume decreases by 24% and it remains empty under pressure. By titrating the pressure of oxygen from 1 to 4 MPa (13), the authors found that this accessory cavity did not fill with oxygen, ruling out its proposed role as an oxygen reservoir for the enzyme. Does this cavity have a purpose?

A recent paper combines SAXS, functional assays, and crystallography at high pressure

(19) to show that Uox undergoes two pressure-induced transitions leading to reduced activity, one of which is most likely dissociation of the functional tetramer into monomers. The nature of the other transition is less clear. The kinetics of the enzyme-catalyzed reaction itself changes little with pressure. However, the substrate affinity for the binding pocket decreases with pressure, reducing overall activity.

The authors (19) made two important observations based on the high-pressure structure. First, thermal *B*-factors among amino acids lining the active site entrance actually increase, consistent with residues gating other cavities (9, 10). Although these side chains become more flexible, the kinetic data suggest that access to the active site is more restricted at high pressure. The authors speculate pressure-induced flexibility might allow a normally open side chain gate to close. Such a gate could be a novel drug target.

Second, the accessory cavity collapses with pressure (**Figure 3**), whereas the active site cavity expands—the accessory cavity acts as ballast for the reaction products, which require a different binding pocket geometry. Pressure disfavors the cavity geometry that binds substrate.

High pressure stabilized a state that likely represents the conformation of the active site after the oxidation reaction has occurred. No complicated chemistry or kinetic trapping was needed to observe this conformation. Moreover, the allosteric interaction between these cavities suggests a general approach to identifying and exploiting allosteric interactions in proteins.

Morphinone Reductase: Multiple Conformational Pathways of an Enzyme

Although pressure was not used directly in morphinone reductase crystallography, we include this example because it suggests how other high-pressure techniques could be used together with crystallography to identify and map out multiple reaction pathways in an enzyme. Capping a remarkable set of papers, Scrutton

and colleagues (21, 46) identified three kinetically distinct pathways in morphinone reductase, a hydride transfer enzyme. An aspartic acid–to-alanine substitution (N189A) increased disorder in the active site and increased usage of the three pathways, whose rates spanned three orders of magnitude.

This mutation demonstrates that enzymes can and do have multiple kinetically distinct reaction pathways. To characterize these pathways, Pudney et al. (46) used pressure to alter the populations of the three different NADH-enzyme conformations.

Crystallographic structures of the wild-type and N189A enzymes identified multiple active site substrate conformations. MD simulations of wild-type enzyme indicated high-pressure-induced changes in active site geometry corresponding to CS of the N189A active site. Further simulations of N189A showed that pressure shifted the equilibrium between these CS, decreasing the hydride donor-acceptor distances in the same way as that observed in simulations of the wild type.

The high-pressure kinetic results are equally intriguing. Pressure essentially has no effect, or speeds up the reaction, depending on the pathway. On the other hand, pressure significantly alters each pathway's utilization. The occupancy of the initial CS changed owing to their different volumes. The overall enzyme activity slows because of the underlying conformational flexibility of the protein, and not because pressure alters the pathways' underlying kinetics.

High-Pressure Crystallography as a Predictive Tool: The Case of Citrine

The most direct suggestion of how to use the ideas developed above comes from Barstow et al. (3, 4), who analyzed high-pressure-induced changes in the pressure-sensitive yellow fluorescent protein, citrine, using the high-pressure cryocooling method (30). Citrine differs from the green fluorescent proteins (GFPs) by a threonine-to-tyrosine substitution at residue 203; the Y203 phenol ring interacts with the main chromophore common to this superfamily of fluorescent proteins. Y203 is anchored to a structurally distinct part of the protein, on the β barrel rather than on the central 3_{10} helix containing the main chromophore, the autocatalytically fused residues called Cro66.

In their first paper, Barstow et al. (3) detailed pressure-induced changes near the active site, where Cro66 and Y203 overlap. The overlap of the two ring structures is thought to result in the distinct yellow color of citrine's fluorescence. If so, then fluorescence should be exquisitely sensitive to small, pressure-induced changes in separation and orientation of the two rings. Indeed, as pressure is increased from 50 to 300 MPa, the fluorescence peak shifts from ∼530 nm to 510 nm, the GFP peak, and the fluorescent yield drops by roughly a factor of two. The effect appears to be reversible, and no such shifts occur in engineered GFP, where the perturbing Y203 is absent.

In experiments on crystals, Barstow et al. unambiguously assigned fluorescence spectral changes to high-pressure protein structures derived from X-ray crystallography. Above 50 MPa, the main chromophore motion is essentially confined to a plane parallel to the Y203 phenol ring. Over ∼300 MPa, it moves in a direction about 45° to the axis formed by the phenolic oxygen and the fourth carbon of the phenyl ring, by almost 1 angstrom. With this small shift, the influence of the Y203 ring is eliminated.

From atmospheric pressure to 50 MPa, the results are in some ways much more interesting. Here, increasing pressure red-shifts the fluorescence peak. This suggests that perturbation by Y203 could be red-shifted, and potentially more efficient, by stabilizing the pressure-induced conformation at ambient pressure. One wonders whether the same could be said for other enzymes such as those discussed above. But how would one take advantage of this knowledge?

Barstow et al. (4) showed a way forward by discussing the global citrine structure. The key finding is that the changes observed in the active site are really due to subtle changes in large, essentially rigid structural domains linked to the active site.

A cluster analysis was used to identify groups of atoms that move together with increasing pressure. The analysis identified two distinct regions, the borders of which are not obvious when looking at the citrine structure. One region contains Y203, with Cro66 and most of the 3_{10} helix in the other region (**Figure 1**).

The two parts of the active site sit in different structural regions, which explains why they move relative to each other. But their relative motion is not what one might first suppose. The centers of mass move at most a couple tenths of an angstrom relative to each other, at the noise limit of the measurement. Instead, small relative rotations of the two structural elements cause Cro66 and Y203 to move relative to each other. The two elements of the active site sit at the ends of long lever arms, an observation beautifully demonstrated by high-pressure crystallography.

This example demonstrates the globally connected nature of proteins. In citrine, pressure resulted in relative shifts of two segments of the protein. These, in turn, pulled on the residues responsible for fluorescent activity, which act as an active site in the molecule. This concrete example gives a plausible explanation for why many enzymes are very pressure sensitive: Similar global distortions of enzymes result in small deformations of active sites. As was the case with citrine, most enzymes involving electrostatic or quantum mechanical interactions will be sensitive even to sub-angstrom deformations of its active site.

SUMMARY: THE FREE ENERGY LANDSCAPE REVISITED

The last few examples in particular illustrate the goal of this review: to map out the free energy landscape of proteins in a way that connects their thermodynamics to their structure. The example of citrine is particularly relevant because it has a conformational substate about which we can ask questions.

What are the conformational substates—those configurations readily grouped together to simplify our representation? For citrine, it is the folded state. The pressure-induced deformations appear smoothly and without jumps, indicating that pressure is simply perturbing this conformational substate from its ambient pressure conformation. What are the coordinates? The analysis by Barstow and coworkers showed that the description of this system can be reduced to two quasi-rigid bodies, with a relative center of mass and a relative orientation. Because this is a pressure experiment, we can determine how much work is done—how much energy is added to the system—from the overall compressibility of the system, determined independently or from the crystallographic experiment. In simplified form, **Figure 1** depicts an illustrative cross section through the free energy landscape, with one position coordinate and one orientation coordinate.

Morphinone reductase is more complicated. MD simulation, X-ray crystallography of the N189A mutant, and high-pressure kinetics experiments indicated the existence of three distinct conformational substates, each having its own distinct kinetics in the enzymatic reaction. These, along with kinetic intermediates, and any final states constitute the set of conformational substates.

One can next attempt to identify a set of coordinates that adequately describes the atomic displacements associated with switching between these conformational substates. With improved structural information, from crystallography, NMR, or the available MD data, one may be able to reduce the number of necessary coordinates using rigid body analysis, as Barstow et al. (3, 4) did with citrine.

The pressure-dependent kinetic data can then be utilized to determine the energy landscape as a function of those coordinates. This case is somewhat unusual in that the branching ratios are here giving information about the relative population of conformational substates. Nonetheless, with that population information in hand, we can determine the relative free energies of those CS. The structural information here allows us to put those into context on their underlying free energy landscape.

In summary, high-pressure structural experiments now allow detailed questions about protein function to be posited and answered. Pressure provides an experimentally accessible variable as potent as, e.g., temperature or pH, but more readily related to measurable atomic positions. Pressure experiments can be used to guide mutation studies and to help engineer proteins. Pressure experiments have the drawback of requiring equipment that, at the moment, is specialized and home constructed. This is likely to change over time.

SUMMARY POINTS

1. High-pressure methods, especially crystallography, are now routine and mature. Any protein that can be crystallized can be subjected to high pressure.

2. High pressure gives access to protein conformations that may not be visible in other ways, providing new insight into the conformational substates and dynamics.

3. High-pressure methods allow us to perturb the structure in a smooth, continuous fashion, making it easier to connect the dots when correlating structural changes to functional changes.

4. A detailed understanding of how protein deforms under high pressure may provide additional insight for advanced structure-based protein engineering.

DISCLOSURE STATEMENT

The authors are not aware of any affiliations, memberships, funding, or financial holdings that might be perceived as affecting the objectivity of this review.

LITERATURE CITED

1. Albright RA, Ibar JL, Kim CU, Gruner SM, Morais-Cabral JH. 2006. The RCK domain of the KtrAB K+ transporter: multiple conformations of an octameric ring. *Cell* 126:1147–59

2. Ando N, Barstow B, Baase WA, Fields A, Matthews BW, Gruner SM. 2008. Structural and thermodynamic characterization of T4 lysozyme mutants and the contribution of internal cavities to pressure denaturation. *Biochemistry* 47:11097–109

3. Barstow B, Ando N, Kim CU, Gruner SM. 2008. Alteration of citrine structure by hydrostatic pressure explains the accompanying spectral shift. *Proc. Natl. Acad. Sci. USA* 105:13362–66

4. Barstow B, Ando N, Kim CU, Gruner SM. 2009. Coupling of pressure-induced structural shifts to spectral changes in a yellow fluorescent protein. *Biophys. J.* 97:1719–27

5. Bridgman PW. 1914. The coagulation of albumin by pressure. *J. Biol. Chem.* 19:511–12

6. Chan HS, Dill KA. 1998. Protein folding in the landscape perspective: chevron plots and non-Arrhenius kinetics. *Proteins* 30:2–33

7. Cho YK, Northrop DB. 1999. Effects of pressure on the kinetics of capture by yeast alcohol dehydrogenase. *Biochemistry* 38:7470–75

8. Cioni P, Strambini GB. 1999. Pressure/temperature effects on protein flexibilty from acrylamide quenching of protein phosphorescence. *J. Mol. Biol.* 291:955–64

9. **Collins MD, Hummer G, Quillin ML, Matthews BW, Gruner SM. 2005. Cooperative water filling of a nonpolar protein cavity observed by high-pressure crystallography and simulation. *Proc. Natl. Acad. Sci. USA* 102:16668–71**

10. Collins MD, Quillin ML, Hummer G, Matthews BW, Gruner SM. 2007. Structural rigidity of a large cavity-containing protein revealed by high-pressure crystallography. *J. Mol. Biol.* 367:752–63

11. Colloc'h N, Girard E, Dhaussy AC, Kahn R, Ascone I, et al. 2006. High pressure macromolecular crystallography: the 140-MPa crystal structure at 2.3 A resolution of urate oxidase, a 135-kDa tetrameric assembly. *Biochim. Biophys. Acta* 1764:391–97

12. Colloc'h N, Sopkova-de Oliveira Santo J, Retailleau P, Vivarès D, Bonneté F, et al. 2007. Protein crystallography under xenon and nitrous oxide pressure: comparison with in vivo pharmacology studies and implications for the mechanism of inhaled anesthetic action. *Biophys. J.* 92:217–24

13. Colloc'h N, Gabison L, Monard G, Altarsha M, Chiadmi M, et al. 2008. Oxygen pressurized X-ray crystallography: probing the dioxygen binding site in cofactorless urate oxidase and implications for its catalytic mechanism. *Biophys. J.* 95:2415–22

14. Davydov DR, Baas BJ, Sligar SG, Halpert JR. 2007. Allosteric mechanisms in cytochrome P450 3A4 studied by high-pressure spectroscopy: pivotal role of substrate-induced changes in the accessibility and degree of hydration of the heme pocket. *Biochemistry* 46:7852–64

15. Day R, Garcia AE. 2008. Water penetration in the low and high pressure native states of ubiquitin. *Proteins* 70:1175–84

16. **Fourme R, Girard E, Kahn R, Dhaussy AC, Ascone I. 2009. Advances in high-pressure biophysics: status and prospects of macromolecular crystallography. *Annu. Rev. Biophys.* 38:153–71**

17. Frauenfelder H, Alberding NA, Ansari A, Braunstein D, Cowen BR, et al. 1990. Proteins and pressure. *J. Phys. Chem.* 94:1024–37

18. **Frauenfelder H, McMahon BH. 2000. Energy landscape and fluctuations in proteins. *Ann. Phys.* 9:665–67**

19. **Girard E, Marchal S, Perez J, Finet S, Kahn R, et al. 2010. Structure-function perturbation and dissociation of tetrameric urate oxidase by high hydrostatic pressure. *Biophys. J.* 98:2365–73**

20. Halabi N, Rivoire O, Leibler S, Ranganathan R. 2009. Protein sectors: evolutionary units of three-dimensional structure. *Cell* 138:774–86

21. Hay S, Pudney CR, McGrory TA, Pang J, Sutcliffe MJ, Scrutton NS. 2009. Barrier compression enhances an enzymatic hydrogen-transfer reaction. *Angew. Chem. Int. Ed. Engl.* 48:1452–54

22. Hebda JA, Miranker AD. 2009. The interplay of catalysis and toxicity by amyloid intermediates on lipid bilayers: insights from type II diabetes. *Annu. Rev. Biophys.* 38:125–52

23. Heremans K. 1982. High pressure effects on proteins and other biomolecules. *Annu. Rev. Biophys. Bioeng.* 11:1–21

9. Provides a good example of how high-pressure crystallography can be used to extend traditional mutagenic studies of protein thermodynamics.

16. Covers many instrumental aspects not discussed here.

18. An excellent introduction to the concepts of energy landscape theory.

19. Summarizes Uox studies and highlights how pressure works in concert with other methods.

24. Hummer G, Garde S, Garcia AE, Paulaitis ME, Pratt LR. 1998. The pressure dependence of hydrophobic interactions is consistent with the observed pressure denaturation of proteins. *Proc. Natl. Acad. Sci. USA* 95:1552–55

25. Inoue K, Yamada H, Imoto T, Akasaka K. 1998. High pressure NMR study of a small protein, gurmarin. *J. Biomol. NMR* 12:535–41

26. Inoue K, Maurer T, Yamada H, Herrmann C, Horn G, et al. 2001. High-pressure NMR study of the complex of a GTPase Rap1A with its effector RalGDS. A conformational switch in RalGDS revealed from non-linear pressure shifts. *FEBS Lett.* 506:180–84

27. Jonas J, Jonas A. 1994. High-pressure NMR spectroscopy of proteins and membranes. *Annu. Rev. Biophys. Biomol. Struct.* 23:287–318

28. Kalbitzer HR, Gorler A, Li H, Dubovskii PV, Hengstenberg W, et al. 2000. 15N and 1H NMR study of histidine containing protein (HPr) from *Staphylococcus carnosus* at high pressure. *Protein Sci.* 9:693–703

29. Katrusiak A, Dauter Z. 1996. Compressibility of lysozyme protein crystals by X-ray diffraction. *Acta Crystallogr. D* 52:607–8

30. Kim CU, Kapfer R, Gruner SM. 2005. High-pressure cooling of protein crystals without cryoprotectants. *Acta Crystallogr. D* 61:881–90

31. Kim CU, Hao Q, Gruner SM. 2007. High-pressure cryocooling for capillary sample cryoprotection and diffraction phasing at long wavelengths. *Acta Crystallogr. D* 63:653–59

32. Kim CU, Chen YF, Tate MW, Gruner SM. 2008. Pressure-induced high-density amorphous ice in protein crystals. *J. Appl. Crystallogr.* 41:1–7

33. Kim CU, Barstow B, Tate MW, Gruner SM. 2009. Evidence for liquid water during the high-density to low-density amorphous ice transition. *Proc. Natl. Acad. Sci. USA* 106:4596–600

34. Kitahara R, Sareth S, Yamada H, Ohmae E, Gekko K, Akasaka K. 2000. High pressure NMR reveals active-site hinge motion of folate-bound *Escherichia coli* dihydrofolate reductase. *Biochemistry* 39:12789–95

35. Kitahara R, Yamada H, Akasaka K. 2001. Two folded conformers of ubiquitin revealed by high-pressure NMR. *Biochemistry* 40:13556–63

36. Kitahara R, Yokoyama S, Akasaka K. 2005. NMR snapshots of a fluctuating protein structure: ubiquitin at 30 bar–3 kbar. *J. Mol. Biol.* 347:277–85

37. Kundrot CE, Richards FM. 1986. Collection and processing of X-ray diffraction data from protein crystals at high pressure. *J. Appl. Crystallogr.* 19:208–13

38. Kundrot CE, Richards FM. 1987. Crystal structure of hen egg-white lysozyme at a hydrostatic pressure of 1000 atmospheres. *J. Mol. Biol.* 193:157–70

39. Kuwata K, Li H, Yamada H, Batt CA, Goto Y, Akasaka K. 2001. High pressure NMR reveals a variety of fluctuating conformers in beta-lactoglobulin. *J. Mol. Biol.* 305:1073–83

40. Macdonald AG. 2002. Experiments on ion channels at high pressure. *Biochim. Biophys. Acta* 1595:387–89

41. Meyer R, Heinemann SH. 1997. Temperature and pressure dependence of Shaker K+ channel N– and C-type inactivation. *Eur. Biophys. J.* 26:433–45

42. Moglich A, Koch B, Gronwald W, Hengstenberg W, Brunner E, Kalbitzer HR. 2004. Solution structure of the active-centre mutant I14A of the histidine-containing phosphocarrier protein from *Staphylococcus carnosus*. *Eur. J. Biochem.* 271:4815–24

43. Mozhaev VV, Heremans K, Frank J, Masson P, Balny C. 1996. High pressure effects on protein structure and function. *Proteins Struct. Funct. Genet.* 24:81–91

44. Paliwal A, Asthagiri D, Bossev DP, Paulaitis ME. 2004. Pressure denaturation of staphylococcal nuclease studied by neutron small-angle scattering and molecular simulation. *Biophys. J.* 87:3479–92

45. Panick G, Malessa R, Winter R, Rapp G, Frye KJ, Royer CA. 1998. Structural characterization of the pressure-denatured state and unfolding/refolding kinetics of staphylococcal nuclease by synchrotron small-angle X-ray scattering and Fourier-transform infrared spectroscopy. *J. Mol. Biol.* 275:389–402

46. Pudney CR, McGrory T, Lafite P, Pang J, Hay S, et al. 2009. Parallel pathways and free-energy landscapes for enzymatic hydride transfer probed by hydrostatic pressure. *Chembiochem* 10:1379–84

47. Rashin AA, Iofin M, Honig B. 1986. Internal cavities and buried waters in globular proteins. *Biochemistry* 25:3619–25

24. Discusses the thermodynamics of burying water in hydrophobic environments at high pressures.

30. The first practical use of high-pressure cryocooling.

40. Summarizes the state of high-pressure functional studies on these important proteins.

48. Refaee M, Tezuka T, Akasaka K, Williamson MP. 2003. Pressure-dependent changes in the solution structure of hen egg-white lysozyme. *J. Mol. Biol.* 327:857–65

49. Schmid G, Ludemann HD, Jaenicke R. 1975. High pressure effects on the activity of glycolytic enzymes. *Biophys. Chem.* 3:90–98

50. Somero GN. 1992. Adaptations to high hydrostatic pressure. *Annu. Rev. Physiol.* 54:557–77

51. Thomanek UF, Parak F, Mössbauer RL, Formanek H, Schwager P, Hoppe W. 1973. Freezing of myoglobin crystals at high pressure. *Acta Crystallogr. A* 29:263–65

52. Unno M, Ishimori K, Ishimura Y, Morishima I. 1994. High-pressure flash photolysis study of hemoprotein: effects of substrate analogues on the recombination of carbon monoxide to cytochrome P450CAM. *Biochemistry* 33:9762–68

53. Urayama P, Phillips GN, Gruner SM. 2002. Probing substates in sperm whale myoglobin using high-pressure crystallography. *Structure* 10:51–60

54. Varo G, Lanyi JK. 1995. Effects of hydrostatic pressure on the kinetics reveal a volume increase during the bacteriorhodopsin photocycle. *Biochemistry* 34:12161–69

55. Williams MA, Goodfellow JM, Thornton JM. 1994. Buried waters and internal cavities in monomeric proteins. *Protein Sci.* 3:1224–35

56. Williamson MP, Akasaka K, Refaee M. 2003. The solution structure of bovine pancreatic trypsin inhibitor at high pressure. *Protein Sci.* 12:1971–79

57. Woenckhaus J, Kohling R, Thiyagarajan P, Littrell KC, Seifert S, Royer CA, R. 2001. Pressure-jump small-angle x-ray scattering detected kinetics of staphylococcal nuclease folding. *Biophys. J.* 80:1518–23

58. Wolynes PG. 2005. Energy landscapes and solved protein-folding problems. *Philos. Trans. A Math. Phys. Eng. Sci.* 363:453–67

59. Xu J, Baase WA, Baldwin E, Matthews BW. 1998. The response of T4 lysozyme to large-to-small substitutions within the core and its relation to the hydrophobic effect. *Protein Sci.* 7:158–77

Nucleosome Structure(s) and Stability: Variations on a Theme

Andrew J. Andrews and Karolin Luger

Howard Hughes Medical Institute and Department of Biochemistry and Molecular Biology, Colorado State University, Fort Collins, Colorado 80523-1870; email: kluger@lamar.colostate.edu

Annu. Rev. Biophys. 2011. 40:99–117

First published online as a Review in Advance on February 14, 2011

The *Annual Review of Biophysics* is online at biophys.annualreviews.org

This article's doi:
10.1146/annurev-biophys-042910-155329

Copyright © 2011 by Annual Reviews.
All rights reserved

1936-122X/11/0609-099$20.00

Keywords

chromatin, nucleosome assembly and disassembly, single-molecule approaches, histone chaperone, kinetics, thermodynamics

Abstract

Chromatin is a highly regulated, modular nucleoprotein complex that is central to many processes in eukaryotes. The organization of DNA into nucleosomes and higher-order structures has profound implications for DNA accessibility. Alternative structural states of the nucleosome, and the thermodynamic parameters governing its assembly and disassembly, need to be considered in order to understand how access to nucleosomal DNA is regulated. In this review, we provide a brief historical account of how the overriding perception regarding aspects of nucleosome structure has changed over the past thirty years. We discuss recent technical advances regarding nucleosome structure and its physical characterization and review the evidence for alternative nucleosome conformations and their implications for nucleosome and chromatin dynamics.

Contents

INTRODUCTION

The DNA of a single mammalian cell fits into the confines of the nucleus by a hierarchical scheme of folding and compaction into chromatin. Nucleosomes, the repeating structural units of chromatin, consist of an octameric histone core (comprising two copies each of H2A, H2B, H3, and H4) around which 147 bp of DNA are wrapped in 1.65 superhelical turns (70). Hundreds of thousands of nucleosomes, together with linker histones, self-organize into higher-order structures of increasing complexity. The compaction of the eukaryotic genome is indeed a seemingly impossible proposition: The nearly 2 m of DNA must be compacted in an ordered manner while still allowing regulated access to the encoded information. At the same time, the DNA has to be protected against physical damage such as breaks and tangles.

In light of the central importance of chromatin architecture, it is not surprising that the cell maintains multiple systems for regulating chromatin structure. In order to understand how DNA accessibility is regulated, we must first understand the thermodynamic and dynamic properties of the nucleosome and how these are affected by DNA sequence, histone variants, and histone posttranslational modifications (PTMs). We must also identify the driving force for nucleosome assembly and disassembly and consider alternative chromatin structures that might be populated during these processes. Here we discuss key aspects of nucleosome structure and recent investigations into nucleosome thermodynamics.

THE QUEST FOR THE NUCLEOSOME STRUCTURE

The quest for the structure of the nucleosome extended over decades (reviewed in References 59 and 87) (**Figure 1**). Modern chromatin history begins with the proposal of the chromatin subunit model in 1973/1974, based on digestion patterns of eukaryotic DNA and electron microscopic visualization of repeating subunits of chromatin in vitro (48, 86, 132). The notion that nucleosomes could perhaps be crystallized was initially met with skepticism. To quote an anonymous reviewer of a paper by Chris Woodcock submitted to *Nature* in 1973: "A eukaryotic chromosome made out of self-assembling 70 Å units, which could perhaps be made to crystallize, would necessitate rewriting our basic textbooks on cytology and genetics! I have never read such a naive paper purporting to be of such fundamental significance. Definitely it should not be published anywhere!" Fortunately, the

---

Histone variant:
nonallelic variant of core histones (mostly of H3 and H2A) with distinct amino acid sequence and distinct expression patterns

PTM:
posttranslational modification

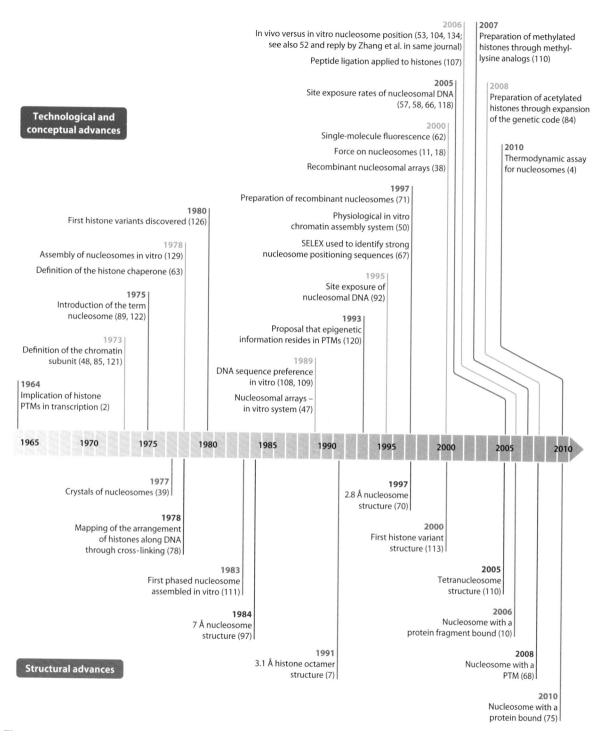

Figure 1

Time line of structural and conceptual advances in our understanding of nucleosome structure and function. Inspired by a similar representation in Reference 87. Abbreviations: PTM, posttranslational modification; SELEX, systematic evolution of ligands by exponential enrichment.

Superhelix: the path
described by the DNA
double helix around
the histone octamer

Histone fold: tertiary
structure motif, first
described in core
histones, consisting of
a long α-helix flanked
by two shorter helices
in a characteristic
U-shaped arrangement

paper was ultimately published, albeit with significant delay (132).

Ten years later, the 7 Å crystal structure by Richmond et al. (97) confirmed previous models in which the DNA was looped around the outside of a disk-shaped histone octamer (49). Although the low resolution made it impossible to correctly assign the histone subunits, important parameters of the nucleosomal DNA supercoil (the superhelix) were derived from this structure and still stand today. In 1991, the structure of the histone octamer without DNA, in the presence of high salt, was determined to 3.1 Å resolution (7). This structure revealed for the first time the intricately intertwined nature of the histone subunits in what was termed the handshake motif of the histone fold. The structure fit perfectly well into the 7 Å nucleosome density that had been assigned to histones, unlike earlier attempts by the same group (125). It was not until 1997, however, that the structure of the nucleosome was determined to a resolution of 2.8 Å (70). The two main obstacles that had to be overcome were increasing the diffrac-

tion limits beyond 7 Å and providing specific sites for heavy-atom attachment, thereby producing sufficient phasing power (71). The ensuing 2.8 Å structure produced unprecedented insight into the intricacies of protein-protein and protein-DNA interactions (**Figure 2**). The conformation of the histone octamer within the nucleosome structure was in excellent agreement with the nonnucleosomal histone octamer structures determined at high ionic strength (7, 22, 131). The later extension of the diffraction limits of nucleosome crystals to 1.9 Å revealed the precise structural parameters of the DNA (28). To date, there are over 40 entries in the Protein Data Bank (PDB) for nucleosome structures. Still missing from the collection of nucleosome structures is a complex of the nucleosome with the linker histone H1, or with any other chromatin architectural protein that affects chromatin higher-order structure.

Another eight years passed after publication of the 2.8 Å nucleosome structure before the first crystallographic model of a tetranucleosome was revealed (102). At 9 Å, this truly

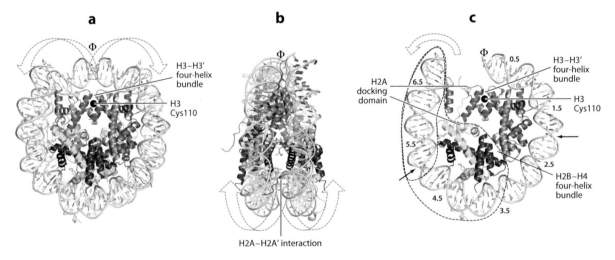

Figure 2

Nucleosome structure. (*a*) The nucleosome viewed down the superhelical axis. Histones H2A, H2B, H3, and H4 are shown in yellow, red, blue, and green, respectively. The dyad axis is indicated (Φ). Arrows indicate possible movements to generate the lexosome (65). (*b*) Viewed as in panel *a* but rotated around the y axis by 90°. (*c*) Viewed as in panel *a*, but only half of the nucleosomal DNA and associated protein is shown. The areas with potential movement to generate an open nucleosome (66) are encircled by dashed magenta lines. The portions of the DNA that are transiently released during site exposure (66) are encircled by dashed gray lines. The location of preferred hot spots for DNA stretching (114) are indicated by arrows; two more exist at the same location on the other half of the DNA. Locations of minor groove–DNA interactions are numbered 0.5–6.5 (70).

impressive structure shows one possible mode of nucleosome-nucleosome interactions, as well as the conformation of linker DNA, although the precise nature of nucleosome-nucleosome interactions still remains speculative. It is clear from a variety of studies that the histone tails contribute to the compaction of chromatin (32, 43, 54, 107). It was recently shown through a combination of mathematical and experimental approaches that several architectures for the 30-nm fiber might coexist in interchangeable forms (44).

NUCLEOSOMES ARE MODULAR MACROMOLECULAR ASSEMBLIES

Many reviews on nucleosome structure have been published (1, 69, 73, 74, 135), and we therefore only briefly review the key features and focus on the aspects of nucleosome structure that are of importance for the ensuing discussion on nucleosome dynamics.

Nucleosome Structure

The four core histones, H2A, H2B, H3, and H4, have two structurally and functionally distinct domains, the tail and fold. The N-terminal ~20–35 residues of each core histone, the histone tail, comprise about 20% of the total amino acids. Histone H2A also has an unstructured C-terminal tail domain. The histone tails are largely unstructured and are not visible in any of the published crystal structures of the nucleosome. About 80–90 residues in each histone comprise the histone fold. Despite barely recognizable sequence similarity, the histone fold regions of the four histones exhibit a high degree of structural homology (73). Together with additional secondary structural elements unique to each histone, the histone fold regions are responsible for mediating interactions with other histones (7) and DNA (28).

Histones are obligate heterodimers in solution. H2A only dimerizes with H2B, and H3 only with H4. When refolded in vitro, and in the absence of interacting factors, H3 and H4 form a $(H3-H4)_2$ tetramer that is best described as a dimer of two H3-H4 dimers in a twisted W-shaped arrangement (**Figure 2a**). This conformation likely exists in vivo (133). H2A-H2B dimers and $(H3-H4)_2$ tetramers do not interact with each other in the absence of DNA under physiological salt conditions, but persist as an octamer composed of two H2A-H2B dimers and one $(H3-H4)_2$ tetramer at 2 M salt (7, 36). In the nucleosome (and in the histone octamer crystallized in the absence of DNA at high salt), each H2A-H2B dimer has two contact points with the $(H3-H4)_2$ tetramer. A four-helix bundle arrangement, similar to that formed by two H3 molecules within the $(H3-H4)_2$ tetramer, connects H2B and H4; a second interaction interface is formed between the H2A docking domain and the other arm of the $(H3-H4)_2$ tetramer (**Figure 2a,c**). Only minor contacts between the two H2A-H2B dimers are observed (**Figure 2b**). Histones exhibit an extremely high degree of conservation between species. Two global mutation studies of all histone residues in yeast have revealed a surprisingly small number (less than 5%) of essential residues in histones (82, 101). Most of these single mutations have no impact on nucleosome structure and likely affect the interaction between nucleosomes or with other nuclear proteins.

Nucleosomal DNA constitutes a tight, 1.65-turn negative superhelix around the histone octamer core. A detailed description of its structure and parameters are given in Reference 96. The 147-bp DNA fragment interacts with histones over its entire length, every time the minor groove faces the histone octamer. Thus, one face of the DNA double helix is heavily constrained by its interaction with protein, while the solvent-facing regions of the DNA are surprisingly mobile even within the constraints of the crystal lattice (70). Compared with other protein-DNA interfaces, histone-DNA interactions are unusual in several important aspects:

■ Not a single base-specific interaction is made between histones and DNA.

FRET: fluorescence
(or Förster) resonance
energy transfer

Nevertheless, sequence dependency for nucleosome positioning has been described (e.g., 53, 134, and references therein).

- The 14 sites of interaction between the individual DNA minor grooves and the histones are independent of each other (**Figure 2c**, numbered 0.5–6.5).
- At each of these sites, at least two (and around the center of the DNA, as many as five) main chain–phosphate interactions (and several side chain-phosphate interactions) are made with both strands of the DNA at the minor groove. Additionally, several salt bridges further contribute to each interaction interface (73).
- An arginine is inserted deep into every one of the minor grooves that face the histone octamer. This is mediated by the negative electrostatic potential of a narrowed minor groove (99, 127).
- Water-mediated hydrogen bonds between protein and DNA are approximately equal in number to the direct hydrogen bonds. The water molecules are thought to provide a means of accommodating intrinsic DNA conformational variation, thus further limiting the sequence dependency of nucleosome positioning while enhancing mobility (28).
- The number of protein-DNA interactions at each of the 14 sites appears to be finely tuned to accommodate intrinsic nucleosome dynamics (see below). The DNA ends, in particular the last ~20 bp, are organized more loosely (as judged from the number of interactions in the crystal structure), whereas the interactions near the center of the DNA are the strongest. Through mechanical unzipping of single molecules of nucleosomal DNA, the smaller number of contacts between DNA and protein is reflected in a significantly lower force required to pull the DNA off the histone octamer surface (46).

Nucleosome Assembly

Given the path of the DNA superhelix (**Figure 2a,b**), it seems unlikely that the histone octamer is deposited onto the DNA as an intact entity. The nucleosome assembles in a sequential manner through the deposition of H3/H4 (either in tetramer or heterodimer form; reviewed in Reference 93) onto DNA, followed by the cooperative addition of two H2A-H2B dimers to organize the peripheral regions of the DNA [**Figure 3**; progressing through states VI → V → (IV or II) → I] (reviewed in References 1 and 98). Disassembly likely follows the reverse pathway. In vitro, the sequential deposition of the histone subcomplexes is facilitated through the use of salt gradients (34, 72, 129) or through recombinant assembly systems pioneered by the Kadonaga laboratory (41); in vivo, histones are safeguarded by a diverse class of proteins, the histone chaperones, and by chromatin assembly factors (1, 37, 90, 93, 98).

EVIDENCE FOR ALTERNATIVE NUCLEOSOME STRUCTURES IN SOLUTION

Nucleosome Crystal Structures: A Biased View of Nucleosomes?

Alternative (or noncanonical) nucleosome structures have been described ever since the nucleosome was first defined structurally (see Reference 135 for a recent review). As discussed below, these range from nucleosomes with partially dissociated DNA ends to split nucleosomes, to nucleosomes with positively (instead of negatively) supercoiled DNA, to nucleosomes with less than a full complement of histones. Technological advances, in particular single-pair fluorescence (or Förster) resonance energy transfer (FRET) approaches, have allowed much progress to be made in identifying transient nucleosomal states (see Reference 19 for a summary on approaches).

Ever since X-ray crystallography has been used to determine the crystal structures of

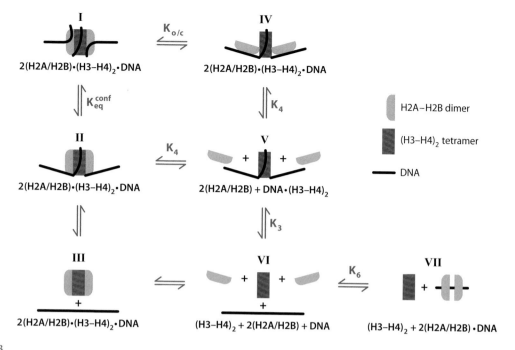

Figure 3

Pathways for nucleosome (dis)assembly, including alternative structural states. The apparent primary pathway for nucleosome assembly is VI → V → (IV or II) → I (17), but several simultaneous pathways are possible. K_3, K_4, and K_6 refer to constants published in Reference 4; $K_{o/c}$ was first described in Reference 17. K_{eq}^{conf} is the equilibrium constant described for transient DNA exposure (128). State VII represents nonnucleosomal H2A/H2B-DNA interactions (4). State III has not been thermodynamically measured in the presence of DNA, and therefore no equilibrium constant is shown.

biological macromolecules, it was questioned to what extent these structures reflect the structure in solution. Crystal formation selects for certain structural states that are conducive to crystallization (indeed, in the early days of protein purification, crystallization was often the last purification step), and therefore alternative solution states might be excluded. Furthermore, structures may be altered from the solution state by crystal lattice contacts and the crystal growth conditions. In many cases, this criticism has been laid to rest by comparisons of crystal structures with NMR structures, and so it is generally viewed as a nonissue (reviewed in Reference 124).

For the nucleosome, however, a comparison between the solution state and the crystal structure requires attention for the following reason. As is the case in many crystals of protein-DNA complexes, base-stacking interactions between

the DNA of neighboring nucleosomes are the major driving force for the formation of nucleosome crystals. The strong dependency on these contacts is the likely reason for the observed stringent length requirement for highly diffracting crystals (145–147 bp of DNA). DNA end-stacking also causes the observed phenomena of DNA stretching and twist diffusion (81, 88, 114). The term stretching alludes to the fact that the 10-bp stretch between DNA-protein contacts exhibits enough plasticity (in particular at four locations along the DNA; two of these are indicated by black arrows in **Figure 2c**) to allow the accommodation of 9, 10, or 11 bp (28, 35, 75, 114). Irrespective of this relatively minor variability, every single nucleosome structure in the PDB has tightly organized DNA ends, because crystal lattice formation selects for this state. However, one notable exception not listed in the database is the 9 Å resolution structure of the nucleosome obtained in a fully hydrated form, in which approximately 10%–15% of the DNA has dissociated from the body of the histone octamer (112).

Photoblinking: during single particle FRET studies, the acceptor fluorophore can become trapped in a dark triplet state for a period of time (photoblinking), which prevents energy transfer from occurring and results in an apparent decrease in FRET signal, leading to misinterpretation of kinetic data

What about nucleosome preparations that do not crystallize? For example, nucleosomes containing the histone variant H2A.Bbd instead of H2A organize less than the canonical 147 bp of DNA (9, 33) and have not yet been crystallized. Similarly, nucleosomes reconstituted with an acetylated form of H3K56 exhibit increased breathing of DNA ends (see below; 83) and form crystals that are fragile and diffract poorly. Finally, all nucleosome structures in the PDB are of nucleosomes reconstituted on either of two DNA sequences. Surely, other examples of failed attempts to crystallize nucleosomes exist in many other laboratories. It is possible that in solution these nucleosomes exhibit structural features that are not compatible with crystal formation. It follows that the structure repository represents a distorted selection of nucleosome structures, while alternative, noncrystallizable structures are underrepresented.

Transient Site Exposure of Nucleosomal DNA

Early studies from the Widom laboratory probing restriction endonuclease accessibility of nucleosomal DNA have put forward the concept of transient site exposure of nucleosomal DNA (3, 92; see Reference 94 for theoretical considerations), which posits that the nucleosomal DNA ends partially dissociate from the surface of the histone octamer (**Figure 2c**, **Figure 3**, state II). This concept holds true for nucleosomal arrays under varying degrees of compaction (91). In 2005, the rates of site exposure on mononucleosomes were measured by single-molecule fluorescence (66, 118). These results were later attributed to photoblinking artifacts (57, 117). [Acceptor blinking is the reversible transitions of the acceptor to an inactive state, resulting in an Förster radius of effectively (100).] Under conditions where photoblinking was mostly

excluded, the van Noort laboratory (57) showed that ~3% of the intact nucleosomes showed intervals with reduced FRET efficiency, clearly distinct from blinking, with a lifetime of 10–120 ms. These approaches were further refined to reveal that nucleosomes are unwrapped with an equilibrium constant of ~0.2–0.6 at nucleosome ends and ~0.1 at a location 27 bp inside the nucleosome, where the histones still remain stably associated (58). As pointed out above, certain histone variants, as well as some histone PTMs may form nucleosomes in which the peripheral DNA spends more time in the open state (9, 33, 83).

Evidence for Open Nucleosome States

The idea of a split nucleosome (or lexosome) as an intermediate during polymerase II transcription emerged relatively early, based on electron microscopy studies (119) and on accessibility of a usually buried cysteine in H3 (65) (**Figure 2a**). However, attempts to demonstrate the existence of this split state have not been successful thus far (95).

More recently, a different open conformational state has been proposed based on single-molecule fluorescence experiments in which FRET between several pairs of judiciously placed fluorophores were measured as a function of ionic strength. This state is characterized by partial dissociation of the H2A-H2B dimers from the H3-H4 tetramer near the dyad while the H2A-H2B dimers remain bound to DNA (**Figure 2c**, **Figure 3**, state IV). It was estimated that at least 0.2% of the nucleosomes adopt this open conformation at physiological salt (17). The transition between the open and closed states has a ΔG of 2–4 kcal mol^{-1}. This model provides an alternative interpretation for several results from other laboratories that have probed the accessibility of nucleosomal DNA. For example, single-molecule studies have investigated the conformational changes that occur in nucleosomes when they are pulled apart from the ends (60, 77). These studies show that the transition from a closed nucleosome to linear DNA is defined by two major steps, the first being reversible and highly sensitive to salt concentration. The interpretation was that the breaking of histone-DNA contacts is monitored while the DNA is pulled off the histone octamer. However, these data can also be explained by the proposed reversible disruption of the H2A-H2B/(H3-H4)$_2$ interface, followed by further unwrapping of the DNA and the irreversible dissociation of the H2A-H2B dimer. Importantly, nucleosome assembly was shown to follow the reverse pathway (including the open complex) as disassembly (17).

The above type of open conformation, if it indeed exists within the more physiological context of a nucleosomal array, has dramatic implications for how mechanistic questions are framed and how results are interpreted in the investigation of histone chaperones, polymerases, and chromatin-remodeling enzymes. For example, the open state could allow a histone chaperone or a remodeling enzyme transient access to the tetramer-dimer interface without completely dissociating the H2A-H2B dimer (90). The results also raise the testable hypothesis that histone variants or certain PTMs of histones alter the equilibrium between the canonical nucleosome conformation and the open conformation.

Nucleosomes and Chromatin with Noncanonical Histone Content

Alternative nucleosome structure can also be characterized by a less than full complement of histones. It can be argued that such structures should not be termed nucleosomes. Indeed, there is no shortage of new terminology (see Reference 135). One controversial example for a noncanonical nucleosome is the centromeric H3 variant CenH3. A variety of models for CenH3-containing nucleosomes have been suggested from a range of in vivo and in vitro data obtained with CenH3 from different species (reviewed in References 12 and 31). Nucleosomes containing only one copy each of CenH3, H4, H2A, and H2B have been proposed for *Drosophila* (26). In yeast, nucleosomes containing one (CenpA-H4)$_2$ tetramer,

together with two copies of a nonhistone protein Scm3, are also discussed (79); however, others conclude that Scm3 is a recruitment factor rather than an integral part of the centromeric nucleosome (20, 21). The direction of the supercoil in centromeric nucleosomes has also been controversial (40, 106). In vitro, nucleosomes with a canonical composition of histones can easily be obtained; however, these have resisted crystallization to date.

A second example has been described as an intermediate during RNA polymerase II transcription through nucleosomes. At physiological strength, the passage of the polymerase causes the loss of one H2A-H2B dimer (56). In vivo, this results in histone dimer exchange (reviewed in Reference 61). It is likely that H2A-H2B dimer depletion in vivo is transient and that histone chaperones such as FACT (facilitates chromatin transcription) aid in nucleosome disassembly and subsequent reassembly (130).

As with any multicomponent system, nucleosome (dis)assembly intermediates as well as nonproductive histone-DNA interactions that are not in the assembly pathway (e.g., state VII in **Figure 3**) are also likely to be populated to a certain extent in vivo, unless prevented or recycled by activities such as those performed by histone chaperones (4). We have recently shown that noncanonical interactions between H2A-H2B dimer and DNA are disfavored in the presence of the histone chaperone Nap1 (nucleosome assembly protein 1) in vitro and in vivo. With respect to DNA organization, such noncanonical chromatin may function in transcription regulation and other DNA-related processes (4). It remains to be seen whether other histone chaperones have similar functions (80).

Biological Implications

It is intuitive that any intrinsic modification that might favor open nucleosome states profoundly affects the accessibility of the underlying DNA, as well as the local compaction of nucleosomes into higher-order structures. Given the rapid technological advances especially in the

FACT: facilitates chromatin transcription

Nap1: nucleosome assembly protein 1

application of single-molecule approaches (reviewed in References 19 and 23), we can now test long-held hypotheses regarding the effects of DNA sequence, histone modifications, and histone variants on the population of these alternative nucleosome states. Several laboratories have already begun to investigate DNA sequence effects (42) and the effects of histone acetylation (83); the pace of such investigations is likely to increase in the coming years.

HOW STABLE IS THE NUCLEOSOME?

It is frequently hypothesized that the incorporation of histone variants or the addition of PTM to histones stabilizes or destabilizes the nucleosome. For reasons explained below, these hypotheses could not be rigorously tested until recently. The nucleosome likely assembles through a combination of two pathways: first, by active promotion of the sequential deposition of components, and second, by the (active) prevention of nonproductive interactions (4). It follows that we need to have a quantitative understanding of the thermodynamics of the nucleosome as well as of competing complexes, to fully understand the thermodynamics of nucleosome (dis)assembly. At the same time, we need to obtain quantitative information on the kinetics of nucleosome assembly and the competing processes.

Nucleosome Thermodynamics– Equilibrium Studies

The term nucleosome stability has multiple definitions that are often used interchangeably in the literature. It has been used to describe any of the nonnucleosomal states (i.e., states II–VI; **Figure 3**) that are mechanistically and conceptually distinct. Other reviews have discussed approaches to measure DNA accessibility (116). Here, we focus on the thermodynamics of nucleosome (dis)assembly.

Any experiment that aims to provide thermodynamic data (either relative or absolute) has to be at equilibrium. An obvious and straightforward approach to measure nucleosome stability is the dilution of preformed nucleosomes to monitor the fraction of free versus nucleosomal DNA. However, this reaction is not at equilibrium because the nucleosome cannot be assembled under these highly dilute conditions (see Reference 115 for a detailed discussion).

The energetic differences of nucleosome formation between two different DNA sequences have traditionally been analyzed by salt jump experiments (67, 108). This is done under conditions in which a labeled DNA fragment is competed against a population of random or defined-sequence DNA fragments with limiting concentrations of histone octamer, while the salt concentration is varied from 1 to 0.1 M. The fraction of labeled DNA incorporated relative to the bulk DNA is then used to calculate a $\Delta\Delta G$ for the sequence of interest. This reaction is easily reversible by changing the direction of the salt titration. This approach was instrumental in determining the rules for nucleosome positioning elements that are intrinsic to the DNA sequence in vitro (67), and allowed a comparison to genomic nucleosome position in vivo (105, 134; see Reference 52 and reply by Zhang et al. in the same issue). An inherent limitation to this approach is that the observed effects may not be relevant under physiological conditions because the numbers obtained reflect the thermodynamic sum of the effects of salt on histone-histone interactions, histone-DNA contacts, and DNA wrapping. Additionally, this approach cannot be used to test the effect of histone variants or PTMs on nucleosome stability because the histone octamer cannot be in excess.

The thermodynamic properties of the nucleosome can be measured under physiological ionic strength through the use of the histone chaperone Nap1 from yeast (yNap1) (6). yNap1 destabilizes nonnucleosomal histone-DNA interactions (4) and binds histones tightly (5). Because both yNap1 and histones can be fluorescently labeled with FRET donor-acceptor pairs, yNap1 can be used as a biosensor for histones by monitoring the amount of

Nap1-histone complex during nucleosome assembly (see Reference 6 for a detailed description of the assay). With this experimental setup, the binding constant of the H2A-H2B dimer for the tetrasome can be measured (representing states I and II in **Figure 3**). Nucleosome assembly can be reversed by the addition of excess yNap1, demonstrating that this step of nucleosome assembly is at equilibrium. This assay utilizes the two defining properties of a histone chaperone: (*a*) histone binding and (*b*) the prevention of nonnucleosomal histone-DNA interactions (63).

In a recent application, we compared the thermodynamic constants for nucleosome formation and competing complexes on two different DNA sequences of the same length. We found that the sequence with a higher positioning propensity forms more stable nucleosomes, but also has a weaker affinity for nonnucleosomal histones (4). Using the system developed by Neumann et al. (83, 84), we have also established that the acetylation of H3K56 (a modification implied in nucleosome assembly and DNA repair; 103) affects tetrasome formation but has little effect on nucleosome stability. Applying this and similar approaches to other histone modifications should allow greater insight into how the stability of the nucleosome is regulated.

The Kinetics of Nucleosome (Dis)Assembly

The importance of understanding the kinetics of nucleosome formation has been exemplified for the PHO5 promoter in yeast, where nucleosome disassembly and subsequent reassembly appear to be the rate-limiting step for gene expression (14–16). It has been proposed recently that the differences between global gene expression rates may be due to the cells' ability to (dis/re)assemble chromatin (30).

Given the limitations set by nonnucleosomal histone-DNA interactions detailed above, we are left with two options for studying nucleosome kinetics. (*a*) We can monitor the off-rates of histone complexes from a preformed

HISTONE CHAPERONES

Histone chaperones are a structurally diverse class of nuclear proteins that are required to prevent the basic core histones from making improper interactions with DNA and other macromolecules prior to assembly into nucleosomes. First defined by Laskey et al. (64), histone chaperones have become important tools for in vitro chromatin assembly (41) and for studying nucleosome thermodynamics (6). In vivo, they have roles in transcription, replication, and DNA repair (reviewed in References 27, 29, and 37).

nucleosome by dilution, assuming an insignificant fraction of reassembly. (*b*) We can use histone chaperones to promote nucleosome (dis)assembly. In both cases, one must make certain that the reaction is driven in the desired direction. For salt gradients, this can be achieved by adding excess DNA to limit nucleosome reassembly on labeled DNA. The use of histone chaperones translates into additional interactions that need to be taken into account, resulting in changes in the free concentration of histones during the reaction and a possible change of the rate-limiting step.

Kinetics of nucleosome disassembly. Two pathways for nucleosome disassembly must be considered: (*a*) the dissociation of octamer from DNA (**Figure 3**, state I → II → III) and (*b*) the sequential release of histone subcomplexes [**Figure 3**, state I → (II or IV) → V → VI]. Many earlier studies have inferred that the histone octamer is released as one entity (8, 18, 45, 61); however, a more widely held and recently verified view is that H2A-H2B dimers dissociate before the $(H3-H4)_2$ tetramer (17). Both pathways are an oversimplification because they do not include likely intermediates. For example, in the first scenario the DNA would need to unwrap almost completely before the octamer could be released as an intact entity. In the second scenario, the DNA would have to unwrap partially and/or the octamer would have to split first, before the first H2A/H2B is released (**Figure 3**).

Tetrasome: $(H3-H4)_2$ tetramer-DNA complex

yNap1: yeast Nap1

Assuming that DNA unwrapping is the first step in nucleosome disassembly, we can use measurements of DNA exposure to estimate the fastest possible rate at which the nucleosome will disassociate. However, the experimentally determined rates of nucleosome disassembly, as judged by the appearance of free DNA under conditions where no reassembly can occur, are slow, with a $t_{1/2}$ of ~0.6–4 h (35, 76). The fact that nucleosomes are stable for much longer times at higher concentrations than under dilute conditions suggests that there is reassembly or exchange of histones. Spontaneous H2A/H2B exchange has in fact been observed at low frequency in vivo (30, 51, 55). However, (H3-H4)$_2$ tetramer exchange is much less likely in the absence of a salt gradient or a chaperone.

Kinetics of nucleosome assembly. As is the case with studying nucleosome thermodynamics in equilibrium, nucleosome assembly kinetics can be studied by varying the ionic strength or through the use of histone chaperones. Irrespective of the approach, the primary difficulty in studying nucleosome assembly kinetics is verifying that a nucleosome has formed in solution. This requires either measuring FRET from multiple judiciously chosen FRET pairs or combining other solution-state approaches such as atomic force microscopy, analytical ultracentrifugation, sucrose gradients, and gel electrophoresis. In addition, nucleosomes, like any nucleoprotein complex, are sensitive to their environment (ionic strength, pH, metal ions, and temperature). If nonnucleosomal histone–DNA interactions are not prevented, the apparent rate of nucleosome formation is slow (reviewed in Reference 128).

Salt jump approaches (rapid change from >0.8 M to <0.2 M salt) have been used by many laboratories to study the kinetics of nucleosome assembly (24, 25). Using fluorescently labeled histone H3, Daban & Cantor (24, 25) demonstrated two-phase kinetics (with time constants of 10^{-1} and ~7×10^{-3} s^{-1}). Intriguingly, these rates are not concentration dependent,

suggesting that the histone-DNA interaction is fast and that the rate-limiting steps are unimolecular conformational changes.

In contrast, single-molecule studies on the rate of nucleosome formation found rates that were both fast and concentration dependent. Wagner et al. (123) used polyglutamic acid, RNA, or the histone chaperone Nap1 to measure an overall rate constant for nucleosome formation that approaches diffusion [~10^6–10^7 (s mol/liter)$^{-1}$]. Nap1-mediated assembly provided data that could be separated into multiple kinetic constants. It was concluded that the rate-limiting step in Nap1-mediated nucleosome assembly is tetrasome formation. Importantly, both studies use histone and Nap1 concentrations well above their equilibrium dissociation constants, further complicating analysis. The discrepancies between the various studies highlight the limitations of each method outlined above.

Thermodynamics and Kinetics of Nucleosome Formation: A Close Relationship

Although we are still far from a complete understanding of the thermodynamic properties governing nucleosome assembly and disassembly, experiments performed to date have provided considerable insight into these processes. However, simply measuring the thermodynamic parameters of the nucleosome under equilibrium conditions might result in a skewed view of nucleosome stability. For example, it is likely that nucleosome formation is intrinsically fast but is slowed significantly by the preponderance of nonnucleosomal interactions. This suggests that nucleosome assembly is essentially prevented by a kinetic trap (108, 128). Once formed, nucleosomes may also be kinetically trapped because of the sequential disassembly pathway. For example, if the rewrapping rates of DNA around the nucleosome were much faster than the off-rate of H2A/H2B or octamer from DNA, then the nucleosome would effectively be trapped in the folded state.

OUTLOOK

Almost 15 years have passed since the publication of the high-resolution crystal structure of the nucleosome. As the structure continues to grace textbook covers, T-shirts, and mouse pads, the picture that is engrained in our minds is that of a static and unchangeable assembly that exists in one state only. Recent technological advances, in single-molecule approaches and in studying the thermodynamic properties of the nucleosome at physiological conditions using histone chaperones, have resulted in a much more nuanced view of this ubiquitous structure. We may now begin to test numerous hypotheses regarding the effect of DNA sequence, PTMs of histones, and histone variant incorporation on the various structural states of the nucleosome.

Ultimately, the nucleosome needs to be viewed in its physiological context—that is, in the structural context of a reasonably compacted nucleosomal array, in the presence of linker histones and other chromatin architectural proteins, and in the presence of a multitude of factors that act upon them. Thus, the approaches described above must be refined to allow further investigations into the effects of chromatin architectural proteins (e.g., linker histone H1) on structural transitions of the nucleosome, as well as to study the nucleosome in the more natural context of a nucleosomal array of increasing complexity. Important advances in dealing with these difficult substrates have already been made (13, 91), and further progress is inevitable.

A multitude of cellular factors are devoted to modifying, remodeling, sliding, assembling, disassembling, and transcribing around the nucleosome. The studies summarized in this review lay the necessary groundwork toward understanding the intricate and diverse mechanisms by which these factors act upon chromatin.

SUMMARY POINTS

1. Histones avoid base-specific interactions with the DNA. The sequence-encoded propensity of DNA to assume superhelical conformation likely provides the positioning signals observed in vitro and in vivo.

2. The nucleosome exists in a variety of alternative states of varying compactness. The available crystal structures of the nucleosome likely represent the most compact state.

3. Nucleosome assembly and disassembly occur in a stepwise manner through the sequential deposition of histone subcomplexes.

4. Nucleosome assembly depends in part on avoiding nonnucleosomal interactions between its components; this is accomplished by the use of salt gradients (in vitro) and histone chaperones (in vivo).

5. In quantitative studies of nucleosome assembly and disassembly, the formation of nonnucleosomal complexes (i.e., between histones and DNA or between histones and histone chaperones and assembly factors) must also be considered.

FUTURE ISSUES

1. To what extent do DNA sequence, histone variants, and PTMs alter the distribution between canonical and alternative states, and how do they affect the thermodynamics of the nucleosome?

2. How are alternative nucleosome states affected by the presence of other nuclear factors, in particular by chromatin architectural proteins such as linker histones?

3. Are these alternative structures significantly populated in the context of nucleosomal arrays and can we devise techniques to observe them in vivo?

DISCLOSURE STATEMENT

The authors are not aware of any affiliations, memberships, funding, or financial holdings that might be perceived as affecting the objectivity of this review.

ACKNOWLEDGMENTS

KL and AJA are supported by the Howard Hughes Medical Institute. Support from the NIH (GM R01GM061909 to KL, F32GM083532 to AJA is gratefully acknowledged. We thank Uma M. Muthurajan for help with **Figure 2** and, together with Duane Winkler and Aaron Hieb for comments and discussion. We thank Kitty Brown for a critical reading of the manuscript.

LITERATURE CITED

1. Akey CW, Luger K. 2003. Histone chaperones and nucleosome assembly. *Curr. Opin. Struct. Biol.* 13:6–14

2. Allfrey VG, Mirsky AE. 1964. Structural modifications of histones and their possible role in the regulation of RNA synthesis. *Science* 144:559

3. Anderson JD, Thastrom A, Widom J. 2002. Spontaneous access of proteins to buried nucleosomal DNA target sites occurs via a mechanism that is distinct from nucleosome translocation. *Mol. Cell. Biol.* 22:7147–57

4. Andrews AJ, Chen X, Zevin A, Stargell LA, Luger K. 2010. The histone chaperone Nap1 promotes nucleosome assembly by eliminating nonnucleosomal histone DNA interactions. *Mol. Cell* 37:834–42

5. Andrews AJ, Downing G, Brown K, Park YJ, Luger K. 2008. A thermodynamic model for Nap1-histone interactions. *J. Biol. Chem.* 283:32412–18

6. Andrews AJ, Luger K. 2011. A coupled equilibrium approach to study nucleosome thermodynamics. *Methods Enzymol.* 488:265–85

7. Arents G, Burlingame RW, Wang BC, Love WE, Moudrianakis EN. 1991. The nucleosomal core histone octamer at 3.1 Å resolution: a tripartite protein assembly and a left-handed superhelix. *Proc. Natl. Acad. Sci. USA* 88:10148–52

8. Ausio J, Seger D, Eisenberg H. 1984. Nucleosome core particle stability and conformational change. Effect of temperature, particle and NaCl concentrations, and crosslinking of histone H3 sulfhydryl groups. *J. Mol. Biol.* 176:77–104

9. Bao Y, Konesky K, Park YJ, Rosu S, Dyer PN, et al. 2004. Nucleosomes containing the histone variant H2A.Bbd organize only 118 base pairs of DNA. *EMBO J.* 23:3314–24

10. Barbera AJ, Chodaparambil JV, Kelley-Clarke B, Joukov V, Walter JC, et al. 2006. The nucleosomal surface as a docking station for Kaposi's sarcoma herpesvirus LANA. *Science* 5762:856–61

11. Bennink ML, Leuba SH, Leno GH, Zlatanova J, de Grooth BG, Greve J. 2001. Unfolding individual nucleosomes by stretching single chromatin fibers with optical tweezers. *Nat. Struct. Biol.* 8:606–10

12. Black BE, Bassett EA. 2008. The histone variant CENP-A and centromere specification. *Curr. Opin. Cell Biol.* 20:91–100

13. Blacketer MJ, Feely SJ, Shogren-Knaak MA. 2010. Nucleosome interactions and stability in an ordered nucleosome array model system. *J. Biol. Chem.* 285:34597–607

14. Boeger H, Griesenbeck J, Kornberg RD. 2008. Nucleosome retention and the stochastic nature of promoter chromatin remodeling for transcription. *Cell* 133:716–26

15. Boeger H, Griesenbeck J, Strattan JS, Kornberg RD. 2003. Nucleosomes unfold completely at a transcriptionally active promoter. *Mol. Cell* 11:1587–98

16. Boeger H, Griesenbeck J, Strattan JS, Kornberg RD. 2004. Removal of promoter nucleosomes by disassembly rather than sliding in vivo. *Mol. Cell* 14:667–73

17. Boehm V, Hieb AR, Andrews AJ, Gansen A, Rocker A, et al. 2010. Nucleosomal DNA accessibility governed by the dimer/tetramer interface. *Proc. Natl. Acad. Sci. USA* In press

18. Brower-Toland BD, Smith CL, Yeh RC, Lis JT, Peterson CL, Wang MD. 2002. Mechanical disruption of individual nucleosomes reveals a reversible multistage release of DNA. *Proc. Natl. Acad. Sci. USA* 99:1960–65

19. Buning R, Van Noort J. 2010. Single-pair FRET experiments on nucleosome conformational dynamics. *Biochimie* 92:1729–40

20. Camahort R, Li B, Florens L, Swanson SK, Washburn MP, Gerton JL. 2007. Scm3 is essential to recruit the histone h3 variant cse4 to centromeres and to maintain a functional kinetochore. *Mol. Cell* 26:853–65

21. Camahort R, Shivaraju M, Mattingly M, Li B, Nakanishi S, et al. 2009. Cse4 is part of an octameric nucleosome in budding yeast. *Mol. Cell* 35:794–805

22. Chantalat L, Nicholson JM, Lambert SJ, Reid AJ, Donovan MJ, et al. 2003. Structure of the histone-core octamer in KCl/phosphate crystals at 2.15 Å resolution. *Acta Crystallogr. D* 59:1395–407

23. Chien FT, van Noort J. 2009. 10 years of tension on chromatin: results from single molecule force spectroscopy. *Curr. Pharm. Biotechnol.* 10:474–85

24. Daban JR, Cantor CR. 1982. Role of histone pairs H2A,H2B and H3,H4 in the self-assembly of nucleosome core particles. *J. Mol. Biol.* 156:771–89

25. Daban JR, Cantor CR. 1982. Structural and kinetic study of self-assembly of nucleosome core particles. *J. Mol. Biol.* 156:749–69

26. Dalal Y, Wang H, Lindsay S, Henikoff S. 2007. Tetrameric structure of centromeric nucleosomes in interphase *Drosophila* cells. *PLoS Biol.* 5:e218

27. Das C, Tyler JK, Churchill ME. 2010. The histone shuffle: histone chaperones in an energetic dance. *Trends Biochem. Sci.* 35:476–89

28. Davey CA, Sargent DF, Luger K, Maeder AW, Richmond TJ. 2002. Solvent mediated interactions in the structure of the nucleosome core particle at 1.9 Å resolution. *J. Mol. Biol.* 319:1097–113

29. De Koning L, Corpet A, Haber JE, Almouzni G. 2007. Histone chaperones: an escort network regulating histone traffic. *Nat. Struct. Mol. Biol.* 14:997–1007

30. Deal RB, Henikoff JG, Henikoff S. 2010. Genome-wide kinetics of nucleosome turnover determined by metabolic labeling of histones. *Science* 328:1161–64

31. Dechassa ML, D'Arcy S, Luger K. 2009. A positive spin on the centromere. *Cell* 138:22–24

32. Dorigo B, Schalch T, Bystricky K, Richmond TJ. 2003. Chromatin fiber folding: requirement for the histone H4 N-terminal tail. *J. Mol. Biol.* 327:85–96

33. Doyen CM, Montel F, Gautier T, Menoni H, Claudet C, et al. 2006. Dissection of the unusual structural and functional properties of the variant H2A.Bbd nucleosome. *EMBO J.* 25:4234–44

34. Dyer PN, Edayathumangalam RS, White CL, Bao Y, Chakravarthy S, et al. 2004. Reconstitution of nucleosome core particles from recombinant histones and DNA. *Methods Enzymol.* 375:23–44

35. Edayathumangalam RS, Weyermann P, Dervan PB, Gottesfeld JM, Luger K. 2005. Nucleosomes in solution exist as a mixture of twist-defect states. *J. Mol. Biol.* 345:103–14

36. Eickbush TH, Moudrianakis EN. 1978. The histone core complex: an octamer assembled by two sets of protein-protein interactions. *Biochemistry* 17:4955–64

37. Eitoku M, Sato L, Senda T, Horikoshi M. 2008. Histone chaperones: 30 years from isolation to elucidation of the mechanisms of nucleosome assembly and disassembly. *Cell Mol. Life Sci.* 65:414–44

38. Fan JY, Gordon F, Luger K, Hansen JC, Tremethick DJ. 2002. The essential histone variant H2A.Z regulates the equilibrium between different chromatin conformational states. *Nat. Struct. Biol.* 19:172–76

39. Finch JT, Lutter LC, Rhodes DS, Brown RS, Rushton B, et al. 1977. Structure of the nucleosome core particles of chromatin. *Nature* 269:29–36

40. Furuyama T, Henikoff S. 2009. Centromeric nucleosomes induce positive DNA supercoils. *Cell* 138:104–13

41. Fyodorov DV, Kadonaga JT. 2003. Chromatin assembly in vitro with purified recombinant ACF and NAP-1. *Methods Enzymol.* 371:499–515

42. Gansen A, Toth K, Schwarz N, Langowski J. 2009. Structural variability of nucleosomes detected by single-pair Förster resonance energy transfer: histone acetylation, sequence variation, and salt effects. *J. Phys. Chem. B* 113:2604–13

43. Gordon F, Luger K, Hansen JC. 2005. The core histone N-terminal tail domains function independently and additively during salt-dependent oligomerization of nucleosomal arrays. *J. Biol. Chem.* 280:33701–6

44. Grigoryev SA, Arya G, Correll S, Woodcock CL, Schlick T. 2009. Evidence for heteromorphic chromatin fibers from analysis of nucleosome interactions. *Proc. Natl. Acad. Sci. USA* 106:13317–22

45. Hagerman TA, Fu Q, Molinie B, Denvir J, Lindsay S, Georgel PT. 2009. Chromatin stability at low concentration depends on histone octamer saturation levels. *Biophys. J.* 96:1944–51

46. Hall MA, Shundrovsky A, Bai L, Fulbright RM, Lis JT, Wang MD. 2009. High-resolution dynamic mapping of histone-DNA interactions in a nucleosome. *Nat. Struct. Mol. Biol.* 16:124–29

47. Hansen JC, Ausio J, Stanik VH, van Holde KE. 1989. Homogeneous reconstituted oligonucleosomes, evidence for salt-dependent folding in the absence of histone H1. *Biochemistry* 28:9129–36

48. Hewish DR, Burgoyne LA. 1973. Chromatin sub-structure. The digestion of chromatin DNA at regularly spaced sites by a nuclear deoxyribonuclease. *Biochem. Biophys. Res. Commun.* 52:504–10

49. Hjelm RP, Kneale GG, Sauau P, Baldwin JP, Bradbury EM, Ibel K. 1977. Small angle neutron scattering studies of chromatin subunits in solution. *Cell* 10:139–51

50. Ito T, Bulger M, Pazin MJ, Kobayashi R, Kadonaga JT. 1997. ACF, an ISWI-containing and ATP-utilizing chromatin assembly and remodeling factor. *Cell* 90:145–55

51. Jamai A, Imoberdorf RM, Strubin M. 2007. Continuous histone H2B and transcription-dependent histone H3 exchange in yeast cells outside of replication. *Mol. Cell* 25:345–55

52. Kaplan N, Moore I, Fondufe-Mittendorf Y, Gossett AJ, Tillo D, et al. 2010. Nucleosome sequence preferences influence in vivo nucleosome organization. *Nat. Struct. Mol. Biol.* 17:918–20

53. Kaplan N, Moore IK, Fondufe-Mittendorf Y, Gossett AJ, Tillo D, et al. 2008. The DNA-encoded nucleosome organization of a eukaryotic genome. *Nature* 458:362–66

54. Kato H, Gruschus J, Ghirlando R, Tjandra N, Bai Y. 2009. Characterization of the N-terminal tail domain of histone H3 in condensed nucleosome arrays by hydrogen exchange and NMR. *J. Am. Chem. Soc.* 131:15104–5

55. Kimura H. 2005. Histone dynamics in living cells revealed by photobleaching. *DNA Repair (Amst.)* 28:939–50

56. Kireeva ML, Walter W, Tchernajenko V, Bondarenko V, Kashlev M, Studitsky VM. 2002. Nucleosome remodeling induced by RNA polymerase II. Loss of the H2A/H2B dimer during transcription. *Mol. Cell* 9:541–52

57. Koopmans WJ, Brehm A, Logie C, Schmidt T, van Noort J. 2007. Single-pair FRET microscopy reveals mononucleosome dynamics. *J. Fluoresc.* 17:785–95

58. Koopmans WJ, Buning R, Schmidt T, van Noort J. 2009. spFRET using alternating excitation and FCS reveals progressive DNA unwrapping in nucleosomes. *Biophys. J.* 97:195–204

59. Kornberg RD, Lorch Y. 1999. Twenty-five years of the nucleosome, fundamental particle of the eukaryote chromosome. *Cell* 98:285–94

60. Kruithof M, van Noort J. 2009. Hidden Markov analysis of nucleosome unwrapping under force. *Biophys. J.* 96:3708–15

61. Kulaeva OI, Gaykalova DA, Studitsky VM. 2007. Transcription through chromatin by RNA polymerase II: histone displacement and exchange. *Mutat. Res.* 618:116–29

62. Ladoux B, Quivy JP, Doyle P, du Roure O, Almouzni G, Viovy JL. 2000. Fast kinetics of chromatin assembly revealed by single-molecule videomicroscopy and scanning force microscopy. *Proc. Natl. Acad. Sci. USA* 97:14251–56

63. Laskey RA, Honda BM, Mills AD, Finch JT. 1978. Nucleosomes are assembled by an acidic protein which binds histones and transfers them to DNA. *Nature* 275:416–20

64. Laskey RA, Mills AD, Morris NR. 1977. Assembly of SV40 chromatin in a cell-free system from *Xenopus* eggs. *Cell* 10:237–43

65. Lee MS, Garrard WT. 1991. Transcription-induced nucleosome 'splitting': an underlying structure for DNase I sensitive chromatin. *EMBO J.* 10:607–15

66. Li G, Levitus M, Bustamante C, Widom J. 2005. Rapid spontaneous accessibility of nucleosomal DNA. *Nat. Struct. Mol. Biol.* 12:46–53

67. Lowary PT, Widom J. 1998. New DNA sequence rules for high affinity binding to histone octamer and sequence-directed nucleosome positioning. *J. Mol. Biol.* 276:19–42

68. Lu X, Simon MD, Chodaparambil JV, Hansen JC, Shokat KM, Luger K. 2008. The effect of H3K79 dimethylation and H4K20 trimethylation on nucleosome and chromatin structure. *Nat. Struct. Mol. Biol.* 15:1122–24

69. Luger K, Hansen JC. 2005. Nucleosome and chromatin fiber dynamics. *Curr. Opin. Struct. Biol.* 15:188–96

70. Luger K, Maeder AW, Richmond RK, Sargent DF, Richmond TJ. 1997. Crystal structure of the nucleosome core particle at 2.8 Å resolution. *Nature* 389:251–59

71. Luger K, Rechsteiner TJ, Flaus AJ, Waye MM, Richmond TJ. 1997. Characterization of nucleosome core particles containing histone proteins made in bacteria. *J. Mol. Biol.* 272:301–11

72. Luger K, Rechsteiner TJ, Richmond TJ. 1999. Preparation of nucleosome core particle from recombinant histones. *Methods Enzymol.* 304:3–19

73. Luger K, Richmond TJ. 1998. DNA binding within the nucleosome core. *Curr. Opin. Struct. Biol.* 8:33–40

74. Luger K, Richmond TJ. 1998. The histone tails of the nucleosome. *Curr. Opin. Genet. Dev.* 8:140–46

75. Makde RD, England JR, Yennawar HP, Tan S. 2010. Structure of RCC1 chromatin factor bound to the nucleosome core particle. *Nature* 467:562–66

76. Mazurkiewicz J, Kepert JF, Rippe K. 2006. On the mechanism of nucleosome assembly by histone chaperone NAP1. *J. Biol. Chem.* 281:16462–72

77. Mihardja S, Spakowitz AJ, Zhang Y, Bustamante C. 2006. Effect of force on mononucleosomal dynamics. *Proc. Natl. Acad. Sci. USA* 103:15871–76

78. Mirzabekov AD, Shick VV, Belyavsky AV, Karpov VL, Bavykin SG. 1978. The structure of nucleosomes: the arrangement of histones in the DNA grooves and along the DNA chain. *Cold Spring Harb. Symp. Quant. Biol.* 42(Pt. 1):149–55

79. Mizuguchi G, Xiao H, Wisniewski J, Smith MM, Wu C. 2007. Nonhistone Scm3 and histones CenH3-H4 assemble the core of centromere-specific nucleosomes. *Cell* 129:1153–64

80. Morillo-Huesca M, Maya D, Munoz-Centeno MC, Singh RK, Oreal V, et al. 2010. FACT prevents the accumulation of free histones evicted from transcribed chromatin and a subsequent cell cycle delay in G1. *PLoS Genet.* 6:e1000964

81. Muthurajan UM, Bao Y, Forsberg LJ, Edayathumangalam RS, Dyer PN, et al. 2004. Crystal structures of histone Sin mutant nucleosomes reveal altered protein-DNA interactions. *EMBO J.* 23:260–71

82. Nakanishi S, Sanderson BW, Delventhal KM, Bradford WD, Staehling-Hampton K, Shilatifard A. 2008. A comprehensive library of histone mutants identifies nucleosomal residues required for H3K4 methylation. *Nat. Struct. Mol. Biol.* 15:881–88

83. Neumann H, Hancock SM, Buning R, Routh A, Chapman L, et al. 2009. A method for genetically installing site-specific acetylation in recombinant histones defines the effects of h3 k56 acetylation. *Mol. Cell* 36:153–63

84. Neumann H, Peak-Chew SY, Chin JW. 2008. Genetically encoding N(epsilon)-acetyllysine in recombinant proteins. *Nat. Chem. Biol.* 4:232–34

85. Noll M. 1974. Internal structure of the chromatin subunit. *Nucleic Acids Res.* 1:1573–78

86. Olins AL, Olins DE. 1974. Spheroid chromatin units (v bodies). *Science* 183:330–32

87. Olins DE, Olins AL. 2003. Chromatin history: our view from the bridge. *Nat. Rev. Mol. Cell Biol.* 4:809–14

88. Ong MS, Richmond TJ, Davey CA. 2007. DNA stretching and extreme kinking in the nucleosome core. *J. Mol. Biol.* 368:1067–74

89. Oudet P, Gross-Bellard M, Chambon P. 1975. Electron microscopic and biochemical evidence that chromatin structure is a repeating unit. *Cell* 4:281–300

90. Park YJ, Luger K. 2008. Histone chaperones in nucleosome eviction and histone exchange. *Curr. Opin. Struct. Biol.* 18:282–89

91. Poirier MG, Bussiek M, Langowski J, Widom J. 2008. Spontaneous access to DNA target sites in folded chromatin fibers. *J. Mol. Biol.* 379:772–86

92. Polach KJ, Widom J. 1995. Mechanism of protein access to specific DNA sequences in chromatin: a dynamic equilibrium model for gene regulation. *J. Mol. Biol.* 254:130–49

93. Polo SE, Almouzni G. 2006. Chromatin assembly: a basic recipe with various flavours. *Curr. Opin. Genet. Dev.* 16:104–11

94. Prinsen P, Schiessel H. 2010. Nucleosome stability and accessibility of its DNA to proteins. *Biochimie* 92:1722–28

95. Protacio RU, Widom J. 1996. Nucleosome transcription studied in a real-time synchronous system: test of the lexosome model and direct measurement of effects due to histone octamer. *J. Mol. Biol.* 256:458–72

96. Richmond TJ, Davey CA. 2003. The structure of DNA in the nucleosome core. *Nature* 423:145–50

97. Richmond TJ, Finch JT, Rushton B, Rhodes D, Klug A. 1984. Structure of the nucleosome core particle at 7 Å resolution. *Nature* 311:532–37

98. Rocha W, Verreault A. 2008. Clothing up DNA for all seasons: histone chaperones and nucleosome assembly pathways. *FEBS Lett.* 582:1938–49

99. Rohs R, West SM, Sosinsky A, Liu P, Mann RS, Honig B. 2009. The role of DNA shape in protein-DNA recognition. *Nature* 461:1248–53

100. Sabanayagam CR, Eid JS, Meller A. 2005. Long time scale blinking kinetics of cyanine fluorophores conjugated to DNA and its effect on Förster resonance energy transfer. *J. Chem. Phys.* 123(22):224708

101. Sakamoto M, Noguchi S, Kawashima S, Okada Y, Enomoto T, et al. 2009. Global analysis of mutual interaction surfaces of nucleosomes with comprehensive point mutants. *Genes Cells* 14:1271–330

102. Schalch T, Duda S, Sargent DF, Richmond TJ. 2005. X-ray structure of a tetranucleosome and its implications for the chromatin fibre. *Nature* 436:138–41

103. Schneider J, Bajwa P, Johnson FC, Bhaumik SR, Shilatifard A. 2006. Rtt109 is required for proper H3K56 acetylation: a chromatin mark associated with the elongating RNA polymerase II. *J. Biol. Chem.* 281:37270–74

104. Segal E, Fondufe-Mittendorf Y, Chen L, Thastrom A, Field Y, et al. 2006. A genomic code for nucleosome positioning. *Nature* 442:772–78

105. Segal E, Widom J. 2009. From DNA sequence to transcriptional behaviour: a quantitative approach. *Nat. Rev. Genet.* 10:443–56

106. Sekulic N, Bassett EA, Rogers DJ, Black BE. 2010. The structure of (CENP-A-H4)(2) reveals physical features that mark centromeres. *Nature* 467:347–51

107. Shogren-Knaak M, Ishii H, Sun JM, Pazin MJ, Davie JR, Peterson CL. 2006. Histone H4-K16 acetylation controls chromatin structure and protein interactions. *Science* 311:844–47

108. Shrader TE, Crothers DM. 1989. Artificial nucleosome positioning sequences. *Proc. Natl. Acad. Sci. USA* 86:7418–22

109. Shrader TE, Crothers DM. 1990. Effects of DNA sequence and histone-histone interactions on nucleosome placement. *J. Mol. Biol.* 216:69–84

110. Simon MD, Chu F, Racki LR, de la Cruz CC, Burlingame AL, et al. 2007. The site-specific installation of methyl-lysine analogs into recombinant histones. *Cell* 128:1003–12

111. Simpson RT, Stafford DW. 1983. Structural features of a phased nucleosome core particle. *Proc. Natl. Acad. Sci. USA* 80:51–55

112. Struck MM, Klug A, Richmond TJ. 1992. Comparison of X-ray structures of the nucleosome core particle in two different hydration states. *J. Mol. Biol.* 224:253–64

113. Suto RK, Clarkson MJ, Tremethick DJ, Luger K. 2000. Crystal structure of a nucleosome core particle containing the variant histone H2A.Z. *Nat. Struct. Biol.* 7:1121–24

114. Suto RK, Edayathumangalam RS, White CL, Melander C, Gottesfeld JM, et al. 2003. Crystal structures of nucleosome core particles in complex with minor groove DNA-binding ligands. *J. Mol. Biol.* 326:371–80

115. Thastrom A, Gottesfeld JM, Luger K, Widom J. 2004. Histone-DNA binding free energy cannot be measured in dilution-driven dissociation experiments. *Biochemistry* 43:736–41

116. Tims HS, Widom J. 2007. Stopped-flow fluorescence resonance energy transfer for analysis of nucleosome dynamics. *Methods* 41:296–303

117. Tomschik M, van Holde K, Zlatanova J. 2009. Nucleosome dynamics as studied by single-pair fluorescence resonance energy transfer: a reevaluation. *J. Fluoresc.* 19:53–62

118. Tomschik M, Zheng H, van Holde K, Zlatanova J, Leuba SH. 2005. Fast, long-range, reversible conformational fluctuations in nucleosomes revealed by single-pair fluorescence resonance energy transfer. *Proc. Natl. Acad. Sci. USA* 102:3278–83

119. Tsanev R, Petrov P. 1976. The substructure of chromatin and its variations as revealed by electron microscopy. *J. Microsc. Biol. Cell* 27:11–19

120. Turner BM. 1993. Decoding the nucleosome. *Cell* 75:5–8

121. Van Holde KE, Sahasrabuddhe CG, Shaw BR. 1974. A model for particulate structure in chromatin. *Nucleic Acids Res.* 1:1579–86

122. Varshavsky AJ, Georgiev GP. 1975. Studies on chromatin. V. A model for the structure of chromatin subunit. *Mol. Biol. Rep.* 2:255–62

123. Wagner G, Bancaud A, Quivy JP, Clapier C, Almouzni G, Viovy JL. 2005. Compaction kinetics on single DNAs: purified nucleosome reconstitution systems versus crude extract. *Biophys. J.* 89:3647–59

124. Wagner G, Hyberts SG, Havel TF. 1992. NMR structure determination in solution: a critique and comparison with X-ray crystallography. *Annu. Rev. Biophys. Biomol. Struct.* 21:167–98

125. Wang BC, Rose J, Arents G, Moudrianakis EN. 1994. The octameric histone core of the nucleosome. Structural issues resolved. *J. Mol. Biol.* 236:179–88

126. West MH, Bonner WM. 1980. Histone 2A, a heteromorphous family of eight protein species. *Biochemistry* 19:3238–45

127. West SM, Rohs R, Mann RS, Honig B. 2010. Electrostatic interactions between arginines and the minor groove in the nucleosome. *J. Biomol. Struct. Dyn.* 27:861–66

128. Widom J. 1999. Equilibrium and dynamic nucleosome stability. *Methods Mol. Biol.* 119:61–77

129. Wilhelm FX, Wilhelm ML, Erard M, Duane MP. 1978. Reconstitution of chromatin: assembly of the nucleosome. *Nucleic Acids Res.* 5:505–21

130. Winkler DD, Luger K. 2011. The histone chaperone FACT: structural insights and mechanisms for nucleosome reorganization. *J. Biol. Chem.* In press

131. Wood CM, Nicholson JM, Lambert SJ, Chantalat L, Reynolds CD, Baldwin JP. 2005. High-resolution structure of the native histone octamer. *Acta Crystallogr. F* 61:541–45

132. Woodcock CL, Safer JP, Stanchfield JE. 1976. Structural repeating units in chromatin. I. Evidence for their general occurrence. *Exp. Cell Res.* 97:101–10

133. Xu M, Long C, Chen X, Huang C, Chen S, Zhu B. 2010. Partitioning of histone H3-H4 tetramers during DNA replication-dependent chromatin assembly. *Science* 328:94–98

134. Zhang Y, Moqtaderi Z, Rattner BP, Euskirchen G, Snyder M, et al. 2009. Intrinsic histone-DNA interactions are not the major determinant of nucleosome positions in vivo. *Nat. Struct. Mol. Biol.* 16:847–52

135. Zlatanova J, Bishop TC, Victor JM, Jackson V, van Holde K. 2009. The nucleosome family: dynamic and growing. *Structure* 17:160–71

Molecular Mechanisms of Ubiquitin-Dependent Membrane Traffic

James H. Hurley[1] and Harald Stenmark[2]

[1]Laboratory of Molecular Biology, National Institute of Diabetes and Digestive and Kidney Diseases, National Institutes of Health, Bethesda, Maryland 20892-0580; email: hurley@helix.nih.gov

[2]Centre for Cancer Biomedicine, Faculty of Medicine, University of Oslo, and Institute for Cancer Research, Oslo University Hospital, Montebello, N-0310 Oslo, Norway; email: stenmark@ulrik.uio.no

Annu. Rev. Biophys. 2011. 40:119–42

First published online as a Review in Advance on February 14, 2011

The *Annual Review of Biophysics* is online at biophys.annualreviews.org

This article's doi: 10.1146/annurev-biophys-042910-155404

Copyright © 2011 by Annual Reviews. All rights reserved

1936-122X/11/0609-0119$20.00

Keywords

lysosome, EGF receptor, ENaC, RING domain, HECT domain, JAMM domain, ubiquitin-binding domain

Abstract

Over the past 14 years, ubiquitination has emerged as a centrally important mechanism governing the subcellular trafficking of proteins. Ubiquitination, interaction with sorting factors that contain ubiquitin-binding domains, and deubiquitination govern the itineraries of cargo proteins that include yeast carboxypeptidase S, the epithelial sodium channel ENaC, and epidermal growth factor receptor. The molecular structures and mechanisms of the paradigmatic HECT and RING domain ubiquitin ligases, of JAMM- and USP-domain-deubiquitinating enzymes, and of numerous ubiquitin-binding domains involved in these pathways have been worked out in recent years and are described.

Contents

Ubiquitination: the covalent conjugation of the C-terminal carboxylate to a protein substrate, usually, but not always, via an isopeptide bond with the e-amino nitrogen of Lys residues

Polyubiquitin: covalently conjugated chains of ubiquitin moieties formed by ubiquitination reactions

INTRODUCTION

Ubiquitin is a small protein with a large footprint in biology. The covalent ubiquitination of proteins is one of the most widespread regulatory posttranslational modifications of proteins. The C terminus of ubiquitin is conjugated to target proteins by the action of three enzymes: a ubiquitin-activating enzyme (E1), a ubiquitin-conjugating enzyme (E2), and a ubiquitin protein ligase (E3) (30, 53, 88, 122). Ubiquitin is normally conjugated to proteins via an isopeptide bond between the C terminus of ubiquitin and specific Lys residues in the ubiquitinated protein. Ubiquitin may be attached to proteins as one or a few monomers (mono- or multiubiquitination) or as a polyubiquitin chain. Ubiquitin polymers are formed when additional ubiquitin molecules are attached to one of the seven Lys residues on a previously attached ubiquitin.

The first demonstration of an involvement of ubiquitin in membrane traffic came with the discovery that ubiquitination of the mating factor receptor Ste2 (32) and the uracil permease Fur4 (26) in budding yeast is required for their efficient endocytosis. Subsequent work has revealed that many (but far from all) membrane proteins in diverse organisms are endocytosed in a ubiquitin-dependent manner, often in response to ligand binding (33, 114). Moreover, even though ubiquitin was first identified as an endocytosis signal, it has turned out to be even more important as a signal for sorting membrane proteins into the intralumenal vesicles (ILVs) of multivesicular bodies (MVBs). The ILVs are essentially vehicles specialized for delivering integral membrane proteins to the lumen of lysosomes for proteolytic processing or degradation, which occurs when MVBs fuse with these degradative organelles (**Figure 1a**). With very few exceptions, almost all integral membrane proteins that are sorted to the lysosome lumen follow ubiquitin-dependent sorting into MVBs.

Ubiquitin-dependent trafficking of membrane proteins is critical for regulating the numbers of transporters, receptors, and adhesion molecules at the plasma membrane, and for correct delivery and activation of certain lysosomal hydrolases, thereby playing a central role in cell signaling and metabolism. In the first part of this review, we discuss three examples of membrane proteins that follow ubiquitin-dependent trafficking pathways. We follow the trafficking of these proteins from their biosynthesis to their degradation and highlight the E3 ubiquitin ligases, deubiquitinating enzymes (DUBs), and sorting components that these cargoes encounter during their journey through the cell. In the second part of the review, we zoom in on the molecular details, down to the atomic level, of how membrane proteins are ubiquitinated, deubiquitinated, and sorted.

ROLE OF UBIQUITINATION IN THE TRAFFICKING OF YEAST CARBOXYPEPTIDASE S

Carboxypeptidase S (Cps1), a vacuolar hydrolase in the budding yeast *Saccharomyces cerevisiae*, is a prototypic example of a biosynthetic cargo that undergoes ubiquitin-mediated sorting. This protein is synthesized as a type II integral membrane protein that represents a precursor form (pCps1) (111). Its activation requires proteolytic clipping into a mature, soluble form (mCps1) (**Figure 1a**). This proteolytic activation occurs in the vacuole, the yeast equivalent of mammalian lysosomes,

through the activities of the vacuolar proteases Pep4 and Prb1. On its way to the vacuole, Cps1 transits directly from the *trans*-Golgi network (TGN) to endosomes, where it undergoes sorting into ILVs of MVBs. This ILV-associated form is delivered to the vacuole for efficient processing upon fusion of the MVB with the vacuole membrane (50).

Ubiquitination of Cps1 by Rsp5

Ubiquitination of Cps1 is crucial for its correct trafficking, and Cps1 was actually the first cargo for which ubiquitin was shown to be the signal for entry into the MVB pathway (50, 92). Pulse-labeling experiments suggest that pCps1 is ubiquitinated after exiting the Golgi complex (50). Cps1 is ubiquitinated on Lys8 by the HECT (homologous to E6AP carboxyl terminus)-domain-containing E3 ubiquitin ligase Rsp5 (50, 51, 76). This ubiquitin ligase is the only of its kind in budding yeast but has multiple mammalian counterparts, including Nedd4, which is described below. Replacement of Lys8 with arginine results in missorting of Cps1 to the limiting membrane (i.e., the membrane delimiting it from the cytosol) of the vacuole (50). In contrast to most known mammalian substrates for HECT E3 ubiquitin ligases, Cps1 does not contain the canonical PPXY motif that serves as a docking site for one of the WW domains of the ubiquitin ligase. Instead, Cps1 recruits Rsp5 through association with an adaptor protein containing a PPXY motif, the endosomal integral membrane protein Bsd2 (31). In addition, Cps1 contains MVB targeting information within the amino acid sequence PVEKAPR, and direct interaction of this motif with the HECT domain of Rsp5 appears to facilitate its ubiquitination and correct sorting (62).

Monoubiquitination Versus Polyubiquitination of Cps1

According to SDS-PAGE (sodium dodecyl sulfate polyacrylamide gel electrophoresis) analyses, the bulk of ubiquitinated Cps1 contains a single ubiquitin moiety, and this has led

DUB:
deubiquitinating enzyme

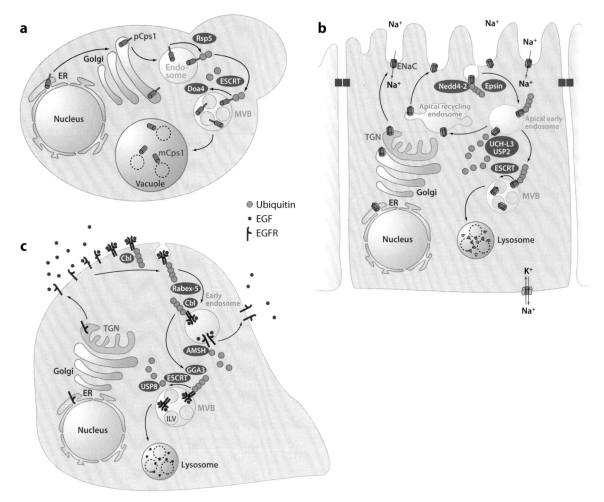

Figure 1

Subcellular itineraries of paradigmatic substrates of ubiquitin-dependent sorting pathways. (*a*) Yeast carboxypeptidase Cps1 is biosynthesized as a transmembrane precursor form (pCps1), which is transported from the biosynthetic pathway to endosomes. Ubiquitination in endosomes, mediated by Rsp5, provides a signal for recognition by the ESCRT machinery. Before the inclusion within the ILVs of MVBs, pCps1 is deubiquitinated by Doa4. When the MVBs fuse with the vacuole, the ILVs and their content are delivered to the vacuole lumen. Here, the ILV membranes are degraded by lipases, whereas pCps1 is processed into the mature form (mCps1) by specific proteases. (*b*) The Na$^+$ channel ENaC is delivered to the apical plasma membrane of epithelial cells, where it channels Na$^+$ from the extracellular space into the cells. Residency of ENaC in the apical plasma membrane is controlled by the E3 ubiquitin ligase, Nedd4-2, which ubiquitinates ENaC and thereby signals its epsin-dependent endocytosis from clathrin-coated pits and delivery to an apical early endosome. Deubiquitination by UCH-L3 or USP2-45 favors recycling of ENaC to the plasma membrane via the apical recycling endosome. ENaC molecules that remain ubiquitinated are recognized by the ESCRT machinery and sorted into the MVB pathway for lysosomal degradation. (*c*) When EGF binds to EGFR at the plasma membrane, this induces a conformation change of the receptor that promotes dimerization. Dimerization triggers tyrosine phosphorylation, which, in addition to initiating signal cascades, promotes receptor ubiquitination by the E3 ubiquitin ligase Cbl. The activated receptor is endocytosed and delivered to an early endosome. If the ligand dissociates in the early endosome, ubiquitination is no longer sustained, and the EGFR becomes deubiquitinated by the DUB AMSH. This promotes recycling to the plasma membrane. If still activated in the endosome, the EGFR remains ubiquitinated and is recognized by GGA3 and the ESCRT machinery. This causes its sorting into ILVs following deubiquitination by the DUB USP8. The receptor and its ligand are degraded when the MVB fuses with a lysosome. Abbreviations: DUB, deubiquitinating enzyme; EGF, epidermal growth factor; EGFR, epidermal growth factor receptor; ENaC, epithelium sodium channel; ER, endoplasmic reticulum; ESCRT, endosomal sorting complex required for transport; ILV, intralumenal vesicle; MVB, multivesicular body; TGN, *trans*-Golgi network; USP, ubiquitin-specific protease.

to the view that a single ubiquitin is sufficient for correct sorting of Cps1 (50). This view was recently challenged, however, by experiments that used a yeast strain that expresses Ub-Lys63Arg (K63R), which cannot form Lys63-linked polyubiquitin, as the only ubiquitin species (61). Cps1 is normally transported from the biosynthetic to the endocytic pathway; surprisingly, transport into the MVB pathway is compromised in the Ub-K63R mutant strain. This suggests that Lys63-linked polyubiquitination is required for Cps1 sorting into the MVB pathway, an interpretation consistent with the finding that Rsp5 preferentially catalyzes Lys63-linked polyubiquitination of its substrates (55). It should be noted, though, that it is difficult to detect any polyubiquitinated Cps1 by immunoblotting—the only higher ubiquitinated form that can be detected convincingly in such experiments is a diubiquitinated species (61). It is therefore conceivable that Lys63-linked diubiquitination is a sufficient signal for sorting Cps1 into the MVB pathway. The difficulties in detecting polyubiquitinated Cps1 might also be explained by the presence of (unidentified) DUBs that may trim the polyubiquitin chains at high rates. At present, one also cannot rule out the possibility that the failure of Cps1 to become correctly sorted in the Ub-K63R yeast strain might be unrelated to an eventual inhibition of Cps1 polyubiquitination, for example, by preventing (unidentified) polyubiquitination of components of the sorting machinery.

Deubiquitination of Cps1 by the DUB Doa4

Whereas only a very minor fraction of Cps1 is ubiquitinated in wild-type yeast, the level of (mono-)ubiquitination increases substantially in yeast strains deleted for the DUB Doa4, indicating that Doa4 is the main DUB for Cps1 (50). Like other MVB cargoes, the vacuolar sorting of Cps1 is dramatically impaired in *doa4* mutants (4). The most straightforward explanation for this is that depletion of Doa4 prevents recycling of monoubiquitin, thereby depleting cellular pools of free ubiquitin. However, Cps1 fails to become correctly sorted to the vacuole lumen in *doa4* mutants even if the cellular levels of ubiquitin are restored experimentally, suggesting that Doa4 could also have an additional function in Cps1 sorting. Doa4 might deubiquitinate components of the sorting machinery, or cargo deubiquitination itself could promote cargo internalization (80, 95).

ENaC: A SODIUM CHANNEL THAT UNDERGOES UBIQUITIN-DEPENDENT TRAFFICKING

The epithelial Na^+ channel, ENaC, is widely expressed in epithelial tissues such as sweat ducts, salivary glands, colon, airway, and kidney nephrons, where it shows a polarized distribution to apical membranes (11). ENaC is a major regulator of salt and water reabsorption, which has been particularly well characterized in the distal nephrons of the kidney.

Subunit Structure and Channel Properties of ENaC

ENaC is composed of three homologous subunits, α, β, and γ. These channels form heterotetramers composed of two α-subunits, one β-subunit, and one γ-subunit (24). They have a high selectivity for Na^+ over K^+ and can be efficiently inhibited by amiloride ($K_i \sim 100$ nM). The α-, β-, and γ-subunits have a similar topology, with two transmembrane domains, a large extracellular region, and cytoplasmic N and C termini (107). The β- and γ-subunits are palmitoylated on specific cytosol-exposed cysteine residues, and this posttranslational modification modulates the gating properties of the channel, possibly by facilitating interactions between the cytoplasmic domains and the plasma membrane (77). ENaC channels are also regulated by channel-activating proteases, which cleave extracellular loops and thereby increase Na^+ conductance by increasing the channel-open probability (11, 56).

Deubiquitination: the hydrolytic cleavage of the abovementioned isopeptide bond

ESCRT: endosomal sorting complex required for transport

USP: ubiquitin-specific protease

Intracellular Trafficking of ENaC

Biosynthesized ENaC molecules transit from the Golgi complex via the TGN to the apical plasma membrane (11). Inactivation of ENaC channels can occur either by transient alteration of gating properties or by internalization of the channels. The latter is triggered by ubiquitination and is discussed here. Endocytosis of ENaC occurs primarily from clathrin-coated pits and is dependent on the clathrin- and ubiquitin-binding endocytosis adaptor Epsin1 (113). The endocytosed ENaC is delivered to early endosomes, where a minor fraction is recycled to the TGN, while the rest is either recycled to the plasma membrane or transported to lysosomes, depending on the ubiquitination status of the channel (11) (**Figure 1b**). Trafficking of ENaC is hormonally controlled, and both the mineralocorticoid aldosterone and the antidiuretic hormone vasopressin cause a rapid redistribution of ENaC from an intracellular pool to the apical plasma membrane, increasing Na$^+$ conductance (110). The molecular mechanisms of this hormone-controlled translocation of ENaC remain to be characterized.

Control of ENaC Endocytosis and Lysosomal Trafficking Through Nedd4-2-Mediated Ubiquitination

ENaC is ubiquitinated at the plasma membrane by the action of the E3 ubiquitin ligase Nedd4-2, which binds to PPXY motifs in the C termini of the ENaC α- and γ-subunits (and possibly also the β-subunit), causing their ubiquitination on N-terminal lysine residues (113). Mutation of these lysines to arginines profoundly inhibits endocytosis from the apical membrane, thereby increasing channel half-time, as do mutations that prevent Nedd4-2 binding (11). Although recent data indicate that multiple ubiquitin molecules are conjugated to the ENaC subunits, it is not yet clear whether this modification represents poly-Ub, multi-mono-Ub, or a combination of these. What is more certain is that ubiquitinated ENaC subunits are recognized by the Hrs subunit of the endosomal sorting complex required for trans-

port (ESCRT)-0 complex and that this entry into the ESCRT pathway mediates lysosomal degradation of ubiquitinated ENaC molecules (127). Given the crucial function of Nedd4-2 in the regulation of ENaC trafficking, it is not surprising that ubiquitination of ENaC by this ubiquitin ligase is under strict control by physiological cues. In particular, phosphorylation of Nedd4-2 by the serum and glucocorticoid kinase, SGK1, or by the AMP-activated kinase, AMPK, prevents its interaction with ENaC, thereby increasing the half-life of ENaC (1, 16). Likewise, phosphorylation of the C-terminus of the β-subunit of ENaC by the G-protein-receptor-coupled kinase, GRK2, causes insensitivity to Nedd4-2-mediated downregulation (18).

Control of ENaC Recycling Through Deubiquitination

ENaC molecules that remain ubiquitinated in endosome membranes appear to be committed to lysosomal degradation by ESCRT-dependent sorting. Conversely, deubiquitination mediated by the DUBs UCH-L3 and USP2 (ubiquitin-specific protease 2) promotes recycling of ENaC to the apical plasma membrane (11). Consistent with this, knockdown of UCH-L3 causes increased ubiquitination of ENaC and a loss of ENaC-mediated Na$^+$ transport (12). Likewise, overexpression of USP2 leads to decreased ubiquitination of ENaC and increases its activity at the plasma membrane (22). These findings illustrate the importance of ENaC ubiquitination for its evasion of the endocytic recycling route and targeting to the degradative MVB pathway.

ENaC Ubiquitination and Genetic Disease

Liddle syndrome is an autosomal dominant salt-sensitive hypertension syndrome caused by gain-of-function mutations in the β- and γ-subunits of ENaC (41, 96). The common theme of these mutations is that they preclude Nedd4-2 binding to ENaC, thereby preventing endocytosis and causing ENaC to reside at the apical plasma membrane for a prolonged period (11). Increased residence time of ENaC in the apical membranes of the distal kidney nephrons results in excessive Na^+ reabsorption and ensuing hypertension.

UBIQUITIN-DEPENDENT TRAFFICKING OF EPIDERMAL GROWTH FACTOR RECEPTORS

Epidermal growth factor (EGF) receptors (EGFRs) are expressed in diverse cell types, where they mediate signaling that results in proliferation, growth, survival, and migration (124). The fact that EGFRs are frequently overexpressed or contain gain-of-function mutations in cancers has made them an attractive target for cancer therapy (123). EGFR has six known ligands, of which EGF and transforming growth factor α (TGFα) are the best characterized. Binding of EGF to EGFR has been characterized at the atomic level, and these analyses show that the monomeric EGFR exists in an autoinhibited state that is relieved by ligand binding (23). Ligand-mediated release of autoinhibition results in the exposure of a dimerization interface, and two ligand-bound EGFR monomers therefore dimerize efficiently. Dimerization of EGFRs draws together the two cytoplasmic tyrosine kinase domains, with each monomer trans-phosphorylating the opposing monomer at multiple cytosolic tyrosine residues. The phosphotyrosine groups recruit a host of important signaling mediators, including the regulatory subunit of class I phosphoinositide 3-kinase; the SH2- and SH3-domain-containing adaptor Grb2; the signal transducer and activator of transcription, STAT3; and phospholipase C γ (124). In addition, phosphorylated EGFR recruits a molecule that profoundly controls its downregulation, namely the E3 ubiquitin ligase c-Cbl (64).

Ubiquitination of EGFR by Ube2D1-4 and c-Cbl

Ubiquitination of activated EGFR takes place at the plasma membrane as a rapid response to ligand binding and continues during endocytosis and early-endosomal trafficking, balanced by deubiquitination (70, 117) (**Figure 1c**). The E2 ubiquitin-conjugating enzymes Ube2D1–4 are involved, as is the E3 ubiquitin ligase c-Cbl (64, 117). c-Cbl belongs to a small subfamily of RING-domain-containing E3 ubiquitin ligases that also comprises two other ubiquitin ligases capable of ubiquitinating EGFR, Cbl-b and Cbl-c. c-Cbl can bind to the human EGFR in two ways, either directly to phosphorylated Tyr1045 or indirectly via binding to Grb2, which in turn binds to phosphorylated Tyr1068 and Tyr1086 of the receptor (70). Ubiquitination of the EGFR is readily detectable at EGF concentrations as low as 1 ng ml^{-1} (52) but is considerably higher above a threshold EGF concentration of about 10 ng ml^{-1} (104). The type of EGFR ubiquitination following EGF stimulation has been determined both by indirect methods and by mass spectrometry (29, 39). The latter analyses have identified six major ubiquitination sites in EGFR, all located within the tyrosine kinase domain. More than half of the ubiquitin molecules attached to the EGFR are in the form of polyubiquitin (mostly Lys63-linked), whereas the remainder represent multi-monoubiquitination (39).

Endocytosis of EGFRs

EGFR endocytosis has been characterized since the 1970s (13). These receptors can be endocytosed via several alternative pathways, possibly depending on the extent of ubiquitination (105), but internalization via clathrin-coated pits is the most predominant (28). The tyrosine kinase activity of EGFRs is required for efficient endocytosis (108), whereas ubiquitination facilitates—but is not strictly required for—EGFR endocytosis (28). A recent study has suggested that at least four partially redundant mechanisms contribute to EGFR endocytosis through clathrin-coated pits. These include ubiquitination of the tyrosine kinase domain, interaction with the clathrin adaptor AP-2, acetylation of three C-terminal lysine residues outside the tyrosine kinase domain,

EGFR: epidermal growth factor receptor

and interaction with Grb2 (28). Consistent with a role for ubiquitination in EGFR endocytosis, knockdown of the ubiquitin-binding endocytosis adaptors Epsin and EPS15 causes a partial inhibition of internalization (52, 105, 112). On the other hand, an EGFR mutant in which all major ubiquitination sites are mutated is internalized at a nearly normal rate (38). The internalization of this mutant is nevertheless sensitive to Cbl depletion, suggesting that Cbl may have roles in endocytosis other than mediating receptor ubiquitination. Indeed, Cbl has multiple interaction partners, including the actin- and ubiquitin-binding protein CIN85, which in turn interacts with endophilin to facilitate clathrin-mediated endocytosis of EGFRs (109).

Ubiquitin-Dependent Activation of GTPase Cascades Downstream of EGFR

In addition to the well-characterized phosphorylation-dependent activation of the RAS GTPase downstream of activated EGFR, resulting in activation of the MAP kinase pathway, there is evidence for ubiquitin-dependent activation of small GTPase cascades downstream of ligand-bound EGFR. The unconventional E3 ubiquitin ligase Rabex-5 binds to EGF-activated EGFR through its two ubiquitin-binding domains (84). Even though the functional implications of this interaction still have to be worked out, it is worth noting that Rabex-5 was originally identified as a guanine nucleotide exchange factor for the small GTPase RAB5, which controls endosomal membrane fusion (36, 83). Moreover, Rabex-5 has also been identified as a downstream effector of a structurally related endosomal GTPase, RAB22 (128). Interestingly, EGFRs are endocytosed in Rabex-5-deficient cells but escape degradation in lysosomes (128). This opens the possibility that one of the consequences of EGFR ubiquitination may be, via recruitment of Rabex-5, to activate RAB5/RAB22-dependent EGFR trafficking through the early-endocytic pathway.

Endosomal Sorting of EGFRs

Endocytosed EGFRs have two alternative fates: recycle to the plasma membrane or transport to lysosomes for degradation by hydrolases such as cathepsin L (25, 35) (**Figure 1c**). Electron microscopy of EGFRs in cells stimulated with EGF has contributed strongly to the paradigm of lysosomal sorting via the MVB pathway, because such studies were the first to demonstrate the ligand-dependent trafficking of a receptor into the ILVs of MVBs (74). Somewhat surprisingly, different ligands cause different trafficking of EGFRs, as best documented in the case of EGF versus TGFα. Whereas stimulation with high concentrations of EGF mainly causes endocytosed EGFRs to be degraded in lysosomes, TGFα causes recycling of the bulk of endocytosed receptors (70). This paradox can be explained by the differential affinities of the two ligands for the receptors at the mildly low pH (5.5–6.0) found in early endosomes. EGF remains bound to EGFR at this pH, resulting in sustained receptor activation, whereas TGFα dissociates (70). The latter leads to a rapid decrease in EGFR ubiquitination and recycling instead of lysosomal sorting. These findings argue that there is a continuous ubiquitination and deubiquitination of EGFRs along the endocytic pathway, and that sustained ubiquitination requires receptor activation through ligand binding. Consistent with this, knockdown of c-Cbl and Cbl-b strongly inhibits ligand-induced degradation of EGFR while having less effect on internalization (39, 63, 85). There is strong evidence that degradative endosomal sorting of ubiquitinated EGFRs is mediated by the ESCRT machinery. EGFRs can be coimmunoprecipitated with Hrs after EGF stimulation (105, 117), and knockdown of various subunits of ESCRT-0, -I, -II, or -III strongly inhibits EGF-mediated lysosomal degradation of EGFRs (6, 7, 9, 71, 91). An involvement of GGA3 in the degradation of endocytosed EGF has also been reported, possibly reflecting alternative entries into the ESCRT pathway, with a GGA3-containing complex serving as an alternative ESCRT-0 (90). In addition, an isoform of EPS15 that associates with Hrs, EPS15B, is required for efficient lysosomal degradation of endocytosed EGFRs (99). The function of this protein in EGFR trafficking is not known, but the fact that it contains two C-terminal ubiquitin-interacting motifs suggests that it may contribute to increase the avidity of ESCRT-0 for ubiquitinated receptors. In keeping with this, biochemical evidence suggests that polyubiquitinated EGFRs interact preferentially with ESCRT-0 and are selectively targeted for degradation (117).

Deubiquitination of EGFRs

Ubiquitinated EGFRs are substrates for the DUBs AMSH and USP8/UBPY (72, 75). These DUBs share the same binding site on the ESCRT-0 subunit STAM through a non-canonical SH3-domain-binding motif, PX(V/I)(D/N)RXXKP (14). AMSH and USP8/UBPY also contain microtubule interaction and transport (MIT) domains that enable them to interact with ESCRT-III. The MIT domains of these DUBs have different but overlapping interactions with ESCRT-III subunits, described in more detail below. AMSH specifically cleaves Lys63-linked polyubiquitin chains and has been proposed to oppose E3 ubiquitin ligase activity on the EGFR through deubiquitination at an early stage of the endosomal sorting process, thereby promoting EGFR recycling (118). USP8, on the other hand, has no preference for any particular type of polyubiquitin chains. It is most similar to yeast Doa4 and might fulfill an equivalent role by deubiquitinating EGFRs prior to their inclusion into ILVs in order to ensure recycling of ubiquitin (2, 98). However, like AMSH, USP8 has also been implicated at an earlier stage, countering the entry of EGFR into the ESCRT pathway (8). The competitive binding of AMSH and USP8 to ESCRT subunits, and the fact that they associate with both ESCRT-0 and ESCRT-III, suggests that they may regulate cargo deubiquitination in complex ways.

STRUCTURE AND FUNCTION OF UBIQUITIN AND UBIQUITIN CHAINS

Having described the cellular itineraries of three archetypal ubiquitin-dependent cargoes, we now consider in structural detail the mechanisms behind these trafficking pathways. Fortunately, ubiquitin is widespread in biology, and many of the relevant molecular mechanisms are conserved between trafficking and other pathways. Thus, the field of trafficking benefits from a number of basic mechanistic and structural studies that were motivated in many cases by the role of ubiquitin in nontrafficking processes.

The ubiquitin fold consists of a five-stranded β-sheet and a single α-helix (121) (**Figure 2a**). The N and C termini are on opposite ends of the three-dimensional structure. The N terminus is well ordered, whereas the last six residues of the C terminus extend beyond the structural core and are flexible. This latter feature of the structure is critically important because the C-terminal carboxylate at Gly76 is the point of covalent attachment to cargo. As such, it must be flexible enough to undergo reactions catalyzed by ubiquitin ligases and DUBs. Ubiquitin has three major hydrophobic patches (106): one surrounding Phe4 near the N terminus; one around Ile36; and the third surrounding Ile44, which is near the C terminus in three dimensions. The latter patch is involved in every recognition event by the membrane trafficking machinery, with only one known exception (42). The ubiquitin core is relatively rigid, but the intrinsic flexibility of the side chains of a subset of the key recognition residues allows ubiquitin to bind a wide range of different partners (60, 86).

Polymeric ubiquitin chains are formed primarily by conjugation of the ubiquitin C terminus to either the α-amino group of the polypeptide chain or the ε-amino groups of any one of the seven Lys residues of ubiquitin (89) (**Figure 2a**). The best-studied ubiquitin chains are linked through Lys48, located next to the Ile44 hydrophobic patch and relatively close to the C-terminal flexible region, and through

MIT: microtubule interaction and transport

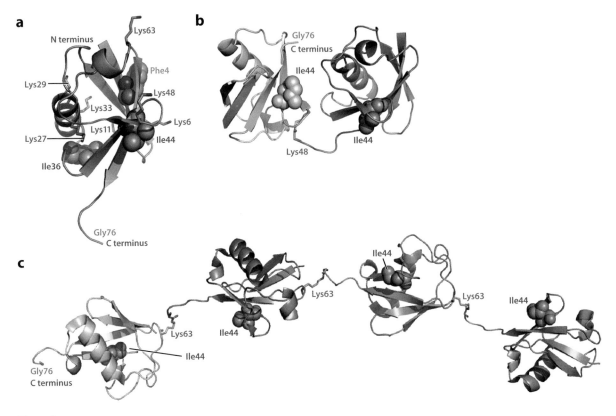

Figure 2

Structure of ubiquitin and selected ubiquitin chains. (*a*) Structure (1UBQ) and major functional features of ubiquitin. The seven Lys residues are shown in a stick model, and the three hydrophobic residues at the center of the main hydrophobic patches are highlighted in space-filling spheres. (*b*) Structure of a K48-linked diubiquitin chain (1TBE). (*c*) Structure of a Lys63-linked tetraubiquitin chain (3HM3). In panels *b* and *c*, the Lys involved in the linkage is shown in a stick model, and Ile44 of each ubiquitin monomer is shown in space-filling spheres. The proximal ubiquitin (the one that would be linked directly to cargo) is colored light orange, while other moieties are colored orange.

Lys63, located near the N terminus in the three-dimensional structure.

Excluding branched chains, eight types of peptide or isopeptide chains occur, and four of these have been characterized structurally (17, 57, 89). The Lys11- and Lys48-linked ubiquitin chains are linked at points close together in the three-dimensional structure and thus have little conformational freedom and prefer closed conformations (**Figure 2***b*). The C- to N-terminally linked or linear ubiquitin chain and the Lys63-linked chain are connected at the two ends of the structure farthest from one another and consequently adopt open conformations (**Figure 2***c*).

MECHANISMS OF CARGO UBIQUITINATION

The Nedd4/Rsp5 Family

The human genome codes nine Rsp5 orthologs, which compose the Nedd4 ubiquitin ligase family (44). Rsp5 and the Nedd4 family members have a domain architecture consisting of an N-terminal C2 domain, two to four WW domains with connecting linkers, and a C-terminal HECT domain (**Figure 3***a*). C2 (protein kinase C conserved region 2) domains are β-sandwiches that come in two different structural permutations (78). The Nedd4 class of C2 domains belongs to type II (**Figure 3***b*). C2

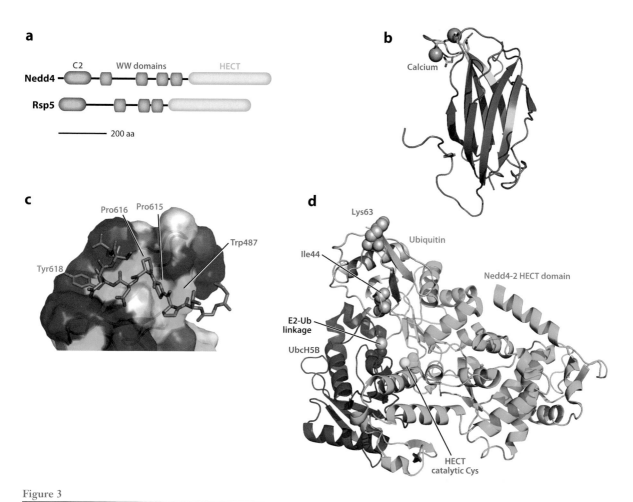

Figure 3

Rsp5 and the Nedd4 family of ubiquitin ligases. (*a*) Domain architecture of Nedd4 and Rsp5. (*b*) Structure of the C2 domain of Nedd4 (2NSQ). Ca^{2+} ions and their ligands are shown as salmon-colored spheres and sticks, respectively. (*c*) Structure of the third WW domain of Nedd4 bound to the PPXY motif of ENaC (1I5H). (*d*) Structure of the Nedd4-2 HECT domain (3JW0; *green*, with catalytic Cys highlighted in space-filling spheres) in complex with an UbcH5B~Ub (*magenta* and *orange*, respectively) adduct. The Ub-UbcH5B bond is highlighted with a space-filling sphere. Ile44 and Lys63 are highlighted to orient the viewer. Abbreviations: ENaC, epithelium sodium channel; Ub, ubiquitin.

domains also come in two flavors: those that bind Ca^{2+} and those that do not (78). Typically, Ca^{2+} binding promotes binding to membranes and sometimes dimerization or binding to other proteins. The Nedd4 family C2 domains bind Ca^{2+} (**Figure 3*b***). Ca^{2+} binding to the Rsp5 C2 domain promotes its binding to acidic lipids (21).

WW domains are compact three-stranded β-sheet structures of just ~40 residues, one of the smallest functional protein domains known

(69). WW domains are best known for binding to PPXY motifs and other Pro-rich recognition sequences, but in some cases they also bind to phosphorylated Ser (pSer)- and Thr (pThr)-containing sequences (66). Nedd4 binds to some substrates, such as the EGFR-associated kinase ACK, via its third WW domain (WW3), which binds to a classical PPXY motif in ACK (65). Nedd4-2 WW3 binds to ENaC via the sequence PPPNYDSL, which comprises a C-terminally extended variant of the PPXY motif (49) (**Figure 3c**). Many other substrates of Nedd4 family ubiquitin ligases do not possess PPXY motifs. In some of these cases, adaptor proteins that possess PPXY motifs may serve as a bridge to Nedd4 family members through their WW domains (103). In many others, alternative combinations of phosphorylated residues and PY elements are responsible for these interactions. The Nedd4-2 WW2 binds to pThr-PY peptides of Smad2/3 with high micromolar affinity (27). The structural mechanism for pThr binding to the WW2 domain is not clear, because a key Arg residue involved in phosphopeptide binding in certain other WW domains (119) is not conserved in Nedd4 family WW domains.

HECT domains are ~350-residue catalytic modules that form a covalent adduct with the C terminus of ubiquitin via a catalytic Cys residue before transferring ubiquitin to the substrate. Structures of three Nedd4 family HECT domains are known: WWP1 (120), Smurf2 (82), and Nedd4 (PDB 2ONI). These structures revealed flexibility between the ~250-residue N-terminal E2-binding domain and the ~100-residue C-terminal Cys-containing domain. A non-Nedd4 family HECT domain from E6-AP has been crystallized in complex with an associated E2 enzyme, UbcH7 (40). This complex structure showed there was a 41 Å gap between the catalytic Cys residues of the E2 and E3 enzymes. When the HECT domain of Nedd4-2 was cocrystallized with UbcH5B~ubiquitin (the covalent adduct between ubiquitin and the E2 enzyme, instead of the E2 enzyme alone), the HECT C-terminal domain rotated about the hinge such that the E2-Cys distance shrank to just 8 Å, nearly close enough for direct ubiquitin transfer (47) (**Figure 3d**).

The Cbl Family

The Cbl proteins responsible for EGFR ubiquitination have similar architectures to one another: Beginning at the N terminus, there is a tyrosine kinase-binding (TKB) domain, followed by a RING ubiquitin ligase domain, a proline-rich domain (PRD), a Ser- and Tyr-rich unstructured region, and finally a UBA domain. The architecture of Cbl-c corresponds to a truncated form of c-Cbl and Cbl-b, ending early in the PRD (**Figure 4a**). The TKB domain is essentially an expanded and divergent variant of the SH2 domain, which is the archetypal phosphotyrosine-binding domain. The TKB domain consists of a divergent SH2 domain augmented by a four-helix bundle subdomain and a Ca^{2+}-binding EF hand subdomain (73) (**Figure 4b**). Although the variant SH2 domain of the TKB is missing the secondary β-sheet that contributes to peptide binding in classical SH2 domains, the pTyr-binding pocket and the overall mode of peptide binding is otherwise similar to that of classical SH2 domains (73). Ca^{2+} binding to the EF hand enables the TKB domain to adopt the right conformation for the four-helix bundle to contribute to high-affinity pTyr peptide binding, and the integrity of all three subdomains is required for function (73). Following phosphorylation of Tyr1069, EGFR residues 1063–1075 fit the canonical Cbl TKB-binding motif (N/D)XpY(S/T)XXP, and the structure of this complex has been determined (79). The EGFR peptide forms an unusual internal salt bridge between the Arg immediately preceding the pTyr and the phosphate group of the pTyr (79).

There are two major classes of ubiquitin ligases, the HECT domain ligases, such as the Nedd4 family discussed above, and the RING domain class, of which Cbl is a member. In contrast to HECT domain ubiquitin ligases, RING domain ubiquitin ligases do not form a covalent adduct with the ubiquitin moiety. Rather, they serve as adaptors to bring together the Ub-bound E2 enzyme and its substrate. The RING domain recruits the E2 enzyme, while other regions of the ligase are typically responsible for substrate binding. The crystal structure of the TKB-RING portion of c-Cbl has been determined in complex with the E2 enzyme UbcH7 and a pTyr peptide from the substrate ZAP70 (126) (**Figure 4b**). The RING domain itself

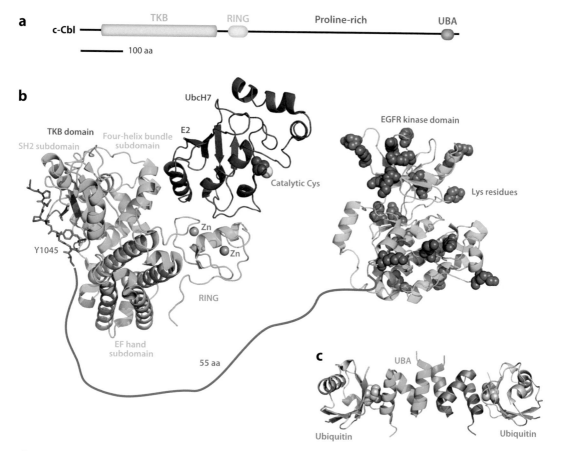

a

c-Cbl ━━━▭━━━TKB━━━▭━━━RING━━━━Proline-rich━━━●━━UBA

━━━━━ 100 aa

b

UbcH7

E2

TKB domain
SH2 subdomain   Four-helix bundle subdomain

Catalytic Cys

EGFR kinase domain

Lys residues

Zn
Zn

Y1045

RING

EF hand subdomain

55 aa

c   UBA

Ubiquitin   Ubiquitin

Figure 4
━━
The Cbl family of ubiquitin ligases. (*a*) Domain architecture of c-Cbl. (*b*) Structural model for Cbl-dependent ubiquitination of the EGFR intracellular region. The structure of the TKB and RING portion of Cbl (*green*) in complex with UbcH7 (*magenta*) and a pTyr peptide (*blue*) from ZAP70 (1FBV) is used as a stand-in for the complex with Ube2D1–4 and the Tyr1045 region of EGFR. The catalytic domain of EGFR (3GT8) is colored blue. Lys residues of the EGFR catalytic domain are highlighted as space-filling spheres, even though not all of these Lys have been directly shown to be ubiquitinated. Structural zinc ions of the Cbl RING domain are shown as salmon-colored spheres. (*c*) Dimeric complex of the Cbl-b UBA domain with ubiquitin (2OOB). Abbreviations: EGFR, epidermal growth factor receptor; EF, elongation factor; TKB, tyrosine kinase binding domain.

is built around two tetrahedrally coordinated zinc ions, which have a structural rather than catalytic role. Cbl has a groove formed by the zinc-binding loops and an α-helix in which the E2 enzyme binds. Five residues from these two loops and the helix control E2 enzyme specificity. In the case of Cbl, these residues are mostly hydrophobic, notably Ile383 and Trp408. This site on the RING domains specifies an E2 that has a Phe residue on its L1 loop, Phe63 in the case of UbcH7. The Ube2D enzymes responsible for Cbl-dependent EGFR

ubiquitination conform to this sequence pattern (117), and thus the UbcH7-Cbl complex probably provides a reasonable model for the structures of the functional Ube2D-Cbl complexes.

c-Cbl functions in EGFR endocytosis by virtue of its coupling to the SH3-domain-containing endocytic proteins CIN85 and endophilin. Residues 902–912 of the c-Cbl PRD form a unique SH3-binding motif in which two SH3 recognition sequences are interwoven with one another (46). A similar mechanism

UBD:
ubiquitin-binding
domain

allows the double-sided ubiquitin interaction motif (DUIM) Hrs to bind two ubiquitin moieties at once, as described below. Two copies of the CIN85 SH3 domain thus bind to a single 11-residue peptide motif from c-Cbl (46). This drives the dimerization of CIN85 and other proteins with similar SH3 domains. Finally, the C-terminal UBA domains of c-Cbl and Cbl-b bind ubiquitin. The Cbl-b UBA domain binds monoubiquitin with ~60 μM affinity and K48-Ub$_4$ with 2 μM affinity (87). The Cbl-b UBA domain dimerizes, but only weakly in the absence of ubiquitin chains (87). Cross-linking of the UBA domains by binding to polyubiquitin promotes dimerization (**Figure 4c**) and, in turn, activation of Cbl. How the ubiquitin-chain-induced dimerization of Cbl-b is integrated into the larger picture of Cbl physiology remains to be discovered.

HOW THE UBIQUITIN TAG IS RECOGNIZED

Virtually all the known effects of ubiquitination on protein sorting are mediated by binding of the ubiquitin tags to specific ubiquitin-binding domains (UBDs) (17, 34, 42). The past decade has seen rapid advances in the structure, function, and biology of UBDs both in trafficking area and in other fields. UBDs as a class have been reviewed elsewhere (17, 34, 42), and the supplemental section of the online version of this article describes the specifics of UBD function in the trafficking of the cargoes described above (follow the **Supplemental Material link** from the Annual Reviews home page at **http://www.annualreviews.org**).

REMOVAL OF THE UBIQUITIN TAG

DUBs cleave the isopeptide bond between the ubiquitin C-terminal Gly and the Lys residue of the ubiquitinated protein (14, 57). DUBs are therefore proteases and use the same chemical reaction mechanisms as conventional proteases. There are five structural classes of DUBs (57). Of these, two families, JAMM

(JAB1/MPN/MOV34) and USP, are the most extensively implicated in regulating the cargoes described earlier in this review.

The JAMM Family Members AMSH and AMSH-LP

The human proteome includes eight JAMM family members. The members AMSH and AMSH-LP are implicated in regulating membrane traffic, and both contain N-terminal MIT domains (43, 116) (**Figure 5a**). Many MIT domains, including the MIT domain of AMSH, bind to MIT-interacting motifs (MIMs), short sequences at the C termini of ESCRT-III proteins that bind to grooves between the three helices of the MIT domain helical bundle (54, 81, 115). The AMSH MIT domain binds to CHMP1A, -1B, and -3 by multiple reports (68, 97, 116, 125). These are MIM1-containing proteins, and the MIM1-binding site of the structurally characterized VPS4-MIT is mostly preserved in AMSH-MIT. The structurally defined Vps4-MIT-MIM1 interaction (81, 115) is a reasonable model for the AMSH MIT-CHMP3 interaction. The experimental structure of the latter complex still requires elucidation, because AMSH-MIT, unlike VPS4 MIT, does not bind to CHMP2 isoforms.

JAMM domain DUBs are zinc metalloproteases that are mechanistically related (although not homologous) to the archetypal metalloprotease, thermolysin. Not all JAMM domain proteins are DUBs. Compared with the non-DUB JAMM domain, the JAMM domain of AMSH-LP has two AMSH-family-specific sequence insertions, Ins-1 and Ins-2 (101). The AMSH family JAMM domains bind two zinc ions: One is a structural zinc ion bound to the conserved JAMM core, and the other is a catalytic zinc ion bound by residues of Ins-2. The structure of a catalytically inactivated AMSH-LP JAMM domain in a stable complex with Lys63-Ub$_2$ has been determined (**Figure 5b**), yielding remarkably detailed insight into the specificity of AMSH and AMSH-LP for Lys63-linked ubiquitin chains (101). Lys63-Ub$_2$ is bound in a highly open

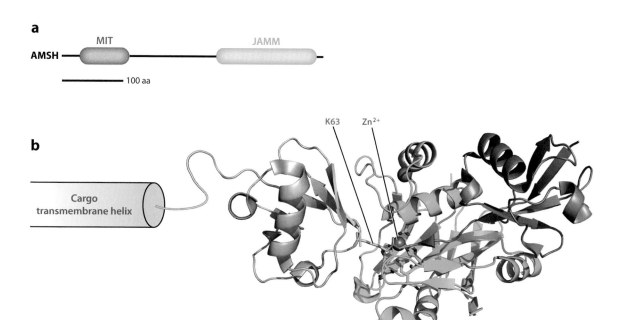

Figure 5

The zinc isopeptidase AMSH. (*a*) Domain architecture of AMSH. (*b*) Structural model for the catalytic complex of AMSH-LP bound to a Lys63-Ub$_2$-modified cargo. The structural model was derived by superimposing the structure of an active, zinc-bound, Ub-free form of AMSH-LP (2ZNR) on the structure of the Lys63-Ub$_2$ complex (2ZVN; the two moieties are depicted in shades of orange) with an inactivated mutant lacking the catalytic zinc ion. To illustrate the positioning of the proximal and distal moieties of the Lys63-Ub, the cargo is modeled as a single-pass transmembrane protein with the ubiquitin conjugated close to the transmembrane domain. Zinc ions are shown as blue-colored space-filling spheres. Abbreviations: JAMM, JAB1/MPN/MOV34; MIT, microtubule interaction and transport; Ub, ubiquitin.

conformation, similar to the open structure of Lys63 chains crystallized by themselves or with Lys63-specific recognition domains (15, 58, 59, 100, 102). The Gln62-Lys63-Glu64 tripeptide of the proximal ubiquitin moiety (corresponding to the moiety attached to a Lys63-Ub$_2$ modified cargo protein) makes multiple hydrogen bonds directly with the Ins-2 region of the JAMM domain (101). These hydrogen bonds align Lys63-Ub for catalytically productive isopeptide bond hydrolysis, while other types of linkages would fail to make these interactions. Thus, Lys63 specificity appears to be hard-wired into the architecture of the AMSH subfamily of JAMM domains.

USP Family Members USP8 and Doa4

With the exception of the JAMM domain family described above, the other four families

of DUBs are all Cys proteases (57). The Cys thiol is activated by participation in a Cys-His-Asp/Asn triad similar to that of other Ser and Cys proteases. The unprotonated Cys thiol group acts as a nucleophile to attack the carbon atom of the isopeptide bond. This leads to formation of an acyl-enzyme intermediate and ejection of the Lys residue. The electrostatic environment of the enzyme active site stabilizes the negative charge on the acyl-enzyme intermediate, which is then hydrolyzed by an attacking water molecule. UCH-L3, responsible for deubiquitination of ENaC, is a ubiquitin C-terminal hydrolase (UCH) family member and has been crystallized (45). The largest family of DUBs in the human proteome is the USP family, of which human USP8 is a member. USP8 contains an N-terminal MIT domain, a rhodanese homology domain,

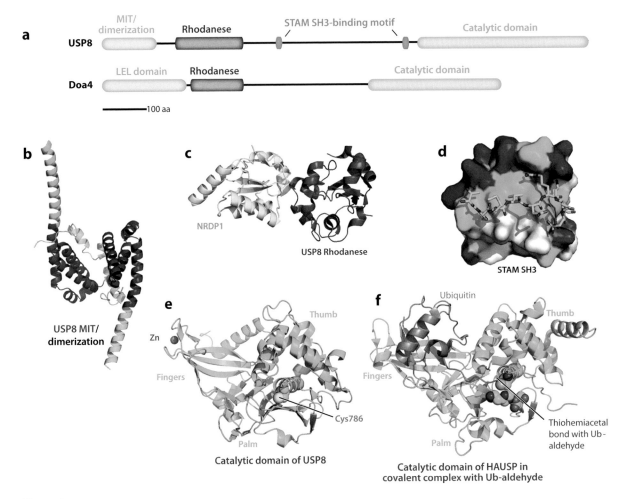

Figure 6

The Cys isopeptidases USP8 and Doa4. (*a*) Domain architecture of USP8 and Doa4. (*b*) Dimeric structure of the N-terminal domain of USP8 (2A9U). The portions of the two monomers that correspond to the MIT sequence motif are colored red and magenta, and the remainder of the subunits are colored green and cyan. (*c*) Structure of the rhodanese domain of USP8 (2GWF) in complex with the noncatalytic C-terminal domain of the ubiquitin ligase NRDP1. (*d*) A surface representation of the SH3 domain of STAM2 (1UJ0). Hydrophobic residues are green, basic residues are blue, acidic residues are red, and uncharged polar residues are white. USP8 residues 699–709 are shown in a stick model colored by atom type. (*e*) Catalytic domain of USP8 in an inactive conformation (2GF0). The noncatalytic (structural) zinc ion is a salmon-colored space-filling sphere. (*f*) Catalytic domain of HAUSP in covalent complex with Ub-aldehyde (1NBF), as a model for the active form of USP8. The residues of the Cys-His-Asp catalytic triad are shown in space-filling spheres and colored by atom type. Abbreviations: MIT, microtubule interaction and transport; Ub, ubiquitin; USP, ubiquitin-specific protease.

a ~500-residue linker, and a catalytic domain (**Figure 6*a***). Yeast Doa4, responsible for deubiquitinating Cps1 and other yeast MVB cargoes, has a domain architecture similar to that of USP8, with both rhodanese and Cys-containing catalytic domains.

The USP8 MIT domain binds tightly to CHMP1B and more weakly to a subset of other CHMPs (97). Unlike other MIT domains, which are stand-alone three-helix bundles, the USP8 MIT is part of a larger dimeric seven-helical domain (5) (**Figure 6*b***). The dimer interface is extensive, but it has not been established that it is required for USP8 function, nor it is known how dimerization might relate to ESCRT-III binding. Doa4 contains an N-terminal domain containing five to seven predicted helices, which is also implicated in recruitment to endosomes by ESCRT-III (3). Although not detectably related by sequence homology, their similar size, function, helical structure, and placement in the domain architecture suggest an underlying structural commonality between these regions of USP8 and Doa4. The USP8 and Doa4 sequences contain a rhodanese domain just C-terminal to their ESCRT-III-binding domain. Classical rhodanese domains use a catalytic Cys to carry out phosphatase or sulfurtransferase reactions, while noncatalytic variants lack the catalytic Cys and have collapsed active sites. The USP8 rhodanese domain belongs to the inactive category and is involved in protein-protein interactions, including a structurally characterized complex with the noncatalytic C-terminal domain of the E3 ligase NRDP1 (5) (**Figure 6c**). Residues 699–709 of the long rhodanese-UCH linker region bind to the STAM SH3 with ~30 μM affinity, and the structure of the complex has been determined (48) (**Figure 6d**). These residues, TPMVN-RENKP, do not conform to an SH3-binding consensus motif, but the nonideal nature of the sequence is partially compensated by electrostatic interactions between the Arg and Lys residues of the motif with acidic pockets on the SH3 domain (48).

The USP8 catalytic domain has been crystallized in an inactive conformation in the absence of ubiquitin (5) (**Figure 6e**). The structure of another USP protein, HAUSP, in a covalent complex with Ub-aldehyde (37) (**Figure 6f**), provides a reasonable model for the mechanism of deubiquitination across the USP family. The catalytic domain has been compared to a hand, with the fingers grasping the bulk of the ubiquitin moiety, while the C-terminal tail of ubiquitin is threaded between the palm and the thumb. The catalytic Cys-His-Asp triad is located between the thumb and

palm such that a covalent bond can be formed between the C-terminal carboxylate of ubiquitin and the catalytic Cys. This places the cargo protein (or a more proximal ubiquitin moiety) on the distal side of the thumb region with respect to the ubiquitin moiety being cleaved (**Figure 6f**).

CONCLUDING REMARKS AND PERSPECTIVES

Since the concept of ubiquitin-dependent membrane traffic was born in the mid-1990s, substantial progress has been made in understanding the molecular details of how cargo proteins are ubiquitinated by E3 ubiquitin ligases, how sorting components such as GGAs and ESCRT subunits recognize the ubiquitinated cargoes, how ubiquitin recognition controls the intracellular itineraries of membrane proteins, and how DUBs are recruited to remove ubiquitin moieties. Even though sufficient data are available to allow a broad overview of ubiquitin as a tag for protein trafficking, several outstanding questions remain to be addressed before such processes can be modeled in detail.

One of the topics that awaits further clarification is the importance of Lys63-linked polyubiquitin chains as a sorting signal. In mammalian systems, there is evidence that Lys63-linked polyubiquitin via the Ube2D1–4 E2 enzymes and the Cbl E3 ligase is crucial for endolysosomal sorting of EGFRs (117), and that Lys63-linked polyubiquitination via the Ubc13 E2 enzyme and the K3 E3 ligase is required for Kaposi's sarcoma-associated and herpes-virus-induced endocytosis and endolysosomal sorting of MHC class I molecules (19, 20). Likewise, Lys63-linked polyubiquitination appears to be required for vacuolar targeting of the yeast amino acid permease, Gap1 (61). On the other hand, there is convincing evidence that monoubiquitin is a sufficient signal for endocytosis of several yeast membrane proteins such as Gap1, the maltose transporter Mal1, and the mating factor receptor Ste2 (61). Moreover, Lys63-linked diubiquitination appears sufficient to target Cps1 to the vacuole (61). It will be important to establish whether the differential requirements for Lys63 polyubiquitination may be related to the types of sorting components that are involved. A few measurements have been made of the affinities and avidities of sorting machinery for ubiquitin chains (59, 94), but more data are needed, particularly in the context of cargo and membranes.

The cargoes described in this review were selected because they have been so well studied compared with most others. Examples such as the specific recognition of phosphorylated EGFR by the TKB domain of Cbl (79) beautifully illustrate the specific targeting and regulation of the ubiquitination machinery. Such clear-cut examples are still more the exception than the rule. Chain specificity remains just as poorly characterized in most cases, with a few spectacular exceptions such as the mechanism of Lys63-ubiquitin-specific cleavage by AMSH (100). More insights into these mechanisms are eagerly awaited.

Ubiquitin signals and ubiquitin receptors operate at multiple sequential steps in trafficking. It remains to be understood how ubiquitinated cargoes are handed off at each step. Simple models, such as the possibility of a gradient of sorting complex UBD ubiquitin affinities that follows the directionality of sorting, do not seem sufficient to account for cargo hand-off. In this connection, it is intriguing that at certain steps in the pathway, ubiquitin ligases, DUBs, and ubiquitin-binding proteins appear to form assemblies with one another. For example, the yeast ESCRTs interact, directly or indirectly, with both the ligase Rsp5 (10, 93) and the DUB Doa4 (3, 67). This raises a host of possibilities for ubiquitin chain remodeling and compartmentalized, complete ubiquitination/deubiquitination cycles that have barely begun to be explored. The molecular gymnastics of the ubiquitin system in membrane trafficking should continue to offer insights and surprises for years to come.

SUMMARY POINTS

1. *S. cerevisiae* Cps1 is the prototypical ubiquitin-dependent biosynthetic cargo. Its sorting from the Golgi complex to the MVB depends on its ubiquitination by the HECT domain E3 ligase Rsp5 and recognition of the ubiquitin moieties by UBDs of the ESCRT complexes. Ubiquitin is recycled by the cysteine isopeptidase DUB Doa4.

2. The level of ENaC at the apical membrane of epithelial cells is controlled by Nedd4-2-dependent ubiquitination and internalization to early endosomes. Nedd4-2 is a HECT family E3 ligase that binds to ENaC through its third WW domain. At the early endosome, deubiquitination by the cysteine isopeptidase DUBs UCH-L3 and USP-45 directs ENaC recycling to the apical membrane. In the absence of DUB action, ENaC is directed to the lysosome for degradation via the ESCRT complexes.

3. Following EGF binding and transphosphorylation of its C terminus, EGFR is ubiquitinated by the RING domain E3 ligase Cbl. At the early endosome, ubiquitin signals are read out by Rabex-5, leading to RAB5 activation, a key event in endosome maturation. EGFR can be recycled following deubiquitination by the zinc-dependent DUB AMSH, or it can bind to GGA3 and the ESCRT complexes and enter the MVB pathway.

4. The key structural features of ubiquitin important for membrane traffic are its Ile44 patch, which binds to nearly all the components of ubiquitin-dependent sorting; its flexible C terminus, which enables acrobatic ubiquitination and deubiquitination reactions and conjugation to diverse protein substrates; and Lys63, which is involved in forming open, flexible polyubiquitin chains.

5. Most ubiquitination events in trafficking are carried out by Nedd4 family E3 ligases, which bind to proline-rich and/or phosphorylated sequences in their cargo substrates.

6. Substrates containing phosphorylated tyrosine are recognized by Cbl through its TKB domain.

7. DUBs are targeted to the trafficking machinery mainly through interactions between their MIT domains and the C termini of ESCRT-III subunits, and between SH3 domains and proline-rich sequences.

8. Certain components of the sorting machinery, including AMSH and the NZF2 domain of *S. cerevisiae* ESCRT-II, are highly specific for Lys63-linked polyubiquitin. Lys63-linked ubiquitin is abundant on substrates of membrane trafficking such as EGFR. However, there is little evidence for Lys63-specific E2 enzymes in trafficking pathways, and definitive evidence that Lys63 linkages are required for trafficking has been elusive.

DISCLOSURE STATEMENT

The authors are not aware of any affiliations, memberships, funding, or financial holdings that might be perceived as affecting the objectivity of this review.

ACKNOWLEDGMENTS

We thank Y. Ye and R. Stanley for comments on the manuscript. This research was supported by the intramural program of the NIH, NIDDK to J.H.H., and by the Research Council of Norway, the Norwegian Cancer Society, and the European Research Council to H.S.

LITERATURE CITED

1. Almaca J, Kongsuphol P, Hieke B, Ousingsawat J, Viollet B, et al. 2009. AMPK controls epithelial Na(+) channels through Nedd4-2 and causes an epithelial phenotype when mutated. *Pflugers Arch.* 458:713–21

2. Alwan HA, van Leeuwen JE. 2007. UBPY-mediated epidermal growth factor receptor (EGFR) de-ubiquitination promotes EGFR degradation. *J. Biol. Chem.* 282:1658–69

3. Amerik A, Sindhi N, Hochstrasser M. 2006. A conserved late endosome-targeting signal required for Doa4 deubiquitylating enzyme function. *J. Cell Biol.* 175:825–35

4. Amerik AY, Nowak J, Swaminathan S, Hochstrasser M. 2000. The Doa4 deubiquitinating enzyme is functionally linked to the vacuolar protein-sorting and endocytic pathways. *Mol. Biol. Cell* 11:3365–80

5. Avvakumov GV, Walker JR, Xue S, Finerty PJ, Mackenzie F, et al. 2006. Amino-terminal dimerization, NRDP1-rhodanese interaction, and inhibited catalytic domain conformation of the ubiquitin-specific protease 8 (USP8). *J. Biol. Chem.* 281:38061–70

6. Babst M, Odorizzi G, Estepa EJ, Emr SD. 2000. Mammalian tumor susceptibility gene 101 (TSG101) and the yeast homologue, Vps23p, both function in late endosomal trafficking. *Traffic* 1:248–58

7. Bache KG, Raiborg C, Mehlum A, Stenmark H. 2003. STAM and Hrs are subunits of a multivalent ubiquitin-binding complex on early endosomes. *J. Biol. Chem.* 278:12513–21

8. Berlin I, Schwartz H, Nash PD. 2010. Regulation of the epidermal growth factor receptor ubiquitination and trafficking by the USP8/STAM complex. *J. Biol. Chem.* 285:34909–21

9. Bishop N, Horman A, Woodman P. 2002. Mammalian class E vps proteins recognize ubiquitin and act in the removal of endosomal protein-ubiquitin conjugates. *J. Cell Biol.* 157:91–101

10. Bowers K, Lottridge J, Helliwell SB, Goldthwaite LM, Luzio JP, Stevens TH. 2004. Protein-protein interactions of ESCRT complexes in the yeast *Saccharomyces cerevisiae*. *Traffic* 5:194–210

11. Butterworth MB. 2010. Regulation of the epithelial sodium channel (ENaC) by membrane trafficking. *Biochim. Biophys. Acta* 1802:1166–77

12. Butterworth MB, Edinger RS, Ovaa H, Burg D, Johnson JP, Frizzell RA. 2007. The deubiquitinating enzyme UCH-L3 regulates the apical membrane recycling of the epithelial sodium channel. *J. Biol. Chem.* 282:37885–893

13. Carpenter G, Cohen S. 1976. 125I-labeled human epidermal growth factor. Binding, internalization, and degradation in human fibroblasts. *J. Cell Biol.* 71:159–71

14. Clague MJ, Urbe S. 2006. Endocytosis: the DUB version. *Trends Cell Biol.* 16:551–59

15. Datta AB, Hura GL, Wolberger C. 2009. The structure and conformation of Lys63-linked tetraubiquitin. *J. Mol. Biol.* 392:1117–24

16. Debonneville C, Flores SY, Kamynina E, Plant PJ, Tauxe C, et al. 2001. Phosphorylation of Nedd4-2 by Sgk1 regulates epithelial Na(+) channel cell surface expression. *EMBO J.* 20:7052–59

17. Dikic I, Wakatsuki S, Walters KJ. 2009. Ubiquitin-binding domains—from structures to functions. *Nat. Rev. Mol. Cell Biol.* 10:659–71

18. Dinudom A, Fotia AB, Lefkowitz RJ, Young JA, Kumar S, Cook DI. 2004. The kinase Grk2 regulates Nedd4/Nedd4-2-dependent control of epithelial Na+ channels. *Proc. Natl. Acad. Sci. USA* 101:11886–90

19. Duncan LM, Nathan JA, Lehner PJ. 2010. Stabilization of an E3 ligase-E2-ubiquitin complex increases cell surfaces MHC class I expression. *J. Immunol.* 184:6978–85

20. Duncan LM, Piper S, Dodd RB, Saville MK, Sanderson CM, et al. 2006. Lysine-63-linked ubiquitination is required for endolysosomal degradation of class I molecules. *EMBO J.* 25:1635–45

21. Dunn R, Klos DA, Adler AS, Hicke L. 2004. The C2 domain of the Rsp5 ubiquitin ligase binds membrane phosphoinositides and directs ubiquitination of endosomal cargo. *J. Cell Biol.* 165:135–44

22. Fakitsas P, Adam G, Daidie D, van Bemmelen MX, Fouladkou F, et al. 2007. Early aldosterone-induced gene product regulates the epithelial sodium channel by deubiquitylation. *J. Am. Soc. Nephrol.* 18:1084–92

23. Ferguson KM, Berger MB, Mendrola JM, Cho HS, Leahy DJ, Lemmon MA. 2003. EGF activates its receptor by removing interactions that autoinhibit ectodomain dimerization. *Mol. Cell* 11:507–17

24. Firsov D, Gautschi I, Merillat AM, Rossier BC, Schild L. 1998. The heterotetrameric architecture of the epithelial sodium channel (ENaC). *EMBO J.* 17:344–52

25. Futter CE, Pearse A, Hewlett LJ, Hopkins CR. 1996. Multivesicular endosomes containing internalized EGF-EGF receptor complexes mature and then fuse directly with lysosomes. *J. Cell Biol.* 132:1011–23

26. Galan JM, Moreau V, Andre B, Volland C, Haguenauer-Tsapis R. 1996. Ubiquitination mediated by the Npi1p/Rsp5p ubiquitin-protein ligase is required for endocytosis of the yeast uracil permease. *J. Biol. Chem.* 271:10946–52

27. Gao S, Alarcon C, Sapkota G, Rahman S, Chen PY, et al. 2009. Ubiquitin ligase Nedd4L targets activated Smad2/3 to limit TGF-beta signaling. *Mol. Cell* 36:457–68

28. Goh LK, Huang F, Kim W, Gygi S, Sorkin A. 2010. Multiple mechanisms collectively regulate clathrin-mediated endocytosis of the epidermal growth factor receptor. *J. Cell Biol.* 189:871–83

29. Haglund K, Sigismund S, Polo S, Szymkiewicz I, Di Fiore PP, Dikic I. 2003. Multiple monoubiquitination of RTKs is sufficient for their endocytosis and degradation. *Nat. Cell Biol.* 5:461–66

30. Hershko A, Ciechanover A, Varshavsky A. 2000. The ubiquitin system. *Nat. Med.* 6:1073–81

31. Hettema EH, Valdez-Taubas J, Pelham HR. 2004. Bsd2 binds the ubiquitin ligase Rsp5 and mediates the ubiquitination of transmembrane proteins. *EMBO J.* 23:1279–88

32. Hicke L. 2001. Protein regulation by monoubiquitin. *Nat. Rev. Mol. Cell Biol.* 2:195–201

33. **Hicke L, Riezman H. 1996. Ubiquitination of a yeast plasma membrane receptor signals its ligand-stimulated endocytosis. *Cell* 84:277–87**

34. Hicke L, Schubert HL, Hill CP. 2005. Ubiquitin-binding domains. *Nat. Rev. Mol. Cell Biol.* 6:610–21

35. Hiwasa T, Sakiyama S, Yokoyama S, Ha JM, Fujita J, et al. 1988. Inhibition of cathepsin L-induced degradation of epidermal growth factor receptors by c-Ha-ras gene products. *Biochem. Biophys. Res. Commun.* 151:78–85

36. Horiuchi H, Lippé R, McBride HM, Rubino M, Woodman P, et al. 1997. A novel Rab5 GDP/GTP exchange factor complexed to Rabaptin-5 links nucleotide exchange to effector recruitment and function. *Cell* 90:1149–59

33. Established the concept of ubiquitination as a determinant of membrane trafficking.

37. Hu M, Li PW, Li MY, Li WY, Yao TT, et al. 2002. Crystal structure of a UBP-family deubiquitinating enzyme in isolation and in complex with ubiquitin aldehyde. *Cell* 111:1041–54

38. Huang F, Goh LK, Sorkin A. 2007. EGF receptor ubiquitination is not necessary for its internalization. *Proc. Natl. Acad. Sci. USA* 104:16904–9

39. Huang F, Kirkpatrick D, Jiang X, Gygi S, Sorkin A. 2006. Differential regulation of EGF receptor internalization and degradation by multiubiquitination within the kinase domain. *Mol. Cell* 21:737–48

40. Huang L, Kinnucan E, Wang GL, Beaudenon S, Howley PM, et al. 1999. Structure of an E6AP-UbcH7 complex: insights into ubiquitination by the E2-E3 enzyme cascade. *Science* 286:1321–26

41. Hummler E, Vallon V. 2005. Lessons from mouse mutants of epithelial sodium channel and its regulatory proteins. *J. Am. Soc. Nephrol.* 16:3160–66

42. Hurley JH, Lee S, Prag G. 2006. Ubiquitin binding domains. *Biochem. J.* 399:361–72

43. Hurley JH, Yang D. 2008. MIT domaina. *Dev. Cell* 14:6–8

44. Ingham RJ, Gish G, Pawson T. 2004. The Nedd4 family of E3 ubiquitin ligases: functional diversity within a common modular architecture. *Oncogene* 23:1972–84

45. Johnston SC, Larsen CN, Cook WJ, Wilkinson KD, Hill CP. 1997. Crystal structure of a deubiquitinating enzyme (human UCH-L3) at 1.8 angstrom resolution. *EMBO J.* 16:3787–96

46. Jozic D, Cardenes N, Deribe YL, Moncalian G, Hoeller D, et al. 2005. Cbl promotes clustering of endocytic adaptor proteins. *Nat. Struct. Mol. Biol.* 12:972–79

47. Kamadurai HB, Souphron J, Scott DC, Duda DM, Miller DJ, et al. 2009. Insights into ubiquitin transfer cascades from a structure of a UbcH5B similar to ubiquitin-HECT(NEDD4L) complex. *Mol. Cell* 36:1095–102

48. Kaneko T, Kumasaka T, Ganbe T, Sato T, Miyazawa K, et al. 2003. Structural insight into modest binding of a non-PXXP ligand to the signal transducing adaptor molecule-2 Src homology 3 domain. *J. Biol. Chem.* 278:48162–68

49. Kanelis V, Rotin D, Forman-Kay JD. 2001. Solution structure of a Nedd4 WW domain-ENaC peptide complex. *Nat. Struct. Biol.* 8:407–12

50. Katzmann DJ, Babst M, Emr SD. 2001. Ubiquitin-dependent sorting into the multivesicular body pathway requires the function of a conserved endosomal protein sorting complex, ESCRT-I. *Cell* 106:145–55

51. Katzmann DJ, Sarkar S, Chu T, Audhya A, Emr SD. 2004. Multivesicular body sorting: Ubiquitin ligase Rsp5 is required for the modification and sorting of carboxypeptidase S. *Mol. Biol. Cell* 15:468–80

52. Kazazic M, Bertelsen V, Pedersen KW, Vuong TT, Grandal MV, et al. 2009. Epsin 1 is involved in recruitment of ubiquitinated EGF receptors into clathrin-coated pits. *Traffic* 10:235–45

53. Kerscher O, Felberbaum R, Hochstrasser M. 2006. Modification of proteins by ubiquitin and ubiquitin-like proteins. *Annu. Rev. Cell. Dev. Biol.* 22:159–80

54. Kieffer C, Skalicky JJ, Morita E, De Domenico I, Ward DM, et al. 2008. Two distinct modes of ESCRT-III recognition are required for VPS4 functions in lysosomal protein targeting and HIV-1 budding. *Dev. Cell* 15:62–73

55. Kim HC, Huibregtse JM. 2009. Polyubiquitination by HECT E3s and the determinants of chain type specificity. *Mol. Cell. Biol.* 29:3307–18

56. Kleyman TR, Carattino MD, Hughey RP. 2009. ENaC at the cutting edge: regulation of epithelial sodium channels by proteases. *J. Biol. Chem.* 284:20447–51

57. Komander D, Clague MJ, Urbe S. 2009. Breaking the chains: structure and function of the deubiquitinases. *Nat. Rev. Mol. Cell Biol.* 10:550–63

58. Komander D, Reyes-Turcu F, Licchesi JDF, Odenwaelder P, Wilkinson KD, Barford D. 2009. Molecular discrimination of structurally equivalent Lys 63-linked and linear polyubiquitin chains. *EMBO Rep.* 10:466–73

59. Kulathu Y, Akutsu M, Bremm A, Hofmann K, Komander D. 2009. Two-sided ubiquitin binding explains specificity of the TAB2 NZF domain. *Nat. Struct. Mol. Biol.* 16:1328–30

60. Lange OF, Lakomek NA, Fares C, Schroder GF, Walter KFA, et al. 2008. Recognition dynamics up to microseconds revealed from an RDC-derived ubiquitin ensemble in solution. *Science* 320:1471–75

61. Lauwers E, Jacob C, Andre B. 2009. K63-linked ubiquitin chains as a specific signal for protein sorting into the multivesicular body pathway. *J. Cell Biol.* 185:493–502

47. Provides the most complete account to date of the molecular gymnastics of NEDD4-family-dependent ubiquitination.

50. Establishes ubiquitin as the key signal for cargo sorting into MVBs, demonstrates ubiquitin-dependent sorting of Cps1, and names and characterizes the first ESCRT complex.

62. Lee JR, Oestreich AJ, Payne JA, Gunawan MS, Norgan AP, Katzmann DJ. 2009. The HECT domain of the ubiquitin ligase Rsp5 contributes to substrate recognition. *J. Biol. Chem.* 284:32126–37

63. Levkowitz G, Waterman H, Ettenberg SA, Katz M, Tsygankov AY, et al. 1999. Ubiquitin ligase activity and tyrosine phosphorylation underlie suppression of growth factor signaling by c-Cbl/Sli-1. *Mol. Cell* 4:1029–40

64. **Levkowitz G, Waterman H, Zamir E, Kam Z, Oved S, et al. 1998. c-Cbl/Sli-1 regulates endocytic sorting and ubiquitination of the epidermal growth factor receptor. *Genes Dev.* 12:3663–74**

65. Lin Q, Wang J, Childress C, Sudol M, Carey DJ, Yang WNA. 2010. HECT E3 ubiquitin ligase Nedd4-1 ubiquitinates ACK and regulates epidermal growth factor (EGF)-induced degradation of EGF receptor and ACK. *Mol. Cell. Biol.* 30:1541–54

66. Lu PJ, Zhou XZ, Shen MH, Lu KP. 1999. Function of WW domains as phosphoserine- or phosphothreonine-binding modules. *Science* 283:1325–28

67. Luhtala N, Odorizzi G. 2004. Bro1 coordinates deubiquitination in the multivesicular body pathway by recruiting Doa4 to endosomes. *J. Cell Biol.* 166:717–29

68. Ma YM, Boucrot E, Villen J, Affar el B, Gygi SP, et al. 2007. Targeting of AMSH to endosomes is required for epidermal growth factor degradation. *J. Biol. Chem.* 282:9805–12

69. Macias MJ, Hyvonen M, Baraldi E, Schultz J, Sudol M, et al. 1996. Structure of the WW domain of a kinase-associated protein complexed with a proline-rich peptide. *Nature* 382:646–49

70. Madshus IH, Stang E. 2009. Internalization and intracellular sorting of the EGF receptor: a model for understanding the mechanisms of receptor trafficking. *J. Cell Sci.* 122:3433–39

71. Malerod L, Stuffers S, Brech A, Stenmark H. 2007. Vps22/EAP30 in ESCRT-II mediates endosomal sorting of growth factor and chemokine receptors destined for lysosomal degradation. *Traffic* 8:1617–29

72. McCullough J, Clague MJ, Urbe S. 2004. AMSH is an endosome-associated ubiquitin isopeptidase. *J. Cell Biol.* 166:487–92

73. Meng WY, Sawasdikosol S, Burakoff SJ, Eck MJ. 1999. Structure of the amino-terminal domain of Cbl complexed to its binding site on ZAP-70 kinase. *Nature* 398:84–90

74. Miller K, Beardmore J, Kanety H, Schlessinger J, Hopkins CR. 1986. Localization of the epidermal growth factor (EGF) receptor within the endosome of EGF-stimulated epidermoid carcinoma (A431) cells. *J. Cell Biol.* 102:500–9

75. Mizuno E, Iura T, Mukai A, Yoshimori T, Kitamura N, Komada M. 2005. Regulation of epidermal growth factor receptor down-regulation by UBPY-mediated deubiquitination at endosomes. *Mol. Biol. Cell* 16:5163–74

76. Morvan J, Froissard M, Haguenauer-Tsapis R, Urban-Grimal D. 2004. The ubiquitin ligase Rsp5p is required for modification and sorting of membrane proteins into multivesicular bodies. *Traffic* 5:383–92

77. Mueller GM, Maarouf AB, Kinlough CL, Sheng N, Kashlan OB, et al. 2010. Cys palmitoylation of the beta subunit modulates gating of the epithelial sodium channel. *J. Biol. Chem.* 285:30453–62

78. Nalefski EA, Falke JJ. 1996. The C2 domain calcium-binding motif: structural and functional diversity. *Protein Sci.* 5:2375–90

79. Ng C, Jackson RA, Buschdorf JP, Sun QX, Guy GR, Sivaraman J. 2008. Structural basis for a novel intrapeptidyl H-bond and reverse binding of c-Cbl-TKB domain substrates. *EMBO J.* 27:804–16

80. Nikko E, Andre B. 2007. Evidence for a direct role of the Doa4 deubiquitinating enzyme in protein sorting into the MVB pathway. *Traffic* 8:566–81

81. Obita T, Saksena S, Ghazi-Tabatabai S, Gill DJ, Perisic O, et al. 2007. Structural basis for selective recognition of ESCRT-III by the AAA ATPase Vps4. *Nature* 449:735–39

82. Ogunjimi AA, Briant DJ, Pece-Barbara N, Le Roy C, Di Guglielmo GM, et al. 2005. Regulation of Smurf2 ubiquitin ligase activity by anchoring the E2 to the HECT domain. *Mol. Cell* 19:297–308

83. Ohya T, Miaczynska M, Coskun U, Lommer B, Runge A, et al. 2009. Reconstitution of Rab- and SNARE-dependent membrane fusion by synthetic endosomes. *Nature* 459:1091–97

84. Penengo L, Mapelli M, Murachelli AG, Confalonieri S, Magri L, et al. 2006. Crystal structure of the ubiquitin binding domains of rabex-5 reveals two modes of interaction with ubiquitin. *Cell* 124:1183–95

85. Pennock S, Wang Z. 2008. A tale of two Cbls: interplay of c-Cbl and Cbl-b in epidermal growth factor receptor downregulation. *Mol. Cell Biol.* 28:3020–37

64. Shows that Cbl-mediated ubiquitination targets EGFR for lysosomal degradation.

86. Perica T, Chothia C. 2010. Ubiquitin: molecular mechanisms for recognition of different structures. *Curr. Opin. Struct. Biol.* 20:367–76

87. Peschard P, Kozlov G, Lin T, Mirza A, Berghuis AM, et al. 2007. Structural basis for ubiquitin-mediated dimerization and activation of the ubiquitin protein ligase Cbl-b. *Mol. Cell* 27:474–85

88. Pickart CM. 2001. Mechanisms underlying ubiquitination. *Annu. Rev. Biochem.* 70:503–33

89. Pickart CM, Fushman D. 2004. Polyubiquitin chains: polymeric protein signals. *Curr. Opin. Chem. Biol.* 8:610–16

90. Puertollano R, Bonifacino JS. 2004. Interactions of GGA3 with the ubiquitin sorting machinery. *Nat. Cell Biol.* 6:244–51

91. Raiborg C, Malerod L, Pedersen NM, Stenmark H. 2008. Differential functions of Hrs and ESCRT proteins in endocytic membrane trafficking. *Exp. Cell Res.* 314:801–13

92. Reggiori F, Pelham HR. 2001. Sorting of proteins into multivesicular bodies: ubiquitin-dependent and -independent targeting. *EMBO J.* 20:5176–86

93. Ren J, Kee Y, Huibergtse JM, Piper RC. 2007. Hse1, a component of the yeast Hrs-STAM ubiquitin sorting complex, associates with ubiquitin peptidases and a ligase to control sorting efficiency into multivesicular bodies. *Mol. Biol. Cell* 18:324–35

94. Ren XF, Hurley JH. 2010. VHS domains of ESCRT-0 cooperate in high-avidity binding to polyubiquitinated cargo. *EMBO J.* 29:1045–54

95. Richter C, West M, Odorizzi G. 2007. Dual mechanisms specify Doa4-mediated deubiquitination at multivesicular bodies. *EMBO J.* 26:2454–64

96. Rotin D. 2008. Role of the UPS in Liddle syndrome. *BMC Biochem.* 9(Suppl. 1):S5

97. Row PE, Lui H, Hayes S, Welchman R, Charalabous P, et al. 2007. The MIT domain of UBPY constitutes a CHMP binding and endosomal localization signal required for efficient EGF receptor degradation. *J. Biol. Chem.* 282:30929–37

98. Row PE, Prior IA, McCullough J, Clague MJ, Urbe S. 2006. The ubiquitin isopeptidase UBPY regulates endosomal ubiquitin dynamics and is essential for receptor down-regulation. *J. Biol. Chem.* 281:12618–24

99. Roxrud I, Raiborg C, Pedersen NM, Stang E, Stenmark H. 2008. An endosomally localized isoform of Eps15 interacts with Hrs to mediate degradation of epidermal growth factor receptor. *J. Cell Biol.* 180:1205–18

100. Sato Y, Yoshikawa A, Mimura H, Yamashita M, Yamagata A, Fukai S. 2009. Structural basis for specific recognition of Lys 63-linked polyubiquitin chains by tandem UIMs of RAP80. *EMBO J.* 28:2461–68

101. Sato Y, Yoshikawa A, Yamagata A, Mimura H, Yamashita M, et al. 2008. Structural basis for specific cleavage of Lys 63-linked polyubiquitin chains. *Nature* 455:358–62

102. Sato Y, Yoshikawa A, Yamashita M, Yamagata A, Fukai S. 2009. Structural basis for specific recognition of Lys 63-linked polyubiquitin chains by NZF domains of TAB2 and TAB3. *EMBO J.* 28:3903–9

103. Shearwin-Whyatt L, Dalton HE, Foot N, Kumar S. 2006. Regulation of functional diversity within the Nedd4 family by accessory and adaptor proteins. *Bioessays* 28:617–28

104. Sigismund S, Argenzio E, Tosoni D, Cavallaro E, Polo S, Di Fiore PP. 2008. Clathrin-mediated internalization is essential for sustained EGFR signaling but dispensable for degradation. *Dev. Cell* 15:209–19

105. Sigismund S, Woelk T, Puri C, Maspero E, Tacchetti C, et al. 2005. Clathrin-independent endocytosis of ubiquitinated cargos. *Proc. Natl. Acad. Sci. USA* 102:2760–65

106. Sloper-Mould KE, Jemc JC, Pickart CM, Hicke L. 2001. Distinct functional surface regions on ubiquitin. *J. Biol. Chem.* 276:30483–89

107. Snyder PM, McDonald FJ, Stokes JB, Welsh MJ. 1994. Membrane topology of the amiloride-sensitive epithelial sodium channel. *J. Biol. Chem.* 269:24379–83

108. Sorkina T, Huang F, Beguinot L, Sorkin A. 2002. Effect of tyrosine kinase inhibitors on clathrin-coated pit recruitment and internalization of epidermal growth factor receptor. *J. Biol. Chem.* 277:27433–41

109. Soubeyran P, Kowanetz K, Szymkiewicz I, Langdon WY, Dikic I. 2002. Cbl-CIN85-endophilin complex mediates ligand-induced downregulation of EGF receptors. *Nature* 416:183–87

110. Soundararajan R, Pearce D, Hughey RP, Kleyman TR. 2010. Role of epithelial sodium channels and their regulators in hypertension. *J. Biol. Chem.* 285:30363–69

101. Provides a definitive structural explanation for the Lys63 linkage specificity of AMSH.

111. Spormann DO, Heim J, Wolf DH. 1992. Biogenesis of the yeast vacuole (lysosome). The precursor forms of the soluble hydrolase carboxypeptidase yscS are associated with the vacuolar membrane. *J. Biol. Chem.* 267:8021–29

112. Stang E, Blystad FD, Kazazic M, Bertelsen V, Brodahl T, et al. 2004. Cbl-dependent ubiquitination is required for progression of EGF receptors into clathrin-coated pits. *Mol. Biol. Cell* 15:3591–604

113. Demonstrates ubiquitin-mediated sorting of ENaC.

113. **Staub O, Gautschi I, Ishikawa T, Breitschopf K, Ciechanover A, et al. 1997. Regulation of stability and function of the epithelial Na+ channel (ENaC) by ubiquitination. *EMBO J.* 16:6325–36**

114. Strous GJ, Van Kerkhof P, Govers R, Ciechanover A, Schwartz AL. 1996. The ubiquitin conjugation system is required for ligand-induced endocytosis and degradation of the growth hormone receptor. *EMBO J.* 15:3806–12

115. Stuchell-Brereton M, Skalicky J, Kieffer C, Karren MA, Ghaffarian S, Sundquist WI. 2007. ESCRT-III recognition by VPS4 ATPases. *Nature* 449:740–44

116. Tsang HTH, Connell JW, Brown SE, Thompson A, Reid E, Sanderson CM. 2006. A systematic analysis of human CHMP protein interactions: Additional MIT domain-containing proteins bind to multiple components of the human ESCRT III complex. *Genomics* 88:333–46

117. Umebayashi K, Stenmark H, Yoshimori T. 2008. Ubc4/5 and c-Cbl continue to ubiquitinate EGF receptor after internalization to facilitate polyubiquitination and degradation. *Mol. Biol. Cell* 19:3454–62

118. Urbe S, McCullough J, Row P, Prior IA, Welchman R, Clague MJ. 2006. Control of growth factor receptor dynamics by reversible ubiquitination. *Biochem. Soc. Trans.* 34:754–56

119. Verdecia MA, Bowman ME, Lu KP, Hunter T, Noel JP. 2000. Structural basis for phosphoserine-proline recognition by group IV WW domains. *Nat. Struct. Mol. Biol.* 7:639–43

120. Verdecia MA, Joazeiro CAP, Wells NJ, Ferrer JL, Bowman ME, et al. 2003. Conformational flexibility underlies ubiquitin ligation mediated by the WWP1 HECT domain E3 ligase. *Mol. Cell* 11:249–59

121. Vijaykumar S, Bugg CE, Cook WJ. 1987. Structure of ubiquitin refined at 1.8 Å resolution. *J. Mol. Biol.* 194:531–44

122. Weissman AM. 2001. Themes and variations on ubiquitylation. *Nat. Rev. Mol. Cell Biol.* 2:169–78

123. Witsch E, Sela M, Yarden Y. 2010. Roles for growth factors in cancer progression. *Physiology* 25:85–101

124. Yarden Y. 2001. The EGFR family and its ligands in human cancer. Signalling mechanisms and therapeutic opportunities. *Eur. J. Cancer* 37(Suppl. 4):S3–8

125. Zamborlini A, Usami Y, Radoshitzky SR, Popova E, Palu G, Gottlinger H. 2006. Release of autoinhibition converts ESCRT-III components into potent inhibitors of HIV-1 budding. *Proc. Natl. Acad. Sci. USA* 103:19140–45

126. This 10-year-old study is still the definitive structural account of Cbl-dependent ubiquitination.

126. **Zheng N, Wang P, Jeffrey PD, Pavletich NP. 2000. Structure of a c-Cbl-UbcH7 complex: RING domain function in ubiquitin-protein ligases. *Cell* 102:533–39**

127. Zhou R, Kabra R, Olson DR, Piper RC, Snyder PM. 2010. Hrs controls sorting of the epithelial Na+ channel between endosomal degradation and recycling pathways. *J. Biol. Chem.* 285:30523–30

128. Zhu H, Liang Z, Li G. 2009. Rabex-5 is a Rab22 effector and mediates a Rab22-Rab5 signaling cascade in endocytosis. *Mol. Biol. Cell* 20:4720–29

The Cyanobacterial Circadian System: From Biophysics to Bioevolution

Carl Hirschie Johnson,[1,2] Phoebe L. Stewart,[2] and Martin Egli[3]

[1]Department of Biological Sciences, [2]Department of Molecular Physiology and Biophysics, and [3]Department of Biochemistry, Vanderbilt University, Nashville, Tennessee 37235; email: carl.h.johnson@vanderbilt.edu

Annu. Rev. Biophys. 2011. 40:143–67

First published online as a Review in Advance on February 14, 2011

The *Annual Review of Biophysics* is online at biophys.annualreviews.org

This article's doi: 10.1146/annurev-biophys-042910-155317

Copyright © 2011 by Annual Reviews. All rights reserved

1936-122X/11/0609-0143$20.00

Keywords

cyanobacteria, Kai, KaiABC, cell division, in vitro oscillators

Abstract

Recent studies have unveiled the molecular machinery responsible for the biological clock in cyanobacteria and found that it exerts pervasive control over cellular processes including global gene expression. Indeed, the entire chromosome undergoes daily cycles of topology/compaction! The circadian system comprises both a posttranslational oscillator (PTO) and a transcriptional/translational feedback loop (TTFL). The PTO can be reconstituted in vitro with three purified proteins (KaiA, KaiB, and KaiC) and ATP. These are the only circadian proteins for which high-resolution structures are available. Phase in this nanoclockwork has been associated with key phosphorylations of KaiC. Structural considerations illuminate the mechanism by which the KaiABC oscillator ratchets unidirectionally. Models of the complete in vivo system have important implications for our understanding of circadian clocks in higher organisms, including mammals. The conjunction of structural, biophysical, and biochemical approaches to this system has brought our understanding of the molecular mechanisms of biological timekeeping to an unprecedented level.

Contents

BACTERIA HAVE CIRCADIAN RHYTHMS

Many biological oscillations have been analyzed biophysically, but most of those oscillators are of relatively high frequency (e.g., millisecond to second) and often involve ionic fluxes across membranes. Circadian rhythms, on the other hand, are ~24-h oscillations in biological processes that are controlled by an endogenous

DD: constant darkness

LL: constant light

biochemical pacemaker. The processes for which activities are choreographed by these clocks range from gene expression, metabolism, and cell division, to development and behavior (12). Circadian rhythms are defined by three diagnostic properties: (*a*) persistence of the oscillations in constant conditions (usually constant darkness, DD, or constant light, LL, at constant temperature), (*b*) temperature compensation (the period length is only slightly affected by temperature changes, i.e., Q_{10} ~0.9–1.1), and (*c*) entrainment of the endogenous pacemaker to the environmental cycle of light and dark (12). Most difficult to explain from the biophysical perspective are the precision of this long time-constant oscillator (~24 h $\pm$ only a few minutes per day) and the temperature compensation property (which is true even for cells and tissues from endothermic animals) (12, 29, 95). However, from an evolutionary perspective, a temperature-dependent or imprecise clock is likely to be useless as an endogenous estimator of environmental time (12, 80). Therefore, a ~24-h clock with the conserved properties of temperature compensation, entrainment, and precision has been the product of natural selection in organisms from bacteria to human. What has not been conserved among cyanobacteria, fungi, plants, and animals are the sequences of the proteins, which are the gears and cogs of these clocks. This implies that circadian clocks have convergently evolved multiple times in response to the selective pressure of an environment with daily cycles (83).

The prokaryotic cyanobacterium *Synechococcus elongatus* PCC7942 has proven advantageous for circadian clock research (10, 43). Cyanobacteria have worldwide importance. The marine cyanobacterium *Prochlorococcus marinus* is possibly the most abundant photosynthetic organism on earth and certainly contributes a large proportion of total global photosynthetic activity (74). Moreover, cyanobacteria are being enlisted as platforms for production of biofuels. This includes *S. elongatus*, which is a unicellular bacterium that depends on photosynthesis autotrophically and therefore could be used to produce biofuels by using sunlight as an inexpensive energy source (4, 5). Until the late 1980s/early 1990s, circadian biologists were reluctant to believe that organisms as simple as prokaryotes could have evolved an elaborate circadian timing mechanism (10, 34); they reasoned that a rapidly dividing bacterium whose lifetime was less than one day had no use for a timing mechanism that extended farther than its lifetime (80). However, *S. elongatus* can divide as rapidly as once every 5–6 h and can still show circadian rhythms without significant perturbation of its circadian pacemaker (32, 42, 55, 59). At this time, there is no question that the ~24-h rhythmic phenomena exhibited by *S. elongatus* are regulated by a bona fide circadian system (10, 43).

The circadian oscillator in *S. elongatus* has uniquely favorable characteristics for biophysical, biochemical, and genetic analyses (10). It is the only organism for which we have full structural information for the key clock proteins (in this case, KaiA, KaiB, and KaiC). *S. elongatus* has a genome size of 2.7 Mbp (smaller than that of *Escherichia coli*), and genetic tools abound (3, 21). Cyanobacteria are one of the few systems in which the adaptive significance of circadian programs has been rigorously tested (12, 31, 73, 103). Most significantly from a biophysical perspective, it is the only circadian system in which a molecular oscillator can be studied in vitro; persistence, precision, and temperature compensation can be reconstituted in vitro with three purified proteins (KaiA + KaiB + KaiC) and ATP (68). Although biophysical analyses of this oscillator have begun (6, 23, 63, 65, 81), we are at the watershed of understanding how this molecular oscillator really works (33, 53).

WHAT'S RHYTHMIC? OUTPUTS OF THE PACEMAKER IN CYANOBACTERIA

The first persuasive evidence for circadian rhythms in a prokaryote came from an investigation that studied the nitrogen-fixing cyanobacterium *Synechococcus* RF1 (10, 16). In this cyanobacterium, nitrogen fixation is regulated by the circadian clock such that it is maximal in the night phase. The enzyme that carries out the reduction of atmospheric nitrogen to ammonia is nitrogenase, an enzyme that is inhibited by oxygen. Because photosynthesis produces oxygen throughout the day and nitrogenase is sensitive to oxygen, turning on nitrogen fixation in the nocturnal phase allows the same cell to perform incompatible metabolic events: photosynthesis during the day and nitrogen fixation during the night (57). This is an example of how the evolutionary emergence of circadian systems may have enhanced fitness by optimizing temporal metabolic programs. In our search for a genetically malleable cyanobacterium, we and our collaborators settled on *S. elongatus* PCC 7942 (which incidentally does not fix nitrogen). The genetic properties of this organism facilitated our discovery of globally regulated gene expression by a circadian timekeeper. We use bacterial luciferase as a reporter of clock-regulated promoter activity; initially we studied the activity of the promoter for the *psbAI* gene (43), but we subsequently discovered that virtually all promoters in the *S. elongatus* genome are regulated by the circadian system (50). **Figure 1a** depicts rhythms based on a few selected promoter::reporter constructs, including the cyanobacterial promoters for the *psbAI*, *kaiA*, *kaiBC*, *purF*, and *ftsZ* genes. The majority of promoters are activated in the subjective day phase. However, the *purF* promoter is activated in the nocturnal phase (**Figure 1a**), an interesting observation considering that its gene product is involved in an oxygen-sensitive pathway (*purF* encodes the enzyme catalyzing the initial step of de novo purine nucleotide biosynthesis; 49). Thus, the circadian regulation of the expression of this *S. elongatus* gene might constitute another example of temporal separation similar to that found for nitrogenase in *Synechococcus* RF1 (10, 16, 49, 50).

As expected from the global control of promoter activity, there is pervasive control by the circadian clock of mRNA abundances in cyanobacteria. In *S. elongatus* and other cyanobacterial species (*Synechocystis* sp. PCC

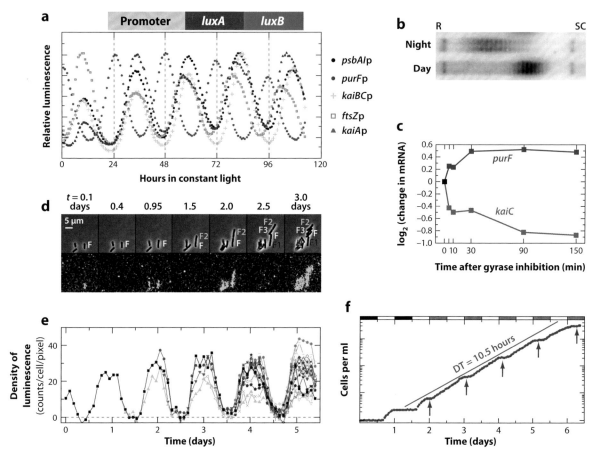

Figure 1

Circadian rhythms in *Synechococcus elongatus*. (*a*) Rhythms of luminescence emanating from cells transformed with bacterial luciferase (*luxAluxB*) fused to the promoters for the *psbAI*, *purF*, *kaiBC*, *ftsZ*, and *kaiA* genes. This plot illustrates circadian rhythms of gene expression. (*b*) Supercoiling of an endogenous plasmid indicates a circadian rhythm in chromosomal topology. In the subjective night topoisomers of the plasmid are more relaxed (R), whereas in the subjective day they are more supercoiled (SC) (104). (*c*) Gyrase inhibition results in an immediate change in gene expression due to drug-induced relaxation. Genes that have higher expression during relaxed circadian times immediately increase in gene expression (*purF*, *red*), whereas genes that have lower expression during relaxed circadian times immediately decrease in gene expression (*kaiC*, *blue*) (from Reference 100 with permission). (*d*) Micrographs of cyanobacterial cells at different times in constant light. Brightfield images (*upper panels*) show growth and cell division as a function of approximate circadian time. Observed luminescence (*lower panels*) reveals circadian rhythms in single cyanobacterial cells. The luminescence reporter was the *psbAI* promoter driving expression of bacterial luciferase. (*e*) Quantification of bioluminescence from a single cell as it divides in constant light. Cell division is indicated by differently colored traces for each daughter cell. (Panels *d* and *e* courtesy of Dr. Irina Mihalcescu from Reference 55). (*f*) Cell division in a population of *S. elongatus* cells is restricted by the circadian system. For the first 36 h the cells are in a light/dark (LD) cycle as indicated by the black and gray bars at the top of the panel. For the remaining time the cells are in constant light (LL). The cell count shows plateaus (*red arrows*) when the cells stop dividing. Plateaus occur during the night in LD cycles as well as the subjective night of LL. The average doubling time (DT) as indicated by the diagonal line was 10.5 h (59).

6803, *Crocosphaera watsonii*, and *Prochlorococcus* MED4), microarray analyses have confirmed that the transcripts of 10%–80% of genes in the genome exhibit circadian and/or daily oscillations of abundance (24, 44, 87, 100, 115). However, whereas promoter activity indicated essentially 100% of the genes are expressed rhythmically (50), microarray analysis in

S. elongatus found that only 30%–60% of mRNA abundances are rhythmic (24, 100). Presumably posttranscriptional regulation accounts for the differences in these two measurements. Transcripts that are either very unstable or very stable may be transcribed rhythmically, but the steady-state levels of the message may nevertheless be nearly constitutive. In initial studies of *S. elongatus* rhythms, the clarity of the *psbAI* promoter activity rhythm was much cleaner than that of *psbAI* mRNA abundance, implying posttranscriptional modulation of transcript abundances (43). These data suggest that in *S. elongatus* measurements of promoter activity are a more accurate indicator of circadian control over gene expression than of transcript abundance. An amusing aspect of studying gene expression in *S. elongatus* is that rhythmic gene expression does not always translate into rhythms of the implicated processes. For example, even though *psbAI* encodes a key photosynthesis protein (the three *psbA* genes encode the D1 protein of photosystem II), implying that photosynthesis is rhythmically regulated, experimental measurements of photosynthetic rate in *S. elongatus* have shown that there is a daily rhythm (in light/dark, LD) of photosynthetic capacity but not a circadian rhythm in LL (112).

How are the global rhythms of promoter activity mediated from the central clockwork? Two alternative scenarios have been proposed: a traditional transcription factor network and an oscillating chromosome model. The transcription factor model is based on studies implicating the putative transcription factor RpaA, which appears to be coupled to the cyanobacterial KaiABC oscillator by the histidine kinase SasA (28, 90). New data implicate other factors in this output pathway, including LabA and CikA (92). CikA may be an important component of both the input (86) and the output pathways. The other model for circadian regulation of global gene expression is the oscillating chromosome hypothesis (60, 104). One argument against mediation of global regulation solely by transcriptional factors that have coevolved with *cis* elements of *S. elongatus*

promoters is that heterologous promoters from *E. coli* such as *conII*p and *trc*p exhibit circadian activity in *S. elongatus* (38, 67). Chromosomal topology/compaction provides a high-level way to influence promoter activity. In *S. elongatus* the circadian clock controls pervasive changes in the compaction and topology of the entire chromosome. Dramatic circadian changes are seen for both compaction/decompaction (visualized by DNA-binding dyes) (88) and DNA topology (indicated by plasmid supercoiling) (104) (**Figure 1***b*). DNA topology and torsion critically affect transcriptional rates, and it is therefore reasonable to hypothesize that such circadian changes in chromosomal topology could be partially responsible for daily modulation of promoter activity (56, 60, 88, 104). We have termed this hypothesis the oscillating nucleoid, or oscilloid model (60, 104).

The role of chromosomal topology in regulating circadian gene expression has always been a "chicken or the egg" dilemma. In other words, is transcription rhythmic because the chromosome is being rhythmically supercoiled (i.e., the oscilloid hypothesis), or is the chromosome cyclically supercoiled because transcription is rhythmic? Recent results favor the oscilloid model (100). Gyrase is a key enzyme that regulates DNA superhelicity, and it is inhibited by the drug novobiocin, which thereby relaxes DNA. When novobiocin is added to *S. elongatus* cells at a phase in which the chromosome is normally negatively supercoiling, there is an immediate change in gene expression, and genes that are normally expressed in antiphase (*kaiBC* versus *purF*, **Figure 1***a*) respond in opposite directions (**Figure 1***c*). This result strongly supports the oscilloid hypothesis that *kaiBC* expression is turned on by negatively supercoiled DNA (and *purF* is turned off), and when the chromosome relaxes 12 h later, the relative expression levels of these two classes of genes flips (100). The two available scenarios for global expression patterns (oscilloid hypothesis versus transcriptional regulation by factors such as RpaA, SasA, LabA, and CikA) do not necessarily exclude each other. An analysis of stochastic gene

LD: light/dark cycle

expression in cyanobacteria (8) provides support for the idea that circadian gene expression is regulated by multiple factors, e.g., changes in both DNA topology and transcriptional factor activity.

Rhythmicity has been recorded both for entire populations of cells (**Figure 1a**) and for a single cyanobacterial cell, as demonstrated by luminescence rhythms from the luciferase reporter fused to the *psbAI* promoter (55) (**Figure 1d**). Remarkably, this study also demonstrated that cell division does not perturb the circadian oscillator (**Figure 1e**). The clock of a daughter cell ticks in phase with the clock of the mother cell. This is consistent with the outcome of analyses of circadian timing in populations of dividing cells. Whether *S. elongatus* cells undergo rapid or slow division (or whether they do not divide at all) does not perturb the intrinsic ~24-h period of the circadian system (42, 59, 61). Moreover, the circadian oscillator specifies a checkpoint for division by regulating the timing at which cell division is permitted. It can be experimentally demonstrated that a population of cells rapidly dividing in LL (average doubling time of 10.5 h) will have cell division restricted by the circadian system (**Figure 1f**). This circadian gating of cell division in *S. elongatus* has been studied recently by the Golden laboratory, and they reported that elevated ATPase activity of KaiC may provide the

circadian checkpoint in cyanobacteria (11). Moreover, recent studies suggest that cell division in *S. elongatus* is influenced by the circadian-implicated genes *cikA* and *cdpA* (11, 51).

Pervasive changes in metabolism, cellular structure, and gene expression that accompany cell division do not perturb circadian timing. This imperturbability appears to be a general property of circadian pacemakers in cyanobacteria and in eukaryotes, and it might be argued that this constitutes evidence for a cyanobacteria-like pacemaking mechanism in eukaryotes (32).

THE TIMEKEEPING MECHANISM: KaiA, KaiB, and KaiC ARE THE CLOCKWORKS' GEARS

The proteins KaiA, KaiB, and KaiC constitute the central components of the clockwork in *S. elongatus* (22). The *kaiA*, *kaiB*, and *kaiC* genes were first identified in 1998, and three-dimensional structures for the proteins they encode became available in 2004 (13, 15, 20, 76, 98, 111). They remain the only core circadian clock proteins for which full-length structures have been determined. KaiA is a domain-swapped dimer with an N-terminal bacterial receiver domain and a C-terminal α-helical

Figure 2

Structures, rhythmic phosphorylation, and associations of KaiA, KaiB, and KaiC. (*a*) Shown from left to right are the crystal structures of the *S. elongatus* KaiA dimer (111), the *Synechocystis* KaiB tetramer (20), and the *S. elongatus* KaiC hexamer (76, 78). Individual subunits of the multimeric proteins are represented in different colors. In the case of KaiC, the subunits are arranged around a central channel that runs vertically (behind the dark blue-colored subunit in this depiction (13). (*b*) Time courses of rhythmic KaiC phosphorylation in vivo and in vitro as assessed by SDS-PAGE (the lowest bands are hypophosphorylated KaiC and the upper bands are various forms of phosphorylated KaiC). Top: KaiC phosphorylation in vivo at different times in constant light (samples were collected every 4 h in constant light and immunoblotted). Bottom: KaiC phosphorylation in the in vitro reaction. Purified KaiA, KaiB, and KaiC were combined with ATP in vitro and samples were collected every 3 h and processed for SDS PAGE and staining. Four bands are obvious in these in vitro samples: hypophosphorylated KaiC and KaiC phosphorylated at the S431, T432, or S431/T432 residues (see labels on the right side of the panel). (*c*) Rhythms of KaiA·KaiB·KaiC complex formation during the in vitro cycling reaction. The color coding of the pie charts indicates the percentage of free KaiC hexamers (*blue*), KaiA·KaiC complexes (*brown*), KaiB·KaiC (*green*) complexes, and KaiA·KaiB·KaiC (*orange*) complexes (63). (*d*) Electron microscopy average images of free KaiC hexamers, KaiA·KaiC complexes, KaiB·KaiC complexes, and presumed KaiA·KaiB·KaiC complexes (63) (color coding is the same as in panel *c*).

bundle, and KaiB adopts a thioredoxin-like fold and forms dimers and tetramers. KaiC is by far the largest of the three and exists as a homohexamer with the appearance of a double-doughnut (**Figure 2a**). The *kaiC* gene is the result of a gene duplication and KaiC displays similar N- and C-terminal domains at the monomeric level (the lobes are referred to as CI

and CII, respectively). Six ATP molecules are bound between subunits in both the CI and the CII rings. The astonishing breakthrough from the Kondo laboratory in 2005 was the report that a molecular oscillator can be reconstituted when the three Kai proteins are combined together with ATP in a test tube (68). This in vitro oscillator ticks with a ~24-h period for at

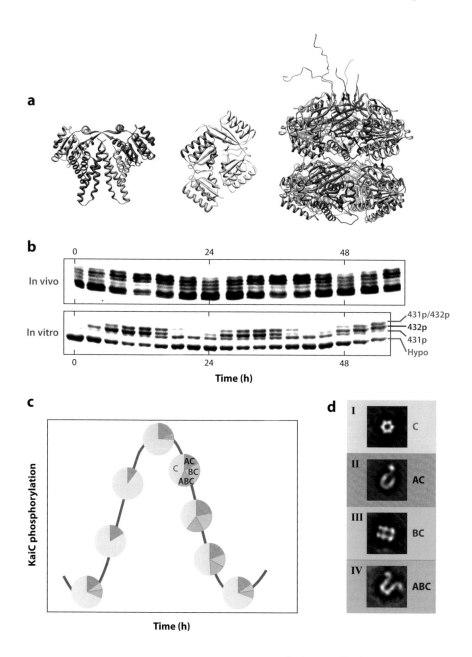

a

b

In vivo

In vitro

431p/432p
432p
431p
Hypo

Time (h)

c

KaiC phosphorylation

C AC
BC
ABC

Time (h)

d

I — C

II — AC

III — BC

IV — ABC

EM: electron
microscopy

SAXS: small angle
X-ray scattering

least 10 days, with KaiC alternating between a hypophosphorylated and a hyperphosphorylated state (**Figure 2b**) (23). Mass spectrometry and X-ray crystallography were used to establish that KaiC is phosphorylated at S431 and T432 in the CII lobe (72, 108). There appear to be no phosphorylation sites in the CI domain. Phosphorylation of T432 and S431 proceeds across the subunit-subunit interface (76, 108).

KaiC exhibits both autokinase and autophosphatase activities (27, 71, 85, 107), and it has been reported that 15 ATP molecules are rhythmically hydrolyzed per subunit during a complete in vitro 24-h cycle (93). KaiA serves as a promoter of the formation of the KaiC hyperphosphorylated state, and KaiB antagonizes KaiA's actions and promotes a return to the hypophosphorylated state. Structural and biophysical studies have provided insight into the KaiA·KaiC (78, 97) and KaiB·KaiC complexes (77). A range of techniques, including gel filtration chromatography, two-dimensional gel electrophoresis, negative-stain electron microscopy (EM), and small angle X-ray scattering (SAXS), have been employed to quantify the relative levels of KaiC versus KaiA·KaiC versus KaiB·KaiC versus KaiA·KaiB·KaiC complexes formed during the in vitro reaction cycle (**Figure 2c,d**) (1, 37, 63). Another key aspect of the in vivo and in vitro KaiC phosphorylation cycle is the strict order of phosphorylation and dephosphorylation of the two P sites, which involves four steps: (*a*) T432 phosphorylation, (*b*) S431 phosphorylation, (*c*) T432 dephosphorylation, and (*d*) S431 dephosphorylation (**Figure 2b**) (71, 85). The vital challenge is to understand the underlying molecular mechanisms of this clockwork.

GETTING TOGETHER: CYCLING INTERACTIONS AMONG THE Kai PROTEINS

KaiA enhances the autokinase activity of KaiC by binding to a C-terminal peptide from a subunit of the latter repeatedly and rapidly (**Figure 3**) (37, 81). A single KaiA dimer appears to be sufficient to upregulate phosphorylation of a KaiC hexamer to saturated levels (18), consistent with the higher abundance of KaiC hexamers in vivo relative to KaiA dimers (40). What is the mechanism underlying KaiA's function? NMR spectroscopy established that KaiA binds to the C-terminal tentacle peptides of KaiC (97). This interaction unravels an S-shaped loop within the contacted KaiC subunit as KaiA pulls the S-loop adjacent to the central channel of the KaiC hexamer (13). The KaiC crystal structure revealed that S-loop residues at amino acids 485–497 form hydrogen bonds across subunits at the periphery of the channel (76). Therefore, one can expect the disruption of the S-shaped loop of a single subunit to weaken the interface between adjacent CII lobes and to promote conformational changes within the CII ring conducive to phosphorylation at T432 and S431. A three-dimensional EM structure of the KaiA·KaiC complex revealed that KaiA assumes at least two orientations above the C-terminal dome of the KaiC hexamer. In one orientation, KaiA is tethered to KaiC via a flexible linker (78). In the second orientation, KaiA is engaged on the KaiC surface, which is suggestive of a transient interaction between an apical loop in the C-terminal domain of a KaiA monomer and the ATP-binding cleft on KaiC.

Unlike KaiA, which binds as a dimer to the C-terminal tentacle portion of KaiC, KaiB does not exhibit any affinity to these tentacles. Moreover, whereas KaiA remains associated with KaiC during the entire phosphorylation cycle, KaiB displays a distinct preference for the phosphorylated form of the hexamer (37, 63, 71, 85). Hybrid structural biology approaches including cryo- and negative-stain EM and X-ray crystallography, along with native PAGE and fluorescence methods, revealed that KaiB dimers bind to the CII ring (77). Thus, EM images are consistent with KaiB dimers forming a third layer on top of CII without obscuring the central channel. This arrangement serves to prevent KaiA from approaching the ATP-binding clefts on KaiC, although the KaiA dimer is still tethered to the C-terminal CII peptide.

More recently, we discovered that although the initial interaction of KaiA with

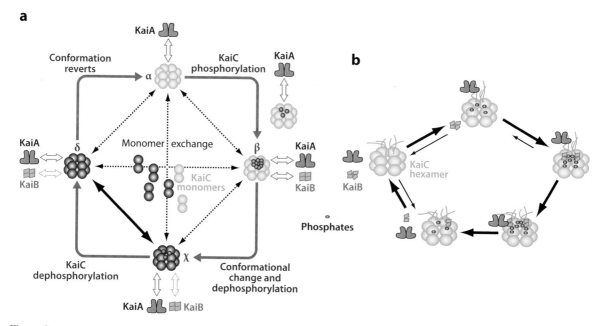

Figure 3

Models of the KaiABC oscillator. (*a*) Diagram representing the mathematical model for the KaiC phosphorylation cycle. The double circle dumbbell shapes in the center represent KaiC monomers. The KaiC hexamer can associate with and dissociate from KaiA and KaiB. KaiC hexamers are shown in light blue and dark blue, representing two conformational states (approximately equivalent to kinase versus phosphatase forms of KaiC). Red dots are phosphates at KaiC phosphorylation sites (residues S431 and T432). Monomer exchange between KaiC hexamers is depicted with the double-headed arrows in the center. The rates of monomer exchange vary among KaiC states, with a solid line indicating a high rate and a dashed line indicating a low rate (from Reference 63). (*b*) Diagram showing the formation of KaiA·KaiB·KaiC complexes. Starting from the leftmost molecular representation and proceeding clockwise: during the phosphorylation phase of the cycling reaction, KaiA (*red dimers*) repeatedly and rapidly interacts with KaiC's C-terminal tentacles. (KaiC molecules are the blue double-donut hexamers.) When KaiC becomes hyperphosphorylated (phosphates on T432 and S431 are depicted as red dots), it first binds KaiB (*green diamonds*) stably. Then, the KaiB·KaiC complex binds KaiA, sequestering it from further interaction with KaiC's tentacles. At that point, KaiC initiates dephosphorylation. When KaiC is hypophosphorylated, it releases KaiB and KaiA, thereby launching a new cycle (from Reference 81).

unphosphorylated KaiC is labile, once KaiC becomes hyperphosphorylated and binds KaiB, KaiA is incorporated into a stable A·B·C complex (81). Amazingly, this complex does not dissociate during a several-hour electrophoresis in native PAGE. Moreover, the formation of this stable complex is not dependent on KaiC's tentacles because KaiA and KaiB form the stable A·B·C complex with the hyperphosphorylated KaiC[489] mutant from which the C-terminal tentacles have been deleted (81). Our current hypothesis is that KaiA first repetitively interacts with the tentacles of hypophosphorylated KaiC to enhance KaiC's autokinase activity until KaiC is hyperphosphorylated, at which time

the KaiC hexamer undergoes a conformational change that allows it to form a stable complex with KaiB (**Figure 3b**) (27, 33, 37, 39, 63, 81, 97). This stable KaiB·KaiC complex exposes a novel binding site for KaiA, which sequesters KaiA in a stable KaiA·KaiB·KaiC complex. The sequestered KaiA is unable to further stimulate the autokinase activity of KaiC, and therefore the autophosphatase activity dominates such that KaiC dephosphorylates to its hypophosphorylated conformation from which KaiB and KaiA dissociate and the cycle begins anew (**Figure 3b**) (81). Using electron spin resonance (ESR) spectroscopy, Mutoh et al. (65) found that spin-labeled Cys mutants

ESR: electron spin resonance

of KaiB from the thermophilic cyanobacterium *Thermosynechococcus elongatus* directly interacted with KaiA from the same species. Fluorescence and native PAGE assays of mixtures of *S. elongatus* KaiA and KaiB in the absence of KaiC did not reveal any interactions between the two proteins (77). Therefore, either the KaiA-KaiB interaction revealed by ESR is too transient to be captured by fluorescence or native gel analyses, or KaiB binding to KaiC creates a new interface for KaiA binding in *Synechococcus*.

KaiC subunits have two critical phosphorylation sites, S431 and T432. The role of a third potentially phosphorylated residue, T426, has been debated. In the crystal structure of the KaiC homohexamer from *S. elongatus*, all six T432 residues and four of the six S431 residues were phosphorylated (76, 108). Closer inspection of the surroundings of the two amino acids revealed that the protein chain curls between residues I425 and I430 so that residues D427, S428, and H429 form a loop in which T426 faces S431. In fact, the last two residues are closely spaced and the side chain of T426 forms a hydrogen bond with the phosphate group of S431-P (76, 108). In the absence of S431 phosphorylation, the two side chains are too far apart to interact. Single T432A, S431A, and T426A mutants are arrhythmic, and we had postulated that T426 might serve as a third phosphorylation site (108). Subsequently, we carried out more detailed biochemical and structural studies on in vitro and in vivo phosphorylation with T426 mutants (75, 109). A central question was whether the amino acid at position 426 must be phosphorylatable, or is it sufficient for it to form a hydrogen bond with S431-P? As a test, an asparagine that should be capable of hydrogen bonding to S431-P was substituted at position 426. This T426N mutant displayed the expected hydrogen bond between 426N and the phosphate on S431 in the crystal structure but was arrhythmic in vivo (75, 109). This observation and other data support the idea that position 426 must be phosphorylatable for the clock to function properly (75, 109). In vivo coexpression studies involving KaiC^{WT} and mutants of KaiC with alternative residues at position 426 demonstrated dramatic effects on dominant/recessive relationships and also revealed that substitutions at T426 alter key properties, such as period, amplitude, robustness, and temperature compensation. Together with the finding that mutations of T426 critically affect the formation of complexes between KaiC and KaiA/KaiB, these observations confirmed that T426 is an important site that regulates the KaiC phosphorylation status in vivo and in vitro.

The association-dissociation process of Kai proteins in vivo and in vitro and the stoichiometry of their complexes in vitro have been analyzed by gel filtration chromatography, native gel assays, and negative-stain EM (36, 37, 63, 81). SAXS has also been used to study the dynamics of untagged Kai proteins as they assemble and disassemble over the in vitro oscillation (1). SAXS data provided support for the idea that the initial phase of the cyanobacterial circadian oscillator is determined largely by the assembly and disassembly of Kai proteins. Further, the period of the clock was resistant to intracellular noise, arising from collisions, crowding, and cytoplasmic viscosity.

WHY TIME IN THE IN VITRO OSCILLATOR IS UNIDIRECTIONAL

In the crystal structure of KaiC from *S. elongatus*, the T432 residues from all six subunits and the S431 residues from four subunits were phosphorylated (76, 108). T432's side chain oxygen atoms are closer on average to the ATP γ-phosphate (7.3 Å), compared with those from the S431 residues (8.4 Å). Once T432 residues are phosphorylated, new stabilizing interactions are formed across the CII subunits because the phosphate group of T432 becomes engaged in a salt bridge to R385 as shown in **Figure 4** (108). While it is reasonable to assume that local conformational fluctuations will be more limited after T432 is phosphorylated, there remains sufficient flexibility at the subunit

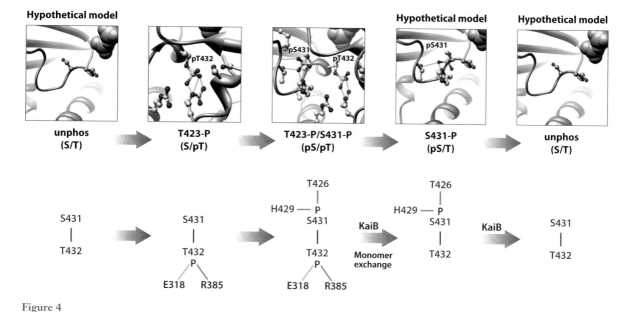

Figure 4

A ratcheting mechanism for unidirectional motion of the KaiABC oscillator. Starting with the unphosphorylated (unphos) form of KaiC (S/T), KaiC is first phosphorylated on T432 (S/pT), leading to the formation of a salt bridge (*solid green line*) to R385 on the adjacent KaiC subunit (*blue chain*; T432-P shows proximity to E318, as shown by the *dashed green line*). KaiC then autophosphorylates on S431, leading to the doubly phosphorylated form (pS/pT) that adds hydrogen bonds (*solid green lines*) to residues T426 and H429 on the same KaiC subunit (*pink chain*). The formation of these hydrogen bonds makes the reverse reactions unfavorable so that the KaiABC oscillator is unidirectional during the phosphorylation phase (33). The hyperphosphorylated KaiC (pS/pT) then interacts with KaiB and initiates monomer exchange and dephosphorylation, forming first pS/T and ultimately unphosphorylated KaiC (S/T) again. The S/T and pS/T forms of KaiC are inferred and labeled as hypothetical models on the figure because no crystal structures of these forms have been reported, whereas the S/pT and pS/pT forms have been successfully crystallized and reported (76, 75, 108).

interface to also allow transfer of a phosphate to S431 in the second step. Once phosphorylated, S431 can engage in additional hydrogen-bonding interactions with amino acids (T426 and H429) in the same subunit (**Figure 4**). These hydrogen-bonding interactions are probably crucial because mutation at T426 to alanine abolishes clock function (108, 109). In general, S431-P residues are more shielded inside a pocket formed by the phosphorylation loop compared with T432-P residues, which may explain the longer survival of the former during the dephosphorylation phase (**Figure 2b**). Overall, the structural information on the phosphorylation events at the KaiCII subunit interfaces and the inter- and intrasubunit interactions formed by the phosphorylated residues indicates that the number of hydrogen bonds increases while

first T432 and subsequently S431 are phosphorylated. This progressive increase in molecular interaction would make the reverse reactions unfavorable, causing a built-in ratcheting mechanism that drives the KaiC oscillator unidirectionally during the phosphorylation phase toward T432-P/S431-P (pS/pT) (33) (**Figure 4**).

We currently have no clear understanding of the mechanism of the autophosphatase activity of KaiC during the dephosphorylation phase. It is possible but not absolutely necessary that a conformational change occurs to drive KaiC forward to the phosphatase state and achieve dephosphorylation first of T432 and then of S431 in all six subunits. This order of phosphorylation and dephosphorylation events has been observed in biochemical assays as mentioned above (**Figure 2b**) (71, 85). It appears

TTFL: transcription/translation feedback loop

that the interaction of KaiB with KaiC facilitates the formation of the phosphatase state. KaiB binds preferentially to a phosphorylated form of KaiC (specifically, the S431-P state; 71, 85). This behavior is different from that exhibited by KaiA, which can bind various forms of KaiC (37). Although the crystal structure of KaiC^{WT} shows the homohexamer captured in the hyperphosphorylated state, with 10 out of 12 phosphate groups present at S431 and T432 sites (76), there are currently no hints from the structure regarding the dephosphorylation mechanism, although Mg^{2+} serves as a cofactor. In fact, to date no crystal structure of unphosphorylated KaiC has been reported.

A TICKING CLOCKWORK IN VITRO

The unanticipated finding that the three Kai proteins together with ATP generate stable oscillations within a ~24-hour period in a test tube (**Figure 2b**) (68) provided a contradiction to the dogma that all circadian oscillators are dependent on a transcription/translation feedback loop (TTFL) (68, 82, 94, 107). Furthermore, the rates of KaiC phosphorylation, dephosphorylation, and ATP hydrolysis, as well as the in vitro rhythm, are all temperature compensated (64, 68, 93, 94). We take this as evidence that temperature compensation is built into the molecular characteristics of the three Kai proteins and the nature of their interactions. Temperature compensation remains an important unresolved question not just for the cyanobacterial system but for circadian clocks in general.

Terauchi et al. (93) have proposed that the rhythm of KaiC ATPase activity constitutes the most fundamental reaction underlying circadian periodicity in cyanobacteria. An alternative (and not exclusive) possibility is that ATP hydrolysis provides the energy needed for conformational changes in KaiC. Our results indicate that intermolecular dynamics of KaiA, KaiB, and KaiC determine the period and amplitude of this in vitro oscillator, leading to the hypothesis that (*a*) the basic timing loop of the KaiABC oscillator and (*b*) its outputs are mediated by conformational changes of KaiC in association with KaiA and KaiB. For example, mutations within KaiB that alter affinity to KaiC modulate the period of this clock in vivo and in vitro as predicted by mathematical modeling (81). Our interpretation is that the formation of Kai protein complexes is coupled with KaiC phosphorylation status; because different KaiB variants modulate the rate of KaiB·KaiC formation, they also affect the period of KaiC phosphorylation (81). At the very least, if the ATPase activity is the basic timing loop as suggested by Terauchi et al. (93), then the intermolecular associations with KaiB must regulate KaiC's ATPase activity in a deterministic way.

Now that an in vitro clock system has been identified, biophysical, biochemical, and structural tactics can be deployed to analyze the molecular nature of a circadian clockwork in a way that was previously impossible. The time-dependent formation of Kai protein complexes has been quantified with EM, gel filtration chromatography, SAXS, and native gel electrophoresis techniques (1, 9, 37, 63). Throughout the in vitro oscillation, KaiC exists in all possible combinations with KaiA and KaiB: free KaiC hexamers, binary KaiA·KaiC and KaiB·KaiC complexes, and ternary KaiA·KaiB·KaiC complexes (**Figures 2c,d**). The proportions of these complexes vary in a phase-dependent manner, with free KaiC hexamers predominating at all phases. About 10% of KaiC hexamers are present as KaiA·KaiC complexes at all phases; by comparison, KaiB·KaiC and KaiA·KaiB·KaiC complexes are clearly rhythmic and are most common during the KaiC dephosphorylation phase (**Figure 2c**) (37, 63). KaiC undergoes rhythmic changes in conformation, phosphorylation status, and interactions with KaiA and KaiB during the in vitro oscillation. Our working hypothesis is that the core of the oscillator is constituted by rhythmic changes in the conformation of KaiC that in turn modulate the interactions with KaiA and KaiB and the activity of transduction factors such as SasA and RpaA (37, 63, 90).

STAYING TOGETHER: MONOMER EXCHANGE AND THE MAINTENANCE OF SYNCHRONY AMONG HEXAMERS

Cyanobacterial cells in populations behave as autonomous oscillators that appear to be unable to communicate phase information intercellularly (2, 55). Once synchronized, however, cells in populations remain in sync for many cycles, implying a robust mechanism for maintaining a precise, high-amplitude rhythm inside each cell. Cellular events such as DNA synthesis, cell division, and metabolic changes generate "noise" or perturbations that pose a significant challenge for circadian oscillators (32, 55, 82). In the case of the in vitro oscillator, KaiC monomer exchange among different hexamers is a process that can potentially synchronize the phosphorylation status of individual hexamers within a population of hexamers, thereby sustaining a high-amplitude oscillation (**Figure 3a**) (23, 37, 63). Monomer exchange was first observed using the technique of pull-down assays with FLAG-tagged KaiC proteins (37; the FLAG tag is an octapeptide

protein tag with the following sequence: DYKDDDDK). Because the pull-down technique can suffer from aggregation/cross-reactivity problems, we investigated monomer exchange by fluorescence resonance energy transfer (FRET). Briefly, a population of KaiC labeled with IAEDANS [5-((((2-iodoacetyl)amino)ethyl)amino)naphthalene-1-sulfonic acid] (EX 336/EM 470 nm) was mixed with a population of KaiC labeled with MTSF [2-((5-fluoresceinyl)aminocarbonyl)ethyl methanethiosulfonate-4-fluorescein] (EX 490/EM 515 nm). **Figure 5a** shows the time-dependent quenching of IAEDANS fluorescence (indicative of FRET) as a gauge of KaiC monomer exchange. Using this method, we confirmed that KaiC hexamers exchange their monomers, but we could not confirm the earlier report that KaiA inhibited KaiC monomer exchange (**Figure 5b**) (37, 63). Subsequent experiments from the Kondo laboratory indicated that KaiC monomer exchange occurs primarily in the dephosphorylation phase of the KaiABC in vitro oscillation (23). Our model simulations show that phase-dependent monomer exchange

FRET: fluorescence resonance energy transfer

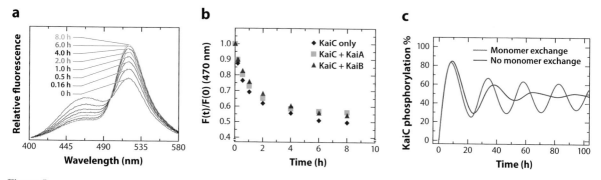

Figure 5

FRET analysis of KaiC monomer exchange. (*a*) A sample of KaiC labeled with IAEDANS (EX 336/EM 470 nm) was mixed with a sample of KaiC labeled with MTSF (EX 490/EM 515 nm). The emission spectrum of the mixture under excitation at 336 nm was recorded at the following times at 30°C: 0 h, 0.16 h, 0.5 h, 1 h, 2 h, 4 h, 6 h, and 8 h. The decrease in fluorescence intensity at 470 nm of IAEDANS-labeled KaiC is indicative of energy transfer due to monomer exchange between the two labeled KaiC populations. (*b*) Effect of KaiA and KaiB on monomer exchange. Measurement of monomer exchange between IAEDANS-labeled and MTSF-labeled KaiC when KaiA (0.05 µg µl^{-1}) or KaiB (0.05 µg µl^{-1}) was added to the mixture of KaiC (0.2 µg µl^{-1} total concentration). The decrease in fluorescence intensity at 470 nm was plotted as a function of time. (*c*) Model prediction of the in vitro KaiABC oscillation in the presence (*blue line*) or absence (*red line*) of phase-dependent monomer exchange (from Reference 63). Abbreviations: FRET, fluorescence resonance energy transfer; IAEDANS, 5-((((2-iodoacetyl)amino)ethyl)amino)naphthalene-1-sulfonic acid; MTSF, 2-((5-fluoresceinyl)aminocarbonyl)ethyl methanethiosulfonate-4-fluorescein.

allows the individual KaiC hexamers within the population of hexamers to maintain an equivalent level of average phosphorylation, such that the hexamers remain synchronized in terms of phospho-status (**Figure 5c**) (63). In this fashion, dynamic rhythms of KaiC phosphorylation can be sustained for at least 10 cycles in vitro (23). Therefore, intracellular synchronicity is achieved by biochemical reactions occurring among thousands of Kai molecules per cell (40) that enable a posttranslational oscillator (PTO) of high precision and synchrony (23, 33, 63).

MODELING THE IN VITRO OSCILLATOR

Since the publication of the in vitro KaiABC rhythm (68), there have been many attempts to model this oscillator (7, 9, 14, 23, 37, 45, 47, 48, 54, 58, 63, 66, 82, 85, 91, 99, 102, 110, 113) and probably others of which we are unaware. As a representative example, we proposed in 2007 a model that stochastically simulates the kinetics of KaiC hexamers and the degree of phosphorylation of each monomer in every hexamer (**Figure 3a**) (63). Beginning with a hypophosphorylated state of KaiC (state α), rapid and repeated association and disassociation of KaiA facilitate phosphorylation until the KaiC hexamer becomes hyperphosphorylated (state β) (27). Association of KaiB with KaiC then brings about a conformational change to a new state (KaiC*, state χ). Eventually, the KaiC* hexamer (state χ) undergoes dephosphorylation, reaches a relatively hypophosphorylated status (state δ), and relaxes to the original conformation (state α). Over the duration of the phosphorylation cycle of a hexamer, monomer exchange between any two hexamers in any of the states can occur. The rate of this subunit exchange reaches a maximum during the KaiC dephosphorylation phase, when KaiB is associated with KaiC (23). KaiB binds to the KaiC hexamer when the total degree of phosphorylation of the KaiC hexamer exceeds a threshold that places it in state β. Although we originally envisioned KaiA stochastically binding and unbinding

rapidly from KaiC hexamers, we now know that when KaiB binds to hyperphosphorylated KaiC, KaiA is sequestered to a novel site to form a stable KaiA·KaiB·KaiC complex (81). The model depicted in **Figure 3a** incorporates phase-dependent KaiC monomer exchange as a mechanism for keeping the phosphorylation state of hexamers synchronized in the population, and accurately predicts observed patterns of in vitro KaiC phosphorylation (63).

It is beyond the scope of this review to describe all the proposed models for the cyanobacterial clockwork (see Reference 7 for an evaluation of many of these models). However, one significant aspect in which the models differ is the mechanism of KaiC hexamer synchronization. In this respect, the various models fall mostly into two groups: synchronization by KaiA sequestration (6, 9, 85, 99) or synchronization by phase-dependent monomer exchange (23, 37, 63, 113). [Interestingly, monomer exchange was predicted by a modeling study before it was experimentally measured (14).] There is clear experimental evidence for phase-dependent monomer exchange (23, 63); however, several studies have also confirmed that KaiA is indeed sequestered into a stable A·B·C complex (6, 81, 85). These data suggest that KaiA sequestration may act in concert with monomer exchange to accomplish the synchrony of KaiC phosphorylation that enables the robust high-amplitude rhythms for many cycles in vitro (23). We have generated a combined model in which monomer exchange is a mechanism for maintaining phase synchrony among KaiC hexamers while KaiA sequestration is involved in the switch from autokinase to autophosphatase mode (81). Finally, modeling studies are beginning to address how the cyanobacterial pacemaker may regulate gene expression (82), metabolism (19), and cell division (110).

PTO AND TTFL: WHO IS DRIVING WHOM?

It is generally assumed that the mechanism of circadian clocks in eukaryotes is dependent on autoregulatory TTFLs (12, 17). Indeed, the discovery of the KaiA, KaiB, and KaiC proteins as key clock components in *S. elongatus* (22) did not initially raise any doubts about the importance of a TTFL at the core of this prokaryotic clockwork. This interpretation was based on the same kind of evidence that currently supports the existence of TTFL oscillators in eukaryotes, namely (*a*) rhythms of abundance for mRNAs and proteins encoded by clock genes, (*b*) feedback of clock proteins on their gene's transcription, and (*c*) phase setting by experimental expression of clock proteins (12, 22, 33). However, a number of more recent observations seemed to be inconsistent with a core TTFL oscillator in cyanobacteria. For example, the circadian rhythm of KaiC phosphorylation appeared unaffected by global inhibition of transcription and translation (94). Moreover, replacement of the promoters driving *kaiBC* gene expression with nonspecific heterologous promoters did not disturb the circadian rhythm (67, 107). When cyanobacterial cells were treated with the protein synthesis inhibitor chloramphenicol for extended periods of time, no phase changes were observed in the circadian system after return to normal conditions (35, 106). The discovery of the in vitro KaiABC oscillator in 2005 proved that a TTFL was not necessary for circadian oscillations in cyanobacteria (68), and that this KaiABC system was likely to act as a PTO in vivo.

The experimental observations that the *kaiABC* gene cluster is essential for rhythms in vivo and that rhythmic KaiC phosphorylation runs without a TTFL in vitro and in vivo (35, 68, 94, 106) implied that the KaiABC PTO was the self-sustained core pacemaker. Therefore, transcription and translation were posited to be involved only in output (68, 94). Subsequently, Kitayama et al. (41) suggested that oscillations in KaiC abundance based on transcription and translation are also important for generating the in vivo circadian rhythm. In particular, those authors reported that constitutive hyperphosphorylation of KaiC (either by overexpression of KaiA or by using a mutant of KaiC that mimicked constitutive hyperphos-

phorylation) allowed rhythmicity to proceed in vivo. Their interpretation was motivated by the observation that cyanobacterial cells apparently exhibited oscillations when the KaiABC oscillator was inactivated by stalling the phosphorylation status of KaiC. In other words, the KaiABC oscillator (PTO) did not seem to be an obligatory core oscillator in cyanobacteria because transcription and translation oscillated even in the absence of the KaiC phosphorylation cycle and the oscillation persisted regardless of the phosphorylation state and kinase activity of KaiC (41). Another group modeled these results and found that a hypothetical pacemaker composed of tightly intertwined PTO and TTFL cycles can generate robust circadian rhythms over a broad range of growth conditions (116).

We decided to extend the experiments by Kitayama et al. and have recently arrived at a different conclusion regarding the hierarchy between the PTO and the TTFL. Our current working model is that the PTO is the core pacemaker and the TTFL is a damped slave oscillator (82) (**Figure 6**). We base this interpretation on our finding that the rhythms generated by cells expressing hyperphosphorylated KaiC have a long period and are clearly damped. Moreover, these damped rhythms are not compensated for by changes in metabolic activity and therefore cannot be considered a bona fide circadian phenomenon. Modeling studies showed that the experimental data were compatible with a core PTO driving the TTFL and that the combined PTO/TTFL system is resilient to noise. Modeling also suggested that de novo synthesis of clock proteins coupled with KaiC monomer exchange results in phase shifts or entrainment of the core PTO pacemaker (**Figure 6**). Our new model of the cyanobacterial oscillator explains how the core pacemaker can be a PTO while receiving input from a TTFL (82). This interpretation has exciting implications for eukaryotic clock systems and suggests that the existence of a common mechanism at the heart of circadian oscillations in all biological systems merits a re-evaluation (32, 82, 83).

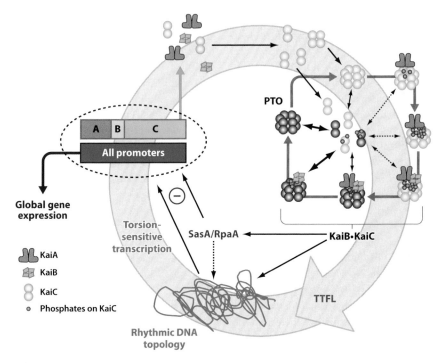

Figure 6

The core posttranslational oscillator (PTO) is embedded in a larger transcription/translation feedback loop (TTFL). The PTO is linked to the damped TTFL by transcription and translation of the *kaiABC* cluster. Global gene expression is mediated by rhythmic modulation of the activity of all promoters, including those driving the expression of the central clock gene cluster, *kaiABC* (ABC in figure). Rhythmic DNA torsion and/or transcriptional factor activity (e.g., SasA/RpaA) modulates global promoter activities. Cyclic changes in the phosphorylation status of KaiC regulate DNA topology/transcriptional factors. The PTO is determined by KaiC phosphorylation as regulated by interactions with KaiA and KaiB (compare with **Figure 3a**). Robustness is maintained by synchronization of KaiC hexameric status via monomer exchange (depicted by dumbbell-shaped KaiC monomers exchanging with KaiC hexamers in the middle of the PTO cycle). The shade of KaiC hexamers (*dark versus light blue*) denotes conformational changes that roughly equate to kinase versus phosphatase forms. New synthesis of KaiC feeds into the KaiABC oscillator as nonphosphorylated hexamers or as monomers that exchange into pre-existing hexamers. If the new synthesis of KaiC occurs at a phase when hexamers are predominantly hypophosphorylated, the oscillation of KaiC phosphorylation is reinforced (enhanced amplitude). If new synthesis of unphosphorylated KaiC happens at a phase when hexamers are predominantly hyperphosphorylated, this leads to an overall decrease in the KaiC phosphorylation status, thereby altering the phase of the KaiABC oscillator (phase shift), reducing its amplitude, or both. Phase shifts accomplished by this mechanism could be partially or totally responsible for entrainment in vivo (from Reference 82).

ENTRAINING THE ENDOGENOUS CLOCK TO ENVIRONMENTAL TIME

One of the three salient properties of circadian rhythms is their ability to be entrained by the daily cycle in the environment so that their endogenous ~24-h period takes on a period of exactly 24 h with the appropriate phase rela-

tionship (12). By virtue of its day-to-day consistency, the LD cycle is the most important environmental signal for circadian entrainment in most organisms. In cyanobacteria, mutagenesis screens designed to find factors involved in phase resetting elicited by pulses of darkness were successful in identifying CikA, a histidine kinase and pseudobacteriophytochrome

(86). Because of its similarity to bacteriophytochromes, CikA initially seemed to be a light-absorbing photoreceptor. However, unlike a classical photoreceptor, CikA does not absorb light, but rather it appears to sense the redox state of the intracellular plastoquinone pool (25, 26). Light drives large changes in redox potential via photosynthesis, and this could be an entraining cue for the cyanobacterial clock (52). Quinone not only directly binds to CikA, it also affects the stability of CikA (26). More recently, redox effects were further implicated in clock entrainment by the observation that KaiA can bind quinone and this binding destabilizes the KaiA protein, thereby reducing its ability to stimulate KaiC autophosphorylation (105). These data suggest that KaiA senses environmental signals as changes in redox state and modulates the circadian clock. Other proteins implicated in light resetting of the cyanobacterial clock include Pex and LdpA, in addition to four new proteins that interact with CikA (NhtA, PrkE, IrcA, and CdpA) (25, 51, 89). These proteins may contribute to the input pathway (51).

However, a new paper has proposed a dramatically different explanation for LD entrainment that is based on the observation that the ratio of ATP to ADP declines in the dark in *S. elongatus* cells, which are obligately photosynthetic (84). When ATP:ADP ratio changes that are similar to the in vivo measurements in darkness were simulated in the in vitro oscillator, they caused phase shifts similar to those observed with dark pulses in vivo. These data were interpreted to mean that exposure of cells to darkness changes the ATP/ADP ratio, and this change directly resets the phase of the PTO in vivo. A mathematical model based on these data effectively modeled entrainment by dark pulses (84). Although this new study has the potential to explain entrainment of the cyanobacterial circadian clock, it is presently unclear how this interpretation will fit together with the other data implicating CikA (and other proteins) in the entrainment pathway.

Although LD cycles are usually the most important entraining agent, cyclic exposure to temperature changes can often entrain circadian clocks (30). There has not yet been a report of temperature cycle entrainment for the cyanobacterial system in vivo (except for a brief mention in Reference 114), but there have been two reports of temperature resetting of the in vitro oscillator (63, 114). In 2007, we reported the first phase response curves for the in vitro system, generated by 6-h pulses of either 16°C or 37°C away from the free-running temperature of 30°C (63). This analysis demonstrated that the in vitro oscillation shares characteristic phase-dependent responses to perturbation with most circadian systems (12, 30). A more extensive analysis of temperature-induced phase resetting that used steps between 30°C and 45°C also found that the temperature stimuli shifted the phase of the in vitro rhythm in a phase-dependent manner suggestive of a nonparametric entrainment mechanism (12, 114).

Nakajima et al. (69) have suggested that daily changes in the intracellular concentrations of KaiB and KaiC could modulate the angular velocity of the pacemaker in vivo and provide an entraining stimulus. KaiB and KaiC abundance levels oscillate in LL, whereas KaiA levels remain rather constant (27, 40, 94, 106). Although some studies have reported that the abundance of Kai proteins is cyclic in LD cycles as well (36), we find considerable variability of Kai cycling in LD; in some experiments, Kai protein levels seem not to oscillate in LD when the KaiC phosphorylation rhythm is robust (82). Therefore, the hypothesis that entrainment in vivo is based on changes of Kai protein abundances warrants more experimental testing (69). Our studies of the relationship between the PTO and TTFL have led us to propose that new synthesis of Kai proteins in the daytime alters and adjusts KaiC phospho-status as needed for proper entrainment to the environment (**Figure 6**) (82).

EVOLUTION OF CIRCADIAN TIMING

Cyanobacteria are an ancient group. They were likely one of the earliest life forms, dating back at least 3.5 billion years. Therefore, over the span of their evolution they have been subjected to the original earth's anoxic atmosphere, which was also characterized by high levels of ultraviolet (UV) radiation. Indeed, cyanobacteria are thought to be the primary organism that transformed earth's atmosphere by photosynthetic oxygen emission into its current oxygenated state with the concomitant creation of the ozone layer, which helps to shield earth's inhabitants from the deleterious effects of UV radiation. Currently, the most plausible model for the evolution of circadian timing—the escape from light hypothesis (32, 70, 79, 80)—strongly implicates daily gating of cell division and DNA damage as the selective forces that resulted in the evolution of circadian pacemakers. In a nutshell, the escape from light hypothesis proposes that the illuminated portion the light/dark cycle had profound and mostly deleterious effects on early life; therefore, a strong initial driving force for the early evolution of a circadian timer could have been the advantage of phasing to the nighttime those cellular processes that are vulnerable to light. For example, UV light produces thymidine dimers in DNA, which if not repaired accurately mutate the DNA. Moreover, even visible light can modulate metabolism and other processes because it is absorbed by omnipresent pigments in cells such as cytochromes. Consequently, cellular events that are hypersensitive to light (such as DNA damage and electron transport) would be best performed at night in organisms that cannot shield themselves from irradiation. The escape from light hypothesis predicts that the vestiges of nocturnal programming of light-sensitive processes might have been retained in present-day organisms. This prediction was tested in the eukaryotic unicell *Chlamydomonas*, where it was found that the cells were indeed most sensitive to UV light in the early night (70).

FUTURE DIRECTIONS FOR CYANOBACTERIAL CLOCKS

As Jacques Monod said in his Nobel Prize acceptance speech in 1965, "The ambition of molecular biology is to interpret the essential properties of organisms in terms of molecular structures. This objective has already been achieved for DNA, and it is in sight for RNA, but it still seems very remote for the proteins." Although we may be on the verge of realizing Monod's ambition with the KaiABC system, many outstanding and unresolved issues remain:

- A perplexing concern about the in vitro oscillator when seen in the context of its in vivo operation relates to the stoichiometry among the KaiA, KaiB, and KaiC proteins for optimal operation. The original publication (68) used concentrations of the Kai proteins that were supposedly based on estimates of in vivo concentrations (40), but in fact significantly more KaiA is needed in the in vitro reaction than appears to be present in vivo. In addition, while early reports suggested that the period and amplitude of the in vitro rhythm were relatively invariant within an allowed range of Kai protein concentrations (37), recent data indicate significant effects on the period of the in vitro oscillator when the ratio of [KaiA] was varied relative to the concentration of [KaiB]+[KaiC] (69). Modeling indicates that the PTO is robust and resilient within a limited scope of Kai protein concentration fluctuations (82), but these recent results (69) raise the concern that the in vitro oscillator might not be a reliable pacemaker in vivo. Perhaps a function of the TTFL is to maintain Kai protein concentrations within a range that maintains a dependable timekeeper.
- What is the configuration of KaiC in the unphosphorylated state? At the present time, no three-dimensional structure of unphosphorylated KaiC has been reported.
- We have crystal structures of the three Kai proteins individually, but not of all three Kai proteins in a complex or interacting with SasA, CikA, etc. Moreover, methods of structural analysis that allow

dynamic measurements of structure (i.e., SAXS, NMR, and EM) should be pursued further because crystal structures are snapshots of proteins whose structures are almost certainly dynamic, and these dynamic changes will be important for fully understanding the mechanism of the KaiABC oscillator.

- Despite progress, a complete mechanistic understanding of phosphorylation (autokinase), dephosphorylation (autophosphatase), and ATPase activity remains obscure.

- Is the KaiC hexamer a propeller? Do all the monomers in the KaiC hexamer act in concert, or is there spatial distribution of activities? Considering that KaiC and F1-ATPase share structural similarities, perhaps there is a rotary phosphorylation/activity within KaiC as in F1-ATPase (96, 101).

- How are the phosphorylation and dephosphorylation rates of KaiC compensated for temperature (68, 94)? The mechanism of temperature compensation, which is a key property of all circadian systems, remains a key mystery.

- How are the Kai proteins coupled to the downstream control of gene expression? One study has proposed that the ATP hydrolytic activity of KaiC is the key output of the KaiABC oscillator (11). Alternatively, an output pathway for KaiC that is based on two-component signaling was proposed whereby KaiC influences the phosphorylation status of the histidine kinase SasA, which in turn regulates the activity of a response regulator, RpaA, that may act as a transcriptional factor (90). Other factors have also become implicated in this output pathway (92). The sequence and structural similarity of KaiC to RecA and DnaB has suggested another possibility, namely that the Kai nanomachine may have a helicase activity (46, 62). This possibility—which does not exclude the participation of a SasA/RpaA or ATPase pathway—is particularly intriguing given the global rhythms of chromosomal topology (60, 100, 104). If KaiC can act on DNA, perhaps it directly mediates the pervasive supercoiling. However, although we found that KaiC has weak binding affinity for forked DNA substrates (62), hitherto attempts to measure helicase activity from KaiC have been unsuccessful.

- Is an oscillating chromosomal topology the basis for rhythmic global gene expression (60, 100, 104)? Systems biology approaches may yield answers.

- What is the mechanism for the competition/selection phenomena that illustrate the adaptive significance of the *S. elongatus* clock in rhythmic environments (19, 31, 73, 103)?

- Will the clockwork in *S. elongatus* ultimately prove to be completely distinct from the clocks of eukaryotes? Or will the insights gleaned from cyanobacteria induce a reassessment of clocks in higher organisms (32, 33, 82, 83)?

SUMMARY POINTS

1. Biological clocks have played an important role in bioevolution as evidenced by their existence in organisms that initially evolved as far back as 3.5 billion years ago.

2. Prokaryotic cyanobacteria have a circadian timekeeping system that enhances fitness.

3. These cells exhibit pervasive circadian regulation of gene expression, possibly by regulation of chromosomal topology.

4. A circadian rhythm of the phosphorylation of the central clock protein KaiC can be reconstituted in vitro with three proteins derived from the cyanobacterium *S. elongatus* (KaiA, KaiB, and KaiC) and ATP.

5. KaiA, KaiB, and KaiC are the only circadian proteins for which the three-dimensional structure of full-length proteins is known.

6. Structural, biochemical, and biophysical methods have been used to study the mechanism by which KaiC is rhythmically phosphorylated and dephosphorylated. Dephosphorylation is temporally coordinated with monomer exchange, which may also function to maintain synchrony of individual Kai molecules in the population of molecules.

7. Modeling has been applied to the in vitro and in vivo systems. Models of the complete in vivo system have indicated the existence of a core biochemical oscillator that controls a larger transcription/translation feedback loop.

8. The *kaiC* gene is widespread among prokaryotes (Eubacteria and Archaea), but it may be performing a nonclock function in prokaryotic species outside of the cyanobacteria. The elucidation of this nonclock function could lead to fascinating clues about the selective pressure(s) that led to the evolution of circadian clocks.

DISCLOSURE STATEMENT

The authors are not aware of any affiliations, memberships, funding, or financial holdings that might be perceived as affecting the objectivity of this review.

ACKNOWLEDGMENTS

We thank our colleagues and coauthors at Vanderbilt University, especially M. Byrne, H. Mchaourab, T. Mori, X. Qin, R. Pattanayek, S. Pattanayek, D. Williams, M. Woelfle, and Y. Xu. We also thank T. Kondo, S. Golden, M. Ishiura, A. Liwang, E. O'Shea, and their laboratory members, whose seminal contributions continue to make the study of cyanobacterial clocks fascinating. Our work is supported by funds from the National Institutes of Health, specifically the NIGMS (GM067152 and GM088595 to CHJ; GM073845 to ME; GM081646 to PLS).

LITERATURE CITED

1. Akiyama S, Nohara A, Ito K, Maéda Y. 2008. Assembly and disassembly dynamics of the cyanobacterial periodosome. *Mol. Cell* 29:703–16

2. Amdaoud M, Vallade M, Weiss-Schaber C, Mihalcescu I. 2007. Cyanobacterial clock, a stable phase oscillator with negligible intercellular coupling. *Proc. Natl. Acad. Sci. USA* 104:7051–56

3. Andersson CR, Tsinoremas NF, Shelton J, Lebedeva NV, Yarrow J, et al. 2000. Application of bioluminescence to the study of circadian rhythms in cyanobacteria. *Methods Enzymol.* 305:527–42

4. Angermayr SA, Hellingwerf KJ, Lindblad P, de Mattos MJ. 2009. Energy biotechnology with cyanobacteria. *Curr. Opin. Biotechnol.* 20:257–63

5. Atsumi S, Higashide W, Liao JC. 2009. Direct photosynthetic recycling of carbon dioxide to isobutyraldehyde. *Nat. Biotechnol.* 27:1177–80

6. Brettschneider C, Rose RJ, Hertel S, Axmann IM, Heck AJ, Kollmann M. 2010. A sequestration feedback determines dynamics and temperature entrainment of the KaiABC circadian clock. *Mol. Syst. Biol.* 6:389

7. Byrne M. 2009. Mathematical modeling of the in vitro cyanobacterial circadian oscillator. In *Bacterial Circadian Programs*, ed. JL Ditty, SR Mackey, CH Johnson, 16:283–300. Berlin: Springer

8. Chabot JR, Pedraza JM, Luitel P, van Oudenaarden A. 2007. Stochastic gene expression out-of-steady-state in the cyanobacterial circadian clock. *Nature* 450:1249–52

9. Clodong S, Düring U, Kronk L, Axmann I, Wilde A, et al. 2007. Functioning and robustness of a bacterial circadian clock. *Mol. Syst. Biol.* 3:90

10. Ditty JL, Mackey SR, Johnson CH, eds. 2009. *Bacterial Circadian Programs*. Berlin: Springer. 333 pp.

11. Dong G, Yang Q, Wang Q, Kim YI, Wood TL, et al. 2010. Elevated ATPase activity of KaiC applies a circadian checkpoint on cell division in *Synechococcus elongatus*. *Cell* 140:529–39

12. Dunlap JC, Loros JJ, DeCoursey PJ, eds. 2004. *Chronobiology: Biological Timekeeping*. Sunderland, MA: Sinauer. 406 pp.

13. Egli M, Stewart PL. 2009. Structural aspects of the cyanobacterial KaiABC circadian clock. In *Bacterial Circadian Programs*, ed. JL Ditty, SR Mackey, CH Johnson, 7:121–40. Berlin: Springer

14. Emberly E, Wingreen NS. 2006. Hourglass model for a protein-based circadian oscillator. *Phys. Rev. Lett.* 96:0383003

15. Garces RG, Wu N, Gillon W, Pai EF. 2004. *Anabaena* circadian clock proteins KaiA and KaiB reveal potential common binding site to their partner KaiC. *EMBO J.* 23:1688–98

16. Grobbelaar N, Huang T-C, Lin HY, Chow TJ. 1986. Dinitrogen-fixing endogenous rhythm in *Synechococcus* RF-1. *FEMS Microbiol. Lett.* 37:173–77

17. Hardin PE, Hall JC, Rosbash M. 1990. Feedback of the *Drosophila* period gene product on circadian cycling of its messenger RNA levels. *Nature* 343:536–40

18. Hayashi F, Ito H, Fujita M, Iwase R, Uzumaki T, Ishiura M. 2004. Stoichiometric interactions between cyanobacterial clock proteins KaiA and KaiC. *Biochem. Biophys. Res. Commun.* 316:195–202

19. Hellweger FL. 2010. Resonating circadian clocks enhance fitness in cyanobacteria in silico. *Ecol. Model.* 221:1620–29

20. Hitomi K, Oyama T, Han S, Arvai AS, Getzoff ED. 2005. Tetrameric architecture of the circadian clock protein KaiB. A novel interface for intermolecular interactions and its impact on the circadian rhythm. *J. Biol. Chem.* 280:19127–35

21. Holtman CK, Chen Y, Sandoval P, Gonzales A, Nalty MS, et al. 2005. High-throughput functional analysis of the *Synechococcus elongatus* PCC 7942 genome. *DNA Res.* 12:103–15

22. Ishiura M, Kutsuna S, Aoki S, Iwasaki H, Andersson CR, et al. 1998. Expression of a gene cluster *kaiABC* as a circadian feedback process in cyanobacteria. *Science* 281:1519–23

23. Ito H, Kageyama H, Mutsuda M, Nakajima M, Oyama T, Kondo T. 2007. Autonomous synchronization of the circadian KaiC phosphorylation rhythm. *Nat. Struct. Mol. Biol.* 14:1084–88

24. Ito H, Mutsuda M, Murayama Y, Tomita J, Hosokawa N, et al. 2009. Cyanobacterial daily life with Kai-based circadian and diurnal genome-wide transcriptional control in *Synechococcus elongatus*. *Proc. Natl. Acad. Sci. USA* 106:14168–73

25. Ivleva NB, Bramlett MR, Lindahl PA, Golden SS. 2005. LdpA: a component of the circadian clock senses redox state of the cell. *EMBO J.* 24:1202–10

26. Ivleva NB, Gao T, LiWang AC, Golden SS. 2006. Quinone sensing by the circadian input kinase of the cyanobacterial circadian clock. *Proc. Natl. Acad. Sci. USA* 103:17468–73

27. Iwasaki H, Nishiwaki T, Kitayama Y, Nakajima M, Kondo T. 2002. KaiA-stimulated KaiC phosphorylation in circadian timing loops in cyanobacteria. *Proc. Natl. Acad. Sci. USA* 99:15788–93

28. Iwasaki H, Williams SB, Kitayama Y, Ishiura M, Golden SS, Kondo T. 2000. A kaiC-interacting sensory histidine kinase, SasA, necessary to sustain robust circadian oscillation in cyanobacteria. *Cell* 101:223–33

29. Izumo M, Johnson CH, Yamazaki S. 2003. Circadian gene expression in mammalian fibroblasts revealed by real-time luminescence reporting: temperature compensation and damping. *Proc. Natl. Acad. Sci. USA* 100:16089–94

30. Johnson CH. 1990. *An Atlas of Phase Response Curves for Circadian and Circatidal Rhythms*. Nashville, TN: Vanderbilt Univ. 715 pp.

31. Johnson CH. 2005. Testing the adaptive value of circadian systems. *Met. Enzymol.* 393:818–37

10. This book is a recent compilation of the knowledge of the cyanobacterial circadian system.

16. First persuasive indication that prokaryotes (cyanobacteria) might have a circadian system comparable to the well-characterized circadian systems of eukaryotes.

22. Reports the identification of the key clock genes *kaiA*, *kaiB*, and *kaiC* that form the core circadian oscillator.

32. Johnson CH. 2010. Circadian clocks and cell division: What's the pacemaker? *Cell Cycle* 9:3864–73

33. Johnson CH, Egli M, Stewart PL. 2008. Structural insights into a circadian oscillator. *Science* 322:697–701

34. Johnson CH, Golden SS, Ishiura M, Kondo T. 1996. Circadian clocks in prokaryotes. *Mol. Microbiol.* 21:5–11

35. Johnson CH, Xu Y, Mori T. 2008. A cyanobacterial circadian clockwork. *Curr. Biol.* 18:R816–25

36. Kageyama H, Kondo T, Iwasaki H. 2003. Circadian formation of clock protein complexes by KaiA, KaiB, KaiC, and SasA in cyanobacteria. *J. Biol. Chem.* 278:2388–95

37. Kageyama H, Nishiwaki T, Nakajima M, Iwasaki H, Oyama T, Kondo T. 2006. Cyanobacterial circadian pacemaker: Kai protein complex dynamics in the KaiC phosphorylation cycle in vitro. *Mol. Cell* 23:161–71

38. Katayama M, Tsinoremas NF, Kondo T, Golden SS. 1999. *cpmA*, a gene involved in an output pathway of the cyanobacterial circadian system. *J. Bacteriol.* 181:3516–24

39. Kim YI, Dong G, Carruthers CW Jr, Golden SS, LiWang A. 2008. The day/night switch in KaiC, a central oscillator component of the circadian clock of cyanobacteria. *Proc. Natl. Acad. Sci. USA* 105:12825–30

40. Kitayama Y, Iwasaki H, Nishiwaki T, Kondo T. 2003. KaiB functions as an attenuator of KaiC phosphorylation in the cyanobacterial circadian clock system. *EMBO J.* 22:2127–34

41. Kitayama Y, Nishiwaki T, Terauchi K, Kondo T. 2008. Dual KaiC-based oscillations constitute the circadian system of cyanobacteria. *Genes Dev.* 22:1513–21

42. Kondo T, Mori T, Lebedeva NV, Aoki S, Ishiura M, Golden SS. 1997. Circadian rhythms in rapidly dividing cyanobacteria. *Science* 275:224–27

43. **Kondo T, Strayer CA, Kulkarni RD, Taylor W, Ishiura M, et al. 1993. Circadian rhythms in prokaryotes: luciferase as a reporter of circadian gene expression in cyanobacteria. *Proc. Natl. Acad. Sci. USA* 90:5672–76**

44. Kucho K, Okamoto K, Tsuchiya Y, Nomura S, Nango M, et al. 2005. Global analysis of circadian expression in the cyanobacterium *Synechocystis* sp. strain PCC 6803. *J. Bacteriol.* 187:2190–99

45. Kurosawa G, Aihara K, Iwasa Y. 2006. A model for the circadian rhythm of cyanobacteria that maintains oscillation without gene expression. *Biophys. J.* 91:2015–23

46. Leipe DD, Aravind L, Grishin NV, Koonin EV. 2000. The bacterial replicative helicase DnaB evolved from a RecA duplication. *Genome Res.* 10:5–16

47. Li C, Chen X, Wang P, Wang W. 2009. Circadian KaiC phosphorylation: a multi-layer network. *PLoS Comput. Biol.* 5:e1000568

48. Li S, Fang YH. 2007. Modelling circadian rhythms of protein KaiA, KaiB and KaiC interactions in cyanobacteria. *Biol. Rhythm Res.* 38:43–53

49. Liu Y, Tsinoremas NF, Golden SS, Kondo T, Johnson CH. 1996. Circadian expression of genes involved in the purine biosynthetic pathway of the cyanobacterium *Synechococcus* sp. strain PCC 7942. *Mol. Microbiol.* 20:1071–81

50. **Liu Y, Tsinoremas NF, Johnson CH, Lebedeva NV, Golden SS, et al. 1995. Circadian orchestration of gene expression in cyanobacteria. *Genes Dev.* 9:1469–78**

51. Mackey SR, Choi JS, Kitayama Y, Iwasaki H, Dong G, Golden SS. 2008. Proteins found in a CikA-interaction assay link the circadian clock, metabolism, and cell division in *Synechococcus elongatus*. *J. Bacteriol.* 190:3738–46

52. Mackey SR, Golden SS. 2007. Winding up the cyanobacterial circadian clock. *Trends Microbiol.* 15:381–88

53. Markson JS, O'Shea EK. 2009. The molecular clockwork of a protein-based circadian oscillator. *FEBS Lett.* 583:3938–47

54. Mehra A, Hong C, Shi M, Loros J, Dunlap J, Ruoff P. 2006. Circadian rhythmicity by autocatalysis. *PLoS Comput. Biol.* 2:e96

55. **Mihalcescu I, Hsing W, Leibler S. 2004. Resilient circadian oscillator revealed in individual cyanobacteria. *Nature* 430:81–85**

56. Min H, Liu Y, Johnson CH, Golden SS. 2004. Phase determination of circadian gene expression in *Synechococcus elongatus* PCC 7942. *J. Biol. Rhythms* 19:103–12

57. Mitsui A, Kumazawa S, Takahashi A, Ikemoto H, Arai T. 1986. Strategy by which nitrogen-fixing unicellular cyanobacteria grow photoautotrophically. *Nature* 323:720–22

43. Reports the use of *S. elongatus* with bacterial luciferase as a reporter of rhythmic gene expression to establish a new model system for extensive circadian clock analyses, this time in a prokaryote!

50. First report of global gene expression regulated by a circadian clock; novel methodology involved the random genomic insertion of a luciferase reporter as a gauge of promoter activity.

55. Reports that circadian timing in *S. elongatus* is resilient to the perturbations of cell metabolism that accompany cell division.

58. Miyoshi F, Nakayama Y, Kaizu K, Iwasaki H, Tomita M. 2007. A mathematical model for the Kai-protein–based chemical oscillator and clock gene expression rhythms in cyanobacteria. *J. Biol. Rhythms* 22:69–80

59. Mori T, Binder B, Johnson CH. 1996. Circadian gating of cell division in cyanobacteria growing with average doubling times of less than 24 hours. *Proc. Natl. Acad. Sci. USA* 93:10183–88

60. Mori T, Johnson CH. 2001. Circadian programming in cyanobacteria. *Semin. Cell Dev. Biol.* 12:271–78

61. Mori T, Johnson CH. 2001. Independence of circadian timing from cell division in cyanobacteria. *J. Bacteriol.* 183:2439–44

62. Mori T, Saveliev SV, Xu Y, Stafford WF, Cox MM, et al. 2002. Circadian clock protein KaiC forms ATP-dependent hexameric rings and binds DNA. *Proc. Natl. Acad. Sci. USA* 99:17203–8

63. **Mori T, Williams DR, Byrne MO, Qin X, Egli M, et al. 2007. Elucidating the ticking of an in vitro circadian clockwork. *PLoS Biol.* 5:e93**

64. Murakami R, Miyake A, Iwase R, Hayashi F, Uzumaki T, Ishiura M. 2008. ATPase activity and its temperature compensation of the cyanobacterial clock protein KaiC. *Genes Cells* 13:387–95

65. Mutoh R, Mino H, Murakami R, Uzumaki T, Takabayashi A, et al. 2010. Direct interaction between KaiA and KaiB revealed by a site-directed spin labeling electron spin resonance analysis. *Genes Cells* 15:269–80

66. Nagai T, Terada TP, Sasai M. 2010. Synchronization of circadian oscillation of phosphorylation level of KaiC in vitro. *Biophys J.* 98:2469–77

67. Nakahira Y, Katayama M, Miyashita H, Kutsuna S, Iwasaki H, et al. 2004. Global gene repression by KaiC as a master process of prokaryotic circadian system. *Proc. Natl. Acad. Sci. USA* 101:881–85

68. **Nakajima M, Imai K, Ito H, Nishiwaki T, Murayama Y, et al. 2005. Reconstitution of circadian oscillation of cyanobacterial KaiC phosphorylation in vitro. *Science* 308:414–15**

69. Nakajima M, Ito H, Kondo T. 2010. In vitro regulation of circadian phosphorylation rhythm of cyanobacterial clock protein KaiC by KaiA and KaiB. *FEBS Lett.* 584:898–902

70. Nikaido SS, Johnson CH. 2000. Daily and circadian variation in survival from ultraviolet radiation in *Chlamydomonas reinhardtii*. *Photochem. Photobiol.* 71:758–65

71. Nishiwaki T, Satomi Y, Kitayama Y, Terauchi K, Kiyohara R, et al. 2007. A sequential program of dual phosphorylation of KaiC as a basis for circadian rhythm in cyanobacteria. *EMBO J.* 26:4029–37

72. Nishiwaki T, Satomi Y, Nakajima M, Lee C, Kiyohara R, et al. 2004. Role of KaiC phosphorylation in the circadian clock system of *Synechococcus elongatus* PCC 7942. *Proc. Natl. Acad. Sci. USA* 101:13927–32

73. Ouyang Y, Andersson CR, Kondo T, Golden SS, Johnson CH. 1998. Resonating circadian clocks enhance fitness in cyanobacteria. *Proc. Natl. Acad. Sci. USA* 95:8660–64

74. Partensky F, Hess WR, Vaulot D. 1999. *Prochlorococcus*, a marine photosynthetic prokaryote of global significance. *Microbiol. Mol. Biol. Rev.* 63:106–27

75. Pattanayek R, Mori T, Xu Y, Pattanayek S, Johnson CH, Egli M. 2009. Structures of KaiC circadian clock mutant proteins: a new phosphorylation site at T426 and mechanisms of kinase, ATPase and phosphatase. *PLoS ONE* 4:e7529

76. **Pattanayek R, Wang J, Mori T, Xu Y, Johnson CH, Egli M. 2004. Visualizing a circadian clock protein: crystal structure of KaiC and functional insights. *Mol. Cell* 15:375–88**

77. Pattanayek R, Williams DR, Pattanayek S, Mori T, Johnson CH, et al. 2008. Structural model of the circadian clock KaiB-KaiC complex and mechanism for modulation of KaiC phosphorylation. *EMBO J.* 27:1767–78

78. Pattanayek R, Williams DR, Pattanayek S, Xu Y, Mori T, et al. 2006. Analysis of KaiA-KaiC protein interactions in the cyano-bacterial circadian clock using hybrid structural methods. *EMBO J.* 25:2017–28

79. Pittendrigh CS. 1965. Biological clocks: the functions, ancient and modern, of circadian oscillations. In *Science and the Sixties. Proceedings of the Cloudcraft Symposium*, pp. 96–111. Air Force Office Sci. Res.

80. Pittendrigh CS. 1993. Temporal organization: reflections of a Darwinian clock-watcher. *Annu. Rev. Physiol.* 55:17–54

63. Combines structural, biochemical, biophysical, and modeling approaches to address the rhythmic regulation of Kai protein complex formation and to derive a model to explain the oscillation and its stability.

68. Reports the landmark discovery that an in vitro circadian oscillator was possible with three purified proteins and ATP.

76. This paper reports the high-resolution structure of KaiC based on X-ray crystallography.

81. Qin X, Byrne M, Mori T, Zou P, Williams DR, et al. 2010. Intermolecular associations determine the dynamics of the circadian KaiABC oscillator. *Proc. Natl. Acad. Sci. USA* 107:14805–10

82. Examines the relationship between the PTO and the TTFL in cyanobacteria and concludes that the PTO is the self-sustained core pacemaker while the TTFL is a slave oscillator that can provide entraining input to the PTO.

82. **Qin X, Byrne M, Xu Y, Mori T, Johnson CH. 2010. Coupling of a core post-translational pacemaker to a slave transcription/translation feedback loop in a circadian system. *PLoS Biol.* 8:e1000394**

83. Rosbash M. 2009. The implications of multiple circadian clock origins. *PLoS Biol.* 7:e62

84. Rust MJ, Golden SS, O'Shea EK. 2011. Light-driven changes in energy metabolism directly entrain the cyanobacterial circadian oscillator. *Science* 331:220–23

85. Rust MJ, Markson JS, Lane WS, Fisher DS, O'Shea EK. 2007. Ordered phosphorylation governs oscillation of a three-protein circadian clock. *Science* 318:809–12

86. Schmitz O, Katayama M, Williams SB, Kondo T, Golden SS. 2000. CikA, a bacteriophytochrome that resets the cyanobacterial circadian clock. *Science* 289:765–68

87. Shi T, Ilikchyan I, Rabouille S, Zehr JP. 2010. Genome-wide analysis of diel gene expression in the unicellular N(2)-fixing cyanobacterium *Crocosphaera watsonii* WH 8501. *ISME J.* 4:621–32

88. Smith RM, Williams SB. 2006. Circadian rhythms in gene transcription imparted by chromosome compaction in the cyanobacterium *Synechococcus elongatus*. *Proc. Natl. Acad. Sci. USA* 103:8564–69

89. Takai N, Ikeuchi S, Manabe K, Kutsuna S. 2006. Expression of the circadian clock-related gene *pex* in cyanobacteria increases in darkness and is required to delay the clock. *J. Biol. Rhythms* 21:235–44

90. Takai N, Nakajima M, Oyama T, Kito R, Sugita C, et al. 2006. A KaiC-associating SasA-RpaA two-component regulatory system as a major circadian timing mediator in cyanobacteria. *Proc. Natl. Acad. Sci. USA* 103:12109–14

91. Takigawa-Imamura H, Mochizuki A. 2006. Predicting regulation of the phosphorylation cycle of KaiC clock protein using mathematical analysis. *J. Biol. Rhythms* 21:405–16

92. Taniguchi Y, Takai N, Katayama M, Kondo T, Oyama T. 2010. Three major output pathways from the KaiABC-based oscillator cooperate to generate robust circadian kaiBC expression in cyanobacteria. *Proc. Natl. Acad. Sci. USA* 107:3263–68

93. Terauchi K, Kitayama Y, Nishiwaki T, Miwa K, Murayama Y, et al. 2007. ATPase activity of KaiC determines the basic timing for circadian clock of cyanobacteria. *Proc. Natl. Acad. Sci. USA* 104:16377–81

94. Tomita J, Nakajima M, Kondo T, Iwasaki H. 2005. No transcription-translation feedback in circadian rhythm of KaiC phosphorylation. *Science* 307:251–54

95. Tosini G, Menaker M. 1998. The tau mutation affects temperature compensation of hamster retinal circadian oscillators. *Neuroreport* 9:1001–5

96. Ueno H, Suzuki T, Kinosita K Jr, Yoshida M. 2005. ATP-driven stepwise rotation of F0F1-ATP synthase. *Proc. Natl. Acad. Sci. USA* 102:1333–38

97. Vakonakis I, LiWang AC. 2004. Structure of the C-terminal domain of the clock protein KaiA in complex with a KaiC-derived peptide: implications for KaiC regulation. *Proc. Natl. Acad. Sci. USA* 101:10925–30

98. Vakonakis I, Sun J, Wu T, Holzenburg A, Golden SS, LiWang AC. 2004. NMR structure of the KaiC-interacting C-terminal domain of KaiA, a circadian clock protein: implications for the KaiA-KaiC interaction. *Proc. Natl. Acad. Sci. USA* 101:1479–84

99. van Zon JS, Lubensky DK, Altena PR, ten Wolde PR. 2007. An allosteric model of circadian KaiC phosphorylation. *Proc. Natl. Acad. Sci. USA* 104:7420–25

100. Vijayan V, Zuzow R, O'Shea EK. 2009. Oscillations in supercoiling drive circadian gene expression in cyanobacteria. *Proc. Natl. Acad. Sci. USA* 106:22564–68

101. Wang J. 2005. Recent cyanobacterial Kai protein structures suggest a rotary clock. *Structure* 13:735–41

102. Wang J, Xu L, Wang E. 2009. Robustness and coherence of a three-protein circadian oscillator: landscape and flux perspectives. *Biophys. J.* 97:3038–46

103. Woelfle MA, Ouyang Y, Phanvijhitsiri K, Johnson CH. 2004. The adaptive value of circadian clocks: an experimental assessment in cyanobacteria. *Curr. Biol.* 14:1481–86

104. Woelfle MA, Xu Y, Qin X, Johnson CH. 2007. Circadian rhythms of superhelical status of DNA in cyanobacteria. *Proc. Natl. Acad. Sci. USA* 104:18819–24

105. Wood TL, Bridwell-Rabb J, Kim YI, Gao T, Chang YG, et al. 2010. The KaiA protein of the cyanobacterial circadian oscillator is modulated by a redox-active cofactor. *Proc. Natl. Acad. Sci. USA* 107:5804–9

106. Xu Y, Mori T, Johnson CH. 2000. Circadian clock-protein expression in cyanobacteria: rhythms and phase setting. *EMBO J.* 19:3349–57

107. Xu Y, Mori T, Johnson CH. 2003. Cyanobacterial circadian clockwork: roles of KaiA, KaiB and the kaiBC promoter in regulating KaiC. *EMBO J.* 22:2117–26

108. Xu Y, Mori T, Pattanayek R, Pattanayek S, Egli M, Johnson CH. 2004. Identification of key phosphorylation sites in the circadian clock protein KaiC by crystallographic and mutagenetic analyses. *Proc. Natl. Acad. Sci. USA* 101:13933–38

109. Xu Y, Mori T, Qin X, Yan H, Egli M, Johnson CH. 2009. Intramolecular regulation of phosphorylation status of the circadian clock protein KaiC. *PLoS ONE* 4:e7509

110. Yang Q, Pando BF, Dong G, Golden SS, van Oudenaarden A. 2010. Circadian gating of the cell cycle revealed in single cyanobacterial cells. *Science* 327:1522–26

111. Ye S, Vakonakis I, Ioerger TR, LiWang AC, Sacchettini JC. 2004. Crystal structure of circadian clock protein KaiA from *Synechococcus elongatus*. *J. Biol. Chem.* 279:20511–18

112. Yen U-C, Huang T-C, Yen T-C. 2004. Observations of the circadian photosynthetic rhythm in cyanobacteria with a dissolved-oxygen meter. *Plant Sci.* 166:949–52

113. Yoda M, Eguchi K, Terada TP, Sasai M. 2007. Monomer-shuffling and allosteric transition in KaiC circadian oscillation. *PLoS ONE* 2:e408

114. Yoshida T, Murayama Y, Ito H, Kageyama H, Kondo T. 2009. Nonparametric entrainment of the in vitro circadian phosphorylation rhythm of cyanobacterial KaiC by temperature cycle. *Proc. Natl. Acad. Sci. USA* 106:1648–53

115. Zinser ER, Lindell D, Johnson ZI, Futschik ME, Steglich C, et al. 2009. Choreography of the transcriptome, photophysiology, and cell cycle of a minimal photoautotroph, *Prochlorococcus*. *PLoS One* 4:e5135

116. Zwicker D, Lubensky DK, ten Wolde PR. 2010. Robust circadian clocks from coupled protein-modification and transcription-translation cycles. *Proc. Natl. Acad. Sci. USA* 107:22540–45

Actin Structure and Function

Roberto Dominguez[1] and Kenneth C. Holmes[2]

[1]Department of Physiology, University of Pennsylvania School of Medicine, Philadelphia, Pennsylvania 19104-6085; email: droberto@mail.med.upenn.edu

[2]Max Planck Institute for Medical Research, D69120 Heidelberg, Germany; email: holmes@mpimf-heidelberg.mpg.de

Annu. Rev. Biophys. 2011. 40:169–86

First published online as a Review in Advance on February 8, 2011

The *Annual Review of Biophysics* is online at biophys.annualreviews.org

This article's doi: 10.1146/annurev-biophys-042910-155359

Copyright © 2011 by Annual Reviews. All rights reserved

1936-122X/11/0609-0169$20.00

Keywords

X-ray crystallography, electron microscopy, fiber diffraction, actin-binding-proteins

Abstract

Actin is the most abundant protein in most eukaryotic cells. It is highly conserved and participates in more protein-protein interactions than any known protein. These properties, along with its ability to transition between monomeric (G-actin) and filamentous (F-actin) states under the control of nucleotide hydrolysis, ions, and a large number of actin-binding proteins, make actin a critical player in many cellular functions, ranging from cell motility and the maintenance of cell shape and polarity to the regulation of transcription. Moreover, the interaction of filamentous actin with myosin forms the basis of muscle contraction. Owing to its central role in the cell, the actin cytoskeleton is also disrupted or taken over by numerous pathogens. Here we review structures of G-actin and F-actin and discuss some of the interactions that control the polymerization and disassembly of actin.

Contents

THE ACTIN MONOMER

Vertebrates express three main actin isoforms, including three α-isoforms of skeletal, cardiac, and smooth muscles and the β- and γ-isoforms expressed in nonmuscle and muscle cells. Actin isoforms differ by only a few amino acids, with most variations occurring toward the N terminus (23). Actin also undergoes various forms of posttranslational modifications. For instance, His73 of skeletal muscle α-actin

G-actin: monomeric (or globular) actin

ABP: actin-binding protein

is methylated, the N-terminal methionine and cysteine residues are acetylated and cleaved, and the resulting N-terminal aspartic acid is then reacetylated.

Since the original determination of the crystal structure of G-actin in complex with DNase I (27), over 80 structures of actin have been reported (**Figure 1**). These are listed in **Supplemental Table 1** (follow the **Supplemental Material link** from the Annual Reviews home page at **http://www.annualreviews.org**). The majority of these structures have been obtained as complexes with actin-binding proteins (ABPs) and small molecules, or by chemically modifying or mutating actin in order to prevent polymerization. Remarkably, irrespective of the bound molecule or nucleotide state, the conformation of the actin monomer is basically the same. Actin belongs to a structural superfamily with sugar kinases, hexokinases, and Hsp70 proteins (3). The Arp proteins (48) and the prokaryotic actin-like proteins MreB (53) and ParM (54) are also now part of this superfamily. Common to these proteins, the 375-amino-acid (aa) polypeptide chain of actin folds into two major α/β-domains (**Figure 1a**). Because of their location within the actin filament, the two major domains of actin are known as the outer and inner domains, and because of their apparently different sizes in electron microscopy (EM) images, they have also been called the small and large domains, respectively. Traditionally, however, a four-subdomain nomenclature has been adopted (27). Subdomains 1 and 3 are structurally related and probably emerged from gene duplication, whereas subdomains 2 and 4 can be viewed as large insertions into subdomains 1 and 3, respectively. For brevity we refer to these as domains 1–4. The actin monomer is rather flat, fitting into a rectangular prism with dimensions 55 Å × 55 Å × 35 Å.

There is relatively little contact between the two major domains of actin; the polypeptide chain passes twice between these domains: at the loop centered at residue Lys336 and at the linker helix Gln137-Ser145, which functions as the axis of a hinge between the domains. As a result, two clefts are formed between the domains. The upper cleft binds the nucleotide and associated divalent cation (Mg^{2+} in cells), which together provide the other important linkage between domains (**Figure 1a**). The lower cleft between domains 1 and 3 is lined by residues Tyr143, Ala144, Gly146, Thr148, Gly168, Ile341, Ile345, Leu346, Leu349, Thr351, and Met355, which are predominantly hydrophobic. This cleft mediates important longitudinal contacts between actin subunits in the filament (16, 35) and constitutes the major binding site for most ABPs, and is thus called the target-binding or hydrophobic cleft (10). The communication between the two clefts provides the structural basis for how nucleotide-dependent conformation changes modulate the binding affinities of ABPs and the strength of intersubunit contacts in the filament.

G-actin is not an effective ATPase, whereas F-actin is, and most crystal structures have been solved in the ATP-bound form. The differences between the ATP- and ADP-bound states are relatively minor and involve primarily two loops (**Figure 1a**): the Ser14 β-hairpin loop and the sensor loop carrying the methylated His73 (21). The Ser14 loop is located in actin subdomain 1 and is structurally equivalent to the loop containing Asp157 in subdomain 3. These two loops engulf the phosphates of the nucleotide. Nucleotide-dependent conformational changes begin with Ser14, which in the ATP state makes a hydrogen-bonding contact with the γ-phosphate. After hydrolysis and γ-phosphate release, Ser14 changes orientation to form a hydrogen-bonding contact with the β-phosphate of the nucleotide. In the ATP state, Ser14 also forms a hydrogen-bonding contact with the main chain of the loop carrying His73. When the side chain of Ser14 rotates in the ADP state, the His73 loop moves toward the nucleotide to occupy some of the space emptied by the γ-phosphate. In this way, this loop appears to sense the state of the nucleotide.

The sensor loop marks the C-terminal end of domain 2, which can be viewed as an insertion into domain 1. At the top of domain 2, residues 39–51 are disordered in most crystal structures. This sequence is referred to as the DNase I-binding loop (or D-loop) because it mediates important interactions in the complex with DNase I (27). When visible in crystal structures, this loop takes on a variety of conformations, including in one case a short α-helix (39). Domain 2, and specifically the D-loop, plays a critical role in longitudinal intersubunit contacts in the filament. Thus, changes in the Ser14 and sensor loops appear to propagate to domain 2, whose conformation changes only slightly between nucleotide states (21), but probably enough to explain the decreased stability of the actin filament in the ADP state. Because the nucleotide sits at the interface between the two major actin domains, another important consequence of nucleotide hydrolysis and γ-phosphate release is a weakening of the linkage between domains, which can then rotate more freely with respect to one another. Two types of rotations have been described: a scissors-like opening and closing of the nucleotide cleft within the plane of the figure (**Figure 1a**) and a perpendicular propeller-twist of one domain with respect to the other (18, 35, 52). Gln137, at the beginning of the hinge helix, is thought to play a critical role in nucleotide hydrolysis (26), which is activated in the filament (8), presumably by interdomain rotation that brings the side chain of Gln137 in contact with the γ-phosphate of the nucleotide (16, 35).

STRUCTURES OF ACTIN BOUND TO ACTIN-BINDING PROTEINS AND SMALL MOLECULES

Nucleotide hydrolysis by F-actin is one of the main factors regulating the G-actin to F-actin transition. In vitro, actin monomers join the fast-growing barbed (or +) end of the filament in the ATP state, hydrolysis takes place in the filament, and ADP-actin monomers dissociate faster from the pointed (or −) end. This steady-state mechanism of actin polymerization/depolymerization is known as actin filament treadmilling (55). However,

D-loop: DNase I-binding loop of actin (residues 39–51)

Propeller-twist: rotation of the outer domain with respect to the inner domain of actin subunits in the filament, which results in a more flat structure of the actin molecule

F-actin: filamentous (or polymeric) actin

G-actin to F-actin transition: refers to the process of actin polymerization

Filament growth: refers to the process of addition of actin monomers to the actin filament, which is faster through the barbed or (+) end of the filament

treadmilling cannot account for the dynamics of filament assembly observed in cells. Numerous ABPs are involved in the regulation of actin cytoskeleton dynamics (44). These proteins have diverse functions, including actin monomer sequestration, filament barbed- and pointed-end capping, filament severing, and filament cross-linking. The structures of various complexes of G-actin with ABPs have been determined by X-ray crystallography (**Figure 1** and **Supplemental Table 1**). Numerous F-actin-binding proteins have been analyzed by EM at limited resolution.

G-actin-binding proteins have diverse functions, belong to different structural families, and present different nucleotide preferences.

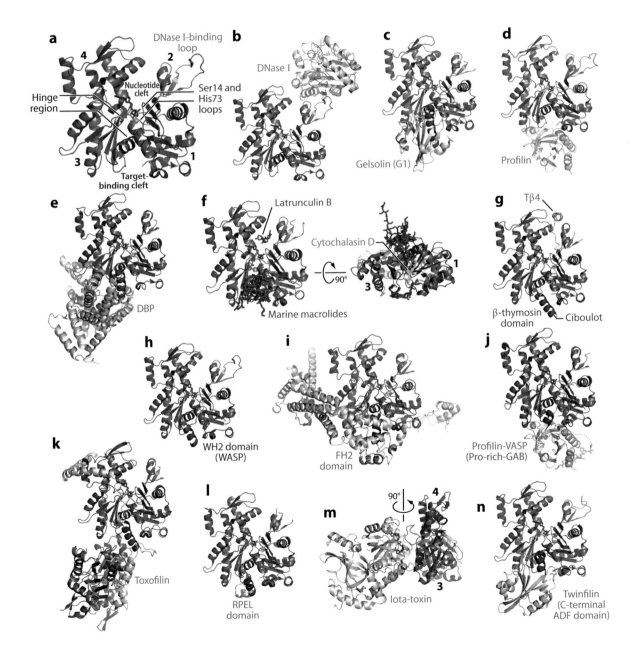

Despite this diversity, most of these proteins bind in the target-binding cleft of actin and typically insert an amphiphilic helix into this cleft (10). Whereas the overall position of the interacting helices is well conserved (**Figure 1**), the directionality of their polypeptide chains is not. The directionality of the helices of toxofilin and of the formin homology 2 (FH2) domain and the (WASP-homology 2) WH2 domain is from back to front (according to the classical view of actin), whereas the helices of gelsolin, vitamin D-binding protein (DBP), ADF/cofilin, and the RPEL domain bind in the opposite direction.

DNase I

Actin and DNase I form a high-affinity complex (33). Although this complex occurs in vivo (49), its physiological significance is still unclear. Nevertheless, it was the actin–DNase I complex that allowed the first crystal structure of actin to be determined (**Figure 1b**; references to the actin structures and PDB accession codes are given in **Supplemental Table 1**).

Gelsolin

Gelsolin belongs to a family of actin-severing and actin-capping proteins, which includes adseverin, villin, capG, advillin, and supervillin (51). Gelsolin consists of six homologous domains (G1–G6) and is regulated by Ca^{2+}, phosphoinositides, and tyrosine phosphorylation. In the inactive, Ca^{2+}-free state, gelsolin adopts a compact conformation stabilized by intramolecular interactions in which actin-binding sites are masked (**Supplemental Figure 1**). Domains G1 and G4 contain the two major actin-binding sites. These domains bind similarly to actin, and as first revealed by the structure of the G1-actin complex (**Figure 1c**), they insert a helix into the target-binding cleft of actin. On the basis of subsequent structures of complexes of G1-G3 and G4-G6 with G-actin (**Supplemental Figure 1**), Robinson and colleagues (5) have proposed a three-step model of filament severing by Ca^{2+}-activated gelsolin. According to this model, Ca^{2+} binding produces a conformational change that partially exposes the

Target-binding cleft: the cleft between domains 1 and 3 of the actin molecule, where most actin-binding proteins interact (also known as the hydrophobic cleft)

FH2: formin-homology domain

WH2: WASP-homology 2 domain

Figure 1

Structures of actin and actin complexes. The structures of actin complexes are shown to scale and in chronological order of publication. (*a*) Classical view of the structure of the actin monomer. The structure shown was derived from the complex with DNase I, with completion of the C terminus from the complex of actin with profilin. Highlighted in orange are the Ser14 and methylated His73 loops, the DNase I-binding loop, and the hinge between domains, consisting of helix Gln137-Ser145 and the loop centered at residue Lys336. Subdomains 1–4 are labeled. (They are also labeled in panels *f* and *m*, which show rotated views of the structure.) Together, subdomains 1 and 2 form the outer (or small) domain, whereas subdomains 3 and 4 constitute the inner (or large) domain. Two large clefts are formed between these domains: the nucleotide- and target-binding clefts. Most actin-binding proteins (ABPs) and small molecules bind in the target-binding cleft, and the interaction frequently involves an α-helix (*magenta*). (*b*) DNase I (1ATN). (*c*) Gelsolin segment 1 (G1) (1EQY). See also **Supplemental Figure 1** for structures of actin with Gelsolin fragments G1–G3 and G4–G6. (*d*) β-actin-profilin (2BTF). (*e*) Vitamin D-binding protein (DBP) (1KXP). (*f*) Two perpendicular views of a superimposition of structures of actin complexes with small molecules, including marine toxins (1QZ5, 1QZ5, 1S22, 1YXQ, 2ASM, 2ASO, 2ASP, 2FXU, 2Q0R, 2Q0U, 2VYP), latrunculin A (2Q0U), and cytochalasin D (3EKS). The marine toxins (*magenta*) bind at the ends of the target-binding cleft, whereas cytochalasin D binds in the middle, and Latrunculin (both A and B) binds in the nucleotide cleft. All these molecules compromise actin polymerization. (*g*) β-thymosin domain. A complete structure of this complex is not available, but combined, the structures of the N-terminal portion of a β-thymosin domain from *Drosophila* ciboulot (1SQK) and the C-terminal end of β-thymosin peptide (Tβ4) (1T44) provide a model of this complex. (*h*) WASP homology domain 2 (WH2) domain of WASP (2A3Z). The WH2 domain, present in many cytoskeletal proteins in the form of tandem repeats, is related to the β-thymosin domain but lacks the C-terminal pointed end capping helix. (*i*) Formin homology 2 (FH2) domain (1Y64). See also **Supplemental Figure 2** for a more detailed representation of this structure. (*j*) Ternary complex with profilin and the Pro-rich G-actin-binding (Pro-rich-GAB) domains of VASP (2PBD). The GAB domain is related to the WH2 domain but presents a shorter N-terminal helix and adopts a slightly different orientation when bound to actin, possibly because it is designed to co-bind with profilin. (*k*) Toxofilin from *Toxoplasma gondii* bound to an antiparallel actin dimer (2Q97). (*l*) RPEL (RPxxEL-containing motif) domain from the serum response factor coactivator MAL (2V52). (*m*) Arginine ADP-ribosylation iota-toxin from *Clostridium perfringens* (3BUZ). View rotated 90° relative to the other complexes. (*n*) C-terminal ADF/cofilin domain of twinfilin (3DAW). See **Supplemental Table 1** for a complete list of references.

Tβ4: β-thymosin 4

actin-binding sites of gelsolin. Proceeding through a transitional complex in which gelsolin binds to the sides of the actin filament, the actin-binding sites of gelsolin become fully exposed, allowing G1 and G4 to engage actin protomers on two different strands of the actin filament and to sever the filament through a coordinated pincer movement.

Gelsolin is primarily cytoplasmic, but a slightly longer isoform circulates in plasma, where it forms part of a homeostatic mechanism termed the actin-scavenger system (30). Under conditions involving cell death or tissue injury, actin can be released into the bloodstream. The actin-scavenger system is responsible for the depolymerization and removal of actin from the circulation. During the first phase of this mechanism, gelsolin severs the actin filaments, followed by the formation of a high-affinity complex of monomeric actin with DBP. This complex is cleared from the circulation rapidly, primarily through the liver. The structure of the DBP-actin complex has also been determined (**Figures 1e**), and despite the absence of any relationship with gelsolin, DBP also inserts a helix into the target-binding cleft of actin.

Profilin

Profilin (together with Tβ4) contributes to maintaining a large fraction (~50%) of the cellular actin in the unpolymerized pool, at a concentration (~20–100 μM) that is 200- to 1000-fold higher than the critical concentration for barbed-end polymerization (~0.1 μM). Whereas free ATP-actin is present in low amounts (0.1–1 μM) and Tβ4-actin is polymerization incompetent, profilin-actin (present at an intracellular concentration of 5–40 μM) constitutes the main source of actin monomers for polymerization (44). Multiple properties allow profilin to play such a central role in filament assembly (13): (*a*) Profilin catalyzes the exchange of ADP for ATP on actin, which replenishes the pool of ATP-actin monomers ready for polymerization; (*b*) it inhibits filament nucleation, which explains in part the need for filament nucleators; (*c*) profilin-bound actin monomers

cannot add to pointed ends but can elongate filament barbed ends at approximately the same rate as free actin monomers; (*d*) owing to its greater affinity for actin, profilin competes effectively with Tβ4 for binding to actin; and (*e*) profilin can bind simultaneously to Pro-rich sequences and actin, and it binds to both with higher affinity as a ternary complex than to either one separately (15). Pro-rich sequences are extremely abundant among cytoskeletal proteins and may serve to channel profilin-actin complexes from the cellular pool onto actin filaments in a regulated manner. Thus, recruitment of profilin-actin complexes through the Pro-rich FH1 domain of formin increases its elongation rate (28). Like most ABPs, profilin binds in the target-binding cleft of actin, but it occupies the back of the cleft (**Figure 1d**), allowing it to bind simultaneously with WH2-related sequences characterized by the presence of a short N-terminal helix, such as the GAB domain of Ena/VASP, which is positioned immediately C-terminal to a Pro-rich profilin-binding sequence (**Figure 1j**).

ADF/Cofilin

Members of this family are expressed in all eukaryotes, where they are generally involved in the recycling of actin monomers, mainly by filament depolymerization, during processes involving rapid cytoskeleton turnover, such as membrane ruffling and cytokinesis (2). This family includes actin-depolymerizing factor (ADF), cofilin-1 (nonmuscle cells), cofilin-2 (muscle cells), and twinfilin, which contains two ADF domains. Despite the importance of this family, the first structure of a complex with actin was determined only recently (40), corresponding to a complex with the C-terminal ADF domain of twinfilin (**Figure 1n**). Contrary to profilin, which as a nucleotide exchange factor has a clear preference for ATP-actin, ADF/cofilin binds ADP-actin with higher affinity than ATP-actin (7). However, the twinfilin-actin structure was determined in the ATP state, and twinfilin has a different function than other members of this family by sequestering actin monomers and capping filament barbed ends. Therefore, this structure may not be fully representative of the ADF/cofilin-ADP-actin complex, but it represents the best model of this interaction currently available.

RPEL Domain

While the cytoplasmic functions of actin are generally well established, its role in the nucleus is less well understood. One such role is the regulation of the myocardin transcriptional coactivator MAL (34, 37). MAL shuttles between the cytoplasm and the nucleus, where it regulates the activity of the transcription factor serum response factor (SRF). The N-terminal portion of MAL contains a tandem repeat of three G-actin-binding RPEL domains (so named because of the presence of the consensus motif RPxxEL). The binding of actin to the RPEL domains inhibits the activity of MAL by preventing its accumulation in the nucleus and thereby repressing transcriptional activation by the MAL-SRF complex. The system is ultimately regulated through RhoA-induced alterations of G-actin to F-actin ratios; increased amounts of cytoplasmic G-actin lead to the formation of the inactive actin-MAL complex, whereas reduced G-actin levels resulting from polymerization may release MAL for nuclear accumulation (34). Structures of actin complexes with two of the RPEL domains of MAL have been determined (**Figure 1*l***). The RPEL motif also binds in the actin-binding cleft, inserting a helix into this cleft, but the orientation of this helix is opposite that of the WH2 domain (which is also the case for DBP, gelsolin, and ADF/cofilin).

Structures of Actin Bound to Small Molecules

Recently, a series of macrolides that interact with actin and disrupt polymerization have been isolated from marine organisms (58). These molecules have diverse structures. Moreover, the strong antitumor activities of some of these compounds have sparked interest in their use

as molecular probes to help elucidate cellular functions of actin and as scaffolds in the design of anticancer drugs. Numerous structures of actin complexes with marine macrolides have been determined (**Supplemental Table 1**). Despite their stereochemical diversity, most of these molecules bind in the target-binding cleft of actin (**Figure 1*f***).

Cytochalasin D is a fungal toxin used extensively in cell biology as an agent that binds to the barbed end of actin filaments and inhibits both the association and dissociation of actin monomers. Cytochalasin D binds in the target-binding cleft of actin (**Figure 1*f***).

Latrunculin A and B are also marine macrolides derived from sponges and nudibranchs (57). Contrary to other marine macrolides, both latrunculin A and B bind in the nucleotide-binding cleft of actin, but they have a similar effect in that they inhibit actin polymerization. Both toxins have been used to prevent actin polymerization in the determination of numerous structures of actin (**Supplemental Table 1** and **Figure 1*f***).

Pathogens and Actin Assembly

Certain human pathogens, such as *Salmonella*, *Shigella*, and *Listeria*, disrupt or hijack the cytoskeleton of host cells during infection (passive invasion mechanism), whereas others, such as *Toxoplasma gondii*, have evolved their own actin cytoskeletal systems (active invasion mechanism) (20). One of the most abundantly expressed proteins of *T. gondii* is toxofilin, a G-actin-binding protein that is secreted into host cells during invasion (59). Toxofilin forms a high-affinity complex with an antiparallel actin dimer and inserts a helix into the target-binding cleft of one of the bound actins (**Figure 1*f***).

Other pathogenic bacteria produce toxins that ADP-ribosylate actin, resulting in cytoskeleton disorganization and cell death (1, 29). Such is the case of *Clostridium botulinum* C2 toxin and *Clostridium perfringens* iota-toxin that modify actin Arg177. The structure of an iota-toxin-actin complex with a nonhydrolyzable NAD analog provides a rare

Filament disassembly: refers to the process of dissociation of actin subunits from the filament

look into an intermediate enzyme-substrate complex implicated in the modification of actin (**Figure 1***m*).

ACTIN-BINDING DOMAINS INVOLVED IN THE DE NOVO POLYMERIZATION OF ACTIN

An important group of ABPs are those that regulate the de novo formation of actin filaments, which include actin filament nucleation and elongation factors (6, 12, 43). Known filament nucleators include the Arp2/3 complex and its large family of nucleation-promoting factors, formins, Spire, Cobl, VopL/VopF, TARP, and Lmod. These molecules control the time and location for polymerization and influence the structures of the actin networks that they generate. Filament nucleators are generally unrelated but, with the exception of formins, they all use the WASP-homology 2 (WH2) domain for interaction with actin. Formins are unique among actin filament nucleators in that they use the FH2 domain for interaction with actin and promote not only nucleation, but also processive barbed-end elongation (19).

The β-Thymosin/WH2 Domain

The WH2 domain is perhaps the most abundant and functionally diverse actin-binding fold (11, 41). It is found in proteins involved in actin monomer sequestration and cytoskeleton scaffolding, but it is particularly abundant among proteins that mediate actin filament nucleation (12). A common architecture, found in Spire, Cobl, and VopL/VopF, consists of tandem WH2 domains that bind three to four actin subunits to form a nucleus. The WH2 domain is short, 17–27 aa long. Its N-terminal portion forms a helix that binds in the target-binding cleft of actin (**Figure 1***b*). After this helix, the WH2 domain presents an extended region that contains the conserved four-residue motif LKKT(V), which is also found in the linker between gelsolin domains G1 and G2.

Tβ4, a short 43-aa polypeptide, is the best-known member of the β-thymosin family and is described in this section because of its structural relationship with the WH2 domain (11, 41). The best model of the Tβ4-actin interaction currently available results from the combination of two structures, corresponding to the complexes of actin with an N-terminal fragment of a β-thymosin domain from ciboulot and with a C-terminal 19-aa fragment of Tβ4 (**Figure 1***g*). Like the WH2 domain, the N-terminal portion of the β-thymosin fold consists of a helix that binds in the target-binding cleft of actin, followed by an LKKT-containing linker. The C-terminal portion, which is missing in the WH2 domain, consists of a helix that binds atop actin subdomains 2 and 4. The combination of N- and C-terminal helices that block monomer addition to both the barbed and pointed ends of the actin filament makes Tβ4 a remarkably simple but effective actin-monomer-sequestering protein. Tβ4-actin complexes cannot participate in actin filament nucleation or elongation. Instead, Tβ4 is thought to function as an actin buffer, losing actin in competitive equilibrium to profilin, which has higher affinity for actin monomers and functions as the main reservoir of actin monomers for filament assembly (44).

The FH2 Domain

In cellular processes ranging from cell locomotion and morphogenesis to cytokinesis, formins mediate the assembly of unbranched actin networks, such as filopodia, stress fibers, and actin cables (19). Diaphanous-related formins, which are the most extensively studied, contain a profilin-binding Pro-rich region known as the FH1 domain toward the middle of their polypeptide chain. Recruitment of profilin-actin through the FH1 domain accelerates formin-mediated filament elongation (28). As often observed among cytoskeletal proteins, the actin-binding FH2 domain is positioned immediately C-terminal to the Pro-rich region. The FH2 domain is responsible for the two filament assembly functions, nucleation and elongation.

The structure of the FH2 dimer consists of two rod-shaped domains connected in head-to-tail fashion by highly flexible linkers (56) (see also **Supplemental Figure 2**). The structure of an actin complex with the FH2 domain of Bni1 was also determined (38) (**Figure 1i**). Formin also contains a helix that inserts into the target-binding cleft of actin (**Figure 1i**).

WH2 Domain and Filament Elongation

While formins promote both nucleation and elongation, none of the WH2-based nucleators seems to promote elongation. However, one family of WH2-containing proteins, Ena/VASP (14), appears to be involved in elongation (4), although it plays no significant role in nucleation. Like formins, Ena/VASP proteins contain a central Pro-rich region and WH2-related sequences, known as the G-actin-binding and F-actin-binding (GAB and FAB) domains. Analogous to the stimulation of formin-mediated elongation produced by profilin (28), various studies indicate that profilin–actin may stimulate Ena/VASP-dependent elongation (reviewed in Reference 12). The ternary complex of profilin–actin with a fragment of VASP comprising a Pro-rich segment and GAB domain suggests that profilin–actin complexes recruited to the Pro-rich region can be delivered directly to the barbed end of elongating filaments (**Figure 1j**). This structure also demonstrated that the GAB domain binds actin in a similar manner to the WH2 domain, but its N-terminal helix is shorter and its orientation is slightly different from that of the WH2 domain, such that it can bind actin simultaneously with profilin.

STRUCTURE OF THE ACTIN FILAMENT

Although closely related, the structures of G-actin and that of the actin subunit in F-actin are in fact different. It has not been possible to crystallize actin filaments, nor do any of the G-actin structures show the conformation of the actin monomer as in the filament. Details of the structure in the filament have been deduced from X-ray fiber diagrams of orientated gels of F-actin and from EM. These studies have now yielded atomic models of the actin filament.

EM of negatively stained actin fibers showed F-actin to be made of two chains that turn gradually around each other to form a right-handed, two-chained long helix (22). The actual symmetry is a single left-handed genetic helix with approximately 13 molecules repeating every six turns in an axial distance of 35.9 nm. The distance between molecules along the helix axis is 27.6 Å. However, because the twist per molecule ($-166.6°$) is close to $180°$, the structure actually appears like two slowly turning right-handed chains (**Figure 2**).

Fiber Diffraction Patterns Give the First Structures of F-Actin

Concentrated solutions of F-actin can be induced to form gels consisting of aligned actin filaments (45). Until recently, the only source of higher-resolution structural data from F-actin was X-ray diffraction from such oriented gels. In oriented gels, the fibers, while parallel, have a random orientation about the fiber axis. Thus, X-ray diffraction yields essentially a section through the diffraction pattern of a single fibrous molecule that has been averaged by spinning it around the fiber axis (cylindrical averaging). Because the structures are periodic in the axial direction, the fiber diagram is limited to layer lines. To interpret a fiber diagram pattern generally one compares the measured data with the computed fiber diagram of a starting model and then uses a refinement procedure to modify the model and achieve a better fit.

Thus, the first near-atomic resolution structure of F-actin (25) was deduced by combining X-ray fiber diffraction data and the crystal structure of G-actin (25). However, this model was based on the G-actin monomer. Because there is in fact a state-change between the monomer in G-actin and F-actin, the fit to the fiber diffraction pattern was not optimal. One consequence of this approximate fit was

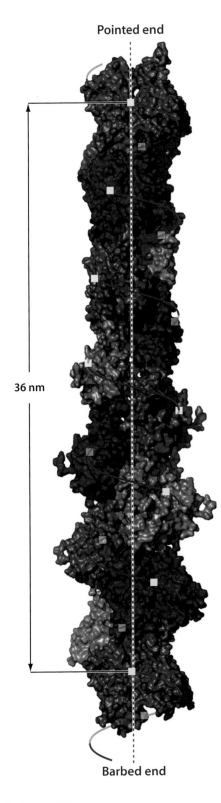

Pointed end

36 nm

Barbed end

that, although the orientation of the actin monomers in the helix was correct, the two long-pitch helices in this early model were about 3 Å too far apart.

Since this initial calculation, a number of attempts have been made to deduce the nature of the G-actin to F-actin transformation by refining models against the data from X-ray fiber diagrams (24, 32, 52). The fiber diagram used had limited resolution (~7 Å). Moreover, disorientation in the sample, which smears out the layer lines, loses information. Unique answers were not obtained.

Actin fibers are diamagnetic. Well-oriented X-ray fiber diffraction patterns can be obtained by placing F-actin gels in a very strong magnetic field (35) (**Supplemental Figure 3**). The resulting higher resolution X-ray fiber diagram (3.3 Å in the radial direction and 5.6 Å along the equator) indeed made it possible to elucidate the nature of the G-actin to F- actin transition (35). The main component of the transition from G-actin to F-actin is a 12°–13° propeller-twist of the outer domain with respect to the inner domain about an axis roughly at right angle to the helix axis (**Figure 3**). Moreover, there are bending movements of domains 2 and 4 in the same direction. The final result is that F-actin is flatter than G-actin by about 17°–18°. Furthermore, the D-loop takes on an open-loop configuration and inserts itself into the target-binding cleft of the subunit immediately above it.

Oda & Maeda (36) have defined the propeller-twist angle (θ) between the two major domains as the angle between two planes,

---

Figure 2

The helical structure of F-actin derived from cryo-electron microscopy (16). The molecules are arranged on a single helix with 13 molecules repeating in almost exactly six left-handed turns. The rise per molecule is 2.76 nm and the twist per molecule is $-166.6 \pm 0.6°$ (for simplicity of drawing, in the figure the value -166.15 has been used to make the structure repeat exactly after 13 residues). Because $-166°$ is close to $180°$, the structure takes on the appearance of a two-start right-handed long-pitch helix.

one containing the Cα atom of residue Gly55 and the axis of rotation, and the other containing Glu207 and the axis of rotation. Gly55 is in domain 2 and Glu207 is in domain 4, so that θ measures the total twist—a rotation of domains 1 and 3 plus a further bending of domains 2 and 4. The average twist angle is about 20° for all G-actin structures (it is the largest for the two monomeric actin structures 1J6Z and 3HBT at 25°), whereas for F-actin structures the twist angle θ is about 7°.

Cryo-Electron Microscopy

Cryo-EM has now yielded an atomic model of F-actin (16). Fujii et al. (16) have produced a 6.6 Å resolution density map of F-actin that allows the disposition of the domains and some of the polypeptide chain to be visualized directly. By controlling humidity and ice thickness during freezing, these authors obtained fields of long, straight actin filaments with nearly perfect helical symmetry. This result made possible the high resolution obtained by these authors. An analysis of their data shows that the angular twist per molecule in the actin helix is $-166.6 \pm 0.6°$ and the rise per residue is 27.6 Å.

Structure of F-Actin

The models produced by Fujii et al. (16) and Oda et al. (35) are similar. The most significant difference is in the D-loop, which Fujii et al. observed directly. Moreover, in the method of refinement used by Fujii et al. coordinates were altered as little as possible from the crystallographic starting model. As a result, the Fujii model is likely to be more precise.

Stereo views of the interface are shown in **Figure 4**. The stronger bonding between molecules is in the direction of the filament axis and takes place between neighboring molecules along the two-start long-pitch helix. Molecules in **Figure 4** are numbered −2, −1, 0, 1, and 2 along the genetic helix (subunit 0 is taken to be the reference). Axial interactions are extensive. One important longitudinal interaction

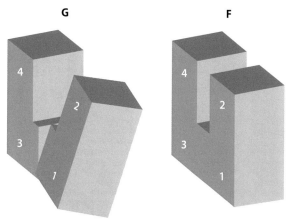

G   **F**

Figure 3

The essence of the G-actin to F-actin transition is a flattening of the actin molecule by a propeller-twist of the outer and inner domains about an axis roughly at right angles to the actin helix axis. The numbers refer to the subdomains (27) (diagram courtesy of Y. Maeda).

is between domains 4 and 3 in neighboring molecules: The loop Pro243-Asp244-Gly245 in domain 4 interacts with two loops in domain 3 of the molecule above, particularly residues Pro322 and Met325 in one loop and Met283, Ile287, and Asp288 in the adjoining loop. Two previous reports have suggested similar interactions (25, 50). The second important longitudinal interaction involves the D-loop. The D-loop of subunit 0 extends into the target-binding cleft of subunit 2 where it interacts with domains 1 and 3. In addition, residue Ile64 forms a bond with domain 3 of molecule 2. Fujii et al. (16) remark that the axial interactions are mostly electrostatic in character, but there are also hydrophobic contacts.

The lateral bonding across the axis between the long-pitch helices is much less extensive than the longitudinal bonding. One interaction is between the plug-like insertion (Gln 263–Gly 273) in domain 3 and the beginning of the D-loop of subunit −1. Interactions of the plug across the helix axis with molecules −1 and 1 stabilize the two-stranded F-actin. The other interchain interaction is just above the plug, where domain 4 contacts domains 1 and 3 of molecule 1. Details are given in the original paper (16).

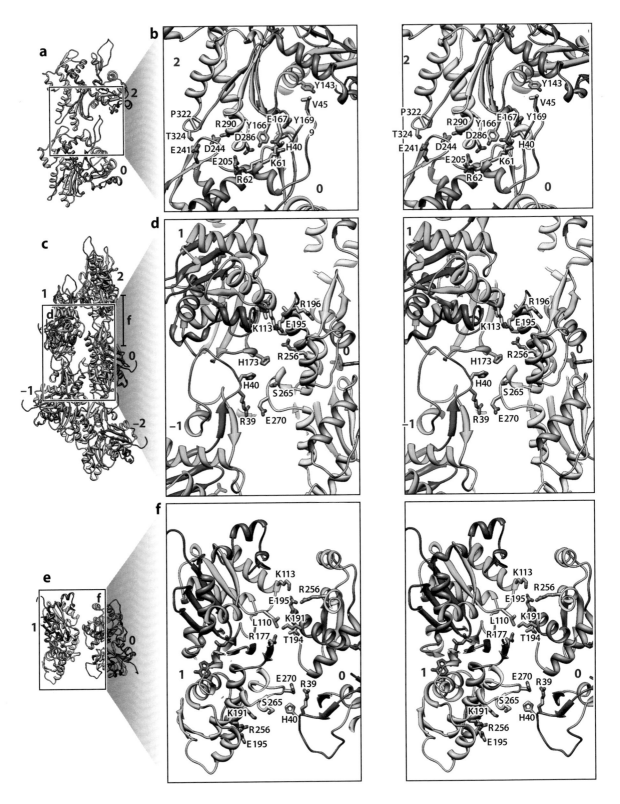

Precursor Actin Helix Stabilized by the Formin FH2 Domain

The precursor helical assembly of G-actin engendered by the FH2 domain (38) is related to the F-actin structure. The asymmetric unit of the crystal structure (1Y64) contains a single FH2-actin complex. The C2 symmetry produces identical FH2-actin complexes at ±28.1 Å and rotated by 180° with respect to one another, with neighboring FH2 subunits tethered by lasso/post interactions (**Figure 5** and **Supplemental Figure 2**). The actin subunits of two successive FH2-actin complexes are thus held in a relative orientation similar to that of the short-pitch F-actin helix (actin subunits of the filament helix are rotated by 166.6° and translated by 27.6 Å), such that the structure found in the crystal looks like a precursor of the F-actin helix (**Figure 5**). However, the longitudinal interactions between actin subunits are not the same as in F-actin and the D-loop is disordered. Nevertheless, the G-actin monomers are positioned by the FH2 scaffold in a way that would facilitate a transition to the F-actin helix. For this transition to occur, the FH2 domain would have to move out to let the D-loop bind in the target-binding cleft of the subunit above it, which is occupied in the complex by a helix of the FH2 knob domain.

Bacterial Actin-Like Homologues

A number of actin-like homologues have been discovered in bacteria. These form filaments (see, e.g., 46). The bacterial actin homology MreB forms filaments (53) that are single stranded and straight. The monomer is flat, similar to the F-actin monomer. However, it does not appear to have a loop equivalent to the D-loop of actin. ParM (47) is a prokaryotic actin homolog that segregates plasmids before bacterial cell division. ParM polymerizes with GTP. The ParM filament has a left-handed long-pitch helix, opposed to the right-handed long-pitch helix of actin (47). It has a sequence that appears to bind to domain 3, similar to the D-loop in actin. Furthermore, cryo-EM has visualized a form in which the nucleotide-binding cleft is open (17). AlfA (42) defines a new class of actins that shows an extremely rapid polymerization rate and assembles with ATP or GTP. The helical symmetry of AlfA filaments differs markedly from the helical symmetry of F-actin. Nevertheless, all these polymers show similar values for the rise per residue, approximately 3.0 nm.

THE G-ACTIN TO F-ACTIN TRANSITION

There is a mass of data on the inhibition of actin polymerization by binding of proteins or small molecules. Sometimes the bound molecules simply impede polymerization by steric hindrance. Thus, the fluorescence probe tetramethylrhodamine-5-maleimide sits exactly in the binding pocket of the D-loop and inhibits D-loop binding. In most cases it appears that the inhibiting molecule competes with the binding of the D-loop to the target-binding cleft of its neighbor along the long-pitch helix. A graded response is sometimes observed. For example, binding of gelsolin leads to severance, whereas binding of cofilin changes the helical pitch of F-actin and increases the filament flexibility (9).

The D-loop breaks into two regions: The sequence 39–43 is involved in stereospecific

---

Figure 4

Intermolecule bonding in F-actin. Shown are five actin molecules labeled −2 to +2. The run of the protein chain is shown as a secondary structure schematic color-coded from blue (N terminus) to red (C terminus). Interacting side chains are shown as sticks. Panels *b*, *d*, and *f* are stereo pairs. Panels *a* and *b* show the main longitudinal interface between molecules 0 and 2. Panels *c* and *d* show the transverse interaction across the helix axis in the neighborhood of the plug. This interaction also has contributions from the D-loop. Panels *e* and *f* show the transverse interaction higher up (see right-hand side of panel *c* for a definition of panel *f*) involving loops 195–198 and 108–113. (Reprinted by permission from Macmillan Publishers Ltd: *Nature* 467:724–728, copyright 2010.)

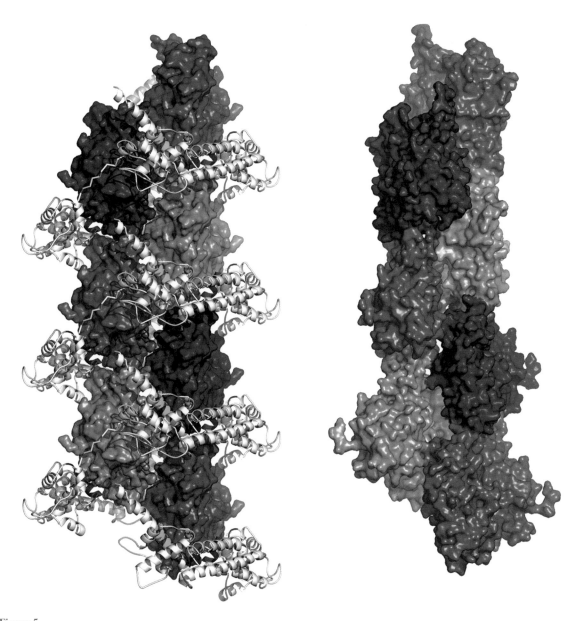

Figure 5

Precursor helix assembled by the formin FH2 domain. The left view shows the precursor helix assembled by FH2 subunits along the crystallographic C2 axis. The formin FH2 domain is shown gray as in **Figure 1i**. Neighboring molecules are positioned by binding respectively to the knob (*magenta*) and the post (*green*) to form a helix with a twist of 180° per molecule and a rise per residue of 28.1 Å (see **Supplemental Figure 2** for details). The right view shows the F-actin helix for comparison. Actin subunits of the filament helix are rotated by −166.6° and translated by 27.6 Å (as in **Figure 2**).

interaction, whereas the sequence 44–51 lies toward the outside of the F-actin structure [this sequence may be a target primarily for F-actin-binding proteins such as myosin (31)].

The WH2 domain sits in the cleft between domains 1 and 3, where it would collide with one end of the D-loop sequence 44–51 but does not interfere with the sequence 39–43. Perhaps differences in response lie in the behavior of the sequence 44–49 and the specific binding site of each protein in the target-binding cleft. For instance, the FAB domain of Ena/VASP is related in sequence to the WH2 domain but binds F-actin and can presumably co-bind in the cleft together with the D-loop of a neighboring actin subunit, whereas the classical WH2 domain cannot (15).

ATP-bound G-actin is the normal substrate for filament formation. It may be induced to polymerize to ATP-containing F-actin, which is energetically favored under physiological conditions. However, F-actin is a slow ATPase, such that the filament soon contains just ADP.

ADP F-actin is less stable than ADP G-actin. This enables actin filaments to be readily disassembled, an essential part of the actin recycling process in cells, but how is this differential stability controlled? It could come about by movements of the D-loop. Studies cited above show that the methylated histidine sensor loop can detect the status of nucleotide and pass this information to the D-loop. Moving the end of the D-loop by 1–2 Å would probably be enough to alter its binding affinity to the domain 1–domain 3 cleft. Unfortunately the data necessary to explore such mechanisms in detail is not available: We still lack the structure of ATP F-actin.

SUMMARY POINTS

1. Actin is of central importance to cell motility.

2. Actin can exist in two forms: G-actin (globular) and F-actin (filamentous).

3. The actin structure is highly conserved. The actin monomer consists of two major domains each of which contains two subdomains. The four subdomains are organized to form a rather flat molecule.

4. Two large clefts are formed between the two major domains of actin. The upper cleft binds the nucleotide. The lower cleft changes conformation depending on the state of the nucleotide and constitutes the binding site for most actin-binding proteins, of which there is a large number.

5. G-actin is not an efficient ATPase.

6. The polymerization of G-actin to F-actin is used in many cellular functions including cell motility and the maintenance of cell shape and polarity. The interaction of F-actin with myosin forms the basis of muscle contraction.

7. Upon polymerization the two major domains of actin undergo a propeller-twist that stimulates F-actin's ATPase activity. F-actin quickly consists of ADP-actin-bound subunits.

FUTURE ISSUES

1. There are no accurate atomic models of proteins bound to F-actin.

2. We do not know why ATP F-actin is more stable than ADP F-actin. We need details of the actin ATPase, detailed structure of ADP F-actin, and detailed structure of ATP F-actin.

3. More accurate structures of the ends of the actin filament (pointed and barbed) and the binding of capping proteins to these ends are needed.

4. How much do F-actin-binding proteins change the structure of the filament?

5. More information is needed regarding the mechano-sensing properties of the actin filament, exerted through interactions with membrane-associated proteins and cytoskeletal proteins.

DISCLOSURE STATEMENT

The authors are not aware of any affiliations, memberships, funding, or financial holdings that might be perceived as affecting the objectivity of this review.

ACKNOWLEDGMENTS

R.D. was supported by NIH grants MH087950, GM073791, and HL086655.

LITERATURE CITED

1. Barth H, Stiles BG. 2008. Binary actin-ADP-ribosylating toxins and their use as molecular Trojan horses for drug delivery into eukaryotic cells. *Curr. Med. Chem.* 15:459–69

2. Bernstein BW, Bamburg JR. 2010. ADF/cofilin: a functional node in cell biology. *Trends Cell Biol.* 20:187–95

3. Bork P, Sander C, Valencia A. 1992. An ATPase domain common to prokaryotic cell cycle proteins, sugar kinases, actin, and hsp70 heat shock proteins. *Proc. Natl. Acad. Sci. USA* 89:7290–94

4. Breitsprecher D, Kiesewetter AK, Linkner J, Urbanke C, Resch GP, et al. 2008. Clustering of VASP actively drives processive, WH2 domain-mediated actin filament elongation. *EMBO J.* 27:2943–54

5. Burtnick LD, Koepf EK, Grimes J, Jones EY, Stuart DI, et al. 1997. The crystal structure of plasma gelsolin: implications for actin severing, capping, and nucleation. *Cell* 90:661–70

6. Campellone KG, Welch MD. 2010. A nucleator arms race: cellular control of actin assembly. *Nat. Rev. Mol. Cell Biol.* 11:237–51

7. Carlier MF, Laurent V, Santolini J, Melki R, Didry D, et al. 1997. Actin depolymerizing factor (ADF/cofilin) enhances the rate of filament turnover: implication in actin-based motility. *J. Cell. Biol.* 136:1307–22

8. Carlier MF, Valentin-Ranc C, Combeau C, Fievez S, Pantoloni D. 1994. Actin polymerization: regulation by divalent metal ion and nucleotide binding, ATP hydrolysis and binding of myosin. *Adv. Exp. Med. Biol.* 358:71–81

9. De la Cruz EM. 2009. How cofilin severs an actin filament. *Biophys. Rev.* 1:51–59

10. Dominguez R. 2004. Actin-binding proteins—a unifying hypothesis. *Trends Biochem. Sci.* 29:572–78

11. Dominguez R. 2007. The beta-thymosin/WH2 fold: multifunctionality and structure. *Ann. N. Y. Acad. Sci.* 1112:86–94

12. Dominguez R. 2010. Structural insights into de novo actin polymerization. *Curr. Opin. Struct. Biol.* 20:217–25

13. dos Remedios CG, Chhabra D, Kekic M, Dedova IV, Tsubakihara M, et al. 2003. Actin binding proteins: regulation of cytoskeletal microfilaments. *Physiol. Rev.* 83:433–73

14. Drees F, Gertler FB. 2008. Ena/VASP: proteins at the tip of the nervous system. *Curr. Opin. Neurobiol.* 18:53–59

15. Ferron F, Rebowski G, Lee SH, Dominguez R. 2007. Structural basis for the recruitment of profilin-actin complexes during filament elongation by Ena/VASP. *EMBO J.* 26:4597–606

16. Fujii T, Iwane AH, Yanagida T, Namba K. 2010. Direct visualization of secondary structures of F-actin by electron cryomicroscopy. *Nature* 467:724–28

17. Galkin VE, Orlova A, Rivera C, Mullins RD, Egelman EH. 2009. Structural polymorphism of the ParM filament and dynamic instability. *Structure* 17:1253–64

18. Galkin VE, VanLoock MS, Orlova A, Egelman EH. 2002. A new internal mode in F-actin helps explain the remarkable evolutionary conservation of actin's sequence and structure. *Curr. Biol.* 12:570–75

19. Goode BL, Eck MJ. 2007. Mechanism and function of formins in the control of actin assembly. *Annu. Rev. Biochem.* 76:593–627

20. Gouin E, Welch MD, Cossart P. 2005. Actin-based motility of intracellular pathogens. *Curr. Opin. Microbiol.* 8:35–45

21. Graceffa P, Dominguez R. 2003. Crystal structure of monomeric actin in the ATP state: structural basis of nucleotide-dependent actin dynamics. *J. Biol. Chem.* 278:34172–80

22. Hanson J, Lowy J. 1963. The structure of F-actin and the actin filaments isolated from muscle. *J. Mol. Biol.* 6:46–60

23. Herman IM. 1993. Actin isoforms. *Curr. Opin. Cell Biol.* 5:48–55

24. Holmes KC, Angert I, Kull FJ, Jahn W, Schroder RR. 2003. Electron cryo-microscopy shows how strong binding of myosin to actin releases nucleotide. *Nature* 425:423–27

25. Holmes KC, Popp D, Gebhard W, Kabsch W. 1990. Atomic model of the actin filament. *Nature* 347:44–49

26. Iwasa M, Maeda K, Narita A, Maeda Y, Oda T. 2008. Dual roles of Gln137 of actin revealed by recombinant human cardiac muscle alpha-actin mutants. *J. Biol. Chem.* 283:21045–53

27. Kabsch W, Mannherz HG, Suck D, Pai EF, Holmes KC. 1990. Atomic structure of the actin:DNase I complex. *Nature* 347:37–44

28. Kovar DR, Harris ES, Mahaffy R, Higgs HN, Pollard TD. 2006. Control of the assembly of ATP- and ADP-actin by formins and profilin. *Cell* 124:423–35

29. Lang AE, Schmidt G, Schlosser A, Hey TD, Larrinua IM, et al. 2010. *Photorhabdus luminescens* toxins ADP-ribosylate actin and RhoA to force actin clustering. *Science* 327:1139–42

30. Lee WM, Galbraith RM. 1992. The extracellular actin-scavenger system and actin toxicity. *N. Engl. J. Med.* 326:1335–41

31. Lorenz M, Holmes KC. The actin-myosin interface. *Proc. Natl. Acad. Sci. USA* 107:12529–34

32. Lorenz M, Popp D, Holmes KC. 1993. Refinement of the F-actin model against X-ray fiber diffraction data by the use of a directed mutation algorithm. *J. Mol. Biol.* 234:826–36

33. Mannherz HG, Goody RS, Konrad M, Nowak E. 1980. The interaction of bovine pancreatic deoxyribonuclease I and skeletal muscle actin. *Eur. J. Biochem.* 104:367–79

34. Miralles F, Visa N. 2006. Actin in transcription and transcription regulation. *Curr. Opin. Cell Biol.* 18:261–66

35. Oda T, Iwasa M, Aihara T, Maeda Y, Narita A. 2009. The nature of the globular- to fibrous-actin transition. *Nature* 457:441–45

36. Oda T, Maeda Y. 2010. Multiple conformations of F-actin. *Structure* 18:761–67

37. Olson EN, Nordheim A. 2010. Linking actin dynamics and gene transcription to drive cellular motile functions. *Nat. Rev. Mol. Cell Biol.* 11:353–65

38. Otomo T, Tomchick DR, Otomo C, Panchal SC, Machius M, Rosen MK. 2005. Structural basis of actin filament nucleation and processive capping by a formin homology 2 domain. *Nature* 433:488–94

39. Otterbein LR, Graceffa P, Dominguez R. 2001. The crystal structure of uncomplexed actin in the ADP state. *Science* 293:708–11

40. Paavilainen VO, Oksanen E, Goldman A, Lappalainen P. 2008. Structure of the actin-depolymerizing factor homology domain in complex with actin. *J. Cell Biol.* 182:51–59

41. Paunola E, Mattila PK, Lappalainen P. 2002. WH2 domain: a small, versatile adapter for actin monomers. *FEBS Lett.* 513:92–97

42. Polka JK, Kollman JM, Agard DA, Mullins RD. 2009. The structure and assembly dynamics of plasmid actin AlfA imply a novel mechanism of DNA segregation. *J. Bacteriol.* 191:6219–30

43. Pollard TD. 2007. Regulation of actin filament assembly by Arp2/3 complex and formins. *Annu. Rev. Biophys. Biomol. Struct.* 36:451–77

44. Pollard TD, Borisy GG. 2003. Cellular motility driven by assembly and disassembly of actin filaments. *Cell* 112:453–65

45. Popp D, Lednev VV, Jahn W. 1987. Methods of preparing well-orientated sols of F-actin containing filaments suitable for X-ray diffraction. *J. Mol. Biol.* 197:679–84

46. Popp D, Narita A, Ghoshdastider U, Maeda K, Maeda Y, et al. 2010. Polymeric structures and dynamic properties of the bacterial actin AlfA. *J. Mol. Biol.* 397:1031–41

47. Popp D, Narita A, Oda T, Fujisawa T, Matsuo H, et al. 2008. Molecular structure of the ParM polymer and the mechanism leading to its nucleotide-driven dynamic instability. *EMBO J.* 27:570–79

48. Robinson RC, Turbedsky K, Kaiser DA, Marchand JB, Higgs HN, et al. 2001. Crystal structure of Arp2/3 complex. *Science* 294:1679–84

49. Rohr G, Mannherz HG. 1978. Isolation and characterization of secretory actin. DNAase I complex from rat pancreatic juice. *Eur. J. Biochem.* 89:151–57

50. Sawaya MR, Kudryashov DS, Pashkov I, Adisetiyo H, Reisler E, Yeates TO. 2008. Multiple crystal structures of actin dimers and their implications for interactions in the actin filament. *Acta Crystallogr. D* 64:454–65

51. Silacci P, Mazzolai L, Gauci C, Stergiopulos N, Yin HL, Hayoz D. 2004. Gelsolin superfamily proteins: key regulators of cellular functions. *Cell Mol. Life Sci.* 61:2614–23

52. Tirion MM, ben-Avraham D, Lorenz M, Holmes KC. 1995. Normal modes as refinement parameters for the F-actin model. *Biophys. J.* 68:5–12

53. van den Ent F, Amos LA, Lowe J. 2001. Prokaryotic origin of the actin cytoskeleton. *Nature* 413:39–44

54. van den Ent F, Møller-Jensen J, Amos LA, Gerdes K, Lowe J. 2002. F-actin-like filaments formed by plasmid segregation protein ParM. *EMBO J.* 21:6935–43

55. Wegner A, Isenberg G. 1983. 12-fold difference between the critical monomer concentrations of the two ends of actin filaments in physiological salt conditions. *Proc. Natl. Acad. Sci. USA* 80:4922–25

56. Xu Y, Moseley JB, Sagot I, Poy F, Pellman D, et al. 2004. Crystal structures of a Formin homology-2 domain reveal a tethered dimer architecture. *Cell* 116:711–23

57. Yarmola EG, Somasundaram T, Boring TA, Spector I, Bubb MR. 2000. Actin-latrunculin A structure and function. Differential modulation of actin-binding protein function by latrunculin A. *J. Biol. Chem.* 275:28120–27

58. Yeung KS, Paterson I. 2002. Actin-binding marine macrolides: total synthesis and biological importance. *Angew. Chem. Int. Ed. Engl.* 41:4632–53

59. Lodoen MB, Gerke C, Boothroyd JC. 2010. A highly sensitive FRET-based approach reveals secretion of the actin-binding protein toxofilin during *Toxoplasma gondii* infection. *Cell Microbiol.* 12(1):55–66

Molecular Origin of the Hierarchical Elasticity of Titin: Simulation, Experiment, and Theory

Jen Hsin,[1,2] Johan Strümpfer,[1,3] Eric H. Lee,[2,3,4] and Klaus Schulten[1,2,3]

[1]Department of Physics, [2]Beckman Institute for Advanced Science and Technology, [3]Center for Biophysics and Computational Biology, and [4]College of Medicine, University of Illinois at Urbana-Champaign, Urbana, Illinois 61801; email: kschulte@ks.uiuc.edu

Annu. Rev. Biophys. 2011. 40:187–203

First published online as a Review in Advance on February 14, 2011

The *Annual Review of Biophysics* is online at biophys.annualreviews.org

This article's doi:
10.1146/annurev-biophys-072110-125325

Copyright © 2011 by Annual Reviews.
All rights reserved

1936-122X/11/0609-0187$20.00

Keywords

molecular dynamics, steered molecular dynamics, elastic protein, mechanical protein

Abstract

This review uses the giant muscle protein titin as an example to showcase the capability of molecular dynamics simulations. Titin is responsible for the passive elasticity in muscle and is a chain composed of immunoglobulin (Ig)-like and fibronectin III (FN-III)-like domains, as well as PEVK segments rich in proline (P), glutamate (E), valine (V), and lysine (K). The elasticity of titin is derived in stages of extension under increasing external force: Ig domain straightening occurs first (termed tertiary structure elasticity), followed by the extension of the disordered PEVK segments. At larger extension and force, Ig domains unfold one by one (termed secondary structure elasticity). With the availability of crystal structures of single and connected Ig domains, the tertiary and secondary structure elasticity of titin was investigated through molecular dynamics simulations, unveiling the molecular origin of titin's elasticity.

Contents

INTRODUCTION

For more than thirty years, molecular dynamics (MD) simulations have been applied in a wide range of biological inquiries, providing dynamic pictures of biomolecules and relating their structure to function (15, 64, 83). One major strength of the MD methodology is its versatility, as simulation methods have extended far beyond the sampling of equilibrium properties and have evolved a diverse set of in silico experiments. The particular MD experiment discussed here is steered molecular dynamics (SMD) (32, 33), in which a force is applied to a biomolecule, mimicking in vivo biological events of cellular mechanics or in vitro atomic force experiments. Such computational testing of the force response of biomolecules is of high value because mechanical forces are associated with a variety of cellular functions, such as cell motion, muscular contraction, regulation of gene expression, cellular communication, and hearing (15, 42, 83).

To showcase the capability of SMD (and other related MD methods), we focus on the muscle protein titin. Titin is responsible for the elasticity and structural stability of muscle in response to stretching forces (21, 86). A series of elegant single-molecule experiments have demonstrated how titin is stretched with application of mechanical force (4, 12, 36, 45, 46, 62, 73, 75, 87, 93). Corroborating the low-resolution experimental evidence, MD studies revealed in chemical detail the molecular mechanisms responsible for the force-bearing characteristics of titin (13, 14, 16, 42, 43, 56–58, 62), providing a description of titin as a complex molecular spring with multiple levels of force response.

The Domain Components of Titin

Titin is the largest known protein in nature (1); its human variant is made of ~38,000 amino acids and is more than 1 μm long (86). Titin spans half of each contractile unit of muscle, called a sarcomere (**Figure 1a**) (38), and is a long linear chain composed of ~300 modular domains, 90% of which are immunoglobulin (Ig)-like or fibronectin III (FN-III)-like (1, 37); the remaining domains are flexible segments such as a PEVK domain [rich in proline (P), glutamate (E), valine (V), and lysine (K)] (51) and a catalytic kinase domain (39, 72) (**Figure 1b**). Crystallographic structures are available for some of titin's domains located at the N terminus (Z1Z2) (23, 60, 89, 94), thick filament-associated A-band (A77–78, A168–170) (2, 65, 66), and flexible I-band (I1, I65–75, I90–91) (31, 63, 86, 90). During muscle contraction and extension, the myosin filament slides along the actin filament, the motion of which is modulated by titin acting as a molecular spring (**Figure 1c**).

The Source of Titin's Extensibility

The role of titin in providing the extensibility and elasticity of muscle has been long recognized, as mutations in titin alter the elastic behavior of muscle and are associated with

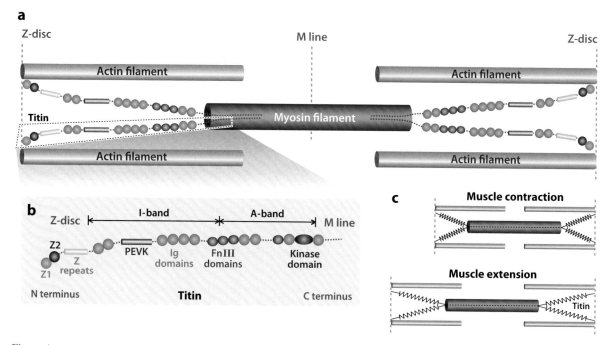

Figure 1

Schematic diagram of a muscle sarcomere. (*a*) The representative protein components in a muscle sarcomere are shown. Besides the well-known actin, or thin filament, and the myosin, or thick filament, muscle contains a third major mechanical component, titin, which gives muscle structural stability and elasticity. (*b*) An overview of the domain components of titin is presented (37). Titin is anchored at the Z-disc with the Z1Z2 domains, followed by the Z-repeats. The PEVK domain is a disordered segment with flexible secondary structure. Connected immunoglobulin (Ig)-like and fibronectin III (FnIII)-like domains make up a large proportion of the rest of the titin chain. (*c*) Shown is the overall mechanic function of titin. When muscle contracts and extends, the myosin filament moves along the actin filament, with the protein titin acting as a molecular spring, protecting muscle from overstretching.

muscle dystrophy (19, 24, 86). Elasticity of titin is derived mainly from the flexible PEVK and Ig segments; here we focus on the elasticity stemming from the Ig domains. We note that different muscle types have slightly different domain composition in their titin isoforms, with the more extensible muscle types (e.g., skeletal muscle) having more Ig domains in titin, and the less extensible muscle types (e.g., cardiac muscle) having fewer Ig domains (86).

Elasticity of titin is thought to arise in multiple regimes (**Figure 2**) (19, 20, 49, 50, 52, 55, 85, 87). When a small force (on the order of a few tens of picoNewtons) is applied, the domains in titin simply align themselves from a zigzag initial orientation and thereby extend titin, providing a tertiary-structure-based elasticity (**Figure 2a**). As force increases, the secondary structure of the titin domains starts to weaken, with the most flexible PEVK region extended first (**Figure 2b**). At physiologically extreme forces (~100 pN), individual Ig domains unravel, losing their β-sandwich secondary structure, and further extend titin (**Figure 2c**). A full picture of the elastic response of titin therefore requires the understanding of both tertiary structure elasticity and secondary structure elasticity.

TERTIARY STRUCTURE ELASTICITY

Despite the physiological relevance of tertiary-structure-based elasticity (it is titin's first mode of force-extension response), investigations of tertiary structure elasticity in the past have been

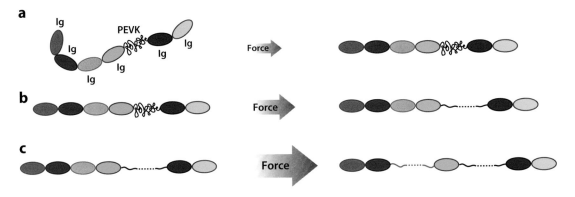

Figure 2

Three modes of elasticity in titin. (*a*) When a small force (on the order of a few tens of picoNewtons) is applied, the Ig domains reorientate themselves, i.e., straighten out, and extend titin, providing a tertiary-structure-based elasticity. (*b*) Continued stretching leads to the unraveling of the structurally flexible PEVK segment. (*c*) At larger forces (above ~100 pN), individual Ig domains unravel, losing their β-sandwich secondary structure, and further extend titin, providing secondary-structure-based elasticity.

difficult (in the case of experiment) or unfeasible (in the case of simulation). Experimentally, the force response stemming from titin's tertiary structure elasticity did not receive as much attention as that from secondary structure elasticity, as the force-extension relation in the latter case is much more pronounced. Also, the lack of resolution in experiments prohibits direct identification of the domain-domain interactions that contribute to tertiary structure elasticity. Simulation of tertiary structure elasticity requires a high-resolution structure of multiple, connected Ig domains (from the functionally elastic titin I-band), which became available only in 2008 (90), whereas structures of single titin Ig domains were published more than a decade earlier (31). In this section, the experimental efforts and simulations probing the tertiary structure elasticity of titin are reviewed.

Stretching Experiments on Titin

Single-molecule force spectroscopy experiments employing methods such as atomic force microscopy (AFM) (4, 12, 35, 45, 46, 62, 75, 93) and optical tweezers (36, 40, 87) are powerful tools for probing the force-bearing properties of titin. These stretching experiments (including also immunofluorescence/immunoelectron microscopy) were carried out with either full

titin molecules (35, 36, 40, 87), or engineered constructs consisting of connected Ig domains (46), or the PEVK segment (51, 67, 79), or a combination of Ig and PEVK segments (46, 53). Because there are many different titin domains included in these experiments, it is not possible to correlate directly the tertiary-structure-based force response with specific domains. Therefore, interpretation of these stretching experiments is often vague.

When force is applied to a titin molecule (or a recombinant segment of connected titin domains), the filament is first extended into a linear chain (**Figure 2a**). The force required for such transformation is on the order of a few tens of picoNewtons, increases slowly with small extension, and rises more quickly at larger extension, resulting in a force-extension profile as shown in **Figure 3a** (87). The shape of the force-extension curve has been described theoretically by three polymer models: the freely jointed chain (FJC) model (10), the worm-like chain (WLC) model (3, 61), and the modified worm-like chain (mWLC) model (91) (**Figure 3b–d**).

Theoretical models interpreting the force-extension profile. The FJC model describes a chain of domains, each of length L_k (known as the Kuhn length), for which there is no

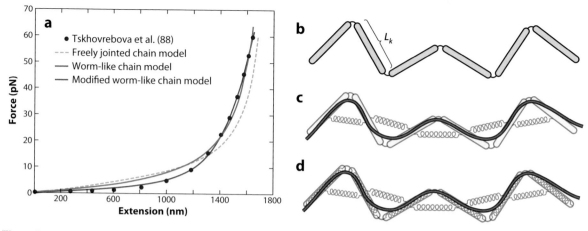

Figure 3

(*a*) A typical force-extension profile from stretching experiments of titin as reported in Reference 87 (*purple dots*). The data is fitted with three commonly employed theoretical models portrayed in panels *b–d*. (*b*) Freely jointed chain (FJC) model is fitted using Equation 1 with parameters $L_k = 1.0$ nm and contour length $L_c = 1,810$ nm. (*c*) Worm-like chain (WLC) model, which is a continuum model that ascribes a bending energy to the chain, is fitted using Equation 2 with parameters $L_p = 0.7$ nm and contour length $L_c = 1,950$ nm. (*d*) Modified worm-like chain (mWLC) model, which introduces a stretching degree of freedom to the chain, is fitted using Equation 3 with parameters $L_p = 1.5$ nm, contour length $L_c = 1,630$ nm, and stretching modulus $K = 520$ pN nm^{-1}. The fittings in panel *a* were performed on stretching data of a full titin molecule, and therefore, elasticity from non-Ig domains, i.e., PEVK, also contributes.

energetic cost to bend neighboring segments relative to each other (**Figure 3*b***). When a freely jointed chain is stretched, the number of available chain conformations is reduced. The net effect is a restoring force of purely entropic origin that pushes the system back to an equilibrium length, which has the most available conformations. As such, a freely jointed chain is often described as an entropic spring. The relation between the applied force (F) and the extension of the chain (X) in the FJC model is given by

$$X = L_c \left[\coth\left(\frac{F L_k}{k_B T} \right) - \frac{k_B T}{F L_k} \right], \quad 1.$$

where L_c is the contour length of the chain (10, 40, 70).

Unlike the freely jointed chain, which is made of well-defined units (each with length L_k), the WLC model is a continuous polymer chain that requires an energetic cost to bend (**Figure 3*c***). Therefore, the WLC model includes an entropic as well as an enthalpic component to its free energy and can be thought of as an entropic-enthalpic spring (76). Its force-

extension relation is

$$F = \frac{k_B T}{L_p} \left[\frac{1}{4(1 - X/L_c)^2} - \frac{1}{4} + \frac{X}{L_c} \right], \quad 2.$$

where L_p is the persistent length of the chain (3, 51, 61). Although a continuum model, one can discretize the WLC into a chain of domains with identical harmonic bending potential for each pair of domains (27, 34), as shown in **Figure 3*c*** (transparent). It can then be demonstrated that the WLC model is equivalent to the FJC model, but with an additional harmonic bending energy (34).

The mWLC model further allows the overall length of the chain to be harmonically stretched (described by another parameter, K, known as the stretching modulus), adding a second enthalpic component (11, 51, 91) (**Figure 3*d***). The corresponding force-extension relationship is

$$F = \frac{k_B T}{L_p} \left[\frac{1}{4(1 - X/L_c + F/K)^2} - \frac{1}{4} + \frac{X}{L_c} - \frac{F}{K} \right]. \quad 3.$$

As shown in **Figure 3a**, the mWLC model fits the force-extension profile of titin best (49, 51).

The primary parameter of the WLC model is the persistence length, L_p, which can be related to the Kuhn length L_k in the FJC model via the relation $2L_p = L_k$ (11). Because the Ig domains of titin are structurally well defined (each being globular and separated by a short linker region), it is expected that they would be the primary components in the various models, such as the Kuhn segment for the FJC model. It can thus be argued that, for the Ig region of titin, one should obtain $L_k = 2L_p =$ 4.4 nm, the approximate length of a single Ig domain. Indeed, L_p is in the range of 3.3 nm to 4.8 nm when fitting the WLC model to experimentally measured force-extension curves of Ig-only domains (40, 87). These values, although not exactly matching the relation above, are satisfactory considering the complexity of titin and the simplicity of the model. Other persistence lengths for titin have also been reported; these values were obtained from experiments that included the force response of non-Ig regions or involved Ig domain unfolding (36, 46, 50, 85).

The good fits of the WLC and mWLC models to the measured force-extension curves imply that there is a non-negligible interaction between the Ig domains and that a multi-Ig chain acts as an entropic-enthalpic spring. However, the fact that the WLC model fits the force-extension relation of titin very well is actually perplexing for several reasons. First, the WLC model assumes homogeneity among the subunits, whereas the domains in titin are quite different from each other, and even for the titin Ig domains alone there exists an appreciable heterogeneity (17, 59, 90). Second, the WLC model assumes a harmonic interaction between each connected domain pair, but the nature of domain-domain interactions in a titin chain is not resolved in experiments and might not necessarily be harmonic. In addition, it remains to be explained why a continuum model such as the WLC model can describe titin, which is made of discrete, well-defined

domains. The full atomic resolution inherent in the MD methodology can be exploited to address these issues.

Molecular Dynamics Simulations of a Multi-Ig Chain

With the availability of the crystal structure of a titin six-Ig domain fragment I65–I70 (Ig6) (90), MD simulations could be carried out to characterize the elastic properties of this construct (43). In an initial 20-ns free dynamics simulation, all domains in the Ig6 segment retained their secondary structure, while significant flexibility was seen in the linker region connecting the Ig domains (43, 88). To probe the force response of Ig6, especially its tertiary-structure-based elasticity, SMD simulations (32, 33) were performed to extend the originally crescent-shaped Ig6 chain into a straight linear chain without disrupting the secondary structure of individual domains (**Figure 4a**) (43). During the SMD simulation, the force applied to the Ig6 chain was measured as a function of chain extension (**Figure 4b**).

The force-extension profile recorded from the SMD simulation (**Figure 4b**) shows that straightening the Ig6 requires only a few tens of picoNewtons. The force remained constant (around 28 pN) during the early stage of extension (<60 Å); the constant force is due to hydrodynamic friction experienced when the Ig6 is dragged through the aqueous environment (water molecules are explicitly described in the simulation) (43). During this early stage of stretching, the intrinsic force required to extend Ig6 is so low that the viscous drag force of 28 pN dominates the force-extension relationship. However, beyond 60 Å extension, the intrinsic force increases above the 28 pN viscous force and becomes discernible in the force-extension profile; with an extension of ~90 Å the intrinsic force increases even more rapidly to a value of about 150 pN. The force-extension profile beyond 60 Å reflects the tertiary structure elasticity intrinsic to Ig6. The hydrodynamic drag force can be

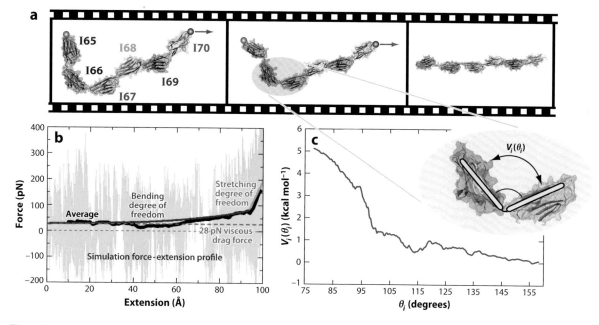

Figure 4

Molecular dynamics (MD) simulation of the force-bearing properties of a chain of six Ig domains (I65–I70; the chain is called Ig6 in this review) (43). (*a*) Snapshots of the steered molecular dynamics (SMD) simulation extending the originally crescent-shaped Ig6. The pulling simulation was terminated when Ig6 was transformed into a linear chain with the secondary structure of all domains preserved. (*b*) Force-extension profile measured in the simulation is plotted in gray, with the average shown in black. Theoretical calculation of the force-extension relation considering the bending degree of freedom is plotted in red, with the motion arising from the stretching degree of freedom shown in orange. Combining these motions fully explains the tertiary-structure-based elasticity of Ig6. (*c*) To compute the energetics associated with the bending degree of freedom, adaptive biasing force calculations implemented in the MD package NAMD (5, 25, 26, 71, 78) were employed (43). The bending motion between linked domains was seen to be soft, requiring only a few kilocalories per mole.

reduced to near zero by decreasing the speed with which the Ig6 chain was stretched, at the expense of higher computing cost. Reducing the stretching speed, for the case of Ig6, also brings the added benefit of pulling the chain under quasi-equilibrium condition (28).

Tertiary structure elasticity is a composite property. As the tertiary structure elasticity of Ig6 originates from domain-domain motions, comprehension of this mode of force response requires investigation of the energetics associated with all forms of domain-domain motions. For the case of Ig6, the most significant motion between two connected domains is the bending motion, in which the domains bend around the linker region (**Figure 4c**). The dynamics of

such motion can be understood based on the energy, $V_i(\theta_i)$, for Ig domain pair i (for Ig6, there are five different pairs of adjacent domains, i.e., $i = 1, 2, \ldots, 5$), which assumes an angle θ_i.

The energy, $V_i(\theta_i)$, can be calculated by using a MD-based protocol called the adaptive biasing force (ABF) method (5, 25, 78) implemented in the MD package NAMD (71). An illustrative result from an ABF calculation of a connected Ig pair is displayed in **Figure 4c**, which shows that the preferred conformation for this Ig pair is an open state, in which the two domains form a large angle, yet transition to a closed state requires only a few kilocalories per mole (43). The low energy cost in bending the Ig domain pairs of Ig6 is in sharp contrast to the energy cost determined for titin's Z1Z2 domain (see **Figure 1b**), for which comparable ABF calculations were performed (41). While the bending motion of Ig domain pairs incurs only a few kilocalories per mole, such motion costs more than 20 kcal mol^{-1} for the Z1Z2 domain (41), indicating inherent rigidity in the structure of Z1Z2. Such rigidity is functionally relevant as Z1Z2 acts as an anchor fixing the N terminus of titin at the Z-line (**Figure 1c**), whereas the Ig domains in the I-band are responsible for titin's extensibility.

Counting all degrees of freedom. The energy, $V_i(\theta_i)$, was computed via ABF for all five domain pairs in the Ig6 chain (43). To combine all the energetic information into a description of the extensibility of an overall chain, first, each $V_i(\theta_i)$ is expressed in terms of the end-to-end distance of the domain pair, $\tilde{V}_i(x_i) = V_i(\theta_i(x_i))$, which is then transformed to a length distribution, $p_i(x_i)$, via the Boltzmann relation $p_i(x_i) \propto \exp[-\tilde{V}_i(x_i)/k_B T]$ (**Figure 5a**). As the overall chain length is the sum of the individual units, i.e., $X = \sum_i x_i$, the length distribution of the full chain, $P(X)$, for a chain made of N-many

domain pairs can be expressed as (41, 43)

$$P(X) = \int_{-\infty}^{+\infty} dx_1 \int_{-\infty}^{+\infty} dx_2 \ldots \int_{-\infty}^{+\infty} dx_N$$
$$\times p_1(x_1)p_2(x_2)\ldots p_N(x_N)\delta\left(\sum_{i=1}^{N} x_i - X\right). \quad 4.$$

The equation above is simply a mathematical expression stating the following: The length distribution of the full chain X is the product of the length distributions of individual domain pairs, but for a given X, one should take into account only those combinations of x_i's that add up to be X.

In the presence of an external force, f_{chain}, the energy landscape, $\tilde{V}_i(x_i)$, is shifted according to $\hat{V}_i(x_i) = \tilde{V}_i(x_i) - f_{\text{chain}}x_i$, resulting in a new equilibrium end-to-end length for each domain pair. The new average end-to-end length, $\langle x_i \rangle$, for domain pair i can be computed using statistical mechanics theory (43) and summed up to calculate the length of the full chain $\langle X \rangle = \sum_i \langle x_i \rangle$, which is a function of the applied force f_{chain} [i.e., $\langle X \rangle = \langle X \rangle(f_{\text{chain}})$]. Thus, the force-extension profile can be numerically computed. The result of this calculation is

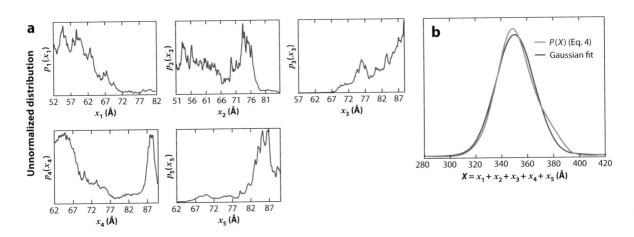

Figure 5

The length distributions of individual adjacent Ig domains are used to compute the overall length distribution of Ig6 by employing Equation 4. (*a*) The length distribution of the five connected Ig domain pairs in Ig6, with the index 1 denoting domain pair I65-I67, 2 denoting domain pair I67-I68, and so on. The length distributions $p_i(x_i)$ were obtained using the adaptive biasing force (ABF) method discussed in the text (43). (*b*) Overall length distribution of the Ig6 chain by inserting the five $p_i(x_i)$ into Equation 4 to obtain $P(X)$ (*orange trace*). Even though the individual $p_i(x_i)$ is non-Gaussian and nonidentical, the overall $P(X)$ can be closely fitted to a Gaussian distribution (*gray trace*).

plotted in **Figure 4b**. As seen in **Figure 4b**, this theoretical treatment describes the Ig6 force response very well up to 90 Å extension.

In order to describe the tertiary structure elasticity over the full extension range, i.e., over the interval [0 Å, 100 Å], all degrees of freedom that permit Ig6 stretching should be accounted for. The orange trace in **Figure 4b** was computed by adding also the small extension involving stretching of all single domains (43). The result fits the simulated average force-extension curve (**Figure 4b**) extremely well.

Emergence of Global Uniformity from Local Heterogeneity

Application of SMD enables direct measurement of the force-bearing properties of a titin Ig chain arising from its tertiary structure elasticity. In such stretching simulations, the most relevant measurement is the extension versus force profile. Owing to the inherent all-atom resolution of MD, one can identify the molecular motions that give rise to the patterns seen in the force-extension profile. For an Ig chain, the major sources of motion are the bending between each connected domain pair and the slight stretching of the domains. Such motions can be parameterized along bending and stretching reaction coordinates, and the characteristics of these motions can be understood by computing the associated free-energy functions, i.e., $\tilde{V}_i(x_i)$, using an MD-based free-energy calculation protocol such as ABF.

For each motion, its energy $\tilde{V}_i(x_i)$ can be computed and translated to a probability distribution $p_i(x_i)$. By accounting for all possible degrees of freedom of the Ig chain that contribute to the overall extension X, one can make use of Equation 4 to calculate the thermodynamics properties of the full chain. The fact that tertiary structure elasticity is a combined property of many smaller-scale motions actually brings upon an important realization. As a result of the central limit theorem (80), the overall distribution $P(X)$ in Equation 4 is nearly Gaussian even if the individual $p_i(x_i)$ are non-Gaussian and nonidentical, provided

that there is a sufficient number of $p_i(x_i)$ combined to yield $P(X)$, as demonstrated in **Figure 5b** for Ig6. Noting that a Gaussian distributed $P(X)$ and harmonic interaction energy $V(X)$ are two sides of the same coin via Boltzmann relation ($P(X) \propto \exp(-V(X)/k_B T)$), the previous sentence can be recast into a more intuitive form: A protein made of repeat units behaves overall like a harmonic spring (at small external force), even if the individual units are nothing like harmonic springs and are different from each other. With this realization, it is worth noticing that many other functionally elastic proteins are also made of repeat domains, e.g., ankyrin, cadherin, and fibrin (42–44, 48, 82–84).

The above contemplation also explains why a continuum model such as the WLC describes the tertiary structure elasticity of titin very well, fitting the experimentally measured force-extension profile quite satisfactorily (40, 46, 85, 87) (**Figure 3a**), despite the heterogeneity of the titin domains (59, 90) and the nonharmonic nature of domain-domain motion contributing to X, owing to the central limit theorem. It also becomes apparent why inclusion of an additional enthalpic component, namely the stretching degree of freedom for each individual domains, captures the tertiary structure elastic response better: All relevant (in regard to contributing to X) motions need to be accounted for in describing the elastic response of a protein chain. In situations where details matter, MD methodology is required to reveal the molecular interactions that combine to manifest the elastic property of titin.

SECONDARY STRUCTURE ELASTICITY

When a straightened titin chain continues to be stretched, the increased applied force eventually unravels one of the domains, which loses its secondary structure and extends, providing relief of tension for the other domains (**Figure 2b,c**). This is the regime of the secondary-structure-based force response of the titin chain.

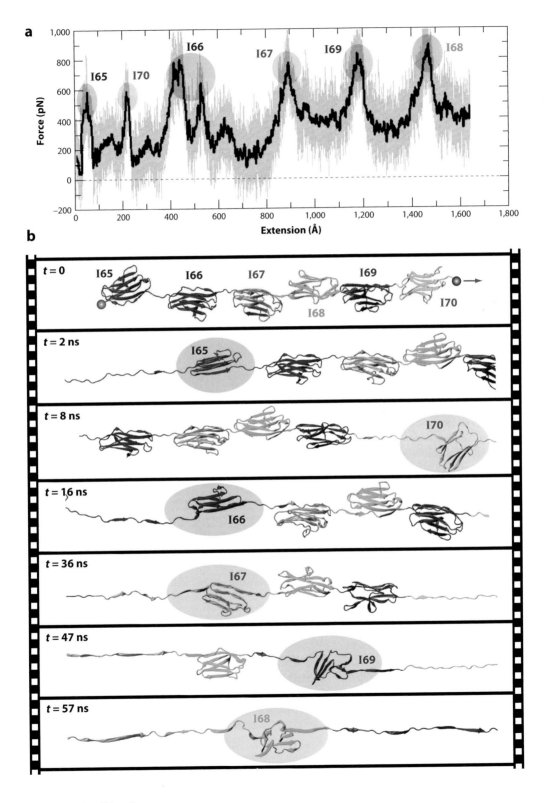

Secondary structure elasticity has been described successfully using MD methods. With the availability of structures for single Ig domains (31, 63, 86), a series of computational investigations was published, revealing in atomic detail how a network of hydrogen bonds holding together the β-sandwich secondary structure of the individual Ig domains provides resistance to external force (13, 14, 16, 42, 43, 56–58, 62). The contribution of MD in elucidating secondary structure elasticity of single Ig domains has been reviewed elsewhere (15, 42, 83); here we discuss again a multi-Ig construct, namely Ig6.

The Mechanical Fingerprint of Multi-Ig Constructs

Complete unfolding of Ig6 was induced in a SMD simulation in which the Ig6 chain was stretched until all six domains ruptured and fully extended (43) (**Figure 6**). In this simulation, the Ig domains were seen to rupture one after another, rather than concurrently. The force-extension curve shows six peak areas (**Figure 6a**) that correspond to domain ruptures in the order I65→I70→I66→I67→I69→I68, i.e., outside in (see **Figure 6b**). In the case of each peak, the force is seen to rise until it reaches a value high enough to rupture the domain spontaneously, i.e., within a nanosecond. Rupturing of an Ig domain leads to an immediate sharp decrease in force as the unraveling domain adds slack to the extension and releases tension. The sequential rupturing of the six Ig domains results in a force-extension profile with a saw-toothed pattern (**Figure 6a**); each peak is attributed to the unfolding/rupturing of one Ig domain (**Figure 6b**). In the particular SMD simulation, the Ig6 chain was stretched with a pulling velocity of 25 Å ns^{-1}, and the full stretching required ∼70 ns and produced nearly 600% of extension.

The saw-toothed force-extension profile of a multi-Ig construct was also observed in single-molecule force spectroscopy experiments employing AFM (4, 12, 45–47, 62, 69, 74, 75, 77, 93). In an AFM experiment, usually recombinant Ig domains are used, i.e., an engineered chain of identical Ig domains is stretched as illustrated in **Figure 7a**, resulting in force traces with evenly spaced peaks sketched in **Figure 7b**. Indeed, so conspicuous is the saw-toothed pattern in the force-extension profile that it has been designated the mechanical fingerprint of titin Ig domains (9).

Comparing Experiment and Simulation

Although the force-extension pattern measured in AFM compares qualitatively well to that seen in simulation (**Figure 6a** and **Figure 7b**), the magnitude of the rupture forces measured is in the range of 100–200 pN, but forces as high as nearly 1000 pN arise in simulations. Among different experiments, there also is no precise agreement on the rupture force of the titin Ig domain (4, 12, 45, 46, 62, 75, 93). The discrepancy is due to different experimental conditions, but due primarily to different stretching velocities employed as shown in **Figure 7c** (7, 8, 29). Different stretching velocities are also the reason for the discrepancy between observed and simulated rupture forces (42). Indeed, experimental and simulation data, as shown in **Figure 7c**, fall on the same theoretical rupture force versus stretching velocity curve (29, 42).

Physiological Functioning of Titin

It is still an open question whether secondary structure elasticity of Ig domains plays a role in the physiological functioning of muscle. Ig domain unfolding may provide the necessary

---

Figure 6

The unfolding of Ig6 induced in a steered molecular dynamics (SMD) simulation (43). (*a*) The measured force-extension profile displays distinct force peaks corresponding to the one-by-one unraveling of the six Ig domains. (*b*) Snapshots of the Ig domains during unfolding.

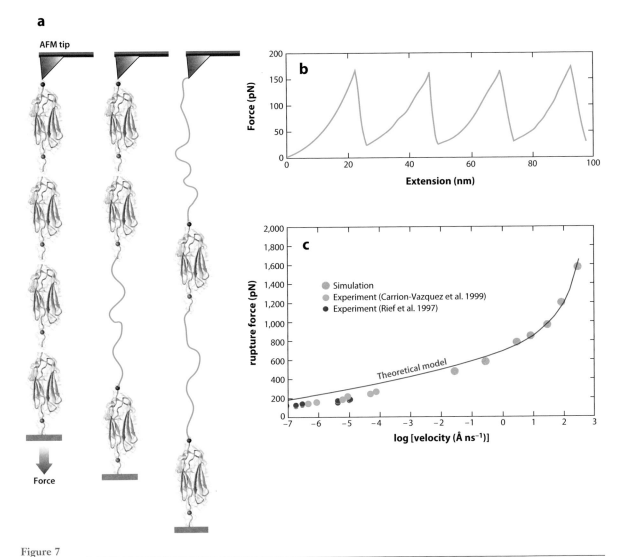

Figure 7

Setup and measured force-extension profile of an atomic force microscopy (AFM) experiment with recombinant Ig domains. (*a*) Ig domains are linked together and attached to an AFM tip; a force is applied at the tip. (*b*) As force is applied, the Ig domains rupture one by one, resulting in the schematic force-extension profile. (*c*) The rupture forces (i.e., the peak forces in panel *b*) are compared across different pulling velocities applied in both experiment and simulation. A single theoretical model (29) (*continuous red curve*) describes both experiment and simulation, suggesting that simulation accounts correctly for single-domain rupture (42).

extension in case of extreme sarcomere stretching (6, 81). Although physiological forces in muscle (0–10 pN) (22, 92) are likely lower than those required to unfold an Ig domain, there may be some titin domains designed to unravel under extreme working conditions of muscle. Furthermore, titin domains can refold within a few seconds, even when still stretched by a low force (18), suggesting that Ig domain unfolding is reversible and does not pose damage to muscle. Indeed, some Ig domains unfold at a relatively low stretching force (17, 46); such Ig domains may serve as shock absorbers, preventing irreparable muscle injury (50, 68).

The sarcomere constitutes a crowded environment with many intermolecular interactions that are not present when in vitro experiments are conducted. Evidence from investigations using myofibrils and whole muscle fibers suggests that, in situ, Ig domain unfolding does not arise (54, 85). In addition, other components in the muscle system, such as tendons in skeletal muscles, also experience stretching forces when a muscle is under strain and, unlike titin, do not possess flexibility; a force strong enough to unfold titin's Ig domains is likely sufficient to break the other less force-resistant muscle components. Complementary investigations using experimental, theoretical, and simulation methods at microscopic to macroscopic levels are required to address the possible physiological function of Ig domain unfolding further and to evaluate the contribution of secondary structure elasticity of the Ig segment to the overall extensibility of titin.

CONCLUSION

The structure of the muscle protein titin is directly related to its function. In particular, the function of titin reflects its hierarchical structural organization (secondary and tertiary). Titin, functioning as a molecular spring, transforms its tertiary structure to buffer small forces and its secondary structure to buffer large forces. By exercising two modes of force response, titin provides muscle with a high degree of structural stability and mechanical elasticity. A combination of crystallography, AFM, simulations, and theory are needed to explain the amazing function of titin.

DISCLOSURE STATEMENT

The authors are not aware of any affiliations, memberships, funding, or financial holdings that might be perceived as affecting the objectivity of this review.

ACKNOWLEDGMENTS

This work was supported by grants from the National Institute of Health (grant no. R01-GM073655 and grant no. P41-RR005969) and the National Science Foundation (PHY0822613). All-atom MD simulations discussed here were performed using the package NAMD (71), with computer time provided by NCSA and TACC via Large Resources Allocation Committee grant MCA93S028. Molecular images in this article were rendered using the molecular visualization software VMD (30). J.H. is supported by the Edelheit Fellowship. Jen Hsin and Johan Strümpfer contributed equally to this work.

LITERATURE CITED

1. Bang ML, Centner T, Fornoff F, Geach AJ, Gotthardt M, et al. 2001. The complete gene sequence of titin, expression of an unusual approximately 700-kDa titin isoform, and its interaction with obscurin identify a novel Z-line to I-band linking system. *Circ. Res.* 89(11):1065–72
2. Bucher RM, Svergun DI, Muhle-Goll C, Mayans O. 2010. The structure of the FnIII tandem A77-A78 points to a periodically conserved architecture in the myosin-binding region of titin. *J. Mol. Biol.* 401:843–53
3. Bustamante C, Marko JF, Siggia ED, Smith S. 1994. Entropic elasticity of λ-phage DNA. *Science* 265:1599–600
4. Carrion-Vazquez M, Oberhauser AF, Fowler SB, Marszalek PE, Broedel SE, et al. 1999. Mechanical and chemical unfolding of a single protein: a comparison. *Proc. Natl. Acad. Sci. USA* 96:3694–99
5. Darve E, Wilson MA, Pohorille A. 2002. Calculating free energies using a scaled-force molecular dynamics algorithm. *Mol. Sim.* 28:113–44

6. Erickson HP. 1994. Reversible unfolding of fibronectin type III and immunoglobulin domains provides the structural basis for stretch and elasticity of titin and fibronectin. *Proc. Natl. Acad. Sci. USA* 91:10114–18

7. Evans E. 2001. Probing the relation between force–lifetime–and chemistry in single molecular bonds. *Annu. Rev. Biophys. Biomol. Struct.* 30:105–28

8. Evans E, Ritchie K. 1997. Dynamic strength of molecular adhesion bonds. *Biophys. J.* 72(4):1541–55

9. Fernandez JM. 2005. Fingerprinting single molecules in vivo. *Biophys. J.* 89:3676–77

10. Flory PJ. 1969. *Statistical Mechanics of Chain Molecules*. New York: Interscience

11. Flory PJ. 2003. *Statistical Mechanics of Polymer Chains*. Cincinnati: Hanser

12. Fowler SB, Best RB, Herrera JLT, Rutherford TJ, Steward A, et al. 2002. Mechanical unfolding of a titin Ig domain: structure of unfolding intermediate revealed by combining AFM, molecular dynamics simulations, NMR and protein engineering. *J. Mol. Biol.* 322(4):841–49

13. Gao M, Lu H, Schulten K. 2001. Simulated refolding of stretched titin immunoglobulin domains. *Biophys. J.* 81:2268–77

14. Gao M, Lu H, Schulten K. 2002. Unfolding of titin domains studied by molecular dynamics simulations. *J. Muscle Res. Cell Motil.* 23:513–21

15. Gao M, Sotomayor M, Villa E, Lee EH, Schulten K. 2006. Molecular mechanisms of cellular mechanics. *Phys. Chem. Chem. Phys.* 8:3692–706

16. Gao M, Wilmanns M, Schulten K. 2002. Steered molecular dynamics studies of titin I1 domain unfolding. *Biophys. J.* 83:3435–45

17. Garcia TI, Oberhauser AF, Braun W. 2009. Mechanical stability and differentially conserved physical-chemical properties of titin Ig-domains. *Proteins Struct. Func. Bioinform.* 75:706–18

18. Garcia-Manyes S, Brujic J, Badilla CL, Fernandez JM. 2007. Force-clamp spectroscopy of single-protein monomers reveals the individual unfolding and folding pathways of I27 and ubiquitin. *Biophys. J.* 93:2436–46

19. Garvey SM, Rajan C, Lerner AP, Frankel WN, Cox GA. 2002. The muscular dystrophy with myositis (mdm) mouse mutation disrupts a skeletal muscle-specific domain of titin. *Genomics* 79:146–49

20. Gautel M, Goulding D. 1996. A molecular map of titin/connectin elasticity reveals two different mechanisms acting in series. *FEBS Lett.* 385:11–14

21. Granzier HL, Labeit S. 2004. The giant protein titin: a major player in myocardial mechanics, signaling, and disease. *Circ. Res.* 94:284–95

22. Granzier HL, Labeit S. 2005. Titin and its associated proteins: the third myofilament system of the sarcomere. *Adv. Protein Chem.* 71:89–119

23. Gregorio CC, Trombitas K, Centner T, Kolmerer B, Stier G, et al. 1998. The NH2 terminus of titin spans the Z-disc: Its interaction with a novel 19-kD ligand (T-cap) is required for sarcomeric integrity. *J. Cell Biol.* 143(4):1013–27

24. Hackman P, Vihola A, Haravuori H, Marchand S, Sarparanta J, et al. 2002. Tibial muscular dystrophy is a titinopathy caused by mutations in *TTN*, the gene encoding the giant skeletal-muscle protein titin. *Am. J. Hum. Genet.* 71:492–500

25. Hénin J, Chipot C. 2004. Overcoming free energy barriers using unconstrained molecular dynamics simulations. *J. Chem. Phys.* 121(7):2904–14

26. Hénin J, Pohorille A, Chipot C. 2005. Insights into the recognition and association of transmembrane alpha-helices. The free energy of alpha-helix dimerization in glycophorin A. *J. Am. Chem. Soc.* 127:8478–84

27. Higuchi H, Nakauchi Y, Maruyama K, Fujime S. 1993. Characterization of beta-connectin (titin 2) from striated muscle by dynamic light scattering. *Biophys. J.* 65:1906–15

28. Hsin J, Schulten K. 2011. Amplification of tertiary structure elasticity in a tandem titin Ig chain. *Biophys. J.* 100:L22–24

29. Hummer G, Szabo A. 2003. Kinetics from nonequilibrium single-molecule pulling experiments. *Biophys. J.* 85(1):5–15

30. Humphrey W, Dalke A, Schulten K. 1996. VMD—visual molecular dynamics. *J. Mol. Graph.* 14:33–38

31. Improta S, Politou AS, Pastore A. 1996. Immunoglobulin-like modules from titin I-band: extensible components of muscle elasticity. *Structure* 4:323–37

32. Isralewitz B, Gao M, Schulten K. 2001. Steered molecular dynamics and mechanical functions of proteins. *Curr. Opin. Struct. Biol.* 11:224–30

33. Izrailev S, Stepaniants S, Isralewitz B, Kosztin D, Lu H, et al. 1998. Steered molecular dynamics. In *Computational Molecular Dynamics: Challenges, Methods, Ideas*. Vol. 4: *Lect. Notes Comput. Sci. Eng.*, ed. P Deuflhard, J Hermans, B Leimkuhler, AE Mark, S Reich, RD Skeel, pp. 39–65. Berlin: Springer-Verlag

34. Jian H, Vologodskii AV, Schlick T. 1997. A combined wormlike-chain and bead model for dynamics simulations of long linear DNA. *J. Comp. Phys.* 136:168–79

35. Kellermayer MSZ, Bustamante C, Granzier HL. 2003. Mechanics and structure of titin oligomers explored with atomic force microscopy. *Biochim. Biophys. Acta* 1604:105–14

36. Kellermayer MSZ, Smith SB, Granzier HL, Bustamante C. 1997. Folding-unfolding transition in single titin modules characterized with laser tweezers. *Science* 276:1112–16

37. Labeit S, Kolmerer B. 1995. Titins: giant proteins in charge of muscle ultrastructure and elasticity. *Science* 270:293–96

38. Lange S, Ehler E, Gautel M. 2006. From A to Z and back? Multicompartment proteins in the sarcomere. *Trends Cell. Biol.* 16:11–18

39. Lange S, Xiang F, Yakovenko A, Vihola A, Hackman P, et al. 2005. The kinase domain of titin controls muscle gene expression and protein turnover. *Science* 308:1599–603

40. Leake MC, Wilson D, Gautel M, Simmons RM. 2004. The elasticity of single titin molecules using a two-bead optical tweezers assay. *Biophys. J.* 87:1112–35

41. Lee EH, Hsin J, Mayans O, Schulten K. 2007. Secondary and tertiary structure elasticity of titin Z1Z2 and a titin chain model. *Biophys. J.* 93:1719–35

42. Lee EH, Hsin J, Sotomayor M, Comellas G, Schulten K. 2009. Discovery through the computational microscope. *Structure* 17:1295–306

43. Lee EH, Hsin J, von Castelmur E, Mayans O, Schulten K. 2010. Tertiary and secondary structure elasticity of a six-Ig titin chain. *Biophys. J.* 98:1085–95

44. Lee G, Abdi K, Jiang Y, Michaely P, Bennett V, Marszalek PE. 2006. Nanospring behaviour of ankyrin repeats. *Nature* 440:246–49

45. Li H, Fernandez JM. 2003. Mechanical design of the first proximal Ig domain of human cardiac titin revealed by single molecule force spectroscopy. *J. Mol. Biol.* 334(1):75–86

46. Li H, Linke WA, Oberhauser AF, Carrion-Vazquez M, Kerkvliet JG, et al. 2002. Reverse engineering of the giant muscle protein titin. *Nature* 418:998–1002

47. Li H, Oberhauser AF, Fowler SB, Clarke J, Fernandez JM. 2000. Atomic force microscopy reveals the mechanical design of a modular protein. *Proc. Natl. Acad. Sci. USA* 97:6527–31

48. Lim B, Lee EH, Sotomayor M, Schulten K. 2008. Molecular basis of fibrin clot elasticity. *Structure* 16:449–59

49. Linke WA. 2000. Stretching molecular springs: elasticity of titin filaments in vertebrate striated muscle. *Histol. Histopathol.* 15:799–811

50. Linke WA, Grützner A. 2008. Pulling single molecules of titin by AFM–recent advances and physiological implications. *Pflug. Arch. Eur. J. Physiol.* 456(1):101–15

51. Linke WA, Ivemeyer M, Mundel P, Stockmeier MR, Kolmerer B. 1998. Nature of PEVK-titin elasticity in skeletal muscle. *Proc. Natl. Acad. Sci. USA* 95(14):8052–57

52. Linke WA, Ivemeyer M, Olivieri N, Kolmerer B, Rügg JC, Labeit S. 1996. Towards a molecular understanding of the elasticity of titin. *J. Mol. Biol.* 261:62–71

53. Linke WA, Kulke M, Li H, Fujita-Becker S, Neagoe C, et al. 2002. PEVK domain of titin: an entropic spring with actin-binding properties. *J. Struct. Biol.* 137:194–205

54. Linke WA, Leake MC. 2004. Multiple sources of passive stress relaxation in muscle fibres. *Phys. Med. Biol.* 49:3613–27

55. Linke WA, Stockmeier MR, Ivemeyer M, Hosser H, Mundel P. 1998. Characterizing titin's I-band Ig domain region as an entropic spring. *J. Cell Sci.* 111:1567–74

56. Lu H, Isralewitz B, Krammer A, Vogel V, Schulten K. 1998. Unfolding of titin immunoglobulin domains by steered molecular dynamics simulation. *Biophys. J.* 75:662–71

57. Lu H, Schulten K. 1999. Steered molecular dynamics simulation of conformational changes of immunoglobulin domain I27 interpret atomic force microscopy observations. *Chem. Phys.* 247:141–53

58. Lu H, Schulten K. 2000. The key event in force-induced unfolding of titin's immunoglobulin domains. *Biophys. J.* 79:51–65

59. Lukman S, Grant GH, Bui JM. 2010. Unraveling evolutionary constraints: a heterogeneous conservation in dynamics of the titin Ig domains. *FEBS Lett.* 584:1235–39

60. Marino M, Zou P, Svergun D, Garcia P, Edlich C, et al. 2006. The Ig doublet Z1Z2: a model system for the hybrid analysis of conformational dynamics in Ig tandems from titin. *Structure* 14:1437–47

61. Marko JF, Siggia ED. 1995. Stretching DNA. *Macromolecules* 28:8759–70

62. Marszalek PE, Lu H, Li H, Carrion-Vazquez M, Oberhauser AF, et al. 1999. Mechanical unfolding intermediates in titin modules. *Nature* 402:100–3

63. Mayans O, Wuerges J, Canela S, Gautel M, Wilmanns M. 2001. Structural evidence for a possible role of reversible disulphide bridge formation in the elasticity of the muscle protein titin. *Structure* 9:331–40

64. McCammon JA, Gelin BR, Karplus M. 1977. Dynamics of folded proteins. *Nature* 267:585–90

65. Mrosek M, Labeit D, Witt S, Heerklotz H, von Castelmur E, et al. 2007. Molecular determinants for the recruitment of the ubiquitin-ligase MuRF-1 onto M-line titin. *FASEB J.* 21:1383–92

66. Müller S, Lange S, Gautel M, Wilmanns M. 2007. Rigid conformation of an immunoglobulin domain tandem repeat in the A-band of the elastic muscle protein titin. *J. Mol. Biol.* 371:469–80

67. Nagy A, Grama L, Huber T, Bianco P, Trombitas K, et al. 2005. Hierarchical extensibility in the PEVK domain of skeletal-muscle titin. *Biophys. J.* 89:329–36

68. Oberhauser AF, Carrion-Vazquez M. 2008. Mechanical biochemistry of protein one molecule at a time. *J. Biol. Chem.* 283:6617–21

69. Oberhauser AF, Hansma PK, Carrion-Vazquez M, Fernandez JM. 2001. Stepwise unfolding of titin under force-clamp atomic force microscopy. *Proc. Natl. Acad. Sci. USA* 98:468–72

70. Oesterhelt F, Rief M, Gaub HE. 1999. Single molecule force spectroscopy by AFM indicates helical structure of poly(ethylene-glycol) in water. *New J. Phys.* 1:6.1–6.11

71. Phillips JC, Braun R, Wang W, Gumbart J, Tajkhorshid E, et al. 2005. Scalable molecular dynamics with NAMD. *J. Comp. Chem.* 26:1781–802

72. Puchner EM, Alexandrovich A, Kho AL, Hensen U, Schäfer LV, et al. 2008. Mechanoenzymatics of titin kinase. *Proc. Natl. Acad. Sci. USA* 105:13385–90

73. Puchner EM, Gaub HE. 2009. Force and function: probing proteins with AFM-based force spectroscopy. *Curr. Opin. Struct. Biol.* 19:605–14

74. Rief M, Fernandez JM, Gaub HE. 1998. Elastically coupled two-level systems as a model for biopolymer extensibility. *Phys. Rev. Lett.* 81:4764–67

75. Rief M, Gautel M, Oesterhelt F, Fernandez JM, Gaub HE. 1997. Reversible unfolding of individual titin immunoglobulin domains by AFM. *Science* 276:1109–12

76. Rief M, Gautel M, Schemmel A, Gaub HE. 1998. The mechanical stability of immunoglobulin and fibronectin III domains in the muscle protein titin measured by AFM. *Biophys. J.* 75:3008–14

77. Rief M, Grubmüller H. 2002. Force spectroscopy of single biomolecules. *Chemphyschem* 3:255–61

78. Rodriguez-Gomez D, Darve E, Pohorille A. 2004. Assessing the efficiency of free energy calculation methods. *J. Chem. Phys.* 120:3563–78

79. Sarkar A, Caamano S, Fernandez JM. 2005. The elasticity of individual titin PEVK exons measured by single molecule atomic force microscopy. *J. Biol. Chem.* 280:6261–64

80. Sornette D. 2006. *Critical Phenomena in Natural Sciences*. Heidelberg: Springer

81. Soteriou A, Clarke A, Martin S, Trinick J. 1993. Titin folding energy and elasticity. *Proc. Biol. Sci.* 254:83–86

82. Sotomayor M, Corey DP, Schulten K. 2005. In search of the hair-cell gating spring: elastic properties of ankyrin and cadherin repeats. *Structure* 13:669–82

83. Sotomayor M, Schulten K. 2007. Single-molecule experiments in vitro and in silico. *Science* 316:1144–48

84. Sotomayor M, Schulten K. 2008. The allosteric role of the Ca^{++} switch in adhesion and elasticity of C-cadherin. *Biophys. J.* 94:4621–33

85. Trombitas K, Greaser M, Labeit S, Jin JP, Kellermayer M, et al. 1998. Titin extensibility in situ: entropic elasticity of permanently folded and permanently unfolded molecular segments. *J. Cell Biol.* 140:853–59

86. Tskhovrebova L, Trinick J. 2003. Titin: properties and family relationships. *Nat. Rev. Mol. Cell Biol.* 4(9):679–89

87. Tskhovrebova L, Trinick J, Sleep JA, Simmons RM. 1997. Elasticity and unfolding of single molecules of the giant muscle protein titin. *Nature* 387:308–12

88. Tskhovrebova L, Walker ML, Grossmann JG, Khan GN, Baron A, Trinick J. 2010. Shape and flexibility in the titin 11-domain super repeat. *J. Mol. Biol.* 397:1092–105

89. Valle G, Faulkner G, De Antoni A, Pacchioni P, Pallavicini A, et al. 1997. Telethonin, a novel sarcomeric protein of heart and skeletal muscle. *FEBS Lett.* 415:163–68

90. von Castelmur E, Marino M, Svergun DI, Kreplak L, Ucurum-Fotiadis Z, et al. 2008. A regular pattern of Ig super-motifs defines segmental flexibility as the elastic mechanism of the titin chain. *Proc. Natl. Acad. Sci. USA* 105:1186–91

91. Wang MD, Yin H, Landick R, Gelles J, Block SM. 1997. Stretching DNA with optical tweezers. *Biophys. J.* 72:1335–46

92. Watanabe K, Nair P, Labeit D, Kellermayer MS, Greaser M, et al. 2002. Molecular mechanics of cardiac titin's PEVK and N2B spring elements. *J. Biol. Chem.* 277:11549–58

93. Williams PM, Fowler SB, Best RB, Toca-Herrera JL, Scott KA, et al. 2003. Hidden complexity in the mechanical properties of titin. *Nature* 422(6930):446–49

94. Zou P, Pinotsis N, Lange S, Song YH, Popov A, et al. 2006. Palindromic assembly of the giant muscle protein titin in the sarcomeric Z-disk. *Nature* 439:229–33

Proton-Pumping Mechanism of Cytochrome *c* Oxidase

Shinya Yoshikawa, Kazumasa Muramoto, and Kyoko Shinzawa-Itoh

Department of Life Science, University of Hyogo, Hyogo 678-1297, Japan;
email: yoshi@sci.u-hyogo.ac.jp, muramoto@sci.u-hyogo.ac.jp, shinzawa@sci.u-hyogo.ac.jp

Annu. Rev. Biophys. 2011. 40:205–23

The *Annual Review of Biophysics* is online at
biophys.annualreviews.org

This article's doi:
10.1146/annurev-biophys-042910-155341

Copyright © 2011 by Annual Reviews.
All rights reserved

1936-122X/11/0609-0205$20.00

Keywords

X-ray structural analyses, membrane proteins, cellular respiration, site-directed mutagenesis for a membrane protein, hemoproteins, O_2 reduction

Abstract

Cytochrome *c* oxidase (CcO), as the terminal oxidase of cellular respiration, coupled with a proton-pumping process, reduces molecular oxygen (O_2) to water. This intriguing and highly organized chemical process represents one of the most critical aspects of cellular respiration. It employs transition metals (Fe and Cu) at the O_2 reduction site and has been considered one of the most challenging research subjects in life science. Extensive X-ray structural and mutational analyses have provided two different proposals with regard to the mechanism of proton pumping. One mechanism is based on bovine CcO and includes an independent pathway for the pumped protons. The second mechanistic proposal includes a common pathway for the pumped and chemical protons and is based upon bacterial CcO. Here, recent progress in experimental evaluations of these proposals is reviewed and strategies for improving our understanding of the mechanism of this physiologically important process are discussed.

Contents

INTRODUCTION

Cytochrome c oxidase (CcO) reduces molecular oxygen (O_2) coupled to the pumping of protons across the mitochondrial or bacterial membrane. This enzyme is one of the most important elements of cellular respiration. Extensive efforts have focused on elucidating the atomic mechanism of the reaction, since the reports 15 years ago on the X-ray structures of CcO from bovine and bacterial origins at 2.8 Å resolution (16, 43).

CcO contains four redox active metal sites, Cu_A, heme a, heme a_3, and Cu_B, each of which reversibly receives one electron equivalent. The dinuclear copper site, Cu_A, receives electrons from cytochrome c in the P-phase (the intermembrane, or periplasmic, phase) and subsequently donates these electrons to heme a. The O_2 reduction site of CcO includes heme a_3 and Cu_B. The two hemes are fixed perpendicularly to the membrane plane by coordination to side chains in the transmembrane α-helices of subunit I, the largest of the 13 different distinct subunits of bovine CcO (44). Each of the four propionate groups of hemes is directed toward the P-phase. The two heme planes are at the same level relative to the membrane plane and are located sufficiently close to each other [at a minimal edge-to-edge distance of 4.7 Å giving an interplane angle of about 100° (44)] to induce electron tunneling between the peripheral groups as shown in **Figure 1**. In fact, the intrinsic electron transfer occurs at the nanosecond level (31).

CcO has two aqueous-exposed surfaces, each facing the P-phase or the N-phase (the matrix or cytoplasmic phase, respectively) (**Figure 1**). The O_2 reduction site is connected to the hydrophilic surface of CcO facing the N-phase by two hydrogen-bond networks, designated the K- and D-pathways. The two pathways are expected to transfer protons (the chemical protons) necessary for forming water at the O_2 reduction site (16, 44). A third possible proton transfer pathway of bovine CcO, the H-pathway, forming a connection between the two hydrophilic surfaces, is for pumping protons (the pumped protons) associated with the formation of a transmembrane proton gradient (51) (**Figure 1**). The H-pathway is located near heme a. The D-pathway is a pathway for both pumped and chemical protons, based on the mutational analyses of bacterial CcOs (20). All

CcO: cytochrome c oxidase

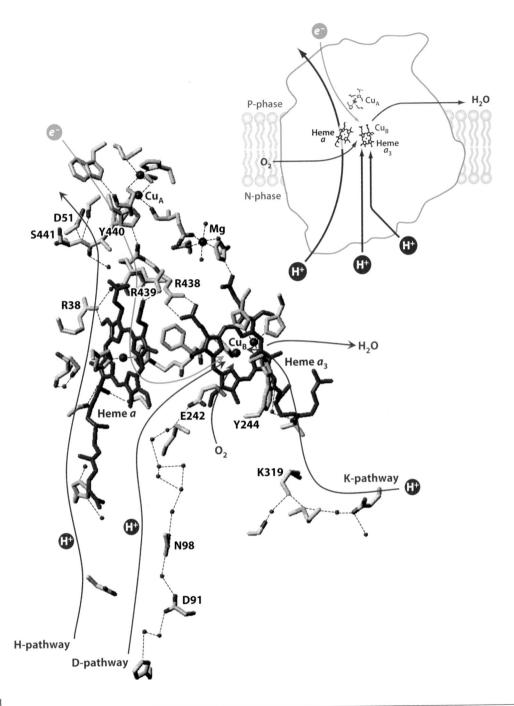

Figure 1

X-ray structure of bovine heart cytochrome *c* oxidase. Structures and locations of redox active sites and possible pathways for protons, O_2, water, and electrons. Inset: Location of the redox active sites in the protein moiety.

P-phase: the space outside the mitochondrial inner membrane or in the periplasmic side of the bacterial cell membrane

N-phase: the region located on the opposite side of the membrane from the P-phase; it has a negative membrane potential relative to the P-phase

Chemical protons: protons required to generate water during the reduction of O_2

Pumped protons: protons translocated across the membrane to store the energy released during the transfer of electrons to O_2 through CcO and by reducing O_2 to water

H-pathway: a proton transfer pathway connecting the N-phase with the P-phase

these networks and metal sites are located in subunits I and II, which are the largest and third largest of the constituent subunits, respectively. The structures and locations of these metal sites and possible pathways for protons and electrons are fairly well conserved in bacterial and eukaryotic aa_3-type CcOs. It is impossible to estimate the scale of the conformational changes sufficient to enable the proton-pumping function. Thus, major efforts have been undertaken to improve the resolution of X-ray structures of CcO. In this review, the basic functions and structures of the above redox active metal sites reported thus far are summarized and the proton-pumping mechanism is discussed.

STRUCTURES AND FUNCTIONS OF THE O_2 REDUCTION SITE

The oxidation and ligand-binding states of intermediate species during CcO reaction discussed in this article are summarized in **Table 1**.

In order to identify the chemical structure of the intermediate species of the O_2 reduction reaction by CcO, the reaction of fully reduced CcO with excess O_2 was followed by resonance Raman spectroscopy using a flow-flash method. The initial intermediate (A) has a band at 571 cm^{-1}. The isotopic shift effects using $^{18}O_2$ and $^{16}O=^{18}O$ indicate that this band corresponds to a Fe-O_2 stretch band similar to those of oxyhemoglobins and oxymyoglobins. The second and third intermediates (P and F) exhibit bands at 804 cm^{-1} and 785 cm^{-1}, respectively (19). The isotope shift effects indicate that both of these bands originate from oxide-bound iron (Fe^{4+}=O^{2-}). The oxide structure of the intermediate P indicates that the O-O double bond has been cleaved at this stage. Thus, a possible structure of the O_2 reduction site in this intermediate is [Fe^{4+}=O^{2-}, Cu$_B^{2+}$-OH$^-$, tyrosine radical]. The tyrosine radical is likely to be that of Y244 located near the O_2 reduction site. The fourth intermediate (O) is characterized by a band at 450 cm^{-1} that is assignable to the Fe-O stretching mode of Fe^{3+}-OH$^-$ (19). The absorption spectra of the A, P, and F intermediates have been characterized as having peaks at 590 (11), 607, and 580 mm (19), respectively. When the fully reduced (or four-electron-reduced) CcO reacts with O_2, the intermediate P is formed significantly faster compared with the reaction between O_2 and the two-electron-reduced enzyme in which Cu$_A$ and heme a are in the oxidized state. The oxidation state of heme a

Table 1 Intermediate species in the cytochrome c oxidase reaction

Abbreviations	States	Oxidation and ligand binding states				
O_r	resting fully oxidized	Fe$_{a3}^{3+}$-O$^-$-O$^-$-Cu$_B^{2+}$		Tyr-OH	^{a}Cu$_A^{3+}$	Fe$_a^{3+}$
O	turnover fully oxidized	Fe$_{a3}^{3+}$-OH$^-$	Cu$_B^{2+}$-OH$^-$	Tyr-OH	Cu$_A^{3+}$	Fe$_a^{3+}$
E	one-electron-reduced	Fe$_{a3}^{3+}$-OH$^-$	Cu$_B^{1+}$	Tyr-OH	Cu$_A^{3+}$	Fe$_a^{3+}$
R	two-electron-reduced	Fe$_{a3}^{2+}$	Cu$_B^{1+}$	Tyr-OH	Cu$_A^{3+}$	Fe$_a^{3+}$
	fully reduced	Fe$_{a3}^{2+}$	Cu$_B^{1+}$	Tyr-OH	Cu$_A^{2+}$	Fe$_a^{2+}$
A	oxygenated^b	Fe$_{a3}^{2+}$-O$_2$	Cu$_B^{1+}$	Tyr-OH	Cu$_A^{3+}$	Fe$_a^{3+}$
					Cu$_A^{2+}$	Fe$_a^{2+}$
P_m	P_m	Fe$_{a3}^{4+}$=O^{2-}	Cu$_B^{2+}$-OH$^-$	Tyr-O$^•$	Cu$_A^{3+}$	Fe$_a^{3+}$
P_r	P_r	Fe$_{a3}^{4+}$=O^{2-}	Cu$_B^{2+}$-OH$^-$	Tyr-O$^-$	^{c}(Cu$_A^{3+}$/Fe$_a^{3+}$) + e$^-$	
F	F	Fe$_{a3}^{4+}$=O^{2-}	Cu$_B^{2+}$-OH$^-$	Tyr-OH	Cu$_A^{3+}$	Fe$_a^{3+}$
					(Cu$_A^{3+}$/Fe$_a^{3+}$) + e$^-$	

^{a}Cu$_A$ is a dinuclear copper site. In the oxidized state, one electron equivalent is distributed between the two cupric (Cu^{2+}) copper sites, while in the reduced state, both copper sites are in the cuprous (Cu^{1+}) state. Thus, Cu$_A$ sites in the oxidized and reduced states are shown by Cu$_A^{3+}$ and Cu$_A^{2+}$, respectively.
^{b}Two types of oxygenated form appear depending on the overall oxidation state of R.
^{c}(Cu$_A^{3+}$/Fe$_a^{3+}$) + e$^-$ denotes that one electron equivalent is distributed between Cu$_A^{3+}$ and Fe$_a^{3+}$.

influences the rate of intermediate P formation (19). Furthermore, when the O_2 reduction reaction is initiated from fully reduced CcO, intermediate P formation is coupled to heme a oxidation. Thus, this intermediate, P_r, is in an overall oxidation state of the O_2 reduction site (including the Y244-OH group) one equivalent lower than that of the intermediate P_m, which is produced from two-electron-reduced CcO. Upon formation of P_m and P_r from A, the O=O double bond is broken to provide O^{2-} and OH^-. This is a nonsequential four-electron reduction process. By this strategy, this enzyme reduces O_2 to water without formation of reactive oxygen species. Extensive X-ray structural analyses of various O_2 analog derivatives of bovine heart CcOs show that the O_2 reduction site induces a conformational change to facilitate the nonsequential donation of three electrons to O_2^- bound at Fe_{a3}^{3+}, which is formed upon O_2 binding to Fe_{a3}^{2+} (26). These results indicate that, in normal enzymatic turnover, when both metals in the O_2 reduction site are in the reduced state (R), CcO binds O_2 to form $Fe_{a3}^{2+} - O_2$ (A). The bound O_2 takes up four electron equivalents to form P_m. Then, P_m sequentially receives four electron equivalents (one at a time) from ferrocytochrome c, generating intermediates F, O, and E (a one-electron-reduced form of O) to regenerate R (**Table 1**).

Fully oxidized bovine heart CcO, as isolated under aerobic conditions (the resting, fully oxidized CcO, O_r), requires six electron equivalents for complete reduction (24). During this reductive titration, the slope of the titration curve monitoring absorption spectral changes during the addition of the initial two electron equivalents is shallow relative to the slope of the curve observed for another four electron equivalents yielding maximal absorption changes. This observation suggests that a majority of the two initial reducing equivalents is consumed by the reduction of an electron acceptor other than the four redox active metal sites of CcO. On the other hand, fully reduced CcO is completely oxidized by four oxidation equivalents donated by O_2. After the oxidative titration, when the oxidized CcO solution is

exposed to excess O_2 for 30 min, the CcO sample, which requires six electron equivalents for complete reduction, is regenerated. These results suggest that the O_r state has a peroxide bound to the O_2 reduction site (24).

The proposed structure of the O_2 reduction site of the O_r state is consistent with its X-ray structure. The F_O-F_C map, obtained from X-ray diffraction experiments carefully conducted to minimize the effect of hydrated electrons from a third-generation synchrotron radiation facility such as SPring-8, shows that a peroxide is bridged between Fe_{a3} and Cu_B ions. The O-O bond distance is 1.7 Å, which is longer than the O-O bond distance of typical peroxide model compounds (1.55 Å), suggesting that the peroxide is in an activated state (22). However, the resting oxidized CcO can be kept at 4°C in the crystalline state for several months without any significant changes to its absorbance spectrum (22). Although the X-ray structural results have been confirmed by a recent resonance Raman investigation showing a band at 755 cm^{-1} assignable to the bridging peroxide (37), it remains desirable to improve the resolution of the X-ray structure to identify the structural origins for the stability of this bridging peroxide.

Fully oxidized CcO generated immediately after complete oxidation, corresponding to the fully oxidized CcO under turnover conditions, the O state, pumps protons upon the addition of one or two electron equivalents, whereas the O_r state does not (47). On the basis of these results, it was proposed that the free energy obtained by O_2 reduction is stored in intermediate P, with half of it released in the oxidative phase. However, the rest of this energy is lost as heat unless the reductive phase starts immediately after complete oxidation of CcO (47). However, the proton pump is detectable upon one-electron reduction of CcO in the E state, prepared by the two-electron reduction of F by CO (35). This E intermediate is surely not prepared from the oxidized form produced immediately after the complete oxidation of CcO. Thus, the proton pump activity of the E intermediate strongly suggests that there are some chemical structural differences between

Oxygenated form (species) (A): O_2-bound intermediate species

F: the third intermediate species next to P_m under turn over conditions

P_r: the second intermediate species next to the O_2-bound form in the reaction between the four-electron-reduced CcO with O_2

P_m: the second intermediate species next to the O_2-bound form in the reaction between the two-electron-reduced CcO with O_2

Resting, fully oxidized CcO (O_r state): fully oxidized CcO as isolated under aerobic conditions

Turnover, fully oxidized CcO (O state): fully oxidized CcO generated immediately after the reduction of O_2 to water, which corresponds to the fully oxidized CcO produced under turnover conditions

E: one-electron-reduced intermediate species of CcO

O_r and O. The structural differences in the O_2 reduction site between O and O_r {$[Fe_{a3}^{3+}$-$OH^-]$ (19) versus $[Fe_{a3}^{3+}$-O^--O^--$Cu_B^{2+}]$ (3)} are likely to cause the difference in the proton-pumping activity.

The O_r state was observed also in a ba_3-type CcO from *Thermus thermophilus*. Thus, the formation of O_r, is not induced by damage during the course of purification as had been suggested (40). The physiological significance of O_r is expected to prevent the formation of active oxygen species under low levels of electron flow in the respiratory chain, in which the fully oxidized CcO is likely to be accumulated. The structure of O $[Fe_{a3}^{3+}$-OH^-, Cu_B^{2+}-OH^-, TyrOH] is expected to react with O_2 spontaneously under aerobic conditions to generate various active oxygen species.

COUPLING BETWEEN THE PROTON-PUMPING AND O_2 REDUCTION CYCLES

A quantitative evaluation of the protons ejected by the reduction of CcO was performed using the O_r state reconstituted into liposomes (6). After quantitative anaerobic addition of electron equivalents by ruthenium (III) hexammine, O_2 was added to initiate the reaction. Significantly, CcO that received two electron equivalents or fewer showed no proton ejection. The number of ejected proton equivalents was essentially proportional to the number of electron equivalents above two equivalents. At four electron equivalents above two equivalents (six electron equivalents in total), the maximal number of protons (four proton equivalents) was ejected. The titration curve (figure 2 in Reference 6) is similar to that of the reductive titration curve of the O_r state under strictly anaerobic conditions (24). The initial lag in the titration curve suggests that the initial two electron equivalents are used for reducing the bridging peroxide in the O_2 reduction site as described above.

In order to confirm that the ratio of protons to electrons is unity in the reductive phase,

the electric potential generated during this process was measured following donation of a single electron via a laser flash system to the O state. Quantitative analyses of the charge separation using the potential generation assignable to the electron transfer from Cu_A to heme a as the internal standard indicate the ejection of one proton equivalent per electron equivalent injected (6). These experimental results indicate that each electron transfer from ferrocytochrome c to CcO under turnover conditions is coupled with pumping of one proton equivalent.

ANALYSES OF THE ELECTRON AND PROTON TRANSFER PROCESSES DURING THE ENZYME REACTION

Flow-Flash Analyses

The CcO reaction has been analyzed extensively by a flow-flash method in which the reaction of fully reduced CcO with an excess amount of O_2 is initiated by a flash-photolysis technique to release CO from the CO-bound, fully reduced CcO in the presence of O_2. Essentially, the same absorbance spectral changes were observed for CcOs isolated from different sources. The time course of this reaction was resolved into at least four phases with approximate time constants (half-lives) of 7, 24, 72, and 800 μs, respectively. These phases are the $R{\to}A$, $A{\to}P_r$, $P_r{\to}F$, and $F{\to}O$ transitions, respectively (1). As described above, oxidation of heme a is coupled to the formation of P_r. Thus, the $P_r{\to}F$ transition does not include electron transfer. The $P_r{\to}F$ and $F{\to}O$ transitions are coupled to both the pumped and the chemical proton transfers. The release and uptake of the protons during these phases have been carefully analyzed using a pH-sensitive dye to measure the change of pH outside and inside proteoliposomes reconstituted with CcO from *Rhodobacter sphaeroides*. One proton equivalent is released to the outside and two proton equivalents are taken up from the inside during these transitions (14).

As described above, hemes a and a_3 are located at the same level relative to the membrane surfaces (**Figure 1**). Thus, the electron transfer between the two hemes during transitions such as A→P_r is unlikely to be electrogenic. However, time-resolved electron and proton transfer (charge translocation) measurements have shown that at neutral pH, the A→P_r transition is superimposeable with the initial phase of the P_r→F transition (5). However, at pH 10.5, at which the P_r→F transition is negligible (27), P_r formation followed by absorbance at 595 mm was coupled to a simultaneous charge separation. The mechanistic implication of the charge separation is discussed below (5).

Single-Electron Injection Analyses

A single-electron-reduction technique using ruthenium bispyridyl was applied to the O state from *Paracoccus denitrificans* to analyze the internal electron and proton transfer reaction (4). After the initial reduction of Cu_A, a rapid electron transfer to heme a with a half-life of about 10 μs occurs concomitantly with membrane potential formation. This 10-μs phase is insensitive to pH and H_2O/D_2O exchange. Therefore, this electron transfer step is not coupled to proton transfer. When the electron movement is finished, 30% and 70% of electron equivalents are distributed between Cu_A and heme a, respectively (31). However, no electron equivalent is distributed in the O_2 reduction site. As described above, hemes a and a_3 are located sufficiently close to each other so that the intrinsic electron transfer rate could be on the nanosecond scale. Thus, the absence of electron transfer from heme a to heme a_3 indicates that the redox potential (E_m) of heme a_3 is much lower than that of heme a (ΔE_m > 100 mV). The amplitude of membrane potential formation is 12% of the total potential formation. The reduction of heme a increases the redox potential of heme a_3 significantly to induce the electron transfer to heme a_3 in the next 150-μs phase. At the end of this phase,

40% and 60% of the electron equivalents are distributed between heme a and the O_2 reduction site (heme a_3/Cu_B), respectively. Namely, Cu_A is completely oxidized during this phase by significant increment of E_m of heme a, induced by reduction of heme a_3. The electron transfer from heme a to heme a_3, although parallel to the membrane surface, contributes to the formation of a large membrane potential (about 42% of the total) due to the proton translocation from the N-phase. The absorbance spectrum at the end of the next phase (800 μs) shows an absorption peak near 640 nm, suggesting that Cu_B is fully reduced and all the other metal sites are in the fully oxidized state (4).

It has been proposed that the absorption peak at 640 nm is assignable to a band due to Cu_B^{1+} because this is blue-shifted from the charge transfer band of the Cu_B^{2+} site at 665 nm (4). However, independent experimental confirmations are needed to validate this proposal. If the final destination of the single electron equivalent is the Cu_B site, the E_m of Cu_B must be much higher than that of all the other metal sites. This 800-μs phase also includes a large membrane potential formation (30% of the total), again suggesting that proton translocation occurs. The charge separation takes place in 800 μs, is induced by the electron transfer parallel to the membrane surface, and is due to proton movement from the N-phase. As the final phase, membrane potential formation with a half-life of 2.6 ms is detectable without any corresponding absorption spectral change. The amplitude is approximately 16% of the total (4). The time constant, electron distribution, and membrane potential generation of each phase in the above single-electron injection process are summarized in **Table 2**.

The incremental increase in E_m of the O_2 reduction site upon protonation of the O_2 reduction site must be higher than that upon protonation of the loading site of the pumped proton, because the loading site must be separated from the O_2 reduction site. Therefore, the 150- and 800-μs phases are assignable to the transfer of pumped protons and chemical protons, respectively. The final 2.6-ms phase

E_m: redox potential of redox active metal site

Table 2 Process of single-electron injection to the turnover fully oxidized intermediate

Phases	Electron distribution				Membrane potential generation	Proton transfer
	Cu_A	Heme a	Heme a_3	Cu_B		
<0.5 μs[a]	100%	0	0	0	0	None
10 μs	30%	70%	0	0	12%	None
150 μs	0	40%	60%	0	42%	To the H^+-loading site (pumped proton)
800 μs	0	0	0	100%	30%	To the O_2 reduction site (chemical proton)
2.6 ms	0	0	0	100%	16%	To the P-phase (pumped proton)

[a]Electron injection process by a laser pulse. The quantum yield is 10%–20%. The experimental details are given in Reference 4.

is for proton release from the proton-loading site, which is driven by conformational changes near the pumped-proton-loading site (4).

The single-electron injection technique has been applied to the transitions F→O and P_m→F using CcOs isolated from bovine and bacteria and for monitoring the formation of membrane potential (20, 40, 46, 52). Essentially, the same results obtained for the O→E transition described above were observed, including a rapid phase followed by a slower phase with at least two components.

The fact that the transitions P_m→F, F→O, and O→E cause similar electron and proton movements strongly suggests that the proton-pumping mechanism is independent of the oxidation and ligand-binding state of the O_2 reduction site. The low-potential site (Cu_A/heme a) transfers each of the four electron equivalents in the catalytic cycle through the same pathway to the high-potential site (heme a_3/Cu_B). Conversely, four types of chemical reactions proceed at the high-potential site upon receipt of each electron equivalent, depending on the overall oxidation state of the high-potential site. Thus, it is likely that the low-potential site drives the proton pump.

The O_r has also been analyzed by the single-electron injection technique. Using carefully prepared reconstituted CcO proteoliposomes to minimize spontaneous reduction, Verkhovsky et al. (48) showed that only the initial membrane potential formation, which is assignable to the electron transfer from Cu_A to heme a, was detectable. As described above, the resting oxidized CcO has a bridging

peroxide in the O_2 reduction site (3). The absence of the second phase in which heme a is oxidized suggests that the increase in E_m of the O_2 reduction site is prohibited by the negative charge of the peroxide. The second flash induces a slower phase (135 μs). However, the slower phase is not sufficiently intense to drive the proton pump (36, 48). These results are consistent with the observation that the O_r state does not have the ability to pump protons upon the addition of one or two electron equivalents as described above (47). A major part of the two electron equivalents available upon the second flash would be used for reduction of the peroxide to yield two oxides or hydroxides (O^{2-} or OH^-). Both heme a_3 and Cu_B are likely to still be essential in the ferric and cupric states, respectively, even after the second flash. Electron donation to the metal sites in the O_2 reduction site seems to be a prerequisite for proton uptake to trigger proton pumping.

ROLES OF PROTON TRANSFER PATHWAYS OF BACTERIAL ENZYMES

Mutational Analyses of the D-Pathway

The site-directed mutagenesis technique has been applied extensively to bacterial aa_3-type CcOs from *P. denitrificans* and *R. sphaeroides* in order to identify the pathways for the chemical and pumped protons. The critical amino acid residues in the D- and K-pathways discussed in this review are conserved in the two bacterial and bovine CcOs. Thus, for the sake of

K-pathway, D-pathway: proton transfer pathways connecting the N-phase with the O_2 reduction site

simplicity, the residue numbers of bovine CcOs are given in this review. Key residues in the D-pathway are E242 near the O_2 reduction site and D91 near the entrance facing the N-phase. The E242Q mutant CcO of *R. sphaeroides* shows no electron transfer or proton-pumping activities under the steady-state turnover conditions. Upon single-electron injection to F state, slower membrane potential generation after rapid potential generation is abolished in the E242Q mutant, suggesting that the F→O transition is blocked by stopping proton transfer through the D-pathway (20). This proton transfer is tightly coupled to electron transfer from heme a to the O_2 reduction site in the F state. The slower potential generation is composed of at least two components that are assignable to the proton translocations of the chemical and pumped protons. Thus, these results suggest that the D-pathway transfers both chemical and pumped protons (20).

A flow-flash analysis of the fully reduced E242Q mutant enzyme from *R. sphaeroides* shows that electron transfer to P_r is impaired without significant oxidation of Cu_A (2). Further, no net proton uptake is detectable. Thus, E242 donates protons to the O_2 reduction site in the P_r→F transition, inducing Cu_A oxidation. The D91N mutant of CcO (which lacks both proton-pumping and O_2 reduction activities) provides protons for the transition of P_r up to the F state without any net proton uptake. In wild-type CcO, the P_r→F transition is coupled to electron transfer from Cu_A to heme a as described above. However, the P_r→F transition in the D91N mutant of CcO does not induce electron transfer, indicative of tight coupling between the D91 and Cu_A sites (2).

A mutation of the *R. sphaeroides* enzyme, N98D, located in the middle of the D-pathway, shows a peculiar phenotype with abolished proton-pumping activity and stimulated O_2 reduction activity (150%–300%). The flow-flash analysis shows that the intermediate species are not perturbed by the mutation. More specifically, the O_2 reduction mechanism has not been modified (29). The N98D mutant CcO generates a membrane potential corresponding to

chemical proton uptake at the O_2 reduction site, consistent with a lack of proton-pumping activity (41). A similar phenotype is reported for N98D and N163D in *P. denitrificans* CcO (30). The X-ray structure of the N98D mutant CcO of *P. denitrificans* at 2.32 Å resolution shows no significant structural change relative to wild-type CcO except for an alternative conformation of the E242 side chain (13).

Mutational Analyses of the K-Pathway

The K319M and T316A mutants in the K-pathway of the CcO of *R. sphaeroides* showed no influence on potential generation due to F→O transitions. These mutant enzymes have neither proton-pumping nor O_2 reduction activities under turnover conditions. Furthermore, reduction of heme a_3 in the mutant CcO is extremely slow. These results suggest that the K-pathway transfers chemical protons coupled to electron transfer to the oxidized O_2 reduction site (Fe_{a3}^{3+}/Cu_B^{2+}) (20, 49). In fact, blocking the K-pathway impairs the electron-coupled transfers of both the chemical and pumped protons, triggered by electron transfer from Cu_A to heme a upon injection of a single electron into the O state (6). The results indicate that the transfer of chemical protons through the K-pathway is coupled to the transfer of pumped protons. An alternative interpretation is that the K-pathway transfers both the chemical and pumped protons.

As a result of these extensive mutagenesis analyses of the K- and D-pathways, the original proposal (that the K- and D-pathways are used for chemical and pumped protons, respectively) (16) has been revised to the present proposal that the K- and D-pathways are used for chemical and chemical/pumped protons, respectively. Most of the experimental results have been interpreted according to the refined proposal.

Mechanism of Proton-Pumping Driven by the D-Pathway

As summarized above, all the phenotypes of the D-pathway mutant CcOs seem consistent

Pumped proton
acceptor site *A* (site
A): the loading site for
the pumped protons in
the D-pathway

with the proposal that the D-pathway transfers both pumped and chemical protons. The proton-pumping site must be completely separated from the O_2 reduction site, because otherwise the pumped protons would be utilized for water formation, thereby generating a short circuit. Thus, the D-pathway must have at least one branching point for separating the pumped protons from the chemical protons. A candidate for the branching point amino acid residue is E242. The X-ray structure of the E242Q mutant of CcO shows that the conformation of Q242 is different from that of E242 in wild-type CcO, suggesting that disruption of a hydrogen bond between the carboxyl group of E242 and the peptide carbonyl group of M107 has occurred (42). Furthermore, the conformational change upon the E/Q exchange induces a translational shift of the indole ring of W126. The indole ring is hydrogen-bonded to the propionate group of the D-ring of heme a_3, which is salt-bridged to R438. The movement of W126 upon E242Q mutation influences the conformation of the propionate to break the salt bridge with R438 through the introduction of a water molecule between the guanidino group and the carboxyl group (42). Based on this structural change and assuming that Q is analogous to deprotonated E, it has been proposed that deprotonation of E242 induces the conformational change of the propionate group to increase its pKa value (9). (It is not clear why Q could be analogous to E in the negatively charged state.)

These structural changes are consistent with the following proton pump mechanism driven by the D-pathway (9). When E242 is in the protonated state, the propionate group salt-bridged to R438 has a low pKa value and is accessible only to the P-phase. Upon one-electron reduction of the O_2 reduction site, a proton is transferred from E242 to the O_2 reduction site for charge compensation. The deprotonation of E242 induces a conformational change in R438 and the D-ring propionate (the pumped proton acceptor site *A*) to increase its pKa value. At this stage, the proton accessibility of site *A* switches from the P-phase to the N-phase. Then, the pumped protons are transferred to site *A* through E242, followed rapidly by the chemical proton transfer to E242. The protonation of E242 induces a conformational change in site *A* to decrease its pKa value and to change its accessibility to the P-phase. Thus, the pumped proton on site *A* is released to the P-phase. See figure 7 in Reference 9 for a more detailed description of this proton pump proposal.

Sufficient X-ray structural basis for this proposal has not been obtained. For example, additional convincing evidence is required for concluding that the conformational change induced by E242Q mutation is analogous to the conformational change induced by deprotonation of E242, as described above. Furthermore, the structural change of the D-ring propionate of heme a_3 by the E/Q exchange does not clearly show the accessibility switching of site *A* from the P-phase to the N-phase.

Under high pH conditions, the $P_r{\rightarrow}F$ transition rate becomes slower, with a pKa of 9.4. The pH dependency of the amplitude of F formation also shows a pKa of 9.4. This pKa value has been ascribed to the ionization of E242 at the branching point of the D-pathway (27). On the other hand, the N98D mutant of *R. sphaeroides* CcO, which has full O_2 reduction activity but no proton-pumping activity, has a pKa of 11 for the pH dependency of the $P_r{\rightarrow}F$ transition rate. These results are consistent with the following proposed function of E242. The pKa value of site *A* of the wild-type enzyme is significantly higher than that of E242 but lower than that of the O_2 reduction site. Thus, the pKa for E242 in the N98D mutant of CcO is 11, higher than that of site *A*. This prevents E242 from transferring protons to site *A*, whereas proton donation to the O_2 reduction site with much higher pKa would be possible (28).

The above proposal that the pKa value determined by the $P_r{\rightarrow}F$ transition is critical for determining proton pump efficiency was confirmed by introducing a second mutation (D91N) into the N98D variant of CcO, which restored both the original pKa (9.5) and proton pumping during steady-state turnover (8).

Direct structural identification of the site used to control pH dependency for the $P_r \rightarrow F$ transition is desirable, because CcO has protonatable functional groups other than E242, such as R38, K319, and the propionate groups of hemes a and a_3.

As described above, the $A \rightarrow P_r$ transition is electrogenic (5). The charge separation during this transition is not blocked by the D91N mutation, which blocks proton uptake to the D-pathway. However, the E242Q mutation does block charge separation. On the basis of these results, it has been proposed that the $A \rightarrow P_r$ transition is coupled to proton transfer from E242 to the pumped-proton-loading site. Thus, the $P_r \rightarrow F$ transition is preceded by protonation of site A driven by electron transfer from heme a to heme a_3 (for P_r formation). The pumped protons at the loading site are released by chemical proton uptake through E242 near the O_2 reduction site. In other words, the proton pump during the $P_r \rightarrow F$ transition, which does not include electron transfer as described above, also is driven by electron transfer to the O_2 reduction site. On the other hand, charge separation is not detectable in K319M mutant CcO, suggesting that the charge separation detectable in the $A \rightarrow P_r$ transition is due to proton movement in the K-pathway (22). It is impossible to identify the charge separation site from only these mutational analyses.

E242 must transfer protons both to the proton acceptor site A and to the O_2 reduction site if it has the proposed branching function in the D-pathway. However, the X-ray structures of CcO reported thus far show no detectable proton transfer pathway to these two sites. E242 is located on the wall of the O_2 transfer channel from the molecular surface of the transmembrane region to the O_2 reduction site. The protons must be transferred to site A across the O_2 transfer channel. No chemical structure that could function as the proton pathway has been identified in the X-ray structure in the region between E242, the site A, and the O_2 reduction site. [Although an extended space near site A is present as a branch of the O_2 path, the vac-uum space cannot function as a controlled proton transfer pathway (45)]. Thus, possible locations of water molecules in this space were calculated by an energy minimization procedure (50), showing that four water molecules could be located in an L-shaped alignment in the region connecting E242 with the O_2 reduction site and site A. These water molecules, placed in a narrow tube with a highly hydrophobic inner surface, are expected to form an oriented single array that facilitates unidirectional proton transfer. The direction of transfer depends on the charge distribution of these metal and proton-accepting sites to appropriately direct protons either to the O_2 reduction or to site A (50). Furthermore, the function of E242 as the valve to prevent reverse proton transfer has been proposed theoretically (17).

However, the following facts are not fully consistent with the presence of the ordered water molecules in this space. In the fully reduced state, water molecules are unlikely to be located in this space, because an O_2 molecule must be transferred though the narrow space. Thus, the four water molecules indicated in the theoretical studies must clear out before introduction of O_2. In fact, no water molecule is detectable in the X-ray structure of the fully reduced CcO, even at 1.8 Å resolution (26).

In X-ray diffraction experiments conducted using the strong X-ray beam of SPring-8, concomitantly with the decrease in peroxide electron density in the O_2 reduction site with increasing X-ray irradiation time, an increase in electron density assigned to a water molecule is detected at Y244OH. Y244 is covalently linked to one of the histidine imidazole groups ligated to Cu_B (3). These results suggest that Y244 functions as a scavenger of water in the O_2 reduction site including the space between E242 and the D-ring propionate. In other words, the absence of water at Y244 in the X-ray structures of bovine heart CcO provides strong evidence for the absence of water molecules in the space. Thus, CcOs in various oxidation and ligand-binding states reported thus far, except for the X-ray-irradiated resting oxidized form and the fully reduced CN$^-$ bound form (26), are unlikely to contain any water molecules in the space because they show no water at Y244.

Possible conformational changes in E242 and the G239 peptide carbonyl group would provide a hydrogen-bond network extending to the OH^- ion on Cu_B from the E242 COOH group, without introducing any water molecules into the O_2 path (S. Yoshikawa, unpublished observations). Thus, the X-ray structure suggests that the role of E242 is to transfer only chemical protons to the O_2 reduction site in the absence of any water molecules in the hydrophobic space.

Although the finding of one water molecule in the hydrophobic space above the top of the D-pathway in the X-ray structure of the fully oxidized CcO from *R. sphaeroides* may represent positive evidence for the presence of these ordered water arrays (34), it is strongly desirable to establish more convincingly the presence of these water molecules experimentally. Randomly oriented water molecules in the space would not be detectable in the high-resolution X-ray structure. Randomly oriented water molecules would also be unable to be involved in controlling the directionality of proton transfer to site A or to the O_2 reduction site.

In order to examine the possible role of the D-ring propionate of heme a_3 and the arginine pair system as the site for the pumped proton acceptor site A, extensive mutational analyses on the arginine pair (R438 and R439) have been performed for bacterial CcO and the homologous *Escherichia coli* quinol oxidase. However, the mutation results are not straightforward to interpret. Mutations of R439 (K, Q, N, and L) and R438 (K and Q) do not impair proton-pumping activity, whereas the R438Q/R439Q double mutant and the R439P and R438 (N and L) mutants do not have the proton-pumping activity (7, 32, 33). The fairly significant mutational changes in this study ($R \rightarrow Q$, N, L) reveal an unexpectedly weak influence on proton-pump activity.

The upper end of the branch of the O_2 path is composed of the R438 and R439 pair and two D-ring propionates of the two hemes. The two guanidino groups of R438 and R439 are hydrogen-bonded to the propionates of heme a_3 and heme a, respectively. This Arg-propionate system (including the salt bridge between R438 and heme a_3 propionate) has been proposed to be the pumped proton loading site, as described above. The X-ray structure indicates that the P-phase side of the arginine pair is protonically equilibrated with P-phase through hydrogen-bond networks inside the protein and that the arginine pair effectively blocks access of the branch of the O_2 path to the P-phase (45). Therefore, the structure of the arginine-propionate system strongly suggests that it functions in reverse proton transfer to maintain mitochondrial membrane potential within an appropriate range, as previously proposed (23). Thus, it is possible that the abolishment of pumping activity by these arginine mutations is due to an increased extent of proton leaking back through the branch to the O_2 reduction site.

In spite of the extensive mutational analyses described thus far, the elements for the proton-pumping function driven by the D-pathway have not been identified experimentally. Even the branching function of E242 for the chemical and pumped protons has not been established. Perhaps the strongest experimental evidence for the proposal that the D-pathway transfers both chemical and pumped protons is provided by the results for the E242Q and D91N mutations of the bacterial D-pathway, which abolish membrane potential generations driving both proton-pumping and O_2 reduction functions (20). However, these results do not identify the location of the pumping pathway but simply indicate that proton transfer through the D-pathway, electron transfer to heme a_3, and proton pumping are all strongly coupled. In fact, the K-pathway mutation (K319M), which abolishes both pumped and chemical proton transfers (6), is also consistent with the proposition that the K-pathway transfers both chemical and pumped protons. Thus, experimental confirmation is indispensable for any proposal based on mutational analyses.

PROTON-PUMPING FUNCTION OF THE H-PATHWAY

The Proton-Pumping Site and Driving Element of the H-Pathway of Bovine Heart CcO

The H-pathway is composed of a hydrogen-bond network connected to the P-phase side and a water channel opening to the N-phase side (**Figure 1** and **Figure 2**). The water molecules in the N-phase are accessible to R38 at the bottom end of the hydrogen-bond network through the water channel. As a result, R38 is protonically equilibrated with the N-phase. D51 of bovine CcO is located at the upper (P-phase) side of the hydrogen-bond network (26). A peptide bond between Y440 and S441 located next to D51 is proposed to facilitate the unidirectional proton transfer by the relative stability of the keto form of the peptide bond compared to the enol form. A theoretical analysis suggests that proton transfer through the keto form of the peptide bond occurs in the physiological timescale in bovine heart CcO because the enzyme has an energetically favorable proton path from the enol OH to the peptide nitrogen, which facilitates an effective enol-keto transition (-(COH) = N- → -(CO)-NH-) (18). The hydrogen-bond network forms two hydrogen bonds with heme *a* from the propionate and formyl groups of heme *a* to a fixed water molecule and R38 in the hydrogen-bond network, respectively. The water channel in the fully reduced CcO has five cavities, each of which is large enough to accommodate at least one water molecule (26, 45).

Figure 3 illustrates that D51, exposed to the bulk water phase of the P-side, becomes buried inside the protein interior upon oxidation of CcO (45). This conformational change of D51 strongly suggests that the protonation state of D51 is dependent on the redox state of CcO. Furthermore, upon oxidation, the accessibility of D51 to the P-phase is essentially eliminated and D51 becomes connected to the N-phase via the H-pathway. Although D51 is expected to function as the proton-pumping site, this residue is not conserved in bacterial and plant

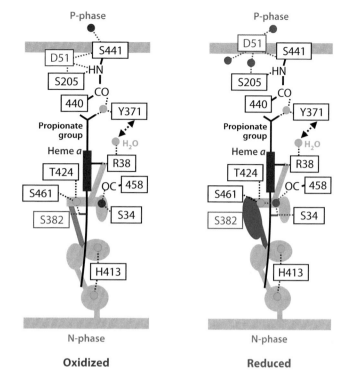

Oxidized   **Reduced**

Figure 2

Schematic representation of the redox-coupled conformational changes in the H-pathway. The water channel is depicted by the gray, dark blue, and red areas. The light blue circles represent the positions of the fixed water molecules. The redox-dependent structural changes are highlighted in red and dark blue.

CcOs. The evolutionary fact is discussed below.

The positive charge created upon oxidation of heme *a* after donating an electron equivalent to the O_2 reduction site is delocalized to the propionate and the formyl groups, each of which forms a hydrogen bond to the hydrogen-bond network of the H-pathway. Thus, protons in the hydrogen-bond network are expected to be actively shuttled through the network, driven by the electrostatic repulsion. These X-ray structures strongly suggest that heme *a* is the driving element of proton pumping.

The Function of D51

The X-ray structures in **Figure 3** suggest that protons are released to the P-phase upon reduction of heme *a*. However, several proton acceptor sites on the protein surface near D51

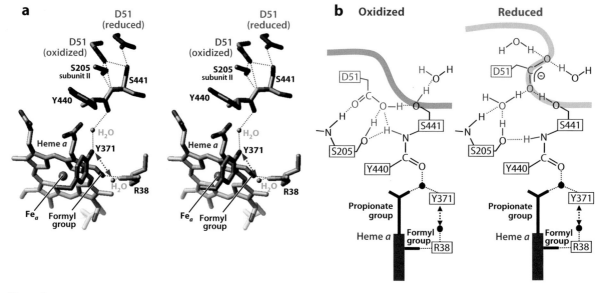

Figure 3

Redox-coupled conformational changes in D51. (*a*) Stereoscopic drawing of the hydrogen-bond network in the fully oxidized and reduced states at 1.8 and 1.9 Å resolution, respectively, viewed from the P-phase side. (*b*) The hydrogen-bonding structure of D51 in the oxidized and reduced states. The smooth, thick curves denote the molecular surface to which the water molecules in P-phase are accessible. The conformational changes upon reduction of CcO are shown by the blue structures on the right. Reprinted from Reference 45.

could trap the pumped protons released from D51 upon reduction. The pumped protons are expected to be distributed over these sites and stabilized electrostatically by the negative charge on D51 (38). The protons are released to the bulk water phase upon oxidation of CcO when the D51COO⁻ is buried inside the protein prior to being protonated by the peptide bond imidic acid, which is a protonated form of the peptide bond (-C(OH)=N⁺H-). Protons are transferred through a peptide bond by forming the imidic acid form as an intermediate state (45). Proton release upon oxidation of heme *a* is consistent with the direct measurement showing the release of the pumped protons upon oxidation of the heme *a*/Cu$_A$ site (10).

The Gating Mechanism for Proton Pumping Through the H-Pathway

Reverse proton transfer from the P-phase is effectively blocked by the peptide bond near the upper end of the hydrogen-bond network.

However, the peptide bond is not able to block reverse proton transfer through the hydrogen-bond network below the peptide bond. Upon oxidation of CcO, one water cavity in the water channel is eliminated by a conformational change of helix X (**Figure 2**) (26, 45). This change narrows the water channel significantly to essentially block water's accessibility to the hydrogen-bond network from the N-phase. Thus, the hydrogen-bond network is not able to equilibrate with the N-phase (the closed state, hereafter) in the oxidized state, at least within the physiological timescale. The same conformational change in helix X occurs also upon binding of strong heme ligands, such as CO or NO, to the fully reduced CcO. Thus, binding of O₂ (a strong heme ligand) is expected to trigger the elimination of the water cavity. Furthermore, preliminary X-ray structural analyses suggest that the P and F intermediates also show the water channel in the closed state. Thus, the water molecules in the N-phase are effectively accessible to the hydrogen-bond network through the water channel (the open state), only when the O_2 reduction site is in the fully reduced state. Thus, the effective blockage of the proton back leakage through the H-pathway is facilitated by the peptide bond near the upper end of the hydrogen-bond network and the water channel located in the N-phase side of the hydrogen-bond network.

Mutational Analyses for the H-Pathway

A stable expression system for bovine CcO genes has been constructed in HeLa cells. Thus far, mutants of three critical sites in the H-pathway, D51N, S441P (the peptide bond in H-pathway), and V386L/M390W (the water channel), have been reported (38, 45). The proline mutation is designed to block formation of the normal protonated imidic acid intermediate $(-(C-OH) = N^+H-)$. The torsion angles of the peptide between Y440 and S441 suggest that the replacement of S441 with proline would have minimal influence on the conformation near the site. The double mutation, V386L/M390W, was generated to block water from accessing the hydrogen-bond network by exchanging the residues on the inside surface of the water channel for the bulkier residues. All three mutant CcOs show the identical phenotype, complete elimination of proton-pumping activity without any significant influence on the function of O_2 reduction. These observations indicate that the functions of these residues are consistent with those proposed by X-ray structural analyses. Furthermore, the X-ray structures of the H-pathway and heme *a* suggest that the coupling between electron transfer and active transport of protons through the H-pathway is not tight. In other words, heme *a* could be oxidized without driving the active transport of protons by electrostatic repulsion, because both the delocalization of the net positive charge at Fe_a and the location of the protons in the hydrogen-bond network are expected to be fairly flexible.

DIVERSITY OF THE PROTON TRANSFER PATHWAYS OF CcO

Bacterial (*R. sphaeroides* and *P. denitrificans*) aa_3-type CcOs have a possible proton transfer pathway analogous to the H-pathway but lack D51 (21). Extensive mutational analyses have not shown any positive evidence for involvement of the bacterial H-pathway in proton pumping (21). These mutational studies, as well as the absence of D51 from bacterial and plant CcOs, may be the strongest evidence against proton pumping driven by the H-pathway. A notable report has indicated that the ba_3-type CcO from a thermophilic bacterium has only one possible proton transfer pathway corresponding to the K-pathway (12, 15). Mutations of the amino acids only in the pathway impair both the oxidative and reductive phases of the catalytic cycle, indicating that no other proton pathway exists in this CcO. In other words, both the H-pathway and D-pathway are not conserved. The common elements across all families of the heme-copper oxygen reductases include the heme a_3/Cu_B dinuclear site, the four histidine imidazole groups that coordinate the metals, and the covalently linked His-Tyr moiety (one of the three imidazoles coordinated to Cu_B is covalently linked to a tyrosine phenol group). Thus, it has been suggested that none of the proton channels plays a central role in the pumping and gating mechanism; that is, roles of these proton channels are only for proton delivery (12).

However, an alternative interpretation of the diversity of proton-pumping systems in all families of the heme-copper oxygen reductases is possible. Reduction of O_2 without releasing active oxygen species is not a simple chemical reaction. No alternative system with comparable or higher efficiency than the Fe_{a3}/Cu_B system has been obtained in the evolution of aerobic organisms. In contrast, proton pumping is a chemically simple reaction, and various amino acid residues can facilitate proton pumping in various ways. Thus, although D51 in bovine CcO is not conserved in bacterial and plant CcOs, this cannot be considered conclusive evidence against the possibility that D51 plays a critical role in the proton pump of animal CcOs. Extensive efforts are required for the structural and functional analyses for each CcO. In this context, mutational analyses for the K- and D-pathways of bovine heart CcO are critical.

CONCLUSIONS

Although these structural comparisons concomitantly with site-directed mutagenesis analyses provide many important insights into understanding the mechanism of proton pumping, these empirical findings must be confirmed by nonempirical experimental results. For example, direct proton movement in CcO during the course of proton pumping must be determined directly, for example, by ultra-high-resolution X-ray crystallographic analyses coupled with time-resolved highly sensitive infrared measurements for elucidation of the proton-pumping mechanism of CcO.

SUMMARY POINTS

1. CcO must transfer protons in the interior of the protein moiety for proton pumping for energy conservation and for making water from molecular oxygen. Three potential proton transfer pathways, D, K, and H, have been identified by X-ray structural analyses of bovine and bacterial CcO. The roles of these pathways have been extensively examined by site-directed mutagenesis analysis.

2. Most of the structural and functional analyses of bacterial CcOs have been interpreted to assume that the D-pathway transfers both pumped and chemical protons. However, the structural elements indispensable for the D-pathway proton-pumping function have not been well characterized, specifically structures relevant for the loading of pumped protons, for switching accessibility between the N- and P-phases, for redox-coupled pKa changes, and for the branching function for the proton transfer direction either to the O_2 reduction site or to the pumped-proton-loading site.

3. Proton pumping through the H-pathway has been proposed based on the X-ray structural analyses of bovine CcO. The structural elements for proton pumping through the H-pathway, including the proton-pumping (loading) site, the structural changes for switching accessibility, and the driving unit, have been demonstrated by X-ray structural analyses at fairly high resolutions.

4. Mutational analyses using the bovine gene expression system have confirmed the functions of some critical H-pathway residues.

FUTURE ISSUES

1. Structural validations for the proposed functions of residues in the D-pathway for evaluations of the D-pathway proposal are needed.

2. Extensive mutagenesis analyses for the D- and K-pathways of bovine CcO would provide important insights into the proton-pumping mechanism.

3. Protons have never been directly observed during the proton-pumping process. In this context, X-ray structural analyses at resolutions sufficient to identify hydrogen and the time-resolved infrared analyses using site-directed stable isotope-labeling techniques are needed for further elucidation of the proton-pumping mechanism of CcO.

DISCLOSURE STATEMENT

The authors are not aware of any affiliations, memberships, funding, or financial holdings that might be perceived as affecting the objectivity of this review.

ACKNOWLEDGMENTS

This work is supported in part by the Grant-in-Aid for Scientific Research 2247012 (S.Y.), the Targeted Protein Research Program (K.M., K.S.-I., and S.Y.), and the Global Center of Excellence Program (S.Y.), each provided by the Japanese Ministry of Education, Culture, Sports, Science and Technology. S.Y. is a Senior Visiting Scientist in the RIKEN Harima Institute.

LITERATURE CITED

1. Adelroth P, Ek M, Brzezinski P. 1998. Factors determining electron-transfer rates in cytochrome c oxidase: investigation of the oxygen reaction in the *R. sphaeroides* enzyme. *Biochim. Biophys. Acta* 1367:107–17
2. Adelroth P, Ek MS, Mitchell DM, Gennis RB, Brzezinski P. 1997. Glutamate 286 in cytochrome aa_3 from *Rhodobacter sphaeroides* is involved in proton uptake during the reaction of the fully-reduced enzyme with dioxygen. *Biochemistry* 36:13824–29
3. Aoyama H, Muramoto K, Shinzawa-Itoh K, Hirata K, Yamashita E, et al. 2009. A peroxide bridge between Fe and Cu ions in the O_2 reduction site of fully oxidized cytochrome c oxidase could suppress the proton pump. *Proc. Natl. Acad. Sci. USA* 106:2165–69
4. Belevich I, Bloch DA, Belevich N, Wikstrom M, Verkhovsky MI. 2007. Exploring the proton pump mechanism of cytochrome c oxidase in real time. *Proc. Natl. Acad. Sci. USA* 104:2685–90
5. Belevich I, Verkhovsky MI, Wikstrom M. 2006. Proton-coupled electron transfer drives the proton pump of cytochrome c oxidase. *Nature* 440:829–32
6. Bloch D, Belevich I, Jasaitis A, Ribacka C, Puustinen A, et al. 2004. The catalytic cycle of cytochrome c oxidase is not the sum of its two halves. *Proc. Natl. Acad. Sci. USA* 101:529–33
7. Bränden G, Bränden M, Schmidt B, Mills DA, Ferguson-Miller S, et al. 2005. The protonation state of a heme propionate controls electron transfer in cytochrome c oxidase. *Biochemistry* 44:10466–74
8. Bränden G, Pawate AS, Gennis RB, Bizezinski P. 2006. Controlled uncoupling and recoupling of proton pumping in cytochrome c oxidase. *Proc. Natl. Acad. Sci. USA* 103:317–22
9. Brzezinski P, Larsson G. 2003. Redox-driven proton pumping by heme-copper oxidases. *Biochim. Biophys. Acta* 1605:1–13
10. Capitanio N, Capitanio G, Boffoli D, Papa S. 2000. The proton/electron coupling at heme a and Cu_A in bovine heart cytochrome c oxidase. *Biochemistry* 39:15454–61
11. Chance B, Saronio C, Leigh JS. 1975. Functional intermediates in the reaction of membrane-bound cytochrome oxidase with oxygen. *J. Biol. Chem.* 250:9226–37
12. Chang HY, Hemp J, Chen Y, Fee JA, Gennis RB. 2009. The cytochrome ba_3 oxygen reductase from *Thermus thermophilus* uses a single input channel for proton delivery to the active site and for proton pumping. *Proc. Natl. Acad. Sci. USA* 106:16169–73
13. Dürr KL, Koepke J, Hellsig P, Muller H, Angerer H, et al. 2008. A D-pathway mutation decouples the *Paracoccus denitrificans* cytochrome c oxidase by altering the side-chain orientation of a distant conserved glutamate. *J. Mol. Biol.* 384:865–77
14. Faxen K, Gilderson G, Adelroth P, Brzezinski P. 2005. A mechanistic principle for proton pumping by cytochrome c oxidase. *Nature* 437:286–89
15. Hemp J, Han H, Roh JH, Kaplan S, Martinez TJ, et al. 2007. Comparative genomics and site-directed mutagenesis support the existence of only one input channel for protons in the C-family (cbb_3 oxidase) of heme-copper oxygen reductases. *Biochemistry* 46:9963–72
16. Iwata S, Ostermeier C, Ludwig B, Michel H. 1995. Structure at 2.8 Å resolution of cytochrome c oxidase from *Paracoccus denitrificans*. *Nature* 376:660–69

17. Kaila VRI, Verkhovsky MI, Hummer G, Wikstrom M. 2008. Glutamic acid 242 is a valve in the proton pump of cytochrome *c* oxidase. *Proc. Natl. Acad. Sci. USA* 105:6255–59

18. Kamiya K, Boero M, Tateno M, Shiraishi K, Oshiyama A. 2007. Possible mechanism of proton transfer through peptide groups in the H-pathway of the bovine cytochrome *c* oxidase. *J. Am. Chem. Soc.* 129:9663–73

19. Kitagawa T, Ogura T. 1997. Oxygen activation mechanism at the binuclear site of heme-copper oxidase superfamily as revealed by time-resolved resonance Raman spectroscopy. *Prog. Inorg. Chem.* 45:431–79

20. Konstantinov A, Siletsky S, Mitchell D, Kaulen A, Gennis RB. 1997. The roles of the two proton input channels in cytochrome *c* oxidase from *Rhodobacter sphaeroides* probed by the effects of site-directed mutations on time-resolved electrogenic intraprotein proton transfer. *Proc. Natl. Acad. Sci. USA* 94:9085–90

21. Lee H-M, Das TK, Rousseau DL, Mills D, Ferguson-Miller S, et al. 2000. Mutations in the putative H-channel in the cytochrome *c* oxidase from *Rhodobacter sphaeroides* show that this channel is not important for proton conduction but reveal modulation of the properties of heme *a*. *Biochemistry* 39:2989–96

22. Lepp H, Svahn E, Faxen K, Brzezinski P. 2008. Charge transfer in the K proton pathway linked to electron transfer to the catalytic site in cytochrome *c* oxidase. *Biochemistry* 47:4929–35

23. Mills, Schmidt B, Hiser C, Westley E, Ferguson-Miller S. 2002. Membrane potential-controlled inhibition of cytochrome *c* oxidase by zinc. *J. Biol. Chem.* 277:14894–901

24. Mochizuki M, Aoyama H, Shinzawa-Itoh K, Usui T, Tsukihara T, et al. 1999. Quantitative reevaluation of the redox-active sites of crystalline bovine heart cytochrome *c* oxidase. *J. Biol. Chem.* 274:33403–11

25. Moody AJ. 1996. As prepared forms of fully oxidized haem/Cu terminal oxidases. *Biochim. Biophys. Acta* 1276:6–20

26. Muramoto T, Ohta K, Shinzawa-Itoh K, Kanda K, Taniguchi M, et al. 2010. Bovine cytochrome *c* oxidase structures enable O_2 reduction with minimization of reactive oxygens and provide a proton-pumping gate. *Proc. Natl. Acad. Sci. USA* 107:7740–45

27. Namslauer A, Angaard A, Katsonouri A, Brzezinski P. 2003. Intramolecular proton-transfer reactions in a membrane-bound proton pump: The effect of pH on the peroxy to ferryl transition in cytochrome *c* oxidase. *Biochemistry* 42:1488–98

28. Namslauer A, Pawate AS, Gennis RB, Brzezinski P. 2003. Redox-coupled proton translocation in biological systems: proton shuttling in cytochrome *c* oxidase. *Proc. Natl. Acad. Sci. USA* 100:15543–47

29. Pawate AS, Morgan J, Namslauer A, Mills D, Brzenzinski P, et al. 2002. A mutation in subunit I of cytochrome oxidase from *Rhodobacter sphaeroides* results in an increase in steady-state activity but completely eliminates proton pumping. *Biochemistry* 41:13417–23

30. Pfitzner U, Hoffmeier K, Harrenga A, Kannt A, Michel H, et al. 2000. Tracing the D-pathway in reconstituted site-directed mutants of cytochrome *c* oxidase from *Paracoccus denitrificans*. *Biochemistry* 39:6756–62

31. Pilet E, Jasaitis A, Liebl U, Vos MH. 2004. Electron transfer between hemes in mammalian cytochrome *c* oxidase. *Proc. Natl. Acad. Sci. USA* 101:16198–203

32. Puustinen A, Wikstrom W. 1999. Proton exit from the heme-copper oxidase of *Escherichia coli*. *Proc. Natl. Acad. Sci. USA* 96:35–37

33. Qian J, Mills DA, Geren L, Wang K, Hoganson CW, et al. 2004. Role of the conserved arginine pair in proton and electron transfer in cytochrome *c* oxidase. *Biochemistry* 43:5748–56

34. Qin L, Liu J, Mills DA, Proshlyakov DA, Hiser C, et al. 2009. Redox-dependent conformational changes in cytochrome *c* oxidase suggest a gating mechanism for proton uptake. *Biochemistry* 48:5121–30

35. Ruitemberg M, Kannt A, Bamberg E, Fendler K, Michel H. 2002. Reduction of cytochrome *c* oxidase by a second electron leads to proton translocation. *Nature* 417:99–102

36. Ruitenberg M, Kannt A, Bamberg E, Ludwing B, Michel H, et al. 2000. Single-electron reduction of the oxidized state is coupled to proton uptake via the K pathway in *Paracoccus denitrificans* cytochrome *c* oxidase. *Proc. Natl. Acad. Sci. USA* 97:4632–36

37. Sakaguchi M, Shinzawa-Itoh K, Yoshikawa S, Ogura T. 2010. A resonance Raman band assignable to the O-O stretching mode in the resting oxidized state of bovine heart cytochrome *c* oxidase. *J. Bioenerg. Biomembr.* 42:241–43

38. Shimokata K, Katayama Y, Murayama H, Suematsu M, Tsukihara T, et al. 2007. The proton pumping pathway of bovine heart cytochrome *c* oxidase. *Proc. Natl. Acad. Sci. USA* 104:4200–5

39. Siletsky S, Kaulen AD, Konstantinov AA. 1999. Resolution of electrogenic steps coupled to conversion of cytochrome c oxidase from the peroxy to the ferryl-oxo state. *Biochemistry* 38:4853–61

40. Siletsky SA, Belevich I, Wikstrom M, Soulimane T, Verkhovsky MI. 2009. Time-resolved OH transition of the aberrant ba_3 oxidase from *Thermus thermophilus. Biochim. Biophys. Acta* 1787:201–5

41. Siletsky SA, Pawate AS, Weiss K, Gennis RB, Konstantinov AA. 2004. Transmembrane charge separation during the ferryl-oxo→oxidized transition in a nonpumping mutant of cytochrome c oxidase. *J. Biol. Chem.* 279:52558–65

42. Svensson-Ek M, Abramson J, Larsson G, Tornroth S, Brzezinski P, et al. 2002. The X-ray crystal structures of wild-type and EQ(1–286) mutant cytochrome c oxidases from *Rhodobacter sphaeroides. J. Mol. Biol.* 321:329–39

43. Tsukihara T, Aoyama H, Yamashita E, Tomizaki T, Yamaguchi H, et al. 1995. Structures of metal sites of oxidized bovine heart cytochrome c oxidase at 2.8 Å. *Science* 269:1069–74

44. Tsukihara T, Aoyama H, Yamashita E, Tomizaki T, Yamaguchi H, et al. 1996. The whole structure of the 13-subunit oxidized cytochrome c oxidase at 2.8 Å. *Science* 272:1136–44

45. Tsukihara T, Shimokata K, Katayama Y, Shimada H, Muramoto K, et al. 2003. The low-spin heme of cytochrome c oxidase as the driving element of the proton-pumping process. *Proc. Natl. Acad. Sci. USA* 100:15304–9

46. Verkhovsky MI, Belevich I, Bloch DA, Wikstrom M. 2006. Elementary steps of proton translocation in the catalytic cycle of cytochrome oxidase. *Biochim. Biophys. Acta* 1757:401–7

47. Verkhovsky MI, Jasaitis A, Verkhovskaya ML, Morgan JE, Wikstrom M. 1999. Proton translocation by cytochrome c oxidase. *Nature* 400:480–83

48. Verkhovsky MI, Tuukkanen A, Backgren C, Puustinen A, Wikstrom M. 2001. Charge translocation coupled to electron injection into oxidized cytochrome c oxidase from *Paracoccus denitrificans. Biochemistry* 40:7077–83

49. Vygodina TV, Pecoraro C, Mitchell D, Gennis R, Konstantinov AA. 1998. Mechanism of inhibition of electron transfer by amino acid replacement K362M in a proton channel of *Rhodobacter sphaeroides* cytochrome c oxidase. *Biochemistry* 37:3053–61

50. Wikstrom M, Verkhovsky MI, Hummer G. 2003. Water-gated mechanism of proton translocation by cytochrome c oxidase. *Biochim. Biophys. Acta* 1604:61–65

51. Yoshikawa S, Shinzawa-Itoh K, Nakashima R, Yaono R, Yamashita E, et al. 1998. Redox-coupled crystal structural changes in bovine heart cytochrome c oxidase. *Science* 280:1723–29

52. Zaslavsky D, Kaulen AD, Smirnova IA, Vygodina T, Konstantinov AA. 1993. Flash-induced membrane potential generation by cytochrome c oxidase. *FEBS Lett.* 336:389–93

SAXS Studies of Ion–Nucleic Acid Interactions

Lois Pollack

School of Applied & Engineering Physics, Cornell University, Ithaca, New York 14853;
email: Lp26@cornell.edu

Annu. Rev. Biophys. 2011. 40:225–42

First published online as a Review in Advance on
February 14, 2011

The *Annual Review of Biophysics* is online at
biophys.annualreviews.org

This article's doi:
10.1146/annurev-biophys-042910-155349

Copyright © 2011 by Annual Reviews.
All rights reserved

1936-122X/11/0609-0225$20.00

Keywords

ASAXS, condensed counterions, electrostatic interactions, RNA,
DNA

Abstract

Positively charged ions, atoms, or molecules compensate the high negative charge of the nucleic acid backbone. Their presence is critical to the
biological function of DNA and RNA. This review focuses on experimental studies probing (*a*) interactions between small ions and nucleic
acids and (*b*) ion-mediated interactions between nucleic acid duplexes.
Experimental results on these simple model systems can be compared
with specific theoretical models to validate their predictions. Small angle
X-ray scattering (SAXS) provides unique insight into these interactions.
Anomalous SAXS reports the spatial correlations of condensed (e.g., locally concentrated) counterions to individual DNA or RNA duplexes.
SAXS very effectively reports interactions between nucleic acid helices,
which range from strongly repulsive to strongly attractive depending
on the ionic species present. The sign and strength of interparticle interactions are easily deduced from dramatic changes in the scattering
profiles of interacting duplexes.

Contents

INTRODUCTION

Small angle X-ray scattering (SAXS) is a remarkably informative probe of macromolecular structure (35, 37). The global structural properties of nucleic acids (NAs) that can be extracted from SAXS data include radius of gyration (50), persistence length (16), cross-sectional radius (58, 95), compactness (83, 84), and arrangement of helical elements (86). All these properties are relevant to biological function. To neutralize the negative charge of the NA sugar-phosphate backbone, charge compensating ions, atoms, or molecules must be present. This review considers the case of charge neutralization by small ions. Although some counterions interact in a site-specific way with the NA, the majority are territorially or diffusively bound (29). In addition to screening the large electrostatic potentials of the NA structure, these ions facilitate (or prevent) interactions between strands (101, 102), depending on their valence, size, concentration, or charge density. These condensed counterions (59) lend additional dimensions to SAXS studies because the counterions contribute in a distinct way to the overall scattering profile.

This review begins with a general discussion of SAXS and then treats the specific case of scattering from a two-component system consisting of a 25-bp DNA or RNA duplex plus counterions. If the counterions are elements with electron binding energies above ∼7 keV (roughly those with atomic number $Z > 25$), anomalous SAXS (91) (ASAXS) provides additional information about their arrangement around the NA and uniquely enables a comparison with theoretical predictions of ion spatial distribution.

This review focuses on the significant insight provided by SAXS into ion-mediated interactions between NA duplexes (4, 72–74). The concentration and valence of counterions surrounding NAs affect forces between NAs, which range from strongly repulsive (43) to strongly attractive (11). These interparticle interactions dramatically alter the measured scattering profiles, notably at the lowest angles (43, 90), reflecting the strength and range of the large electrostatic potentials that exist in the presence of NA. Studying the role of ions in NA interactions enables a more thorough understanding of counterion-driven processes such as DNA condensation or RNA folding.

Finally, we consider the information SAXS conveys about the role of ions in large-scale conformational changes of individual macromolecules. This topic has been treated in several recent reviews (27, 29, 71, 75, 101, 102) and is discussed only briefly here.

SAXS BACKGROUND

General Considerations

SAXS provides information about the size, shape, compactness, and molecular weight of macromolecules in solution. In contrast to X-ray crystallography, its structural information is low resolution, ∼10 Å. The power of this approach lies in reporting functional (e.g., unconstrained) conformations and dynamics.

A schematic of a typical SAXS experiment is shown in **Figure 1**. Each electron in the sample reradiates a spherical wave upon excitation by the electric field of the incident X-ray beam. For

---

Small angle X-ray scattering (SAXS): an experimental technique that provides information about the global structure of macromolecules in solution

NA: nucleic acid

Anomalous SAXS (ASAXS): exploits element-specific, energy-dependent scattering strengths to emphasize the contribution to the overall scattering signal of one particular species of counterion

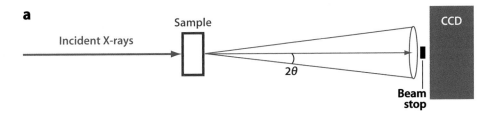

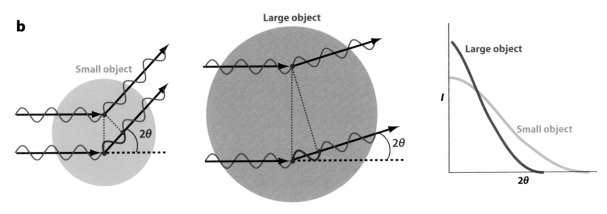

Figure 1

Shows a schematic of a SAXS experiment. (*a*) An X-ray beam is incident on the sample, typically macromolecules in solution. Interaction with the sample scatters or redirects the beam away from its incident direction. Information about the size and shape of the molecules is acquired by examining the angular distribution of the scattered radiation. This is carried out by performing an integral at fixed θ over the face of the cone. Distance from the center of the circle corresponds to different scattering angles. Intensity is usually plotted as a function of $q = 4\pi \sin\theta/\lambda$. Typical scattering profiles are shown in panel *b*. Smaller objects (*green*) have scattering profiles that extend to larger angles, whereas the scattering profiles of larger objects (*brown*) fall off more rapidly. Abbreviations: CCD, charge coupled device; SAXS, small angle X-ray scattering.

regular SAXS, we consider elastic scattering into angles within a few degrees of the θ = 0, or forward, direction. The amplitude of the scattered wave is computed by considering phase differences between all waves scattered into a particular direction, characterized by the angle 2θ or the momentum transfer $q = 4\pi \sin\theta/\lambda$. In the small-angle regime of interest, $q \sim \theta$. For the collection of N electrons (or N atoms treated as localized clusters of electrons) that comprise the macromolecule of interest, the scattering amplitude $F(q)$ has the following form:

$$F(q) = \sum_{n=1}^{N} e^{iq \cdot r_n} f_n(q). \qquad 1.$$

The position of the nth electron (or atom) relative to an origin is denoted r_n. The scattering factor for the electron (atom) is $f_n(q)$. At

$q = 0$, it is proportional to atomic number, Z; its variation with q is slight in the angular range of interest.

Because experimental detectors measure intensity, not amplitude, $F(q)$ is multiplied by its complex conjugate to yield $I(q)$. The spherically averaged intensity $\langle I(q) \rangle$ is computed to account for the random orientation of the macromolecule in solution relative to the direction of the incident beam (17):

$$\langle I(q) \rangle \propto \sum_{n=1}^{N} \sum_{m=1}^{N} f_n(q) f_m(q)$$
$$\times (\sin q r_{nm} / q r_{nm}). \qquad 2.$$

This equation applies when considering discrete (as opposed to continuous) distributions of electrons and/or atoms. The scattering factor of the jth electron/atom is denoted f_j and the distance between the *m*th and *n*th electron/atom is described by $r_{nm} = r_n - r_m$.

Because the waves scattered from all electrons and/or atoms have the same phase in the forward ($q = 0$) direction, $I(q = 0)$ or I_0 is proportional to the square of the total number of electrons in the sample. Scattering from a collection of C noninteracting molecules in solution is simply C times the scattering from an individual particle. Changes in I_0 report aggregation or association (30). For example, I_0 doubles if a given sample transitions from all monomers to all dimers.

At angles larger than zero, the scattering from larger objects falls off more rapidly than scattering from smaller objects because of the greater path length (hence phase) difference of X-rays scattered from widely separated sources (**Figure 1**). Thus, the shape of the scattering profile reflects the arrangement of scattering units within the macromolecule.

Molecular size can be quantified by radius of gyration R_g, the mean square distance between all pairs of electrons. Changes of up to a factor of 2 in R_g can accompany large-scale conformational changes such as folding (83). The R_g is computed using the so-called Guinier approximation. At the lowest scattering angles (for $q < 1.3/R_g$), $I(q)$ can be approximated by a Gaussian form:

$$\langle I(q) \rangle = I_0 e^{-q^2 R_g^2 / 3}. \qquad 3.$$

A linear fit to $ln(I(q))$ versus q^2 yields a slope that is proportional to R_g^2. As an alternative, indirect methods (93) derive R_g from data acquired at scattering angles up to π/D_{max}, where D_{max} is the maximum particle diameter. For rod-like particles, $I(q)$ varies as the product of $1/q$ and a function that reflects the particle's cross-sectional geometry (35). At low angles, the radius of gyration of the cross-section can be extracted from a linear fit to $ln(qI(q))$ versus q^2.

Scattering profiles contain much more information than R_g. Analytical (or simple computational) forms describe the scattering of objects with regular shapes, such as spheres, discs, or cylinders (35). For nonregular objects, e.g., RNA with tertiary structure, different strategies are employed to interpret SAXS data. One popular approach for folding studies (87) relies on Kratky plots of $I(q)q^2$ versus q to emphasize the shape of $I(q)$ for q values above the Guinier cutoff (35). Kratky plots of globular, or compact, states display a broad peak in this mid-q range. For the more extended states characteristic of unfolded structures, peaks in the Kratky plots are less pronounced or vanish altogether (25) (**Figure 2**).

Most recently, sophisticated mathematical algorithms (18, 39, 92, 96) have been applied to reconstruct macromolecular shapes from scattering profiles. For example, (continuous) distributions of beads placed in a carefully sized box are adjusted to yield scattering profiles that coincide with measurement. A recent review summarizes many of the resources available for SAXS reconstructions (53).

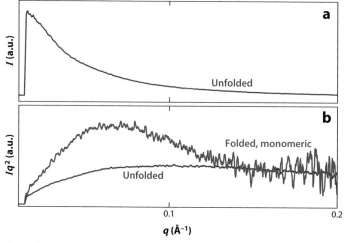

Figure 2

(*a*) The intensity versus q from a small RNA domain, P4-P6 from the *Tetrahymena* ribozyme (Pollack laboratory, unpublished data). (*b*) Kratky plots. The red curve shows the scattering profile acquired in low salt where the domain is unfolded. The blue curve was acquired under conditions where P4-P6 is folded and monomeric, 160 ms after the addition of 10 mM Mg^{2+} (86).

The Form Factor

Both the NA and its associated counterions contribute to the measured scattering profile of DNA and RNA. The scattering amplitude for each component can be computed from Equation 1. For a DNA duplex plus associated ions (64),

$$F_{DNA\text{-}ion}(q) = f_{DNA}F_{DNA}(q) + f_{IONS}F_{IONS}(q). \quad 4.$$

In this expression, f_{DNA}/f_{IONS} represents the solvent-corrected or excess scattering factor of DNA or ions, e.g., the effective number of electrons that contributes to the scattering relative to the background solvent. For the ions $f_{IONS} = N_{ions}*f_{ion}$, where N_{ions} is the number of ions correlated to the NA and f_{ion} is the solvent-corrected scattering factor associated with each ion. A more detailed derivation and explanation of each term is provided in Reference 64. By definition (26), $F(0) = 1$; thus the contribution to $I(q)$ from the DNA or the ions is determined by the relative magnitude of f_{DNA} and f_{IONS}. The intensity $I(q)$ equals the product of $F_{DNA\text{-}ion}$ and its complex conjugate (64).

Anomalous Scattering

ASAXS takes advantage of contrast variation methods to highlight the small angle scattering signal from a single elemental component correlated with a larger system (91). Using ASAXS we can derive information about ion distribution. All elements possess a unique set of characteristic energies, corresponding to electron binding energies, and can be individually targeted by tuning the energy of an X-ray beam until it is exactly equal to or is resonant with a specific electronic transition. Because the contrast or scattering strength of an element changes when probed near these binding energies, energy-dependent changes in scattering profiles provide additional information about counterion correlation to the NA. If a scattering profile is first measured far from a resonant edge and a second scattering profile is measured close to a resonant edge, only the scattering of the resonant element will be altered. By subtraction of carefully normalized signals, information can be obtained about the spatial distribution of the resonant elements arranged around nonresonant structures. If appropriately

selected ions are employed, it is possible to target the ions around NAs. To date, this approach has been applied to study the distribution of select monovalent, divalent, and trivalent ions around short DNA and RNA duplexes (2, 3, 26, 64, 66).

For counterions with moderate Z, $f_{DNA} > f_{IONS}$. In this regime it is sufficient to measure scattering profiles at two carefully selected energies close to (but below) the ion absorption edge (see Reference 64 for a complete discussion). The measured difference in intensity at these two energies is

$$\Delta I(q) \propto f_{DNA}F_{DNA}F_{IONS} + f_{IONS}F_{IONS}^2. \quad 5.$$

The second term in this expression is always smaller than the first, but it is not necessarily negligible. Therefore, Fourier inversion of the difference signal does not strictly provide the interference term $F_{DNA}F_{IONS}$ (66). Instead, data are analyzed by direct comparison with models. For example, numerical solutions based on the nonlinear Poisson-Boltzmann (NLPB) equation can be obtained from solvers such as the adaptive Poisson-Boltzmann solver (APBS) (8) and used to generate counterion distributions around atomically detailed NA structures. Scattering profiles of these model systems are computed at the two different energies, and these theoretically generated curves are compared with the experimental curves to assess the validity of the model's assumptions.

The Structure Factor

In addition to reporting the size and shape of isolated (noninteracting) objects in solution, SAXS effectively reports interactions between them (22, 38). For concentrated solutions, $\langle I(q) \rangle$ is the product of two functions: the form factor $P(q) = F(q)F^*(q)$ described above and the structure factor $S(q)$. The latter relates to the spatial arrangement of macromolecules. Three regimes are of interest to this work: noninteraction, repulsive interaction, and attractive interaction. In the simplest case, noninteraction, the relative positions of noninteracting particles are uncorrelated so the total scattering from C

NLPB: nonlinear Poisson-Boltzmann

Form factor [$P(q)$]: the spherically averaged small angle scattering profile associated with a single macromolecule

Structure factor [$S(q)$]: a function that modulates the SAXS form factor and reflects inteparticle interactions.

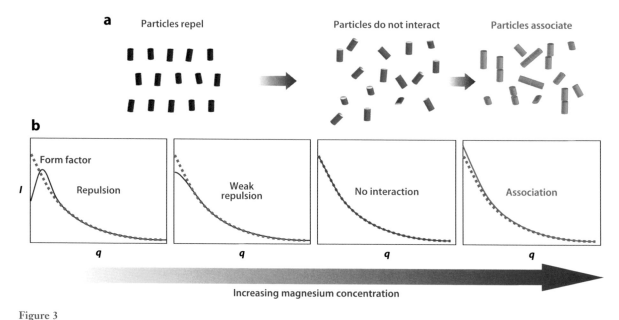

Figure 3

(*a*) Schematic represents interparticle interference. (*Left*) The cylinders are charged and self-avoid by forming a loose lattice. (*Center*) The cylinders do not interact and are randomly distributed. (*Right*) Some cylinders stack in end-to-end configurations. (*b*) Illustrates scattering profiles from the interacting cylinders shown in panel *a*. From left to right, strong electrostatic repulsion between cylinders results in a peak in the structure factor that is visible in a plot of $I(q)$ versus q. When weak repulsion is present, the peak softens to a downturn (*second from left*). When the particles do not interact (*third from left*), the measured intensity is identical in shape to the form factor. Finally, the plot at right corresponds to the case of weak attraction. The presence of dimers (e.g., end-to-end stacked monomers) results in an increase in I_0 relative to the form factor (*gray dashed line*).

particles is just C times the scattering from one particle. Repulsive interaction is easily realized if the particles are charged, e.g., DNA or RNA, in a low-ionic-strength solution where screening distances exceed particle separations. Particles self-avoid and form a loose lattice as suggested in **Figure 3*a***. This ordering introduces an additional length scale to the problem. When strong interparticle interference is present, the structure factor displays a pronounced peak and the intensity decreases relative to the form factor at the lowest angles (**Figure 3*b***). When weak repulsion is present, the peak softens and a downturn or decrease at the lowest q is measured in $I(q)$ relative to the form factor (**Figure 3*b***). In **Figure 3*b***, the second plot from the right illustrates the case of noninteraction. Upturns at low q reflect increases in I_0 and indicate macromolecular association (**Figure 3*a***, right; **Figure 3*b***, far right).

These methods have been applied by others to quantify interactions between charged macromolecules (22, 38).

The structure factor provides relevant information about internucleic acid interactions. There are two ways to interpret the information provided by the structure factor. First, an effective structure factor $S_M(q)$ can be computed that enables direct comparison with potentials. Second, $S_M(q)$ can be extrapolated to $q = 0$ to yield the second virial coefficient. The first method provides a full curve for $S_M(q)$ for comparison with theoretical predictions, but it relies on a decoupling approximation because most NA particles are not spherical. For nonisotropic particles, the structure factor does not decouple from the form factor. Fortunately, for short NA duplexes, a decoupling approximation can be applied to simplify the calculation (63). Within this approximation, the effective structure factor is

$$S_M(q) = \left(1 + \frac{\langle F(q)\rangle^2}{\langle F(q)^2\rangle}(S(q) - 1)\right). \quad 6.$$

Here, $F(q)$ is the particle scattering amplitude and the brackets indicate an orientational average. The isotropic structure factor $S(q)$ can be computed assuming that the particles interact via a distance-dependent potential, such as the two Yukawa potentials (12, 55, 89).

The second approach relies on computation of second virial coefficients and avoids much of the complexity introduced by anisotropy. Here, only the $q = 0$ value is used. $S_M(0)$ is proportional to the osmotic compressibility of the system (93) and

$$\frac{P(q=0)}{I(c, q=0)} = \frac{1}{S_M(q=0)} = 1 + (2MA_2)c. \quad 7.$$

In this expression c is the macromolecular concentration and A_2 is the second virial coefficient, a term in the series expansion of osmotic pressure that describes interparticle interactions (93). When $A_2 > 0$, interparticle interactions are repulsive and particles are more evenly distributed. A negative A_2 indicates a lower osmotic pressure and net attractive interactions, which can result in interparticle association. To use this method, the form factor $P(q)$ must be accurately determined, either by computation or measurement at a very low sample concentration where particles do not interact. The structure factor is experimentally derived by dividing the measured $I(q)$ by the form factor. Measurements are made at a number of NA concentrations, c. Extrapolation of each curve to $q = 0$ yields $S(c, 0)$. Finally, the slope of $1/S(0, c)$ versus c provides the value of A_2 for a given bulk solvent condition. Computation of A_2 enables straightforward comparison of interparticle interaction strength as ionic conditions are varied (72).

Reconstructions

SAXS also probes ion-induced conformational changes of both DNA and RNA (27). To date, measurements of this type have emphasized changing NA conformations, e.g., the initiation of RNA folding by the addition of divalent ions

to solutions containing low concentrations of monovalent ions. The more localized screening provided by divalent ions allows formerly extended molecules to relax into compact ensembles. Tertiary contacts form within these compact ensembles and lock the molecule into folded functional structures. SAXS studies of ion-induced folding provide information about the global structure of these molecules, at all times even before specific contacts are made. These global measurements are essential complements to other techniques that sense local interactions (33, 49, 86, 94).

Analysis of RNA conformations during folding can be carried out in different ways, e.g., measuring the radius of gyration (49), computing Fourier transforms of $I(q)$ (93), using Kratky plots to assess molecular compaction (84), or employing singular value decomposition (SVD) methods to determine the number of independent states present (21). Recently, reconstruction methods have been applied to visualize the structures of transient, folding intermediates (51).

Ion-induced ordering of DNA usually leads to liquid crystalline forms (recently reviewed in Reference 101). These structures are amenable to analysis by standard diffraction techniques (47, 76, 77).

SAXS STUDIES OF NUCLEIC ACID DUPLEXES

The first SAXS measurements of dilute solutions of DNA were made in the early 1960s (58) and reported the radius of gyration of the DNA cross section as well as the linear mass. These measurements did not reveal the associated counterions, because light (Na$^+$) ions were used. The counterions remained silent to SAXS until 1990 (19), when heavy-atom replacement enabled detection of counterions with larger atomic numbers, e.g., Tl$^+$ and Ba^{2+}. The Poisson-Boltzmann distribution was employed to model the spatial distribution of these ions around DNA, and the total scattering profile was computed. However, there was no attempt to experimentally distinguish ion scattering from DNA scattering. Stuhrmann (91) was the first to suggest using ASAXS to measure the distribution of Cs^+ ions around DNA.

Probing the Ion Atmosphere with ASAXS

Knowledge of the number and spatial distribution of counterions is important in computing the detailed free energies of systems containing NAs (57). The sensitivity of NA interactions to the number and valence of counterions is further emphasized by the observation that inter-DNA interactions can be tuned from highly repulsive to strongly attractive when even small numbers of trivalent ions are added to DNA in Na^+-containing solutions (67). It is of great interest from both theoretical and practical (biotechnology) viewpoints to understand how ions facilitate such dramatic changes in NA interaction.

Anomalous SAXS enables measurement of the nonlocalized counterions that evade crystallographic detection. Because of the increasing absorption of X-rays by water at energies below ~7 keV, ASAXS is presently limited to experiments on ions with absorption edges above 7 keV. This precludes ASAXS measurements of low-Z, physiological elements such as Na, Mg,

or K. The lightest monovalent ion that is easily probed is Rb^+, with a K edge at 15.2 keV; the lightest divalent ion is Sr^{2+} with a K edge at 16.1 keV. Fortunately, many in vitro studies have been carried out using trivalent cobalt hexammine (11, 67, 100), an effective condensing agent for DNA. The cobalt edge at 7.7 keV is readily accessible to ASAXS.

Twenty-five-bp DNA in different ionic solutions. The distributions of monovalent, divalent, and trivalent ions around duplex DNA and monovalent and divalent ions around RNA have been measured with ASAXS (2, 3, 26, 66). Short duplexes, 25 bp in length, were employed for all measurements. These molecules are much shorter than duplex persistence lengths and therefore can be accurately modeled with atomic detail (26).

The first ASAXS experiment probed Rb ions in solutions of DNA in Rb-acetate solution (26). The ASAXS signal was derived by measuring two carefully selected energies just below the Rb K edge, where there is a small decrease in excess scattering length of Rb^+ alone. Typical ASAXS data are shown in **Figure 4**. As validation of the technique, the anomalous signal measured in Rb was compared to a signal derived by heavy-atom replacement: The

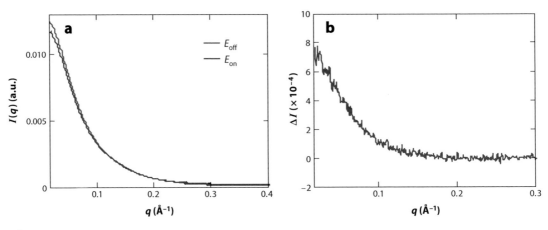

Figure 4

(*a*) Scattering profiles of DNA in Rb-acetate acquired at two different X-ray energies just below the Rb edge. The blue curve was acquired far below the edge, where resonant effects are small. The red curve was acquired just below the edge, where the scattering strength of the Rb^+ ions is decreased due to resonant effects. (*b*) The difference between these curves, the anomalous signal, is shown.

scattering profile of DNA in Na$^+$-containing solution was acquired and compared with the scattering profile of DNA in Rb$^+$-containing solution. Subtraction of the two profiles (carefully DNA concentration matched) yields a curve that has the same shape as the anomalous difference signal. A Rb$^+$ distribution around an atomically detailed model of DNA was computed according to NLPB predictions. To compare with data, scattering profiles derived from simulation were computed at two energies near the Rb edge using the appropriate energy-dependent scattering factors for Rb. Good agreement with the experimental data was obtained, validating the model. A similar experiment was carried out comparing the anomalous signal derived from DNA in Sr^{2+}-containing solutions acquired at energies near the Sr absorption edge with a heavy-atom difference signal comparing DNA in Sr^{2+}-containing solutions to DNA in Mg^{2+}-containing solutions. The similarity in shapes obtained by these two different methods validates the ASAXS approach for the ion-DNA system.

Follow-up experiments (2) exploited counterion competition to monitor the relative fraction of charge screened by monovalent or divalent ions in mixed-valence ionic solution. When ions of comparable size and mixed valence are present, a larger fraction of the overall charge is compensated by the more highly charged ions. A smaller number of divalent (relative to monovalent) cations can compensate the backbone charge of the NA while maintaining a more favorable entropy (fewer ions must be spatially restricted). DNA charge is neutralized more effectively by divalent ions than by monovalent ions.

The relative composition of the counterion cloud was measured by monitoring anomalous signals from both Rb$^+$ and Sr^{2+} ions in a single sample. The experimental results match predictions of NLPB models of competition only if a correction is made to account for finite ion size (80–82). The importance of ion size and charge density has been confirmed by other experiments on NA systems (5, 44–46). ASAXS reveals that the spatial distribution of one ion species remains largely unchanged in the presence of the other (2).

The competition of trivalent cobalt hexammine with monovalent Rb was probed with ASAXS in subsequent experiments (3). For reasons stated above, trivalent ions effectively displace monovalent ions (14, 69). When the trivalent ion concentration is very low relative to the phosphate concentration, ion competition is adequately described by mean-field, NLPB models with corrections for finite ion size. However, attractive forces consistent with DNA condensation can be initiated when even small numbers of cobalt hexammine ions are introduced (67, 100). For the short (25-bp) DNA duplexes probed, attractive interactions occur when the cobalt hexammine concentration exceeds ~0.8 mM (on a background of 100-mM monovalent ions) (3). These attractive forces preclude the application of mean-field models, which treat ions as point particles and by default do not account for correlations between ions and hydration effects. ASAXS measurements of ion distributions provide a framework for comparing more complete theoretical models with experiment. Important issues can be addressed, for example, identification of the physical origin of attractive forces between like-charged objects.

25-bp RNA in different ionic solutions. Recent interest in biophysical characterization of RNA has increased dramatically (40) because of the increasing recognition of RNA's numerous and varied biological roles (88). Short duplexes are important as structural units for folded RNAs, in RNA interference, and for associated therapeutic applications (13). These current topics motivated a parallel series of ASAXS experiments to probe the ion atmosphere around RNA. To facilitate comparison with DNA, similarly sequenced molecules were used (the sequence is identical if dT is replaced by rU). Different spatial distributions were measured for both monovalent Rb and divalent Sr ions around RNA relative to DNA (66).

Structural variations between RNA and DNA account for at least some of these differences. DNA duplexes assume the B form, while RNA duplexes assume the A form (85). The latter helix is shorter and wider, with a higher linear charge density, which should be screened more effectively by ions (29, 59, 60). However, the measured differences result more likely from the vastly different topology of the A form surface relative to the B form, especially the geometry of the major grooves (66). The high, negative potential in the very deep RNA major groove (24) draws the ions in more tightly when compared with DNA. ASAXS confirms this expectation, reporting more ions at shorter distances from the RNA helical axis than from the DNA helical axis.

To summarize, ASAXS probes the delocalized ions in the counterion cloud. The spatial arrangement of monovalent ions is consistent with appropriately modified predictions of NLPB models. Deviations are measured for larger valences, notably trivalent ions. The development of all atom models of ion distribution around 25-bp helices with the same sequence as those probed experimentally (42) is an important step forward in treating ion-specific effects, including size, hydration, and correlations with counterions or co-ions. Recently ASAXS has been applied to determine N_{ions}, providing ion number in addition to spatial correlation to the NA (65).

Ion-Mediated Interactions Between Duplexes

Previous sections considered SAXS characterization of an NA duplex together with its associated counterions. A second series of experiments focuses on understanding ion-mediated interactions between closely spaced NA strands. The 25-bp duplex discussed above continues to serve as an excellent model system for these studies. To probe interparticle interactions, SAXS profiles of duplexes are acquired in a series of carefully prepared solutions. Because the helices themselves do not change conformation, the form factor $P(q)$ remains constant for all conditions probed. Interparticle interactions are monitored through changes in

the structure factor $S(q)$. The ratio of cylinder height to diameter (4:1) is relatively modest and compatible with the decoupling approximation discussed above. SAXS measurements of the structure factor or, alternatively, the second virial coefficient are a valuable tool for quantifying interactions between duplexes.

We first consider interactions between DNA duplexes. These studies were carried out below the entanglement concentration, c^*. At these low concentrations interactions are isotropic (99). In this regime, modeling is straightforward and much insight can be gained into the sign and strength of interparticle interactions. Above c^* (~2.6 mM for 25-bp DNA), the rod-like shape of the NA affects interparticle interactions (97).

DNA duplexes in different ionic solutions. To measure ion-mediated interactions, short DNA duplexes were dialyzed against solutions with ionic strength ranging from low millimolar to nearly molar. (Detailed protocols for sample preparation can be found in Reference 74). For a given bulk salt solution, samples were prepared at several DNA concentrations below c^*. At the lowest concentration, an approximately 0.05 mM duplex, the distance between adjacent duplexes exceeds the screening length. The charge from one duplex is therefore fully screened, hence undetectable at the position of its neighbor. Under these conditions, SAXS profiles reproduce the form factor (74). [In the language of the structure factor, $S(q) = 1$ when interparticle interactions are absent.] At higher DNA concentrations, full scattering profiles were acquired and divided by the form factor to extract the structure factor. The decoupling approximation was applied to correct for the anisotropy of the duplexes (10). A functional form can be derived for the structure factor, assuming a two-Yukawa potential as in References 55 and 56. By examining only the repulsive potential, it is possible to determine the effective charge on the duplex at a specific ionic condition (74). An important (but not unexpected) result is the lower effective charge or enhanced NA screening by divalent Mg^{2+}

relative to monovalent Na^+ at the same bulk ionic strength.

At the lowest ionic strengths, $S(q)$ for more concentrated DNA in solution displays a clear interparticle interference peak. The position of this peak indicates the distance between loosely ordered duplexes and scales with DNA concentration to the $-1/3$ power, consistent with an isotropic and dilute system [the peak position varies with $c^{-1/2}$ for $c > c^*$ (97)]. The interparticle distance should be compared with the screening length, the length scale over which the electrostatic potential of the duplex is screened. This is related to (but is not necessarily equal to) the Debye screening length, the distance at which the potential decays to $1/e$ of its strength at the surface of a charged particle. Interparticle repulsion results if the potential exceeds kT halfway between neighbors; in this case the charge on the first duplex is not fully screened before the neutralizing ions from its neighbor are encountered. An intuitive picture is that the ions required to fully screen the duplex must pile up at this boundary. For a three-dimensional system, the ions form a cage around the DNA and localize it at the center. However, if the charge is fully screened on this length scale, no charge accumulates and the DNAs are unlocalized. The size of this cell decreases as the ionic strength increases, so interparticle ordering weakens.

The second virial coefficient, A_2, provides an alternative method for quantifying interactions (9) and does not rely on a specific choice of potential. The procedure for measuring A_2 was previously described and is illustrated in **Figure 5** for measurements in solutions containing 9 mM NaCl. At low NaCl concentrations, A_2 is large and positive, indicating net repulsion. As the ionic strength increases and the length scale for screening decreases, A_2 decreases. Twenty-five-bp duplexes experience no apparent interactions when monovalent ion concentrations increase above about 0.1 M. Here, A_2 approaches zero.

Similar behavior is observed at low ionic strength when the bulk salt solutions contain divalent ions: Interparticle interactions are

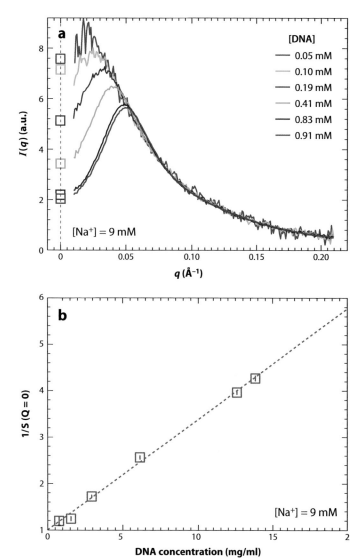

Figure 5

The procedure for extracting the second virial coefficient for DNA in 9 mM NaCl. (*a*) Six scattering profiles were acquired from samples with DNA concentrations ranging from 0.05 to 0.91 mM duplex. The structure factor (peak) becomes more pronounced as the DNA concentration increases: The lattice formed by the particles becomes stiffer (see **Figure 3a**, *left panel*). For each curve the y-intercept is determined and divided by the y-intercept of the form factor. This procedure yields $S(0)$ as a function of DNA concentration. (*b*) The inverse of $S(0)$ is plotted as a function of concentration for each of the six curves shown. The slope of the dashed line yields the second virial coefficient, A_2, for DNA in 9 mM NaCl.

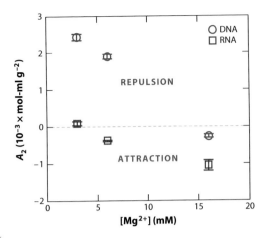

Figure 6

The second virial coefficients for both DNA and RNA as a function of Mg^+ concentration. Positive A_2 indicates net interparticle repulsion, whereas negative A_2 indicates net attraction or association. Scattering profiles measured for DNA and RNA when A_2 is below zero are consistent with end-to-end stacking. The second virial coefficient for RNA is always smaller than that for DNA; interparticle repulsion vanishes at lower bulk ionic strength, and the onset of end-to-end stacking also occurs at lower ionic strength. These observations are consistent with the more effective screening of RNA than DNA under identical bulk solution conditions.

repulsive. However when the divalent ion concentration exceeds ~10 mM (for 25-bp DNA duplexes), the second virial coefficient changes sign, indicating attractive interactions between duplexes (**Figure 6**). Careful examination of the scattering profiles shows that DNA duplexes stack end-to-end. This model was validated by capping the ends with a loop of T residues (52), essentially converting a duplex into a dumbbell. It is interesting that repulsive interactions persist to high ionic strength when end-to-end stacking is prevented by capping. End-to-end stacking has also been observed in low-ionic-strength but highly concentrated NA solutions (62).

RNA duplexes in different ionic solutions. Second virial coefficients were measured for short, similarly sequenced RNA duplexes following the protocols applied to DNA. Data for divalent ions are shown in **Figure 6**. At any given $[Mg^{2+}]$ when A_2 is positive, its value is smaller for RNA than for DNA: RNA is screened more effectively than DNA is. End-

to-end stacking, indicated by the sign reversal of A_2, also initiates at lower ionic strength (66). A similar trend was measured for monovalent-ion-containing solutions. Short helices stack during RNA folding, so end-to-end attraction may be a driving force in assembling structures that take advantage of stacking interactions, for example, in four-way junctions (54).

Tethered DNA helices. SAXS has also been applied to study the conformations of a more complex system consisting of two short duplexes linked by a flexible, uncharged tether (4). This geometry accurately models duplex interactions at the high local concentrations found in biological structures, e.g., folded RNA. The relative arrangement of the short duplexes was measured as a function of ionic strength. Conformations were extended at low ionic strength, consistent with strong electrostatic repulsion between duplexes. Random conformations were measured at higher ionic strengths, consistent with electrostatic relaxation. Notably, no evidence for side-by-side attractive forces was found.

Ion-Induced Conformational Changes

The stabilization of folded or compact NA conformations is an important biological role for ions (23, 28, 103). Many studies on RNA have been carried out because of the variety of biologically relevant tertiary structures that have been identified. SAXS has been applied to study RNAs ranging in size from individual domains through functional molecules (6, 7, 16, 20, 25, 31–34, 41, 49, 51, 61, 68, 70, 78, 84, 86, 94). This work has also been highlighted in recent reviews (27, 71, 75, 101, 102).

Because ion contributions to scattering profiles are small compared with contributions from (ion-induced) conformational changes, e.g., compaction, folding, or unfolding, treatment in this review is brief. The diffusively bound ions have the greatest impact during the earliest stages of folding, where the large-scale conformational changes they trigger are readily detectable by SAXS. As structures form, their changing charge densities attract more ions (29), further enhancing SAXS signals. Although behavior is far from universal, a variety of folding strategies have been reported. At one extreme, ions induce nonspecific collapse in the full-length *Tetrahymena* ribozyme (49, 84). Significant reduction in R_g is measured prior to the formation of specific tertiary contacts (assayed by direct comparison with hydroxyl radical footprinting). However, ions do not always induce such dramatic global changes. Although electrostatic repulsion is an important barrier to folding, other factors must be considered, such as the stiffness of non-base paired regions (86) or the formation of nonnative contacts that favor extended instead of compact states (6). When folding occurs through a series of sequential steps, SAXS can elucidate small changes in compaction that correspond to critical steps, such as rearrangement around prebound metal ions (33). A recent time-resolved SAXS study of the early stages of *Azoarcus* group I intron folding (78) identifies Mg^{2+}-dependent partitioning between parallel pathways. These pathways are distinguished by the formation of either native or nonnative contacts after the initial electrostatic relaxation.

CONCLUSIONS AND OUTLOOK

Interactions with ions are essential to the biological function of both RNA and DNA. The SAXS experiments described here focus on model systems in which interaction parameters can be quantified. Information about counterions is accessible by numerous other techniques (1, 5, 15, 36, 48, 79, 98); however, SAXS alone provides concordant structural information about ions and the underlying NA. Knowledge gained from studies of model systems can be applied to solving fundamentally important problems, such as RNA folding. In addition, it can be used to develop optimal strategies for designing or packaging NAs for therapeutic applications (L. Li, S. Pabit, S. Meisburger & L. Pollack, unpublished data).

Advances in instrumentation and data interpretation favor the continued application of SAXS to these problems. Next-generation X-ray sources will enable high-quality measurements on small-volume samples, offering new possibilities for very rapid time-resolved studies; reconstruction methods, touched on only briefly here, also hold great promise for interpreting SAXS data.

SUMMARY POINTS

1. ASAXS enables measurement of ion number and correlation to DNA and RNA.

2. NLPB predictions of spatial distribution of monovalent ions around DNA or RNA are consistent with results of ASAXS experiments when finite ion size is included.

3. NA helices repel strongly at low ionic strength where screening exceeds half the distance between neighbors.

4. Both DNA and RNA helices stack end-to-end when electrostatic interactions are sufficiently screened on short length scales.

DISCLOSURE STATEMENT

The author is not aware of any affiliations, memberships, funding, or financial holdings that might be perceived as affecting the objectivity of this review.

ACKNOWLEDGMENTS

The author thanks Dr. Suzette Pabit for valuable comments, suggestions, and assistance with some figures. Dr. Xiangyun Qiu also provided valuable assistance with figures. L.P. gratefully acknowledges NIH support through GM085062.

LITERATURE CITED

1. Anderson CF, Record MT. 1990. Ion distributions around DNA and other cylindrical polyions—theoretical descriptions and physical implications. *Annu. Rev. Biophys. Biophys. Chem.* 19:423–65

2. Andresen K, Das R, Park HY, Smith H, Kwok LW, et al. 2004. Spatial distribution of competing ions around DNA in solution. *Phys. Rev. Lett.* 93:248103

3. Andresen K, Qiu X, Pabit SA, Lamb JS, Park HY, et al. 2008. Mono- and trivalent ions around DNA: a small-angle scattering study of competition and interactions. *Biophys. J.* 95:287–95

4. **Bai Y, Das R, Millett IS, Herschlag D, Doniach S. 2005. Probing counterion modulated repulsion and attraction between nucleic acid duplexes in solution. *Proc. Natl. Acad. Sci. USA* 102:1035–40**

5. Bai Y, Greenfeld M, Travers KJ, Chu VB, Lipfert J, et al. 2007. Quantitative and comprehensive decomposition of the ion atmosphere around nucleic acids. *J. Am. Chem. Soc.* 129:14981–88

6. Baird NJ, Gong HP, Zaheer SS, Freed KF, Pan T, Sosnick TR. 2010. Extended structures in RNA folding intermediates are due to non-native interactions rather than electrostatic repulsion. *J. Mol. Biol.* 397:1298–306

7. Baird NJ, Westhof E, Qin H, Pan T, Sosnick TR. 2005. Structure of a folding intermediate reveals the interplay between core and peripheral elements in RNA folding. *J. Mol. Biol.* 352:712–22

8. Baker NA, Sept D, Joseph S, Holst MJ, McCammon JA. 2001. Electrostatics of nanosystems: application to microtubules and the ribosome. *Proc. Natl. Acad. Sci. USA* 98:10037–41

9. Barone G, Sayers Z, Svergun D, Koch MHJ. 1999. A synchrotron radiation X-ray scattering study of aqueous solutions of native DNA. *J. Synch. Rad.* 6:1031–34

10. Bendedouch D, Chen SH. 1983. Structure and interparticle interactions of bovine serum-albumin in solution studied by small-angle neutron-scattering. *J. Phys. Chem.* 87:1473–77

11. Bloomfield VA. 1997. DNA condensation by multivalent cations. *Biopolymers* 44:269–82

12. Bocquet L, Trizac E, Aubouy M. 2002. Effective charge saturation in colloidal suspensions. *J. Chem. Phys.* 117:8138–52

13. Bonetta L. 2009. RNA-based therapeutics: Ready for delivery? *Cell* 136:581–84

14. Braunlin WH, Anderson CF, Record MT. 1987. Competitive interactions of $Co(NH_3)_6^{3+}$ and Na^+ with helical B-DNA probed by Co-59 and Na-23 NMR. *Biochemistry* 26:7724–31

15. Braunlin WH, Xu QW. 1992. Hexaamminecobalt (III) binding environments on double-helical DNA. *Biopolymers* 32:1703–11

16. Caliskan G, Hyeon C, Perez-Salas U, Briber RM, Woodson SA, Thirumalai D. 2005. Persistence length changes dramatically as RNA folds. *Phys. Rev. Lett.* 95:268303

17. Cantor CR, Schimmel PR. 1980. *Biophysical Chemistry Part II: Techniques for the Study of Biological Structure and Function*. New York: Freeman

18. Chacon P, Diaz JF, Moran F, Andreu JM. 2000. Reconstruction of protein form with X-ray solution scattering and a genetic algorithm. *J. Mol. Biol.* 299:1289–302

19. Chang SL, Chen SH, Rill RL, Lin JS. 1990. Measurements of monovalent and divalent counterion distributions around persistence length DNA fragments in solution. *J. Phys. Chem.* 94:8025–28

20. Chauhan S, Caliskan G, Briber RM, Perez-Salas U, Rangan P, et al. 2005. RNA tertiary interactions mediate native collapse of a bacterial group I ribozyme. *J. Mol. Biol.* 353:1199–209

21. Chen LL, Hodgson KO, Doniach S. 1996. A lysozyme folding intermediate revealed by solution X-ray scattering. *J. Mol. Biol.* 261:658–71

22. Chen SH, Sheu EY, Kalus J, Hoffmann H. 1988. Small-angle neutron-scattering investigation of correlations in charged macromolecular and supramolecular solutions. *J. Appl. Crystallogr.* 21:751–69

23. Chen SJ. 2008. RNA folding: conformational statistics, folding kinetics, and ion electrostatics. *Annu. Rev. Biophys.* 37:197–214

24. Chin K, Sharp KA, Honig B, Pyle AM. 1999. Calculating the electrostatic properties of RNA provides new insights into molecular interactions and function. *Nat. Struct. Biol.* 6:1055–61

25. Das R, Kwok LW, Millett IS, Bai Y, Mills TT, et al. 2003. The fastest global events in RNA folding: electrostatic relaxation and tertiary collapse of the *Tetrahymena* ribozyme. *J. Mol. Biol.* 332:311–19

26. **Das R, Mills TT, Kwok LW, Maskel GS, Millett IS, et al. 2003. The counterion distribution around DNA probed by solution X-ray scattering. *Phys. Rev. Lett.* 90:188103**

4. **Introduces tethered helices as a model system to probe side-by-side attraction of DNA duplexes.**

26. **First application of ASAXS to probe the distribution of counterions around DNA.**

27. Doniach S, Lipfert J. 2009. Use of small angle X-ray scattering to characterize conformational states of functional RNAs. *Methods Enzymol.* 469:237–51

28. Draper DE. 2008. RNA folding: thermodynamic and molecular descriptions of the roles of ions. *Biophys. J.* 95:5489–95

29. Draper DE, Grilley D, Soto AM. 2005. Ions and RNA folding. *Annu. Rev. Biophys. Biomol. Struct.* 34:221–43

30. Eliezer D, Chiba K, Tsuruta H, Doniach S, Hodgson KO, Kihara H. 1993. Evidence of an associative intermediate on the myoglobin refolding pathway. *Biophys. J.* 65:912–17

31. Fang X, Littrell K, Yang XJ, Henderson SJ, Siefert S, et al. 2000. Mg2+-dependent compaction and folding of yeast tRNA- and the catalytic domain of the *B. subtilis* RNase P RNA determined by small-angle X-ray scattering. *Biochemistry* 39:11107–13

32. Fang XW, Golden BL, Littrell K, Shelton V, Thiyagarajan P, et al. 2001. The thermodynamic origin of the stability of a thermophilic ribozyme. *Proc. Natl. Acad. Sci. USA* 98:4355–60

33. Fang XW, Thiyagarajan P, Sosnick TR, Pan T. 2002. The rate-limiting step in the folding of a large ribozyme without kinetic traps. *Proc. Natl. Acad. Sci. USA* 99:8518–23

34. Fang XW, Yang XJ, Littrell K, Niranjanakumari S, Thiyagarajan P, et al. 2001. The *Bacillus subtilis* RNase P holoenzyme contains two RNase P RNA and two RNase P protein subunits. *RNA* 7:233–41

35. Glatter O. 1982. *Small Angle X-ray Scattering*. London: Academic

36. Grilley D, Soto AM, Draper DE. 2006. Mg^{2+}-RNA interaction free energies and their relationship to the folding of RNA tertiary structures. *Proc. Natl. Acad. Sci. USA* 103:14003–8

37. Guinier A, Fournet G. 1955. *Small-Angle Scattering of X-Rays*. New York: Wiley

38. Hayter JB, Penfold J. 1981. An analytic structure factor for macroion solutions. *Mol. Phys.* 42:109–18

39. Heller WT, Abusamhadneh E, Finley N, Rosevear PR, Trewhella J. 2002. The solution structure of a cardiac troponin C-troponin I-troponin T complex shows a somewhat compact troponin C interacting with an extended troponin I-troponin T component. *Biochemistry* 41:15654–63

40. Herschlag D, ed. 2009. *Methods in Enzymology, Vol. 469: Biophysical, Chemical, and Functional Probes of RNA Structure, Interactions and Folding, Pt. B*. San Diego, CA: Elsevier/Academic

41. Kilburn D, Roh JH, Guo L, Briber RM, Woodson SA. 2010. Molecular crowding stabilizes folded RNA structure by the excluded volume effect. *J. Am. Chem. Soc.* 132:8690–96

42. Kirmizialtin S, Elber R. 2010. Computational exploration of mobile ion distributions around RNA duplex. *J. Phys. Chem. B* 114:8207–20

43. Koch MHJ, Sayers Z, Sicre P, Svergun D. 1995. A synchrotron-radiation electric-field X-ray solution scattering study of DNA at very-low ionic-strength. *Macromolecules* 28:4904–7

44. Koculi E, Hyeon C, Thirumalai D, Woodson SA. 2007. Charge density of divalent metal cations determines RNA stability. *J. Am. Chem. Soc.* 129:2676–82

45. Koculi E, Lee NK, Thirumalai D, Woodson SA. 2004. Folding of the *Tetrahymena* ribozyme by polyamines: importance of counterion valence and size. *J. Mol. Biol.* 341:27–36

46. Koculi E, Thirumalai D, Woodson SA. 2006. Counterion charge density determines the position and plasticity of RNA folding transition states. *J. Mol. Biol.* 359:446–54

47. Kornyshev AA, Lee DJ, Leikin S, Wynveen A. 2007. Structure and interactions of biological helices. *Rev. Mod. Phys.* 79:943–96

48. Krakauer H. 1971. Binding of Mg^{2+} ions to polyadenylate, polyuridylate, and their complexes. *Biopolymers* 10:2459–90

49. Kwok LW, Shcherbakova I, Lamb JS, Park HY, Andresen K, et al. 2006. Concordant exploration of the kinetics of RNA folding from global and local perspectives. *J. Mol. Biol.* 355:282–93

50. Lake JA, Beeman WW. 1967. Yeast transfer RNA—a small-angle X-ray study. *Science* 156:1371–73

51. Lamb JS, Kwok L, Qiu X, Andresen K, Park H, Pollack L. 2008. Reconstructing three dimensional shape envelopes from time resolved small angle x-ray scattering data. *J. Appl. Crystallogr.* 41:1046–52

52. Li L, Pabit SS, Lamb JS, Park HY, Pollack L. 2008. Closing the lid on DNA end-to-end stacking interactions. *Appl. Phys. Lett.* 92:223901

53. Lipfert J, Doniach S. 2007. Small-angle X-ray scattering from RNA, proteins, and protein complexes. *Annu. Rev. Biophys. Biomol. Struct.* 36:307–27

29. Excellent review summarizing the properties of ions associated to RNA.

54. Liu J, Declais AC, Lilley DMJ. 2004. Electrostatic interactions and the folding of the four-way DNA junction: analysis by selective methyl phosphonate substitution. *J. Mol. Biol.* 343:851–64

55. Liu Y, Chen WR, Chen SH. 2005. Cluster formation in two-Yukawa fluids. *J. Chem. Phys.* 122:44507

56. Liu Y, Fratini E, Baglioni P, Chen WR, Chen SH. 2005. Effective long-range attraction between protein molecules in solutions studied by small angle neutron scattering. *Phys. Rev. Lett.* 95:118102

57. Lohman TM, Dehaseth PL, Record MT. 1980. Pentalysine-deoxyribonucleic acid interactions: a model for the general effects of ion concentrations on the interactions of proteins with nucleic-acids. *Biochemistry* 19:3522–30

58. Luzzati V, Masson F, Nicolaieff A. 1961. Structure of deoxyribonucleic acid in solution. Study by the diffusion of x-rays at small angles. *J. Mol. Biol.* 3:185–201 (In French)

59. Manning GS. 1969. Limiting laws and counterion condensation in polyelectrolyte solutions. I. Colligative properties. *J. Chem. Phys.* 51:924–33

60. Mills PA, Rashid A, James TL. 1992. Monte Carlo calculations of ion distributions surrounding the oligonucleotide d(ATATATATAT)$_2$ in the B, A, and wrinkled D conformations. *Biopolymers* 32:1491–501

61. Moghaddam S, Caliskan G, Chauhan S, Hyeon C, Briber RM, et al. 2009. Metal ion dependence of cooperative collapse transitions in RNA. *J. Mol. Biol.* 393:753–64

62. Nakata M, Zanchetta G, Chapman BD, Jones CD, Cross JO, et al. 2007. End-to-end stacking and liquid crystal condensation of 6- to 20-base pair DNA duplexes. *Science* 318:1276–79

63. Nossal R, Glinka CJ, Chen SH. 1986. SANS studies of concentrated protein solutions. I. Bovine serum-albumin. *Biopolymers* 25:1157–75

64. Pabit SA, Finkelstein KD, Pollack L. 2009. Using anomalous small angle X-ray scattering to probe the ion atmosphere around nucleic acids. *Methods Enzymol.* 469:391–410

65. Pabit SA, Meisburger SP, Li L, Blose JM, Jones CD, Pollack L. 2010. Counting ions around DNA with anomalous small angle X-ray scattering. *J. Am. Chem. Soc.* 132:16334–36

66. Pabit SA, Qiu XY, Lamb JS, Li L, Meisburger SP, Pollack L. 2009. Both helix topology and counterion distribution contribute to the more effective charge screening in dsRNA compared with dsDNA. *Nucleic Acids Res.* 37:3887–96

67. Pelta J, Livolant F, Sikorav JL. 1996. DNA aggregation induced by polyamines and cobalthexamine. *J. Biol. Chem.* 271:5656–62

68. Perez-Salas UA, Rangan P, Krueger S, Briber RM, Thirumalai D, Woodson SA. 2004. Compaction of a bacterial group I ribozyme coincides with the assembly of core helices. *Biochemistry* 43:1746–53

69. Plum GE, Bloomfield VA. 1988. Equilibrium dialysis study of binding of hexammine cobalt(III) to DNA. *Biopolymers* 27:1045–51

70. Pollack L, Doniach S. 2009. Time-resolved X-ray scattering and RNA folding. *Methods Enzymol.* 469:253–68

71. Putnam CD, Hammel M, Hura GL, Tainer JA. 2007. X-ray solution scattering (SAXS) combined with crystallography and computation: defining accurate macromolecular structures, conformations and assemblies in solution. *Q. Rev. Biophys.* 40:191–285

72. Qiu XY, Andresen K, Kwok LW, Lamb JS, Park HY, Pollack L. 2007. Inter-DNA attraction mediated by divalent counterions. *Phys. Rev. Lett.* 99:038104

73. Qiu XY, Andresen K, Lamb JS, Kwok LW, Pollack L. 2008. Abrupt transition from a free, repulsive to a condensed, attractive DNA phase, induced by multivalent polyamine cations. *Phys. Rev. Lett.* 101:228101

74. Qiu XY, Kwok LW, Park HY, Lamb JS, Andresen K, Pollack L. 2006. Measuring inter-DNA potentials in solution. *Phys. Rev. Lett.* 96:138101

75. Rambo RP, Tainer JA. Bridging the solution divide: comprehensive structural analyses of dynamic RNA, DNA, and protein assemblies by small-angle X-ray scattering. *Curr. Opin. Struct. Biol.* 20:128–37

76. Raspaud E, Durand D, Livolant F. 2005. Interhelical spacing in liquid crystalline spermine and spermidine-DNA precipitates. *Biophys. J.* 88:392–403

64. Recent resource describing experimental requirements for ASAXS and data interpretation.

66. First ASAXS study probing the distribution of monovalent and divalent ions around RNA.

72. First use of second viral coefficients to monitor repulsion or end-to-end stacking of short DNA duplexes.

77. Rau DC, Parsegian VA. 1992. Direct measurement of the intermolecular forces between counterion-condensed DNA double helices: evidence for long-range attractive hydration forces. *Biophys. J.* 61:246–59

78. **Roh JH, Guo L, Kilburn JD, Briber RM, Irving T, Woodson SA. 2010. Multistage collapse of a bacterial ribozyme observed by time-resolved small-angle X-ray scattering. *J. Am. Chem. Soc.* 132:10148–54**

79. Romer R, Hach R. 1975. Transfer-RNA conformation and magnesium binding: study of yeast phenylalanine-specific transfer-RNA by a fluorescent indicator and differential melting curves. *Eur. J. Biochem.* 55:271–84

80. Rouzina I, Bloomfield VA. 1996. Competitive electrostatic binding of charged ligands to polyelectrolytes: planar and cylindrical geometries. *J. Phys. Chem.* 100:4292–304

81. Rouzina I, Bloomfield VA. 1996. Influence of ligand spatial organization on competitive electrostatic binding to DNA. *J. Phys. Chem.* 100:4305–13

82. Rouzina I, Bloomfield VA. 1997. Competitive electrostatic binding of charged ligands to polyelectrolytes: practical approach using the nonlinear Poisson-Boltzmann equation. *Biophys. Chem.* 64:139–55

83. Russell R, Millett IS, Doniach S, Herschlag D. 2000. Small angle X-ray scattering reveals a compact intermediate in RNA folding. *Nat. Struct. Biol.* 7:367–70

84. **Russell R, Millettt IS, Tate MW, Kwok LW, Nakatani B, et al. 2002. Rapid compaction during RNA folding. *Proc. Natl. Acad. Sci. USA* 99:4266–71**

85. Saenger W. 1984. *Principles of Nucleic Acid Structure.* New York: Springer-Verlag

86. Schlatterer JC, Kwok LW, Lamb JS, Park HY, Andresen K, et al. 2008. Hinge bending: a barrier to RNA folding. *J. Mol. Biol.* 379:859–70

87. Semisotnov GV, Kihara H, Kotova NV, Kimura K, Amemiya Y, et al. 1996. Protein globularization during folding. A study by synchrotron small-angle X-ray scattering. *J. Mol. Biol.* 262:559–74

88. Sharp PA. 2009. The centrality of RNA. *Cell* 136:577–80

89. Shukla A, Mylonas E, Di Cola E, Finet S, Timmins P, et al. 2008. Absence of equilibrium cluster phase in concentrated lysozyme solutions. *Proc. Natl. Acad. Sci. USA* 105:5075–80

90. Skibinska L, Gapinski J, Liu H, Patkowski A, Fischer EW, Pecora R. 1999. Effect of electrostatic interactions on the structure and dynamics of a model polyelectrolyte. II. Intermolecular correlations. *J. Chem. Phys.* 110:1794–800

91. Stuhrmann HB. 1981. Anomalous small-angle scattering. *Q. Rev. Biophys.* 14:433–62

92. Svergun DI. 1999. Restoring low resolution structure of biological macromolecules from solution scattering using simulated annealing. *Biophys. J.* 76:2879–86

93. **Svergun DI, Koch MHJ. 2003. Small-angle scattering studies of biological macromolecules in solution. *Rep. Prog. Phys.* 66:1735–82**

94. Takamoto K, Das R, He Q, Doniach S, Brenowitz M, et al. 2004. Principles of RNA compaction: insights from the equilibrium folding pathway of the P4-P6 RNA domain in monovalent cations. *J. Mol. Biol.* 343:1195–206

95. Timasheff S, Witz J, Luzzati V. 1961. Structure of high molecular weight ribonucleic acid in solution: a small-angle X-ray scattering study. *Biophys. J.* 1:525–37

96. Walther D, Cohen FE, Doniach S. 2000. Reconstruction of low-resolution three-dimensional density maps from one-dimensional small-angle X-ray solution scattering data for biomolecules. *J. Appl. Crystallogr.* 33:350–63

97. Wang LX, Bloomfield VA. 1991. Small-angle X-ray-scattering of semidilute rodlike DNA solutions: polyelectrolyte behavior. *Macromolecules* 24:5791–95

98. Wensel TG, Meares CF, Vlachy V, Matthew JB. 1986. Distribution of ions around DNA, probed by energy transfer. *Proc. Natl. Acad. Sci. USA* 83:3267–71

99. Weyerich B, Daguanno B, Canessa E, Klein R. 1990. Structure and dynamics of suspensions of charged rod-like particles. *Faraday Dis.* 90:245–59

100. Widom J, Baldwin RL. 1980. Cation-induced toroidal condensation of DNA studies with $Co^{3+}(NH_3)_6$. *J. Mol. Biol.* 144:431–53

78. Recent time-resolved SAXS study reporting parallel pathways within milliseconds of the initiation of RNA folding.

84. First time-resolved SAXS study of RNA folding.

93. Excellent overview of SAXS applications to macromolecules.

101. Wong GC, Pollack L. 2010. Electrostatics of strongly charged biological polymers: ion-mediated interactions and self-organization in nucleic acids and proteins. *Annu. Rev. Phys. Chem.* 61:171–89

102. Woodson SA. 2005. Metal ions and RNA folding: a highly charged topic with a dynamic future. *Curr. Opin. Chem. Biol.* 9:104–9

103. Woodson SA. 2010. Compact intermediates in RNA folding. *Annu. Rev. Biophys.* 39:61–77

P-Type ATPases

Michael G. Palmgren[1] and Poul Nissen[2]

[1]Center for Membrane Pumps in Cells and Disease – PUMPKIN, Danish National Research Foundation, University of Copenhagen, DK-1871 Frederiksberg C, Denmark; email: palmgren@life.ku.dk

[2]Center for Membrane Pumps in Cells and Disease – PUMPKIN, Danish National Research Foundation, Aarhus University, DK-8000 Århus C, Denmark; email: pn@mb.au.dk

Annu. Rev. Biophys. 2011. 40:243–66

First published online as a Review in Advance on February 22, 2011

The *Annual Review of Biophysics* is online at biophys.annualreviews.org

This article's doi: 10.1146/annurev.biophys.093008.131331

Copyright © 2011 by Annual Reviews. All rights reserved

1936-122X/11/0609-0243$20.00

Keywords

cation pumps, lipid flippases, structure, regulation

Abstract

P-type ATPases form a large superfamily of cation and lipid pumps. They are remarkably simple with only a single catalytic subunit and carry out large domain motions during transport. The atomic structure of P-type ATPases in different conformations, together with ample mutagenesis evidence, has provided detailed insights into the pumping mechanism by these biological nanomachines. Phylogenetically, P-type ATPases are divided into five subfamilies, P1–P5. These subfamilies differ with respect to transported ligands and the way they are regulated.

Contents

INTRODUCTION

Phosphorylated intermediate-type ATPase (P-type ATPase): a family of membrane-embedded biological pumps with a single catalytic subunit that form a phosphorylated intermediate during each catalytic cycle

When analyzing the ATP hydrolytic activity of crab nerve membranes, Skou (134) observed that the activity "seems to fulfill a number of conditions that must be imposed on an enzyme which is thought to be involved in the active extrusion of sodium ions". Later it became apparent that the Na^+/K^+ pump he had described belongs to a large superfamily of integral membrane proteins now referred to as P-type ATPases (24)—a name derived from the fact that a transient phosphorylated intermediate is formed at a conserved Asp residue during catalysis.

The P-type ATPases constitute a large protein family that pump ions and lipids across cellular membranes. From detailed structural information on representative members we now know them as sophisticated biological nanomachines. Pumps of various subfamilies with different transport specificities are essential for almost all life forms, e.g., by interconverting metabolic energy into electrochemical gradients that can be harnessed for cellular uptake processes, by mediating cellular signaling, by providing metalloenzymes with their ligands, and by initiating vesicle budding as a basis for eukaryotic compartmentalization. The aim of this review is first to describe the basic principles of the general pump mechanism and then to provide a comprehensive overview of the P-type ATPase super family. Several recent reviews give a more detailed picture of individual subfamilies (17, 30, 33, 71, 73, 82, 87, 89, 149, 163).

DOMAIN STRUCTURE

The Ca^{2+}-ATPase SERCA1a [sarco (endo)plasmic reticulum calcium pump] of rabbit muscles was the first P-type ATPase to have its crystal structure determined (156), and today we have ample information on the structure of this pump in several conformations (97, 98, 138, 155, 157, 158). Later, structures of other P-type pumps, including the Na^+/K^+-ATPase (88, 133) and plasma membrane H^+-ATPase (106), have also emerged. Despite a low degree of sequence conservation, these pumps are strikingly similar and we now have a detailed picture of the general three-dimensional layout of P-type ATPases (19, 87, 97, 151, 152).

P-type ATPases have an elongated form, three times longer than they are wide, with one end embedded in the membrane and only a minor fraction exposed to the lumenal/extracellular side. The other end forms a large cytoplasmic headpiece. Two large cytoplasmic

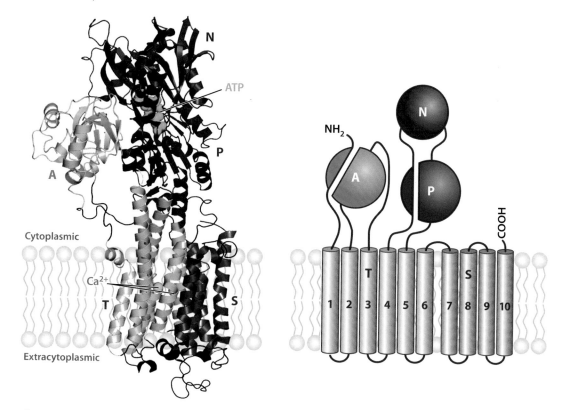

Figure 1

Schematic overview of the structural organization of P-type ATPases. Left: Ribbon model of the crystal structure of the sarco(endo)plasmic reticulum Ca^{2+}-ATPase SERCA2a (PDB accession number 1T5S; Reference 138). The pump is organized into three cytoplasmic domains (A, P, and N) and two membrane-embedded domains (T and S). The N-domain binds ATP and serves as a built-in protein kinase, which phosphorylates the P-domain. The A-domain is an intrinsic protein phosphatase, which dephosphorylates the P-domain once during each catalytic cycle. The transported ions (in this case Ca^{2+}) bind in the middle of the T-domain and alternate between being exposed to the cytoplasm and the extracytoplasmic side. Right: A schematic diagram of the same structure. The different domains are indicated with capital letters. The A-domain is formed by the N-terminal extension of the first cytoplasmic loop. The N-domain is an insertion into the P-domain, both of which are formed by the second and larger cytoplasmic loop. The membrane-embedded region consists of 10 transmembrane-spanning segments. The first six segments form the T-domain, whereas the S-domain is formed by the remaining four transmembrane helices. See text for further details. This figure is kindly provided by Dr. Maike Bublitz.

loops together with the N- and C-terminal ends form the cytoplasmic domains.

P-type ATPases contain five functional and structurally distinct domains as illustrated in **Figure 1**: three cytoplasmic domains (A, actuator; N, nucleotide binding; P, phosphorylation) and two membrane-embedded domains (T, transport; S, class specific support domain). A fourth domain is often present, the regulatory (R) domain, which is always situated at the N or C terminus (or both). During each catalytic cycle the P-domain is phosphorylated at an invariant Asp residue by the N-domain and subsequently dephosphorylated by the A-domain. In this manner the N-domain operates as a protein kinase, the A-domain acts as a protein phosphatase, and the P-domain acts as the substrate for both the protein kinase and phosphatase.

The Actuator Domain

The A-domain is a built-in protein phosphatase formed as a globular module connected to the

A-domain: actuator domain

N-domain: nucleotide binding domain

P-domain: phosphorylation domain

T-domain: transport domain

transmembrane region by two (P1-ATPases) or three (P2-P5-ATPases) rather long linker sequences attached to the transmembrane segments M1 (not in P1-ATPases), M2, and M3. The linkers are flexible (145) and allow for rotation of the A-domain. The A-domain contains a signature motif of P-type ATPases, Thr-Gly-Glu (TGE).

The Phosphorylation Domain

Transient phosphorylation of an invariant Asp is the hallmark of the P-type ATPase family. The phosphorylation site is situated in the P-domain in the middle of a flattened surface pointing away from the membrane. The P-domain is highly conserved among all P-type ATPases and contains no large insertions or deletions. The phosphorylated Asp is situated in the conserved sequence Asp-Lys-Thr-Gly (DKTG). Two other sequences in this domain, Thr-Gly-Asp-Asn (TGDN) and Gly-Asp-Gly-x-Asn-Asp (GDGXND), are involved in Mg^{2+} coordination associated with ATP binding at the phosphorylation site. The structure of the P-domain is homologous to that of members of the haloacid dehalogenase superfamily (6, 118), which are mainly soluble proteins that carry out hydrolysis.

The P-domain of several P-type pumps harbors a cytoplasmic K^+ site (22, 36, 124, 139). The nontransported K^+ ion coordinated by this site facilitates dephosphorylation by stabilizing the docking of the A-domain as it catalyzes the dephosphorylation reaction.

The Nucleotide Binding Domain

The N-domain performs ATP binding and phosphorylates the P-domain. It is a modular insertion into the P-domain (following its first beta-strand with the phosphorylation site) and is connected to it by a narrow hinge. In the N-domain residues that interact with ATP make contacts foremost with the adenosine part, which leaves the three phosphate groups protruding away from the binding site. Among the three cytoplasmic domains, the N-domain

is the most variable. It can carry large insertions and deletions. Nevertheless, the core structure of the N-domain is conserved (31, 52, 55, 56, 123).

The Transport Domain

The T-domain consists of six membrane-spanning segments and harbors the ion binding site(s), located halfway through the membrane. The T-domain is highly flexible and moves during the catalytic cycle as ions are allowed to associate and dissociate. Ions enter and leave the ion binding sites through half-channels formed by three membrane-spanning segments (M2, M4, and M6 in SERCA; 138, 156, 157). Two of the α-helical segments (M4 and M6 in SERCA) are unwound in the middle of the plane of the membrane. The result is an open space and the exposure of peptide bond carbonyl oxygens that are employed for ion coordination. Unwinding of the α-helix corresponding to M4 in SERCA is preceded by a Pro residue, which is conserved in all P-type ATPases (in SERCA in the sequence motif Pro-Glu-Gly-Leu; 81). This feature suggests that all P-type ATPases bind ions (or water) by essentially the same mechanism. The number of ion binding sites varies from one (P3- and P2B-ATPases) to two (P2A-ATPase) to three (P2C-ATPase). In the latter, the third site is formed mainly by the S-domain (see below).

The Support Domain

The S-domain is an auxiliary unit that provides structural support to the T-domain and can have specialized functions such as providing ion-coordinating side chains for additional ion binding sites in some P-type ATPases (Ca^{2+}- and Na^+/K^+-ATPases). Whereas the T-domain is flexible, the S-domain of P2- and P3-ATPases is more rigid and does not change much during the catalytic cycle (157). The membrane-spanning segments of the S-domain show little sequence conservation among the various P-type ATPase subfamilies. They can be inserted into the N-terminal region (P1B-ATPases), the C-terminal end (P2-, P3-, and

P4-ATPases), or both (P5-ATPases). In P1A-ATPases the S-domain consists of a single additional helix found at the C-terminal end.

The Regulatory Domain

In many, if not most, P-type ATPases one of the terminal ends act as an R-domain. It can be the N terminus (as in many heavy-metal pumps and plant calmodulin binding Ca^{2+} pumps), the C-terminus (Na^+/K^+-ATPase and animal calmodulin binding pumps), or both (plasma membrane H^+ pumps; 35). R-domains may act as autoinhibitors by interacting with the rest of the pump molecule, for example, to inhibit domain movements, as sensors for transported cations or as regulators of pump cation affinities.

Other Subunits

Some P-type ATPases operate by the catalytic subunit alone, others are dependent on additional subunits. An accessory subunit is commonly required for functional maturation and endoplasmic reticulum (ER) exit of the pump. In other cases, small subunits with extracytoplasmic mass interact with extracytoplasmic pump loops to stabilize high- or low-affinity conformations.

CATALYTIC MECHANISM

The structural conservation between pumps suggests a common mechanism for pumping for all members of the P-type ATPase family. Structures of the Ca^{2+}-ATPase in several different conformations can be accessed on the homepage of the Toyoshima laboratory (**http://www.iam.u-tokyo.ac.jp/StrBiol/resource/res.html**) and of the PUMPKIN Research Centre in Aarhus, Denmark (**http://www.pumpkin.au.dk/research/download-gallery//**). A mutation database for this pump is maintained by Jens Peter Andersen (**http://sercamutation.au.dk/**). As a result of combining the spatial information with detailed mutagenesis studies, we can now

propose a reasonable model for the whole catalytic cycle.

Skou (135) found that by adding radioactively labeled ADP to the Na^+/K^+ pump assay (with Mg^{2+} and ATP included), small amounts of radioactively labeled ATP could be obtained. He rationalized that the phosphate originating from the ATP-ADP exchange reaction had to come from the ATPase protein and concluded that "one may assume that an intermediate step in the breakdown of ATP is the formation of a phosphorylated enzyme in which the phosphate is bound" (135). Later it was shown that Na^+/K^+-ATPase preparations indeed incorporate radioactive phosphate from ATP in a Na^+- and K^+-dependent manner (111).

Subsequent work suggested the presence of two distinct forms of the phosphorylated Na^+/K^+-ATPase intermediate (37, 110). One form could be dephosphorylated by the addition of ADP (to release ATP) but was insensitive to K^+. The other form was sensitive to K^+ but insensitive to ADP. These two forms were named E_1-P ($Enzyme_1$-P) and E_2-P ($Enzyme_2$-P) (166). The unifying theory defined in the Post-Albers model (5, 110) proposes that the pump is phosphorylated in the E_1 form to produce E_1-P that spontaneously is converted to E_2-P and later is dephosphorylated to E_2 that subsequently reverts to E_1. Ion movement across the membrane occurs simultaneously with the E_1-P to E_2-P and with the E_2 to E_1 transitions. When the Na^+/K^+ pump was digested with proteases, in the presence of either Na^+ or K^+, different breakdown products were obtained (60). It was proposed that different pump conformations, depending on the cation present, exposed different surface loops to proteases. This discovery gave rise to a model in which the enzyme changes overall shape when it alternates between E_1 and E_2.

The classical theory for ion pumping by P-type ATPases proposes that ion binding sites present in the center of the membrane-embedded domain alternate between having high and low affinity. In the E_1 conformation, high-affinity sites are exposed to the cytoplasmic site, whereas in the E_2 conformation, the

E_1: a major conformation of P-type ATPases with ion binding site(s) facing the cytoplasm

E_2: the second major conformation of P-type ATPases with ion binding site(s) facing the extracytoplasmic side of the membrane

Figure 2

Schematic overview of P-type ATPases with transport and countertransport. E_1 states are blue, with phosphorylation stimulated by binding of the light-blue ligand. E_2 states are burgundy, with dephosphorylation stimulated by binding of the orange ligand. Exchange of ligands takes place through half-channels toward the cytoplasmic side (E_1-ATP) or the extracellular/luminal side (E_2-P).

same sites are converted into low-affinity sites exposed to the other side of the membrane. In this manner, the ion is not pushed through the membrane but, rather, at a fixed position alternates by being present at either side of the membrane (**Figure 2**). Structural studies have now shown that this model is correct. The detailed mechanism of pumping by P-type ATPases (exemplified by the SERCA pump) has been described in a number of recent authoritative reviews (82, 87, 154) and the reaction cycle as depicted in **Figure 2** is explained only briefly here.

Binding of ion(s) to the ion binding site induces a helix rearrangement that is transmitted to the P-domain and generates a Mg^{2+} site near the Asp. Owing to electrostatic repulsion from the negatively charged γ-phosphate of ATP, the conserved Asp can be phosphorylated only if Mg^{2+} is bound at this site, and Mg^{2+} is firmly bound to the P-domain only if the transmembrane ion binding sites are occupied. In this way, tight coupling is obtained between ion binding and phosphorylation of the pump.

Phosphorylation of the P-domain breaks the ATP-mediated linkage to the N-domain and causes the domain to bend forward in a manner that results in stretching of the ropelike linker sequence between M3 and the A-domain. This creates a tension in the pump, which is the driving force for the subsequent dephosphorylation reaction. As a result, the A-domain starts to move, carries out a stunning 90° rotation, and slides onto the top of the bended P-domain. The rotation has two consequences: (*a*) It destroys the high-affinity ion binding sites in the

T-domain and opens an ion exit channel on the other side of the membrane; and (*b*) it positions the TGE signature motif of the A-domain in close proximity to the phosphorylated Asp. The conformational changes associated with the rotation of the A-domain result in transition of the E_1-P form of the pump to the E_2-P form. This is the slowest part of the catalytic cycle and the rate-limiting step.

When the ions have diffused away on the other side of the membrane, the pump has to become dephosphorylated in order to proceed further. Binding of counterions to the E_2-P conformation causes the exit pathway of the T-domain to close. This movement is transmitted to the P-domain, where it causes a slight further rotation of the A-domain. This shifts the TGE motif toward the phosphorylated Asp at the P-domain. A water molecule can now be co-ordinated between the Glu residue of the TGE motif and the phosphorylated Asp, and the water is allowed to carry out a nucleophilic attack on the phosphate bond. In this way, counterion binding and occlusion stimulate dephosphorylation of the pump, and ATP binding chases the liberated inorganic phosphate out.

The dephosphorylated pump in the E_2 conformation has the counterions occluded in the center of the T-domain. As a result of dephosphorylation and phosphate release, the A-domain rotates away from the P-domain, which bends upward again, and the pump returns to a relaxed E_1 conformation. This is associated with helical movements in the T-domain that destroy the high-affinity counterion binding sites and open an exit channel for these ions at the cytoplasmic side of the membrane.

PHYLOGENY OF P-TYPE ATPases

Even though P-type ATPases are divergent with respect to primary structure, individual P-type ATPase sequences can be recognized in genomes by more or less conserved core sequences and signature motifs that contribute to the common pump catalytic machinery as discussed above (9).

Phylogenetic analysis of available sequences in completed genomes has shown that the superfamily of P-type ATPase can be divided into five distinct subfamilies (P1–P5), each of which can be further divided into subgroups (A, B, etc.) (9). The most important groups are the following: P1A-ATPases are part of bacterial K^+ transport systems, P1B-ATPases are heavy-metal pumps, P2A-ATPases and P2B-ATPases are assumed Ca^{2+} pumps, P2C-ATPases are Na^+/K^+ and H^+/K^+ pumps of animals, P2D-ATPases are Na^+ pumps of fungi, P3A-ATPases are plasma membrane H^+ pumps, and P4-ATPases are putative lipid flippases. P5-ATPase pumps have so far no assigned specificity.

Below we outline the general structural organization and catalytic mechanism of P-type ATPases. However, there is great individual variation between the different subfamilies and the main structural differences relating to the terminal R-domains and accessory subunits (**Figure 3**). We therefore focus on special characteristics of representative members of the various subfamilies.

SPECIFIC FEATURES AMONG P-TYPE ATPase SUBFAMILIES

P1A-ATPases: The KdpFABC Complex

A most peculiar P-type ATPase is the bacterial KdpB-ATPase, which is only found in about one-third of eubacterial genomes analyzed so far and in a minor number of archaeal genomes. It was first identified in a mutant screen as part of an operon involved in high-affinity uptake of K^+ (65). The KdpB-ATPase is a highly unusual ATPase for three reasons: (*a*) With a molecular

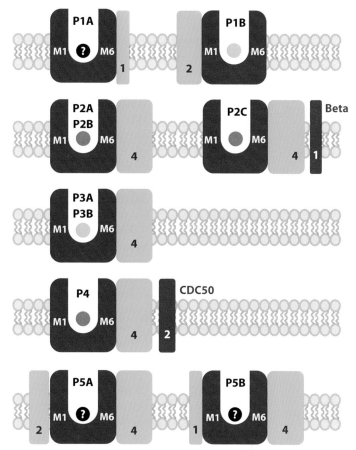

Figure 3

Schematic overview of the P-type ATPase family. The core structure (including the cytoplasmic domains N, P, and A and the transmembrane helices M1 through M6) is shown in burgundy, facing outward in the E_2-P conformation (see also **Figure 2**). Additional S-domains (with the number of transmembrane segments indicated) are green, and subunits (beta for P2C, CDC50 for P4) are magenta. Several P-type ATPases also display autoregulation by N- and/or C-terminal R-domains (not included, but see text for details). Ligands have been included for pumps with identified countertransport (*orange sphere*) or only transport (*blue sphere*). The question mark indicates that the transported ligand is not identified.

mass of 72 kDa, it is the smallest P-type ATPase known; (*b*) Because KdpB is associated in a complex (KdpFABC) with three additional subunits (KdpF, KdpA, KdpC), among which KdpA and KdpC are encoded by the same operon as KdpB, it has a larger number of subunits than any other P-type ATPase; and (*c*) it does not appear to pump any ligand, as K^+ seems to move through KdpA (see below).

Contrary to most P-type ATPases, KdpB has an odd number of transmembrane segments (seven), with an additional transmembrane segment placed after the T-domain. The N-domain is reduced in size and has little sequence homology with other P-type ATPases. Nevertheless, the basic fold of the N-domain of KdpB is close to that of other P-type pumps (55) and shows highest similarity to that of P1B-ATPases (31).

Although KdpB itself has all the signature motifs of a P-type ATPase, all mutations affecting K^+ specificity locate to the 59-kDa KdpA subunit (15, 127). Could it be that KdpB serves as a motor for K^+ transport through KdpA? If so, KdpB may represent a primitive P-type ATPase that has not yet fused with its membrane-bound carrier entity.

There are some problems with this model, however. First, the KdpB-ATPase comprises all six helices of the ubiquitous T-domain. Conserved T-domain features include a conserved Pro in M4 (in the sequence LIPTTIG), which in other P-type ATPases is involved in helix breaking in order to make backbone carbonyl oxygens available for ligand coordination in the ion binding cavity (156). Second, mutations of two charged residues in the T-domain, Asp583 and Lys586, completely abolish the pumping activity of the combined complex (12, 16). Third, KdpA does not resemble the T-domain or S-domain of other P-type ATPases but rather is reminiscent of a K^+ channel (34).

According to an alternative model (101), the KdpB-ATPase pumps an unknown cation out of the cell, in this way creating a local electrochemical gradient that is harvested by a channel protein (KdpA) carrying out K^+ uptake. In support of this model, the reconstituted Kdp complex generates a current in the absence of K^+ (39). In the P3-type plasma membrane H^+-ATPase, the transmembrane proton acceptor/donor is a negatively charged Asp, the pK_a of which is controlled by a nearby positively charged Arg (106). Given that in the putative M5 of the membrane domain of KdpB an Asp (Asp583) and a positively charged Lys (Lys586)

are essential for catalysis (12, 16), it is therefore tempting to speculate that KdpB is a proton pump (101). However, in the presence of K^+ and ATP, the reconstituted Kdp complex generates a negative potential inside proteoliposomes (38). This is to be expected if a positive charge is transported from the lumen of the vesicles outward and argues against an electroneutral exchange of K^+ with another cation. In support of the former model, the Asp-Lys pair might rather act as a strong ion pair that indirectly governs K^+ transport through the KdpA subunit (12).

The roles of the two additional subunits, 3-kDa KdpF and 20-kDa KdpC, still have not been fully elucidated. KdpF has a stabilizing effect on the complex (44). In this manner it resembles the β-subunit of P2C- and P4-ATPases. KdpF is not essential for pump function (44), and in archaeal genomes encoding other members of the Kdp-ATPase complex, KdpF is often absent. KdpC, which is an essential part of the KdpFABC complex (43), has only a single transmembrane span with a hydrophilic extension into the cytoplasm. In this manner, KdpC shows resemblance to small membrane proteins regulating P2A- and P2C-ATPases such as phospholamban and phospholemma. KdpC binds ATP with low affinity, possibly via its ribose moiety, and might be involved in regulating the KdpFABC enzyme complex (3).

P1B-ATPases: HEAVY-METAL PUMPS

Heavy-metal pumps (P1B-ATPases) are omnipresent in all life forms and are the most common P-type ATPases in bacteria and archaea. Phylogenetically, the P1B subfamily is divided into monovalent (Cu^+, Ag^+) and divalent (Cu^{2+}, Zn^{2+}, Co^{2+}, Pb^{2+}, Cd^{2+}) metal pumps (9, 75, 116). Well-characterized heavy-metal pumps include the bacterial CadA (91), ZntA (117), and CopA (93), pumping Cd^{2+}, Zn^{2+}, and Cu^+, respectively; the plant HMA4 Zn^{2+} pump (57); and the human ATP7A and ATP7B Cu^+ pumps (147, 164). Cu^+ pumps appear to be the most abundant heavy-metal pumps in nature (7).

P1B-ATPases serve two purposes. They provide heavy metals (most commonly Cu and Zn) required for maturation of metalloproteins, which in prokaryotes typically occur in the periplasmic space and in eukaryotes occur in the lumen of secretory vesicles. They extrude toxic heavy metals across the plasma membrane (both Cu and Zn can be harmful at elevated concentrations).

The N-domain of P1B-ATPases has a low degree of amino acid similarity to other pumps except for the bacterial P1A-ATPase KdpB, underlining the close relationship between P1A- and P1B-ATPases (159). At the primary structure level, the N-domain in P1B-ATPases is special in that the conserved residues that make contacts to ATP (55) are different from those in other P-type ATPases including KdpB (31, 122, 159). A similar discrepancy between conservation of primary and tertiary structures is seen for the A-domain of P1B-ATPases, which has low sequence homology to other P-type ATPases but has a three-dimensional structure that very much resembles other pumps (123).

P1B-ATPases have peculiar terminal extensions rich in Cys and sometimes also His residues. These residues coordinate metals well, and indeed the terminal domains of P1B-ATPases bind metals with high affinity. Such metal binding domains (MBDs) are found at either the N terminus (in bacteria, fungi, plants, and animals) or the C terminus (in plants). P1B-ATPases transport heavy metals even in the absence of their MBDs (10, 49), arguing against an essential role of these domains in metal transport.

MBDs typically have several metal binding sites and therefore are ideally suited as metal sensors. A Cd pump in yeast, Pca1p, is a short-lived protein with a half-life of less than 5 min owing to a sequence in its N-terminal MBD region that serves as a degradation signal for ubiquitin-mediated protein degradation in the proteasome (1, 2). Binding of Cd to the N-terminal MBDs apparently masks this signal, probably as an effect of a more compact folding, and as a result Pca1 can escape degradation. Sensing of Cd occurs in the ER, and once bound by the MBDs, the pump can escape the ER and is secreted to the plasma membrane as a stable long-living pump. In humans, a related system linked to the N-terminal MBDs, but apparently not involving proteolytic degradation of the protein, redirects P1B-ATPases from internal membranes to the plasma membrane in response to an increased amount of Cu in the cell (107, 162). These results suggests that a role of MBD metal sensors in response to changing heavy-metal concentrations is to regulate the amount of pump protein at the various cellular locations, i.e., allowing the pump protein to switch between a predominant role in metalloprotein maturation to one where it is responsible for heavy-metal detoxification.

A peculiar feature with P1B-ATPase is their low turnover rates when expressed in heterologous systems. The specific ATPase activity of human Cu-ATPase expressed in yeast (40) is at least two orders of magnitude lower than that of SERCA pumps, and that of bacterial CopA is an order of magnitude lower. This fact has slowed down the biochemical characterization of these pumps and could indicate that something is lacking in the heterologous systems typically employed for characterization of these pumps, i.e., an accessory subunit, a specific metallochaperone, or specific lipids.

P2A-ATPases: Sarco(endo)plasmic Reticulum Ca^{2+}-ATPase (Ca^{2+}/H$^+$-ATPase)

P2A-ATPases are common in bacteria, where they most likely have evolved as a mechanism for exporting Ca^{2+} against steep transmembrane gradients (29). During muscle contraction in animals, Ca^{2+} is released from the sarcoplasmic reticulum of muscle cells to establish cross-links between actin and myosin. In order for muscles to relax, Ca^{2+} has to be removed from the environment of the myofibrils and pumped back into the sarcoplasmic reticulum, which is the main function of P2A-ATPases in animal cells (14). P2A-ATPases are also found in the ER, where they have additional functions, and in animals, these pumps are named sarco(endo)plasmic reticulum calcium (SERCA) pumps. The P2A Ca^{2+}-ATPase exports two Ca^{2+} per ATP hydrolyzed and countertransports H^+. The pump is slightly electrogenic, which would suggest that fewer than four H^+ are released into the cytoplasm per two Ca^{2+} pumped into the sarco(endo)plasmic reticulum lumen (13, 90).

Three genes encode SERCA pumps in mammals (SERCA1–3 or ATP2A1–3), but the presence of tissue-dependent splice variants raises the number of possible SERCA pumps to 11 (50). The SERCA1a splice variant is dominant in adult fast-twitch skeletal muscles. Due to its high abundance here (up to 90% of total sarcoplasmic reticulum protein), SERCA1a has since long been a favorite P-type ATPase for biochemical and structural studies. The SERCA2 pump is widely distributed in cells other than muscle cells and is the best studied P2A Ca^{2+}-ATPase with respect to regulation. SERCA2 exists in two splice variants, SERCA2a and SERCA2b (51).

A number of small transmembrane-bound proteins bind to and regulate the activity of P2A-ATPases. Phospholamban (a 55-amino-acid residue polypeptide) regulates the Ca^{2+} affinity of SERCA1a and SERCA2a apparently by interfering with the movement of M2, which negatively affects the Ca^{2+} entry pathway to the pump (63, 142, 144, 153). Phospholamban inhibition is relieved by phosphorylation of phospholamban. Sarcolipin (35 amino acid residues) is a phospholamban homolog that inhibits SERCA1a and SERCA2a in skeletal muscles (8, 92, 165). According to modeling studies, phospholamban and sarcolipin could bind to the same transmembrane binding groove with their lumenal extensions interacting with SERCA lumenal loops (8, 153), although the fact that sarcolipin and phospholamban apparently can bind at the same time argues against this proposal (83). Phospholamban and sarcolipin both bind and stabilize SERCA in its E_2 conformational state, which has low affinity for Ca^{2+} (8, 26, 83).

Compared with the SERCA2a pump, in the SERCA2b pump, the four C-terminal amino acid residues are replaced by 49 mostly hydrophobic residues. This creates an extra transmembrane segment (the eleventh) in the SERCA2b pump and moves the C terminus from the cytoplasmic side to the lumenal side of the membrane (11, 23). Several observations suggest that the additional transmembrane segment of SERCA2b represents a unique R-domain. The eleventh transmembrane segment of SERCA2b appears to dock between M5 and M10, and the three C-terminal residues interact with four lumenal loops of the pump (160). As a result of this intramolecular interaction, the conformational equilibrium of the pump is stabilized toward the Ca^{2+} binding E_1 conformation, which imposes a high, apparent Ca^{2+} affinity (160). Calnexin, a chaperone protein that assists in protein folding in the ER, interacts specifically with SERCA2b at the lumenal C terminus, keeping the pump in an inhibited state (119). When cytosolic Ca^{2+} is increased, calnexin dissociates from SERCA2b as a result of phosphorylation by a protein kinase. This relieves the constraint on SERCA2b and allows the pump to refill the stores in the ER. Interaction between SERCA2b and other lumenal ER chaperones might further regulate Ca^{2+}-filling of empty stores.

A number of specific SERCA inhibitors have been isolated from natural resources. By far the most potent inhibitor is thapsigargin, which is isolated from a rare Mediterranean umbelliferae, *Thapsia garganica*. Thapsigargin increases cytoplasmic concentrations of Ca^{2+} and ultimately causes apoptosis as a result of SERCA inhibition (148). Cyclopiazonic acid is produced by species of the fungi *Aspergillus* and *Penicillum* and inhibits SERCA by intercalating with a divalent ion into the Ca^{2+} entrance pathway (66).

P2B-ATPases: Calmodulin Binding Ca^{2+}-ATPase

Ca^{2+} is an important cytoplasmic messenger that rises in concentration from the nanomolar to the micromolar range in response to extra-cellular stimuli. Calcium that flows into the cell during such events in order to resensitize the cell has to be pumped out again. Calmodulin binding Ca^{2+} pumps (P2B-ATPases) carry out this job and establish a steep steady state Ca^{2+} gradient of $\approx$ 10,000-fold across the plasma membrane of eukaryotic cells (18, 126).

Four features distinguish P2B- and P2A-ATPases from each other: (*a*) P2B Ca^{2+}-ATPases bind calmodulin at a long terminal extension that functions as a pump autoinhibitor. In animal cells, these extensions are in the C-terminal end, whereas in plant cells they localize to the N terminus (53, 76). Binding of calmodulin neutralizes the constraint imposed by the terminal domain and results in an activation of Ca^{2+} pumping. (*b*) Depending on the P2B-ATPases in question, they can reside in almost every membrane of the cell. In animal cells, they are only sitting in the plasma membrane, but in plant and fungal cells, they localize further to the vacuolar membrane, the plastid envelope, and the ER. P2B-ATPases are often referred to as plasma membrane Ca^{2+}-ATPases (PMCA), but this name is unfortunate owing to their wide intracellular distribution in organisms other than animals. (*c*) P2B-ATPases have only a single Ca^{2+} binding site in the T-domain (corresponding to the P2A-ATPases site II located on M4), whereas the space corresponding to site I of P2A-ATPases in P2B-ATPases apparently is occupied by a built-in cation, Arg. P2B-ATPases are partly electrogenic, as it seems that only a single H^+ is countertransported for each Ca^{2+} pumped out (90). (*d*) The steady-state level of phosphorylation in P2B-ATPases is increased by La^{3+}, whereas in other P-type ATPases, including P2A pumps, La^{3+} decreases steady-state phosphorylation levels.

In humans, there are four isoforms of P2B-ATPases, or plasma membrane Ca^{2+} pump 1–4 (PMCA1–4 or ATP2B1–4). For each isoform a number of mRNA splice variants exist, which increases the number of potential gene products to more than 20. The physiological function of each splice variant is not known in detail, but most likely the variants are important for adjust-ing the pumps to the specific needs of individual cell types or extracellular signals. It is striking that most splice events do not affect residues involved in the basic catalytic machinery but rather result in variations of the C-terminal regulatory domains.

The calmodulin binding R-domain regulates the speed of P2B-ATPase catalysis and the affinity of Ca^{2+} binding. For example, the apparent Ca^{2+} affinity of the autoinhibited erythrocyte pump has a K_d of $\sim$30 μM, whereas calmodulin binding to the R-domain causes the apparent affinity to rise to a K_d of $\sim$0.5 μM (125). The affinity for calmodulin depends on the P2B-ATPase isoform or splice variant. In cellular Ca^{2+}-signaling, the combined kinetics of Ca^{2+} influx and efflux determine the magnitude and length of Ca^{2+} spikes (or transients), which are believed to contain information about the nature of the incoming signal. By representing the efflux component, individual P2B-ATPase variants might therefore be adapted to transmit specific cellular signals.

A large number of molecules interact with P2B-ATPases (18). Most of these bind to the terminal R-domain. Calmodulin has already been mentioned. For animal P2B-ATPases, phospholipids [primarily phosphatidylserine and phosphatidylinositol bisphosphate (PIP2)] and polyunsaturated fatty acids also bind to the R-domain and cause an increased Ca^{2+} affinity of the pump. Phosphorylation of the R-domain by protein kinase A downstream of the calmodulin binding region increases V_{max} and the Ca^{2+} affinity. Further, the blood protease calpain cleaves the R-domain just upstream of the calmodulin binding domain to separate it irreversibly from the pump. A number of other proteins bind to the animal P2B pumps via a PDZ (PSD-95, discs-large, and Z0–1 proteins) domain present in the R-domain of the pump, including members of the MAGUK (membrane-associated guanylate kinase, SAP) family of kinases. The physiological significance of these interactions is not clear.

Another group of Ca^{2+} pumps are the secretory pathway Ca^{2+} pumps (SPCA). Besides Ca^{2+}, the secretory pathway pump in Golgi membranes transports Mn^{2+} (77), which is used as an essential cofactor of many enzymes in the lumen of Golgi compartments. Countertransport has not been demonstrated in these pumps (32).

P2C-ATPases: The Sodium-Potassium Pump

The history of the discovery of the sodium pump is described by Skou (136) and Glynn (48). The Na^+/K^+-ATPase extrudes three Na^+ and imports two K^+ for every ATP split (129). As a result, it establishes chemical gradients of Na^+ and K^+; further, due to the imbalance in cation exchange and the activity of ion-specific channels, an electrical gradient across the plasma membrane is generated. The combined electrochemical gradient is essential for a large number of vital processes in animal cells, such as action potential formation, secondary transport across the plasma membrane, signaling, and volume regulation (61).

The Na^+/K^+-ATPase also consists of a β-subunit with a single transmembrane span and most of its mass exposed to the extracellular side of the membrane. This subunit is essential for the newly synthesized α-subunit to exit the ER and also influences the K^+ affinity of the pump (74). The $\alpha\beta$-complex in some cell types interacts with members of the FXYD family, which are small hydrophobic polypeptides with a single transmembrane span, e.g., the γ-subunit in kidney outer medulla (47), that appear to regulate the cytoplasmic Na^+ affinity of the pump. In humans, four different α-subunit isoforms (70), three β-subunit isoforms (46), and at least five FXYD γ-subunits are present (47). This isoform diversity adds to the complexity of Na^+/K^+-ATPase regulation.

In the crystal structures of the Na^+/K^+-ATPase, the two K^+ ions are coordinated in the T-domain in the E_2-P_i-like state at positions equivalent to those of the two Ca^{2+} ions bound by the SERCA Ca^{2+}-ATPase in the E_1-P-like states (88, 133). Only two conserved amino acid residues differ between the K^+ and Ca^{2+} binding sites of these pumps. A possible explanation

for the observed strict ion specificities of each pump is that these are dictated by only minor differences in the positioning of side chains that could originate from, e.g., a differential degree of tilting or twisting of helices (9). The inhibitor ouabain inserts from the extracellular side deep in between two transmembrane segments of the T-domain in such a way that it closes the K^+ entrance pathway (95, 169).

In order to bind Na^+, the K^+-coordinating residues are anticipated to rearrange to accommodate binding of two of the three Na^+ ions (87, 96). In the conformation with bound K^+, an H^+ from the cytoplasm is likely to be bound in place of the third Na^+ and is subsequently released again to the cytoplasm in exchange for Na^+ (112). Mutagenesis and modeling studies suggest that the third Na^+ site is formed mainly by residues in the S-domain (from transmembrane helices M8 and M9) and in the T-domain (from transmembrane helix M5) (69, 96). A central Na^+-coordinating residue at site III appears to be an Asp in M8 (Asp930 in $\alpha2$ and Asp923 in $\alpha3$) (112) that, when mutated in vivo, causes the neurological disorder rapid-onset dystonia parkinsonism (170).

The short C-terminal extension of the Na^+/K^+-ATPase does not extrude into the cytoplasm but rather is inserted into the membrane domain (88) and plugs a polar pathway leading to the Asp residue of site III (112). Deletion of the C terminus causes a strong reduction in general Na^+ affinity of the pump (88, 150), whereas removal of only the two C-terminal Tyr-Tyr residues specifically affects site III of the Na^+/K^+-ATPase (78, 112). Molecular dynamics simulations support a model in which the C terminus serves as a plug that blocks the hydrophilic entrance pathway leading to the third Na^+ site (112).

The gastric H^+/K^+-ATPase is closely related to the Na^+/K^+-ATPase and is only mentioned here. It is an electroneutral pump secreting large amounts of acid into the gastric lumen (120, 132). Like the Na^+/K^+-ATPase, it associates with a β-subunit. In this pump, the three Na^+ sites are replaced by two H^+ sites. Amino acid residues in the two pumps that are important for cation pumping are essentially the same in both pumps, suggesting that Na^+ and H^+ are transported by a similar mechanism. The space for the third Na^+ site in the Na^+/K^+-ATPase is apparently occupied by an Arg in the H^+/K^+-ATPase, which might serve as a built-in and nontransported cation (112).

P3-ATPases: The Plasma Membrane H^+-ATPase

The plasma membrane of fungal and plant cells are energized by an electrochemical gradient of H^+, acidic on the extracellular side, and in these organisms secondary active transport systems as a rule are H^+-coupled cotransport systems. When the plasma membrane H^+-ATPase was cloned, first from the yeast *Saccharomyces cerevisiae* (131) and later from the model plant *Arabidopsis thaliana* (54, 104), it became evident that it was a P-type ATPase related to the Na^+/K^+-ATPase, which is absent from plant and fungal cells.

The plasma membrane H^+-ATPase is unique in several aspects. (*a*) It is a powerful electrogenic pump, and it does not seem to countertransport any ion to neutralize the electrogenic effect of H^+ export. (*b*) Only a single charge (an H^+) is transported per ATP split. As a result, the energy of ATP hydrolysis is converted into transporting a single H^+ against the electrochemical gradient, which makes it possible to establish steep gradients. (*c*) Plasma membrane H^+-ATPase has an extended C-terminal region that serves as an autoinhibitory regulatory domain. In this way it resembles calmodulin binding P2B Ca^{2+}-ATPases, but instead of binding calmodulin it interacts with other effector proteins, most notably the 14-3-3 protein.

The crystal structure of a plant plasma membrane H^+-ATPase (*A. thaliana* AHA2) devoid of its C-terminal regulatory domain has been determined at 3.6 Å resolution (106). The structure is similar to that of the Ca^{2+}-ATPase and the Na^+/K^+-ATPase, which would suggest that the basic transport mechanism is the same for these three pumps.

As a unique feature, the T-domain of the plasma membrane H^+-ATPase has a central cavity, presumably filled with water in the E_1 state. Two conserved charged residues from membrane segments are located around the lumen of this cavity: an aspartate residue (Asp684 in AHA2), which is essential for H^+ transport, and an arginine residue (Arg655 in AHA2), which is important but not essential for transport (20). The aspartate is proposed to be the central H^+ acceptor/donor and that in the protonated form it interacts with an equally conserved Asn106 residue (21, 106). As the arginine moves relative to the aspartate during conformational changes, it alters the pK_a of the aspartate and in this way facilitates either binding or release of the H^+ to the acceptor/donor group. Following release of the H^+ on the extracellular side, the positively charged arginine might serve as a gatekeeper restricting the H^+ from moving backward (106) and as a built-in counterion to promote dephosphorylation.

H^+ pumping by the plasma membrane H^+-ATPase is activated when the hydrophilic C terminus is either removed at the gene level (109) or cleaved off by proteases (102, 103). In vivo, the C-terminal constraint on pump activity is relieved reversibly in response to environmental signals. Addition of glucose to the growth medium of the yeast *S. cerevisiae* causes a rapid increase in H^+ secretion resulting from activation of the plasma membrane H^+-ATPase (130) via phosphorylation of its C-terminal R-domain (67). In plants, blue light activates the plasma membrane H^+-ATPase of stomatal guard cells via phosphorylation of its C-terminal R-domain (62). The resulting electrochemical gradient of H^+ causes the guard cells to take up ions and water, and the resulting swelling leads to opening of the stomatal pore (79), which allows plants to breathe carbon dioxide from the atmosphere. Several other environmental signals result in activation/inactivation of the plant plasma membrane H^+-ATPase via its R-domain (33, 45).

Many proteins interact with the plasma membrane H^+-ATPase and the preferred interaction site is the R-domain. The most prominent protein that binds to the R-domain of plant plasma membrane H$^+$-ATPase is 14-3-3 protein (58, 94, 99), which was first characterized as an abundant protein in animal brain tissue (4). The 14-3-3 protein binding site in the R-domain is created by protein kinase-mediated phosphorylation of the penultimate C-terminal residue, Thr947, in AHA2 (41, 143). Following phosphorylation or binding of the fungal toxin fusicoccin, 14-3-3 protein binds strongly to the extreme C terminus and causes activation of the pump, probably by inhibiting interaction of the R-domain with its receptor site in the H$^+$-ATPase pump molecule (167, 100).

P4-ATPases: Putative Phospholipid Flippases

P4-ATPases are phylogenetically diverse from other P-type ATPases and, for this reason, were first characterized as an ancient group of pumps (146). This does not hold true, however, as they have not been identified in a single prokaryote so far, whereas they are ubiquitous in eukaryotes (9). In most eukaryotes, P4-ATPases are by far the largest P-type ATPase subfamily. In humans, 14 genes encode P4-ATPases in contrast to only 4 genes encoding Na$^+$/K$^+$-ATPase (105). Current models propose two important functions for the P4-ATPase complexes with their β-subunit: First, they establish phospholipid asymmetry in biological membranes by promoting phospholipid flipping across the membrane bilayer. Second, they assist in budding of transport vesicles in the endocytic and secretory pathways.

Erythrocytes rapidly internalize exogenously added labeled phosphatidylserine and phosphatidylethanolamine to generate phospholipid asymmetry in favor of the inner leaflet. This flippase activity is strictly ATP dependent and inhibited by the P-type ATPase inhibitor vanadate (128). A similar aminophospholipid flippase activity (ATPase II, ATP8B1) was partially purified from adrenal chromaffin granules (84, 86), clathrin-coated vesicles (168), and

human erythrocytes (86) and hydrolyzed ATP in a phosphatidylserine-dependent manner (85, 86, 168). Peptides derived from a partially purified ATPase II preparation showed sequence homology to a yeast gene, *DRS2*, encoding a P-type ATPase (146). *drs2* mutant cells were deficient in uptake of labeled phosphatidylserine from the outer leaflet to the inner leaflet of the plasma membrane (146), adding credibility to the assumption that the sequenced peptides originated from a flippase protein. Bioinformatic analysis of all available P-type ATPase sequences revealed that Drs2p belongs to a distinct eukaryotic subfamily of P-type ATPases called P4 ATPase (9).

Graham and coworkers (25) identified yeast *DRS2* in an independent screen for null alleles that are synthetic lethal in the absence of ADP-ribosylation factor (*ARF1*) or clathrin heavy chain (*CHC1*), both of which are involved in budding of secretory vesicles from the Golgi stacks. Drs2p protein localizes to the Golgi complex, where Arf1p and Chc1p are localized as well (25). Clathrin-coated vesicles will not form in *drs2* cells, and as a result *drs2* and *chc1* null mutations have similar phenotypes (42). Taken together, these results suggest that Drs2p has an important role in forming vesicles in the secretory pathway. Among the four P4-ATPases in yeast, Dnf1p and Dnf2p play a role in another vesicle-producing machinery, the endocytic pathway (108). Additional strong evidence for the involvement of P4-ATPases in vesicle budding came from the study of *A. thaliana* plants carrying a knockout in the P4-ATPase gene *ALA3*. This gene is expressed in slime-producing cells of the root tip, and formation of slime vesicles at the Golgi complex is completely blocked in *ala3* plants (113).

Genetic studies by Tanaka and coworkers (121) showed that mutations in a yeast gene, *CDC50* (cold-sensitive cell-division-cycle 50), which is completely unrelated to *DRS2*, cause the same phenotypes as mutating *DRS2*, and that the *CDC50* gene product colocalizes with and binds to Drs2p in the trans-Golgi network. Further, Cdc50p is required for Drs2p in order for the pump to leave the ER after its biosynthesis. Cdc50p is part of a small gene family that in yeast comprises three members. Genes related to Cdc50 form small families in all eukaryotes, and accumulating evidence suggests that they all function as P4-ATPase β-subunits required for ER exit of the pumps.

In the higher plant *A. thaliana*, ALA proteins (plant P4-ATPases) with different phospholipid specificity retain their specificity no matter which ALIS protein (equivalents to yeast Cdc50p) they are associated with (72). This suggests that ALIS proteins do not establish a specific interaction with the transported phospholipid. Like Cdc50p, ALIS proteins appear to promote the exit of P4-ATPase proteins from the ER (72). Reconstitution experiments with purified Drs2p including Cdc50p in substoichiometric amounts have revealed that the complex promotes flipping of fluorescently labeled phosphatidylserine molecules but not phosphatidylcholine (171). Cdc50p preferentially associates with Drs2p in the E_2 conformation (like phospholamban and sarcolipin to SERCA) (68).

P5-ATPases: Orphan Transporters

As genome sequences began to accumulate in the late 1990s, results from data mining, using conserved P-type ATPase sequence motifs as search strings, made it apparent that a group of eukaryotic sequences clustered together as the fifth P-type ATPase subfamily (9). However, none of these genes had been characterized by then. Still today, the substrate specificity of any P5 pump has not been elucidated.

Characteristic for P5-ATPases is a PPxxP motif in the transmembrane segment immediately preceding the P-domain. Further comparison of genome sequences has revealed that the P5-ATPase subfamily should be divided into at least two subfamilies, P5A and P5B (80, 137). The main differences are that (*a*) P5A sequences contain a PP(E/D)xPx(E/D) motif with two conserved negatively charged residues, whereas P5B sequences have a PP(A/V)xP(A/V)x motif with two conserved

hydrophobic residues. In other pumps, amino acid residues in this motif are located around one of the ion binding pockets (site II in SERCA1). As the conserved residues in this motif are entirely different, the ion specificities of the two subgroups can hardly be the same. (*b*) The number of predicted transmembrane segments differ. P5A has 12 transmembrane segments, whereas P5B has 11. (*c*) The intracellular localization of the pumps differs. P5A pumps are found exclusively in the ER, whereas P5B pumps are found in the lysosome (humans), the vacuole (in yeast), and the plasma membrane (*Caenorhabditis elegans*) (137).

The most well-characterized P5A-ATPase is Spf1p in the yeast *S. cerevisiae*. Deletion mutants of this pump have a complex phenotype. *Spf1* (Sensitivity to *Pichia farinosa* 1) was first identified in a screen for mutants showing insensitivity to the killer toxin SMKT secreted by the competing yeast *Pichia farinosa* (141). As Spf1p protein is in the ER and the toxin interacts with the plasma membrane, the effect is most likely indirect. Deletion of *SPF1* further causes upregulation of Kar2p, an ER stress marker, and abnormal distribution of Sec12p, a guanine nucleotide exchange factor (GEF) involved in vesicle budding (140). Other phenotypes of *spf1* cells (27, 28, 161) make it likely that Spf1p exerts its function in the ER, where loss of its function leads to impairment of basic ER functions such as protein folding and processing. Likewise, in the plant *A. thaliana*, disruption of the single P5A-ATPase gene, *MIA*, results in disturbed ER homeostasis (59). Because Spf1p interacts with a protein involved in budding of vesicles, much like P4-ATPases in the Golgi complex (see above), it has been proposed that P5A-ATPases might be the lipid flippases of the ER, just as P4-ATPases are the lipid flippases of the Golgi complex, plasma membrane, and endosomes (114).

Little is known concerning P5B-ATPases, but genetic lesions in members of this subgroup cause neurological diseases in humans. Among the four human P5-ATPase genes, mutations in *ATP13A2* are associated with the neurodegenerative disease Kufor-Rakeb syndrome (115) and mutations in *ATP13A4* are associated with speech-language problems (64). How these relate to P5B-ATPase function is currently unknown.

SUMMARY POINTS

1. P-type ATPases perform active transport through a mechanism exploiting the formation and breakdown of a phosphoenzyme intermediate.

2. There are five distinct subfamilies of P-type ATPases and five subtypes, typically associated with different transport specificities.

3. P-type ATPases display modular structures, where cytoplasmic domains (A, P, N) perform ATP hydrolysis and a transmembrane transport domain (T) and a class-specific support domain (S) perform transport, coupled with conformational changes mediated via linkers and tertiary contacts.

4. P1A-ATPases drive potassium uptake in the bacterial KdpABCF complex. P1B-ATPases are heavy-metal and transition-metal pumps found in all kingdoms of life. For P1-ATPases, only crystal structures of soluble domains are known.

5. Ion pumps of the P2 and P3 family are well characterized at the structural and functional levels. Ca^{2+}-ATPases are ubiquitous in eukaryotes. Na^{+}/K^{+}-ATPase is found almost exclusively in animals, whereas plasma membrane H^{+}-ATPases maintain a similarly critical role in other eukaryotes.

6. P4- and P5-ATPases are also found in all eukaryotes, but not in prokaryotes. P4-ATPases are critical for lipid bilayer homeostasis and for vesicle-mediated transport schemes. The function of P5-ATPases is still unknown.

FUTURE ISSUES

1. How does the KdpB subunit—the ATPase—relay its activity to the KdpA potassium channel subunit to establish an active transport system, and how do the related P1B metal pumps work at the molecular level?

2. What is the molecular mechanism of sodium transport in Na^{+}/K^{+}-ATPase?

3. How do P3 H^{+}-ATPases establish and maintain large electrochemical gradients of H^{+}?

4. What is the molecular role and mechanism of P4-ATPases in lipid flipping, and what is the role and activity of P5-ATPases?

5. How do P-type ATPases work as scaffolds or even receptors in signaling complexes and networks?

DISCLOSURE STATEMENT

The authors are not aware of any affiliations, memberships, funding, or financial holdings that might be perceived as affecting the objectivity of this review.

ACKNOWLEDGMENTS

We are indebted to Peter Leth Jørgensen, Rosa Lopez-Marques, Maria Dalgaard Mikkelsen, and Danny Mollerup Sørensen for useful comments on the manuscript.

LITERATURE CITED

1. Adle DJ, Lee J. 2008. Expressional control of a cadmium-transporting P1B-type ATPase by a metal sensing degradation signal. *J. Biol. Chem.* 283:31460–68
2. Adle DJ, Wei W, Smith N, Bies JJ, Lee J. 2009. Cadmium-mediated rescue from ER-associated degradation induces expression of its exporter. *Proc. Natl. Acad. Sci. USA* 106:10189–94
3. Ahnert F, Schmid R, Altendorf K, Greie JC. 2006. ATP binding properties of the soluble part of the KdpC subunit from the *Escherichia coli* K^+-transporting KdpFABC P-type ATPase. *Biochemistry* 45:11038–46
4. Aitken A, Collinge DB, van Heusden BP, Isobe T, Roseboom PH, et al. 1992. 14-3-3 proteins: a highly conserved, widespread family of eukaryotic proteins. *Trends Biochem. Sci.* 17:498–501
5. Albers RW. 1967. Biochemical aspects of active transport. *Annu. Rev. Biochem.* 36:727–56
6. Aravind L, Galperin MY, Koonin EV. 1998. The catalytic domain of the P-type ATPase has the haloacid dehalogenase fold. *Trends Biochem. Sci.* 23:127–29
7. Argüello JM. 2003. Identification of ion-selectivity determinants in heavy-metal transport P1B-type ATPases. *J. Membr. Biol.* 195:93–108
8. Asahi M, Sugita Y, Kurzydlowski K, De Leon S, Tada M, et al. 2003. Sarcolipin regulates sarco(endo)plasmic reticulum Ca^{2+}-ATPase (SERCA) by binding to transmembrane helices alone or in association with phospholamban. *Proc. Natl. Acad. Sci. USA* 100:5040–45
9. Axelsen KB, Palmgren MG. 1998. Evolution of substrate specificities in the P-type ATPase superfamily. *J. Mol. Evol.* 46:84–101
10. Baekgaard L, Mikkelsen MD, Sørensen DM, Hegelund JN, Persson DP, et al. 2010. A combined Zn/Cd sensor and Zn/Cd export regulator in a heavy metal pump. *J. Biol. Chem.* 285:31243–52
11. Bayle D, Weeks D, Sachs G. 1995. The membrane topology of the rat sarcoplasmic and endoplasmic reticulum calcium ATPases by in vitro translation scanning. *J. Biol. Chem.* 270:25678–84
12. Becker D, Fendler K, Altendorf K, Greie JC. 2007. The conserved dipole in transmembrane helix 5 of KdpB in the *Escherichia coli* KdpFABC P-type ATPase is crucial for coupling and the electrogenic K^+-translocation step. *Biochemistry* 46:13920–28
13. Beeler TJ. 1980. Ca^{2+} uptake and membrane potential in sarcoplasmic reticulum vesicles. *J. Biol. Chem.* 255:9156–61
14. Berchtold MW, Brinkmeier H, Müntener M. 2000. Calcium ion in skeletal muscle: its crucial role for muscle function, plasticity, and disease. *Physiol. Rev.* 80:1215–65
15. Bertrand J, Altendorf K, Bramkamp M. 2004. Amino acid substitutions in putative selectivity filter regions III and IV in KdpA alter ion selectivity of the KdpFABC complex from *Escherichia coli. J. Bacteriol.* 186:5519–22
16. Bramkamp M, Altendorf K. 2005. Single amino acid substitution in the putative transmembrane helix V in KdpB of the KdpFABC complex of *Escherichia coli* uncouples ATPase activity and ion transport. *Biochemistry* 44:8260–66
17. Bramkamp M, Altendorf K, Greie J-C. 2007. Common patterns and unique features of P-type ATPases: a comparative view on the KdpFABC complex from *Escherichia coli. Mol. Membr. Biol.* 24:375–86
18. Brini M, Carafoli E. 2009. Calcium pumps in health and disease. *Physiol. Rev.* 89:1341–78
19. Bublitz M, Poulsen H, Morth JP, Nissen P. 2010. In and out of the cation pumps: P-type ATPase structure revisited. *Curr. Opin. Struct. Biol.* 20:431–39
20. Buch-Pedersen MJ, Palmgren MG. 2003. Conserved Asp684 in transmembrane segment M6 of the plant plasma membrane P-type proton pump AHA2 is a molecular determinant of proton translocation. *J. Biol. Chem.* 278:17845–51
21. Buch-Pedersen MJ, Pedersen BP, Veierskov B, Nissen P, Palmgren MG. 2009. Protons and how they are transported by proton pumps. *Pflugers Arch.* 457:573–79

22. Buch-Pedersen MJ, Rudashevskaya EL, Berner TS, Venema K, Palmgren MG. 2006. Potassium as an intrinsic uncoupler of the plasma membrane H$^+$-ATPase. *J. Biol. Chem.* 281:38285–92

23. Campbell AM, Kessler PD, Fambrough DM. 1992. The alternative carboxyl termini of avian cardiac and brain sarcoplasmic reticulum/endoplasmic reticulum Ca^{2+}-ATPases are on opposite sides of the membrane. *J. Biol. Chem.* 267:9321–25

24. Carafoli E, Pedersen PL. 1987. Ion motive ATPases. I. Ubiquity, properties, and significance to cell function. *Trends Biochem. Sci.* 23:394–401

25. Chen CY, Ingram MF, Rosal PH, Graham TR. 1999. Role for Drs2p, a P-type ATPase and potential aminophospholipid translocase, in yeast late Golgi function. *J. Cell. Biol.* 147:1223–36

26. Chen Z, Akin BL, Stokes DL, Jones LR. 2006. Cross-linking of C-terminal residues of phospholamban to the Ca^{2+} pump of cardiac sarcoplasmic reticulum to probe spatial and functional interactions within the transmembrane domain. *J. Biol. Chem.* 281:14163–72

27. Cronin SR, Khoury A, Ferry DK, Hampton RY. 2000. Regulation of HMG-CoA reductase degradation requires the P-type ATPase Cod1p/Spf1p. *J. Cell Biol.* 148:915–24

28. Cronin SR, Rao R, Hampton RY. 2002. Cod1p/Spf1p is a P-type ATPase involved in ER function and Ca^{2+} homeostasis. *J. Cell Biol.* 157:1017–28

29. Devés R, Brodie AF. 1981. Active transport of Ca^{2+} in bacteria: bioenergetics and function. *Mol. Cell. Biochem.* 36:65–84

30. Di Leva F, Domi T, Fedrizzi L, Lim D, Carafoli E. 2008. The plasma membrane Ca^{2+} ATPase of animal cells: structure, function and regulation. *Arch. Biochem. Biophys.* 476:65–74

31. Dmitriev O, Tsivkovskii R, Abildgaard F, Morgan CT, Markley JL, Lutsenko S. 2006. Solution structure of the N-domain of Wilson disease protein: distinct nucleotide-binding environment and effects of disease mutations. *Proc. Natl. Acad. Sci. USA* 103:5302–7

32. Dode L, Andersen JP, Raeymaekers L, Missiaen L, Vilsen B, Wuytack F. 2005. Functional comparison between secretory pathway Ca^{2+}/Mn^{2+}-ATPase (SPCA) 1 and sarcoplasmic reticulum Ca^{2+}-ATPase (SERCA) 1 isoforms by steady-state and transient kinetic analyses. *J. Biol. Chem.* 280:39124–34

33. Duby G, Boutry M. 2009. The plant plasma membrane proton pump ATPase: a highly regulated P-type ATPase with multiple physiological roles. *Pflugers Arch.* 457:645–55

34. Durell SR, Bakker EP, Guy HR. 2000. Does the KdpA subunit from the high affinity K$^+$-translocating P-type KDP-ATPase have a structure similar to that of K$^+$ channels? *Biophys. J.* 78:188–99

35. Ekberg K, Palmgren MG, Veierskov B, Buch-Pedersen MJ. 2010. A novel mechanism of P-type ATPase autoinhibition involving both termini of the protein. *J. Biol. Chem.* 285:7344–50

36. Ekberg K, Pedersen BP, Sørensen DM, Nielsen AK, Veierskov B, et al. 2010. Structural identification of cation binding pockets in the plasma membrane proton pump. *Proc. Natl. Acad. Sci. USA* 107:21400–5

37. Fahn S, Koval GJ, Albers RW. 1968. Sodium-potassium-activated adenosine triphosphatase of *Electrophorus* electric organ. V. Phosphorylation by adenosine triphosphate-32P. *J. Biol. Chem.* 243:1993–2002

38. Fendler K, Dröse S, Altendorf K, Bamberg E. 1996. Electrogenic K$^+$ transport by the Kdp-ATPase of *Escherichia coli. Biochemistry* 35:8009–17

39. Fendler K, Dröse S, Epstein W, Bamberg E, Altendorf K. 1999. The Kdp-ATPase of *Escherichia coli* mediates an ATP-dependent, K$^+$-independent electrogenic partial reaction. *Biochemistry* 38:1850–56

40. Forbes JR, Cox DW. 1998. Functional characterization of missense mutations in ATP7B: Wilson disease mutation or normal variant? *Am. J. Hum. Genet.* 63:1663–74

41. Fuglsang AT, Visconti S, Drumm K, Jahn T, Stensballe A, et al. 1999. Binding of 14-3-3 protein to the plasma membrane H$^+$-ATPase AHA2 involves the three C-terminal residues Tyr^{946}-Thr-Val and requires phosphorylation of Thr^{947}. *J. Biol. Chem.* 274:36774–80

42. Gall WE, Geething NC, Hua Z, Ingram MF, Liu K, et al. 2002. Drs2p-dependent formation of exocytic clathrin-coated vesicles in vivo. *Curr. Biol.* 12:1623–27

43. Gassel M, Altendorf K. 2001. Analysis of KdpC of the K$^+$-transporting KdpFABC complex of *Escherichia coli. Eur. J. Biochem.* 268:1772–81

44. Gassel M, Möllenkamp T, Puppe W, Altendorf K. 1999. The KdpF subunit is part of the K$^+$-translocating Kdp complex of *Escherichia coli* and is responsible for stabilization of the complex in vitro. *J. Biol. Chem.* 274:37901–7

45. Gaxiola RA, Palmgren MG, Schumacher K. 2007. Plant proton pumps. *FEBS Lett.* 581:2204–14
46. Geering K. 2001. The functional role of beta subunits in oligomeric P-type ATPases. *J. Bioenerg. Biomembr.* 33:425–38
47. Geering K. 2006. FXYD proteins: new regulators of Na-K-ATPase. *Am. J. Physiol. Renal Physiol.* 290:F241–50
48. Glynn IM. 2002. A hundred years of sodium pumping. *Annu. Rev. Physiol.* 64:1–18
49. Gonzalez-Guerrero M, Arguello JM. 2008. Mechanism of Cu^+-transporting ATPases: Soluble Cu^+ chaperones directly transfer Cu^+ to transmembrane transport sites. *Proc. Natl. Acad. Sci. USA* 105:5992–97
50. Grover AK, Khan J. 1992. Calcium pump isoforms: diversity, selectivity and plasticity. *Cell Calcium* 13:9–17
51. Gunteski-Hamblin AM, Greeb J, Shull GE. 1988. A novel Ca^{2+} pump expressed in brain, kidney, and stomach is encoded by an alternative transcript of the slow-twitch muscle sarcoplasmic reticulum Ca-ATPase gene. Identification of cDNAs encoding Ca^{2+} and other cation-transporting ATPases using an oligonucleotide probe derived from the ATP-binding site. *J. Biol. Chem.* 263:15032–40
52. Håkansson KO. 2003. The crystallographic structure of Na,K-ATPase N-domain at 2.6 Å resolution. *J. Mol. Biol.* 332:1175–82
53. Harper JF, Hong B, Hwang I, Guo HQ, Stoddard R, et al. 1998. A novel calmodulin-regulated Ca^{2+}-ATPase (ACA2) from *Arabidopsis* with an N-terminal autoinhibitory domain. *J. Biol. Chem.* 273:1099–106
54. Harper JF, Surowy TK, Sussman MR. 1989. Molecular cloning and sequence of cDNA encoding the plasma membrane proton pump (H^+-ATPase) of *Arabidopsis thaliana*. *Proc. Natl. Acad. Sci. USA* 86:1234–38
55. Haupt M, Bramkamp M, Heller M, Coles M, Deckers-Hebestreit G, et al. 2006. The holo-form of the nucleotide binding domain of the KdpFABC complex from *Escherichia coli* reveals a new binding mode. *J. Biol. Chem.* 281:9641–49
56. Hilge M, Siegal G, Vuister GW, Güntert P, Gloor SM, Abrahams JP. 2003. ATP-induced conformational changes of the nucleotide-binding domain of Na,K-ATPase. *Nat. Struct. Biol.* 10:468–74
57. Hussain D, Haydon MJ, Wang Y, Wong E, Sherson SM, et al. 2004. P-type ATPase heavy metal transporters with roles in essential zinc homeostasis in *Arabidopsis*. *Plant Cell* 16:1327–39
58. Jahn T, Fuglsang AT, Olsson A, Brüntrup IM, Collinge DB, et al. 1997. The 14-3-3 protein interacts directly with the C-terminal region of the plant plasma membrane H^+-ATPase. *Plant Cell* 9:1805–14
59. Jakobsen MK, Poulsen LR, Schulz A, Fleurat-Lessard P, Møller A, et al. 2005. Pollen development and fertilization in *Arabidopsis* is dependent on the MALE GAMETOGENESIS IMPAIRED ANTHERS gene encoding a type V P-type ATPase. *Genes Dev.* 19:2757–69
60. Jorgensen PL. 1975. Purification and characterization of (Na^+, K^+)-ATPase. V. Conformational changes in the enzyme transitions between the Na-form and the K-form studied with tryptic digestion as a tool. *Biochim. Biophys. Acta* 401:399–415
61. Jorgensen PL, Håkansson KO, Karlish SJ. 2003. Structure and mechanism of Na,K-ATPase: functional sites and their interactions. *Annu. Rev. Physiol.* 65:817–49
62. Kinoshita T, Shimazaki K. 1999. Blue light activates the plasma membrane H^+-ATPase by phosphorylation of the C-terminus in stomatal guard cells. *EMBO J.* 18:5548–58
63. Kirchberger MA, Tada T, Katz AM. 1975. Phospholamban: a regulatory protein of the cardiac sarcoplasmic reticulum. *Rec. Adv. Stud. Card. Struct. Metab.* 5:103–15
64. Kwasnicka-Crawford DA, Carson AR, Roberts W, Summers AM, Rehnström K, et al. 2005. Characterization of a novel cation transporter ATPase gene (ATP13A4) interrupted by 3q25-q29 inversion in an individual with language delay. *Genomics* 86:182–94
65. Laimins LA, Rhoads DB, Altendorf K, Epstein W. 1978. Identification of the structural proteins of an ATP-driven potassium transport system in *Escherichia coli*. *Proc. Natl. Acad. Sci. USA* 75:3216–19
66. Laursen M, Bublitz M, Moncoq K, Olesen C, Møller JV, et al. 2009. Cyclopiazonic acid is complexed to a divalent metal ion when bound to the sarcoplasmic reticulum Ca^{2+}-ATPase. *J. Biol. Chem.* 284:13513–18
67. Lecchi S, Nelson CJ, Allen KE, Swaney DL, Thompson KL, et al. 2007. Tandem phosphorylation of Ser-911 and Thr-912 at the C terminus of yeast plasma membrane H^+-ATPase leads to glucose-dependent activation. *J. Biol. Chem.* 282:35471–81

68. Lenoir G, Williamson P, Puts CF, Holthuis JC. 2009. Cdc50p plays a vital role in the ATPase reaction cycle of the putative aminophospholipid transporter Drs2p. *J. Biol. Chem.* 284:17956–67

69. Li C, Geering K, Horisberger JD. 2006. The third sodium binding site of Na,K-ATPase is functionally linked to acidic pH-activated inward current. *J. Membr. Biol.* 213:1–9

70. Li Y, Camacho P. 2004. Ca^{2+}-dependent redox modulation of SERCA 2b by ERp57. *J. Cell Biol.* 164:35–46

71. Li Z, Xie Z. 2009. The Na^+/K^+-ATPase/Src complex and cardiotonic steroid-activated protein kinase cascades. *Pflugers Arch.* 457:635–44

72. López-Marqués RL, Poulsen LR, Hanisch S, Meffert K, Buch-Pedersen MJ, et al. 2010. Intracellular targeting signals and lipid specificity determinants of the ALA/ALIS P4-ATPase complex reside in the catalytic ALA alpha-subunit. *Mol. Biol. Cell* 21:791–801

73. Lutsenko S, Barnes N, Bartee MY, Dmitriev OY. 2007. Function and regulation of human copper-transporting ATPases. *Physiol. Rev.* 87:1011–46

74. Lutsenko S, Kaplan JH. 1993. An essential role for the extracellular domain of the Na,K-ATPase beta-subunit in cation occlusion. *Biochemistry* 32:6737–43

75. Lutsenko S, Kaplan JH. 1995. Organization of P-type ATPases: significance of structural diversity. *Biochemistry* 34:15607–13

76. Malmström S, Askerlund P, Palmgren MG. 1997. A calmodulin-stimulated Ca^{2+}-ATPase from plant vacuolar membranes with a putative regulatory domain at its N-terminus. *FEBS Lett.* 400:324–28

77. Mandal D, Rulli SJ, Rao R. 2003. Packing interactions between transmembrane helices alter ion selectivity of the yeast Golgi Ca^{2+}/Mn^{2+}-ATPase PMR1. *J. Biol. Chem.* 278:35292–98

78. Meier S, Tavraz NN, Dürr KL, Friedrich T. 2010. Hyperpolarization-activated inward leakage currents caused by deletion or mutation of carboxy-terminal tyrosines of the Na^+/K^+-ATPase α subunit. *J. Gen. Physiol.* 135:115–34

79. Merlot S, Leonhardt N, Fenzi F, Valon C, Costa M, et al. 2007. Constitutive activation of a plasma membrane H^+-ATPase prevents abscisic acid-mediated stomatal closure. *EMBO J.* 26:3216–26

80. Møller AB, Asp T, Holm PB, Palmgren MG. 2008. Phylogenetic analysis of P5 P-type ATPases, a eukaryotic lineage of secretory pathway pumps. *Mol. Phylogenet. Evol.* 46:619–34

81. Møller JV, Juul B, le Maire M. 1996. Structural organization, ion transport, and energy transduction of P-type ATPases. *Biochim. Biophys. Acta* 1286:1–51

82. Møller JV, Olesen C, Winther AM, Nissen P. 2010. The sarcoplasmic Ca^{2+}-ATPase: design of a perfect chemi-osmotic pump. *Q. Rev. Biophys.* 43:501–66

83. Morita T, Hussain D, Asahi M, Tsuda T, Kurzydlowski K, et al. 2008. Interaction sites among phospholamban, sarcolipin, and the sarco(endo)plasmic reticulum Ca^{2+}-ATPase. *Biochem. Biophys. Res. Commun.* 369:188–94

84. Moriyama Y, Nelson N. 1988. Purification and properties of a vanadate- and N-ethylmaleimide-sensitive ATPase from chromaffin granule membranes. *J. Biol. Chem.* 263:8521–27

85. Moriyama Y, Nelson N, Maeda M, Futai M. 1991. Vanadate-sensitive ATPase from chromaffin granule membranes formed a phosphoenzyme intermediate and was activated by phosphatidylserine. *Arch. Biochem. Biophys.* 286:252–56

86. Morrot G, Zachowski A, Devaux PF. 1990. Partial purification and characterization of the human erythrocyte Mg^{2+}-ATPase. A candidate aminophospholipid translocase. *FEBS Lett.* 266:29–32

87. Morth JP, Pedersen BP, Buch-Pedersen MJ, Andersen JP, Vilsen B, et al. 2011. A structural overview of the plasma membrane Na^+,K^+-ATPase and H^+-ATPase ion pumps. *Nat. Rev. Mol. Cell. Biol.* 12:60–70

88. Morth JP, Pedersen BP, Toustrup-Jensen MS, Sørensen TL, Petersen J, et al. 2007. Crystal structure of the sodium-potassium pump. *Nature* 450:1043–49

89. Muthusamy B-P, Natarajan P, Zhou X, Graham TR. 2009. Linking phospholipid flippases to vesicle-mediated protein transport. *Biochim. Biophys. Acta* 1791:612–19

90. Niggli V, Sigel E. 2008. Anticipating antiport in P-type ATPases. *Trends Biochem. Sci.* 33:156–60

91. Nucifora G, Chu L, Misra TK, Silver S. 1989. Cadmium resistance from *Staphylococcus aureus* plasmid pI258 *cadA* gene results from a cadmium-efflux ATPase. *Proc. Natl. Acad. Sci. USA* 86:3544–48

92. Odermatt A, Becker S, Khanna VK, Kurzydlowski K, Leisner E, et al. 1998. Sarcolipin regulates the activity of SERCA1, the fast-twitch skeletal muscle sarcoplasmic reticulum Ca^{2+}-ATPase. *J. Biol. Chem.* 273:12360–69

93. Odermatt A, Suter H, Krapf R, Solioz M. 1993. Primary structure of two P-type ATPases involved in copper homeostasis in *Enterococcus hirae*. *J. Biol. Chem.* 268:12775–79

94. Oecking C, Piotrowski M, Hagemeier J, Hagemann K. 1997. Topology and target interaction of the fusicoccin-binding 14-3-3 homologs of *Commelina communis*. *Plant J.* 12:441–53

95. Ogawa H, Shinoda T, Cornelius F, Toyoshima C. 2009. Crystal structure of the sodium-potassium pump (Na^+, K^+-ATPase) with bound potassium and ouabain. *Proc. Natl. Acad. Sci. USA* 106:13742–47

96. Ogawa H, Toyoshima C. 2002. Homology modeling of the cation binding sites of Na^+K^+-ATPase. *Proc. Natl. Acad. Sci. USA* 99:15977–82

97. Olesen C, Picard M, Winther AM, Gyrup C, Morth JP, et al. 2007. The structural basis of calcium transport by the calcium pump. *Nature* 450:1036–42

98. Olesen C, Sørensen TL, Nielsen RC, Møller JV, Nissen P. 2004. Dephosphorylation of the calcium pump coupled to counterion occlusion. *Science* 306:2251–55

99. Olivari C, Meanti C, De Michelis MI, Rasi-Caldogno F. 1998. Fusicoccin binding to its plasma membrane receptor and the activation of the plasma membrane H^+-ATPase. IV. Fusicoccin induces the association between the plasma membrane H^+-ATPase and the fusicoccin receptor. *Plant Physiol.* 116:529–37

100. Ottmann C, Marco S, Jaspert N, Marcon C, Schauer N, et al. 2007. Structure of a 14-3-3 coordinated hexamer of the plant plasma membrane H^+-ATPase by combining X-ray crystallography and electron cryomicroscopy. *Mol. Cell* 25:427–40

101. Palmgren MG. 2001. Plant plasma membrane H^+-ATPases: powerhouses for nutrient uptake. *Annu. Rev. Plant Physiol. Plant Mol. Biol.* 52:817–45

102. Palmgren MG, Larsson C, Sommarin M. 1990. Proteolytic activation of the plant plasma membrane H^+-ATPase by removal of a terminal segment. *J. Biol. Chem.* 265:13423–26

103. Palmgren MG, Sommarin M, Serrano R, Larsson C. 1991. Identification of an autoinhibitory domain in the C-terminal region of the plant plasma membrane H^+-ATPase. *J. Biol. Chem.* 266:20470–75

104. Pardo JM, Serrano R. 1989. Structure of a plasma membrane H^+-ATPase gene from the plant *Arabidopsis thaliana*. *J. Biol. Chem.* 264:8557–62

105. Paulusma CC, Elferink RP. 2010. P4 ATPases—the physiological relevance of lipid flipping transporters. *FEBS Lett.* 584:2708–16

106. Pedersen BP, Buch-Pedersen MJ, Morth JP, Palmgren MG, Nissen P. 2007. Crystal structure of the plasma membrane proton pump. *Nature* 450:1111–14

107. Petris MJ, Mercer JF, Culvenor JG, Lockhart P, Gleeson PA, Camakaris J. 1996. Ligand-regulated transport of the Menkes copper P-type ATPase efflux pump from the Golgi apparatus to the plasma membrane: a novel mechanism of regulated trafficking. *EMBO J.* 15:6084–95

108. Pomorski T, Lombardi R, Riezman H, Devaux PF, van Meer G, Holthuis JC. 2003. Drs2p-related P-type ATPases Dnf1p and Dnf2p are required for phospholipid translocation across the yeast plasma membrane and serve a role in endocytosis. *Mol. Biol. Cell* 14:1240–54

109. Portillo F, de Larrinoa IF, Serrano R. 1989. Deletion analysis of yeast plasma membrane H^+-ATPase and identification of a regulatory domain at the carboxyl-terminus. *FEBS Lett.* 247:381–85

110. Post RL, Kume S, Tobin T, Orcutt B, Sen AK. 1969. Flexibility of an active center in sodium-plus-potassium adenosine triphosphatase. *J. Gen. Physiol.* 54:306–26

111. Post RL, Sen AK, Rosenthal AS. 1965. A phosphorylated intermediate in adenosine triphosphate-dependent sodium and potassium transport across kidney membranes. *J. Biol. Chem.* 240:1437–45

112. Poulsen H, Khandelia H, Morth JP, Bublitz M, Mouritsen OG, Egebjerg J, Nissen P. 2010. Neurological disease mutations compromise a C-terminal ion pathway in the Na^+/K^+-ATPase. *Nature* 467:99–102

113. Poulsen LR, López-Marqués RL, McDowell SC, Okkeri J, Licht D, et al. 2008. The *Arabidopsis* P4-ATPase ALA3 localizes to the Golgi and requires a beta-subunit to function in lipid translocation and secretory vesicle formation. *Plant Cell* 20:658–76

114. Poulsen LR, López-Marqués RL, Palmgren MG. 2008. Flippases: still more questions than answers. *Cell. Mol. Life Sci.* 65:3119–25

115. Ramirez A, Heimbach A, Gründemann J, Stiller B, Hampshire D, et al. 2006. Hereditary parkinsonism with dementia is caused by mutations in ATP13A2, encoding a lysosomal type 5 P-type ATPase. *Nat. Genet.* 38:1184–91

116. Rensing C, Ghosh M, Rosen BP. 1999. Families of soft-metal-ion-transporting ATPases. *J. Bacteriol.* 181:5891–97

117. Rensing C, Mitra B, Rosen BP. 1997. The *zntA* gene of *Escherichia coli* encodes a Zn(II)-translocating P-type ATPase. *Proc. Natl. Acad. Sci. USA* 94:14326–31

118. Ridder IS, Dijkstra BW. 1999. Identification of the Mg^{2+}-binding site in the P-type ATPase and phosphatase members of the HAD (haloacid dehalogenase) superfamily by structural similarity to the response regulator protein CheY. *Biochem. J.* 339:223–26

119. Roderick HL, Lechleiter JD, Camacho P. 2000. Cytosolic phosphorylation of calnexin controls intracellular Ca^{2+} oscillations via an interaction with SERCA2b. *J. Cell Biol.* 149:1235–48

120. Sachs G, Chang HH, Rabon E, Schackman R, Lewin M, Saccomani G. 1976. A nonelectrogenic H^+ pump in plasma membranes of hog stomach. *J. Biol. Chem.* 251:7690–98

121. Saito K, Fujimura-Kamada K, Furuta N, Kato U, Umeda M, Tanaka K. 2004. Cdc50p, a protein required for polarized growth, associates with the Drs2p P-type ATPase implicated in phospholipid translocation in *Saccharomyces cerevisiae*. *Mol. Biol. Cell* 15:3418–32

122. Sazinsky MH, Agarwal S, Argüello JM, Rosenzweig AC. 2006. Structure of the actuator domain from the *Archaeoglobus fulgidus* Cu^+-ATPase. *Biochemistry* 45:9949–55

123. Sazinsky MH, Mandal AK, Argüello JM, Rosenzweig AC. 2006. Structure of the ATP binding domain from the *Archaeoglobus fulgidus* Cu^+-ATPase. *J. Biol. Chem.* 281:11161–66

124. Schack VR, Morth JP, Toustrup-Jensen MS, Anthonisen AN, Nissen P, et al. 2008. Identification and function of a cytoplasmic K^+ site of the Na^+, K^+-ATPase. *J. Biol. Chem.* 283:27982–90

125. Scharff O, Foder B. 1978. Reversible shift between two states of Ca^{2+}-ATPase in human erythrocytes mediated by Ca^{2+} and a membrane-bound activator. *Biochim. Biophys. Acta* 509:67–77

126. Schatzmann HJ. 1983. The red cell calcium pump. *Annu. Rev. Physiol.* 45:303–12

127. Schrader M, Fendler K, Bamberg E, Gassel M, Epstein W, et al. 2000. Replacement of glycine 232 by aspartic acid in the KdpA subunit broadens the ion specificity of the K^+-translocating KdpFABC complex. *Biophys. J.* 79:802–13

128. Seigneuret M, Devaux PF. 1984. ATP-dependent asymmetric distribution of spin-labeled phospholipids in the erythrocyte membrane: relation to shape changes. *Proc. Natl. Acad. Sci. USA* 81:3751–55

129. Sen AK, Post RL. 1964. Stoichiometry and localization of adenosine triphosphate dependent sodium and potassium transport in the erythrocyte. *J. Biol. Chem.* 239:345–52

130. Serrano R. 1983. In vivo glucose activation of the yeast plasma membrane ATPase. *FEBS Lett.* 156:11–14

131. Serrano R, Kielland-Brandt MC, Fink GR. 1986. Yeast plasma membrane ATPase is essential for growth and has homology with ($Na^+ + K^+$), K^+- and Ca^{2+}-ATPases. *Nature* 319:689–93

132. Shin JM, Munson K, Vagin O, Sachs G. 2009. The gastric HK-ATPase: structure, function, and inhibition. *Pflugers Arch.* 457:609–22

133. Shinoda T, Ogawa H, Cornelius F, Toyoshima C. 2009. Crystal structure of the sodium-potassium pump at 2.4 Å resolution. *Nature* 459:446–50

134. Skou JC. 1957. The influence of some cations on an adenosine triphosphatase from peripheral nerves. *Biochim. Biophys. Acta* 23:394–401

135. Skou JC. 1960. Further investigations on a $Mg^{3+}Na^+$-activated adenosinetriphosphatase, possibly related to the active, linked transport of Na^+ and K^+ across the nerve membrane. *Biochim. Biophys. Acta* 42:6–23

136. Skou JC. 1998. Nobel lecture. The identification of the sodium pump. *Biosci. Rep.* 18:155–69

137. Sørensen DM, Buch-Pedersen MJ, Palmgren MG. 2010. Structural divergence between the two subgroups of P5 ATPases. *Biochim. Biophys. Acta* 1797:846–55

138. Sørensen TL, Clausen JD, Jensen AM, Vilsen B, Møller JV, et al. 2004. Localization of a K^+-binding site involved in dephosphorylation of the sarcoplasmic reticulum Ca^{2+}-ATPase. *J. Biol. Chem.* 279:46355–58

139. Sørensen TL, Møller JV, Nissen P. 2004. Phosphoryl transfer and calcium ion occlusion in the calcium pump. *Science* 304:1672–75

140. Suzuki C. 2001. Immunochemical and mutational analyses of P-type ATPase Spf1p involved in the yeast secretory pathway. *Biosci. Biotechnol. Biochem.* 65:2405–11

141. Suzuki C, Shimma YI. 1999. P-type ATPase *spf1* mutants show a novel resistance mechanism for the killer toxin SMKT. *Mol. Microbiol.* 32:813–23

142. Suzuki T, Wang JH. 1986. Stimulation of bovine cardiac sarcoplasmic reticulum Ca^{2+} pump and blocking of phospholamban phosphorylation and dephosphorylation by a phospholamban monoclonal antibody. *J. Biol. Chem.* 261:7018–23

143. Svennelid F, Olsson A, Piotrowski M, Rosenquist M, Ottman C, et al. 1999. Phosphorylation of Thr-948 at the C terminus of the plasma membrane H^+-ATPase creates a binding site for the regulatory 14-3-3 protein. *Plant Cell* 11:2379–91

144. Tada M, Inui M. 1983. Regulation of calcium transport by the ATPase-phospholamban system. *J. Mol. Cell. Cardiol.* 15:565–75

145. Takahashi M, Kondou Y, Toyoshima C. 2007. Interdomain communication in calcium pump as revealed in the crystal structures with transmembrane inhibitors. *Proc. Natl. Acad. Sci. USA* 104:5800–5

146. Tang X, Halleck MS, Schlegel RA, Williamson P. 1996. A subfamily of P-type ATPases with aminophospholipid transporting activity. *Science* 272:1495–97

147. Tanzi RE, Petrukhin K, Chernov I, Pellequer JL, Wasco W, et al. 1993. The Wilson disease gene is a copper-transporting ATPase with homology to Menkes disease gene. *Nat. Genet.* 5:344–50

148. Thastrup O, Cullen PJ, Drøbak BK, Hanley MR, Dawson AP. 1990. Thapsigargin, a tumor promoter, discharges intracellular Ca^{2+} stores by specific inhibition of the endoplasmic reticulum Ca^{2+}-ATPase. *Proc. Natl. Acad. Sci. USA* 87:2466–70

149. Tidow H, Aperia A, Nissen P. 2010. How are ion pumps and agrin signaling integrated? *Trends Biochem. Sci.* 35:653–59

150. Toustrup-Jensen MS, Holm R, Einholm AP, Schack VR, Morth JP, et al. 2009. The C terminus of Na^+,K^+-ATPase controls Na^+ affinity on both sides of the membrane through Arg935. *J. Biol. Chem.* 284:18715–25

151. Toyoshima C. 2008. Structural aspects of ion pumping by Ca^{2+}-ATPase of sarcoplasmic reticulum. *Arch. Biochem. Biophys.* 476:3–11

152. Toyoshima C. 2009. How Ca^{2+}-ATPase pumps ions across the sarcoplasmic reticulum membrane. *Biochim. Biophys. Acta* 1793:941–46

153. Toyoshima C, Asahi M, Sugita Y, Khanna R, Tsuda T, MacLennan DH. 2003. Modeling of the inhibitory interaction of phospholamban with the Ca^{2+} ATPase. *Proc. Natl. Acad. Sci. USA* 100:467–72

154. Toyoshima C, Inesi G. 2004. Structural basis of ion pumping by Ca^{2+}-ATPase of the sarcoplasmic reticulum. *Annu. Rev. Biochem.* 73:269–92

155. Toyoshima C, Mizutani T. 2004. Crystal structure of the calcium pump with a bound ATP analogue. *Nature* 430:529–35

156. Toyoshima C, Nakasako M, Nomura H, Ogawa H. 2000. Crystal structure of the calcium pump of sarcoplasmic reticulum at 2.6 Å resolution. *Nature* 405:647–55

157. Toyoshima C, Nomura H. 2002. Structural changes in the calcium pump accompanying the dissociation of calcium. *Nature* 418:605–11

158. Toyoshima C, Nomura H, Tsuda T. 2004. Lumenal gating mechanism revealed in calcium pump crystal structures with phosphate analogues. *Nature* 432:361–68

159. Tsuda T, Toyoshima C. 2009. Nucleotide recognition by CopA, a Cu^+-transporting P-type ATPase. *EMBO J.* 28:1782–91

160. Vandecaetsbeek I, Trekels M, De Maeyer M, Ceulemans H, Lescrinier E, et al. 2009. Structural basis for the high Ca^{2+} affinity of the ubiquitous SERCA2b Ca^{2+} pump. *Proc. Natl. Acad. Sci. USA* 106:18533–38

161. Vashist S, Frank CG, Jakob CA, Ng DT. 2002. Two distinctly localized p-type ATPases collaborate to maintain organelle homeostasis required for glycoprotein processing and quality control. *Mol. Biol. Cell* 13:3955–66

162. Veldhuis NA, Gaeth AP, Pearson RB, Gabriel K, Camakaris J. 2009. The multi-layered regulation of copper translocating P-type ATPases. *Biometals* 22:177–90

163. Voelker DR. 2009. Genetic and biochemical analysis of non-vesicular lipid traffic. *Annu. Rev. Biochem.* 78:827–56

164. Vulpe C, Levinson B, Whitney S, Packman S, Gitschier J. 1993. Isolation of a candidate gene for Menkes disease and evidence that it encodes a copper-transporting ATPase. *Nat. Genet.* 3:7–13

165. Wawrzynow A, Theibert JL, Murphy C, Jona I, Martonosi A, Collins JH. 1992. Sarcolipin, the "proteolipid" of skeletal muscle sarcoplasmic reticulum, is a unique, amphipathic, 31-residue peptide. *Arch. Biochem. Biophys.* 298:620–23

166. Whittam R, Wheeler KP. 1970. Transport across cell membranes. *Annu. Rev. Physiol.* 32:21–60

167. Würtele M, Jelich-Ottmann C, Wittinghofer A, Oecking C. 2003. Structural view of a fungal toxin acting on a 14-3-3 regulatory complex. *EMBO J.* 22:987–94

168. Xie XS, Stone DK, Racker E. 1989. Purification of a vanadate-sensitive ATPase from clathrin-coated vesicles of bovine brain. *J. Biol. Chem.* 264:1710–14

169. Yatime L, Buch-Pedersen MJ, Musgaard M, Morth JP, Winther AML, et al. 2009. P-type ATPases as drug targets: tools for medicine and science. *Biochim. Biophys. Acta* 1787:207–20

170. Zanotti-Fregonara P, Vidailhet M, Kas A, Ozelius LJ, Clot F, et al. 2008. [123I]-FP-CIT and [99mTc]-HMPAO single photon emission computed tomography in a new sporadic case of rapid-onset dystonia-parkinsonism. *J. Neurol. Sci.* 273:148–51

171. Zhou X, Graham TR. 2009. Reconstitution of phospholipid translocase activity with purified Drs2p, a type-IV P-type ATPase from budding yeast. *Proc. Natl. Acad. Sci. USA* 106:16586–91

Kinesin Assembly and Movement in Cells

Kristen J. Verhey,[1] Neha Kaul,[2]
and Virupakshi Soppina[1]

[1]Department of Cell and Developmental Biology, [2]Department of Mechanical Engineering,
University of Michigan, Ann Arbor, Michigan 48109; email: kjverhey@umich.edu,
nehakaul@umich.edu, vsoppina@umich.edu

Annu. Rev. Biophys. 2011. 40:267–88

First published online as a Review in Advance on
February 14, 2011

The *Annual Review of Biophysics* is online at
biophys.annualreviews.org

This article's doi:
10.1146/annurev-biophys-042910-155310

Copyright © 2011 by Annual Reviews.
All rights reserved

1936-122X/11/0609-0267$20.00

Keywords

microtubule, molecular motor, intracellular transport, autoinhibition,
bidirectional transport, posttranslational modifications

Abstract

Long-distance transport in eukaryotic cells is driven by molecular motors that move along microtubule tracks. Molecular motors of the kinesin superfamily contain a kinesin motor domain attached to family-specific sequences for cargo binding, regulation, and oligomerization. The biochemical and biophysical properties of the kinesin motor domain have been widely studied, yet little is known about how kinesin motors work in the complex cellular environment. We discuss recent studies on the three major families involved in intracellular transport (kinesin-1, kinesin-2, and kinesin-3) that have begun to bridge the gap in knowledge between the in vitro and in vivo behaviors of kinesin motors. These studies have increased our understanding of how kinesin subunits assemble to produce a functional motor, how kinesin motors are affected by biochemical cues and obstacles present on cellular microtubules, and how multiple motors on a cargo surface can work collectively for increased force production and travel distance.

Contents

INTRODUCTION

Microtubules are cylindrical cytoskeletal filaments whose polymerization from heterodimeric α- and β-tubulin subunits results in a structural and functional polarity of the two ends. Microtubules and their associated proteins carry out a variety of important functions in cells, from structural support to cell division to movement within and of cells. Movement along microtubule tracks is driven by molecular motors in the kinesin and dynein families (**Figure 1**), and abnormalities in motor-driven transport have been linked to a wide range of human diseases including neurodegenerative diseases, polycystic kidney disease, and cancer (46, 117, 153).

Kinesin motors are enzymes that convert the chemical energy of ATP hydrolysis into mechanical work along microtubule tracks. Most of our understanding of kinesin mechanics

KIF: kinesin family

comes from work on kinesin-1 (formerly conventional kinesin or KIF5) (88), the founding member of the kinesin superfamily. Minimal kinesin-1 motors, which contain only the motor domain and dimerization sequences, move processively (maintain an interaction with the microtubule for many rounds of catalysis) toward the plus end of the microtubule in 8-nm steps (the distance between adjacent β-tubulin subunits) and generate speeds of 0.6–0.8 μm s^{-1}. Processivity is due to a hand-over-hand stepping mechanism in which the two motor domains alternate their catalytic cycles in order to maintain the motor-microtubule interaction (reviewed in References 45 and 133).

Molecular, genetic, and biochemical analyses of organisms from fungi to animals have revealed a superfamily of proteins that have in common a kinesin motor domain ($\sim$350 amino acids, with $\sim$40% amino acid identity within the superfamily). On the basis of homology between motor domains, the kinesin-related proteins have been placed in 14 families (kinesin-1 to kinesin-14) (63, 88, 98). One of the biggest surprises in studying the mechanics of kinesin motors is that the catalytic motor domain has been adapted over evolution for functions other than movement to microtubule plus ends. For example, kinesin-14 motors undergo directional motility to the minus ends of microtubules. And in kinesin-8 and kinesin-13 motors, catalytic activity results in destabilization or depolymerization of microtubules. Outside the motor domain, each kinesin family contains distinct sequences for cargo binding, oligomerization, and regulation (**Figure 2**).

Motor-dependent transport along microtubules takes place with remarkable efficiency despite the crowded environment in cells. Indeed, a major challenge in the field is to understand how the in vitro motility properties of kinesin motors relate to their cargo transport properties in vivo. In this review, we focus on recent work on the assembly and movement of kinesin motors that drive anterograde trafficking events. The reader is referred to recent reviews on kinesin motors that participate in mitosis (48, 146) and cytoplasmic dynein (45, 74).

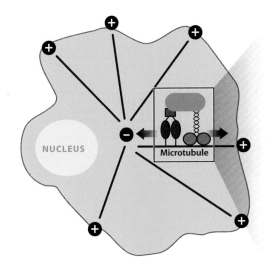 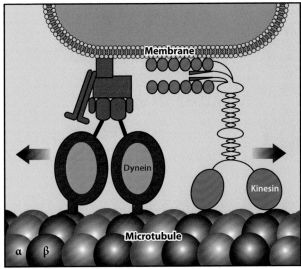

Figure 1

Intracellular transport in interphase mammalian cells. Microtubules are polarized cytoskeletal filaments polymerized from heterodimeric α- and β-tubulin subunits. In most cell types, microtubules (*dark blue lines*) are organized with their plus (growing, +) ends extending into the cell periphery and their minus (anchored, −) ends in the middle of the cell. Transport cargoes (*purple*) such as vesicles, organelles, and protein complexes are carried along microtubule tracks by kinesin (*green*) and dynein (*red*) motors, which read and respond to microtubule polarity to carry out directed motility. In general, kinesin motors carry out long-distance transport to the plus ends of microtubules in the cell periphery (anterograde transport), whereas cytoplasmic dynein carries cargoes in the retrograde direction, toward the minus ends of microtubules in the cell center.

SUBUNIT COMPOSITION AND ASSEMBLY

The Kinesin-1 Family: A Heterotetrameric Motor

The kinesin-1 holoenzyme is a heterotetramer consisting of two kinesin heavy-chain (KHC) and two kinesin light-chain (KLC) subunits (**Figure 2**). The KHC subunit contains an N-terminal kinesin motor domain followed by a neck domain, a coiled-coil stalk, and a globular tail domain (**Figure 2**). The neck domain consists of a neck linker and a neck coil, which are critical for motor processivity and homodimerization, respectively. The coiled-coil stalk contains hinge segments that enable the motor to adopt a folded, autoinhibited state (see below) and also impart flexibility necessary for multiple motors to work collectively when attached to the same surface (10, 25, 26, 42, 70). The globular tail domain participates in regulation of motor activity and cargo binding. The

KLC subunit contains a heptad repeat region for oligomerization with KHC followed by six tetratricopeptide repeat (TPR) motifs that participate in cargo binding (3, 63, 135).

In mammals, the KHC subunit is encoded by three genes (KIF5A, KIF5B, and KIF5C), whereas the KLC subunit is encoded by four genes (KLC1–4). It seems likely that all kinesin-1 transcripts can undergo alternative splicing. Thus, there is the potential to generate many biochemically distinct kinesin-1 motors by differential association of the gene products. Recent work using well-defined antibodies that specifically recognize the different kinesin-1 gene products demonstrated that both KHC and KLC polypeptides exclusively form homodimers and that each KHC homodimer can associate with each KLC homodimer (30). The biochemically distinct kinesin-1 motors assembled from homodimers and/or splice forms can have functionally distinct roles in cells (18, 21, 41, 51, 147).

KHC: kinesin heavy chain

KLC: kinesin light chain

TPR: tetratricopeptide repeat

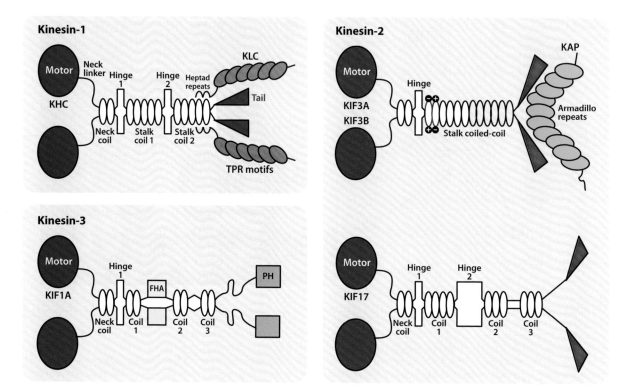

Figure 2

Subunit composition of kinesin motors. Schematic of the subunit composition and organization of kinesin-1, kinesin-2, and kinesin-3 motors. All these kinesins contain a kinesin motor domain (*dark green oval*) at their N terminus for ATP-dependent processive motion toward the plus ends of microtubules. These kinesins also have a neck domain (neck linker and neck coil) and varying amounts of coiled-coil stalk regions for oligomerization. Several of these kinesins have protein-protein or protein-lipid interaction domains such as TPR, Armadillo, FHA, and PH domains. Note that the kinesin-3 motor KIF1A is depicted as a dimeric molecule as this appears to be the state of the processive motor, although it is still unclear whether KIF1A motors are monomeric or dimeric in solution. Abbreviations: FHA, forkhead associated; KAP, kinesin-associated protein; KHC, kinesin heavy chain; KIF, kinesin family; KLC, kinesin light chain; PH, pleckstrin homology; TPR, tetratricopeptide repeat.

Although the kinesin-1 motor is generally considered a heterotetramer, there is evidence that the KHC gene product can exist biochemically and act functionally in the absence of KLC. KHC motors lacking KLC can be isolated during purification of kinesin-1 from the brain, although this was thought to be due to degradation of the KLC subunits (54). In fungi, the kinesin-1 motor lacks KLC subunits (126), whereas in cultured cells, the kinesin-1 pool contains both tetrameric molecules and KHC dimers (32, 52). Functionally, KHC can have distinct roles that do not require the presence of the KLC subunit, including (*a*) transport of mRNA granules (13, 73, 94), (*b*) mitochondrial transport (47), (*c*) cytoplasmic streaming (106), and (*d*) microtubule-microtubule sliding (72, 128).

The Kinesin-2 Family: Heterotrimeric and Homodimeric Motors

The kinesin-2 family contains two subfamilies that are either heterotrimeric or homodimeric assemblies. The heterotrimeric kinesin-2 motors are composed of two motor-domain-containing subunits from the KIF3 family and a nonmotor accessory protein known as kinesin-associated protein (KAP) (**Figure 2**). The homodimeric subfamily includes

KAP:
kinesin-associated protein

CeOSM-3 (*Caenorhabditis elegans*), MmKIF17 (*Mus musculus*), and TtKin5 (*Tetrahymena thermophila*) (**Figure 2**). Both kinesin-2 subfamilies have been implicated in a wide variety of transport events, from transport along axonemal microtubules during the construction and maintenance of cilia and flagella, to transport along cytoplasmic microtubules of organelles, melanosomes, mRNA granules, and membrane-bound vesicles (63, 118). Whereas the assembly of heterotrimeric kinesin-2 motors has been studied (below), much less is known about the assembly of homodimeric kinesin-2 motors, although recent work has shown that the neck coil is sufficient for homodimerization of KIF17 (58).

KIF3 polypeptides are composed of a kinesin motor domain, a neck domain, a coiled-coil stalk, and a globular tail domain (**Figure 2**). Three KIF3 genes have been identified in mammals, KIF3A, KIF3B, and KIF3C. Whereas KIF3A and KIF3B are widely expressed, KIF3C is expressed mainly in neural tissues. A variety of studies have shown that KIF3A forms heterodimers with either KIF3B or KIF3C (8, 27, 99, 109, 112, 151), although KIF3C may also exist in a homodimeric complex (99). The KAP subunit, composed largely of armadillo repeats (**Figure 2**), binds only to the KIF3A/KIF3B heterodimer (14, 36, 150).

How do the KIF3 subunits assemble via heterodimerization and avoid the formation of homodimeric molecules? A unique feature of the KIF3A and KIF3B polypeptides is the highly charged segments at the proximal region of the stalk (**Figure 2**). As KIF3A exhibits opposite charges of KIF3B and KIF3C (**Figure 2**), these charged segments were proposed to specify heterodimer formation. Using synthetic peptides in solution, Chana et al. (19, 20) showed that the KIF3A neck coil is capable of homodimer formation and that the charged segments drive heterodimer formation. In contrast, De Marco et al. found that the C-terminal region of the *Xenopus laevis* Xklp3A and Xklp3B stalk domains is critical for heterodimerization based on coimmunoprecipitation of truncated motors (27) and on solution studies of a stalk frag-

ment (28). Thus, although it seems plausible that both the charged region and the C terminus of the stalk can contribute to heterodimer formation, the exact mechanisms and contribution of each segment remain to be clarified.

Why nature combined two different polypeptide chains to create a heterodimeric motor is still unclear. In general, both native and recombinant kinesin-2 motors display processive motility in microtubule gliding assays or when attached to beads in motility assays (8, 9, 14, 23, 100, 107, 109). That heterodimeric kinesin-2 motors are processive indicates that both motors in the complex are contributing to motility. Indeed, when expressed as monomeric motor domains, the CeKLP-20 (homolog of MmKIF3A and SpKRP85) and CeKLP-11 (homolog of MmKIF3B and SpKRP95) motors display similar microtubule binding properties and ATP hydrolysis rates (14), indicating that both motors have equivalent enzymatic activities. Kinesin-2 motors have a shorter run length than do kinesin-1 motors (8, 9, 100), and recent work has shown that this is due to an extended neck linker sequence in kinesin-2 that relieves the strain between the two motor domains (100, 120). Because of the shorter run length, kinesin-2 motors may be better adapted to participate in bidirectional motility with cytoplasmic dynein.

To dissect the contribution of the KIF3A and KIF3B polypeptides to the motility of kinesin-2, several groups have engineered dimeric constructs that contain two KIF3A heads or two KIF3B heads. Engineered homodimeric molecules of murine KIF3A/KIF3A and KIF3B/KIF3B (100) or *C. elegans* KLP-20/KLP-20 and KLP-11/KLP-11 (107) were found to have motility properties similar to those of heterodimeric wild-type motors, indicating that the ability of the motor to take multiple steps before detaching does not require the presence of two different motor domains. Yet there does appear to be functional differences between the two nonidentical motor domains of the kinesin-2 heterotrimer. Pan et al. (107) found that an engineered *C. elegans* KLP-11/KLP-11 homodimer drives the

KLP: kinesin-like protein

KRP: kinesin-related protein

rotation of microtubules in a motility assay, indicating that the KLP-11 subunit can produce torque. In addition, Brunnbauer et al. (14) determined that engineered homodimeric *C. elegans* KLP-11/KLP-11 molecules have limited activity in ATPase and motility assays, suggesting that the KLP-11 motor exhibits different kinetic properties depending on the identity of its partner head. Differences between these studies, such as the linkages used to engineer homodimeric molecules and the species of KIF3 polypeptide, will need to be resolved to thoroughly understand the contribution of the two different motor domains to the motility of heterotrimeric kinesins.

The Kinesin-3 Family: Monomeric and/or Dimeric Motors

The kinesin-3 family was first identified in *C. elegans* because of mutations in the *unc-104* gene that result in a severe defect in axonal transport of synaptic vesicles, and the mouse homolog, KIF1A, was characterized soon thereafter. Further work identified a variety of gene products across species with homology to the motor domains of CeUNC-104/MmKIF1A, and the kinesin-3 family now comprises five subfamilies (KIF1, KIF13, KIF14, KIF16, and KIF28), named for the corresponding murine gene products (98).

Sequence analysis demonstrated that kinesin-3 motors contain only short α-helical segments with a weak propensity for coiled-coil formation (**Figure 2**) (37, 59, 104, 111, 121, 148). Indeed, oligomerization of kinesin-3 motors has been experimentally challenging, and thus the monomer/dimer state of kinesin-3 motors has been controversial. The initial characterization of MmKIF1A suggested that it is a globular protein whose behavior in biochemical and biophysical assays is most consistent with a monomeric molecule (104). How a monomeric motor can drive processive motility has intrigued the field for many years. Subsequent work raised several interesting possibilities.

The first model for kinesin-3 motility used analysis of engineered MmKIF1A motor domains and showed that, in single-molecule assays, KIF1A moves by a biased diffusion mechanism in which a basic stretch on the KIF1A motor domain (the K-loop) interacts with the acidic C-terminal tails of α- and β-tubulin (the E-hooks) to maintain the motor-microtubule interaction after ATP hydrolysis (75, 101–103). Although this elegant and convincing work provided a molecular mechanism for motility, the speeds of the engineered constructs were too slow (140 nm s^{-1}) compared with the fast motilities of kinesin-3 constructs in multiple-motor gliding assays and kinesin-3 cargoes in vivo (91, 104, 109, 155). In addition, other kinesin-3 family members have shorter K-loops and do not show diffusive motion (4, 109, 131). Thus, motility via a biased diffusion mechanism may not be a general feature of kinesin-3 motors.

A second model was put forth that kinesin-3 motors must work in teams to produce processive motility (109, 115, 131). Indeed, forced dimerization of CeUNC-104 resulted in a motor with high velocity and processivity (131). Teamwork on a vesicle surface could be accomplished by multiple monomeric motors cooperating during motility or by dimerization of motors at high local concentrations. Interestingly, CeUNC-104/MmKIF1A motors contain a lipid binding pleckstrin homology (PH) domain at their C terminus (**Figure 2**) that could recruit motors to the vesicle surface. Klopfenstein et al. (76, 77) provided strong support for this model by showing that the PH-domain-dependent recruitment of CeUNC-104 motors to vesicles or liposomes produces fast motilities in vitro and in vivo. Fast motility is likely due to dimerization on the membrane surface as mutation of the neck coil abolished processive motility but not membrane binding (76).

More recently, the possibility has been raised that kinesin-3 motors may indeed be homodimeric molecules. Analysis of MmKIF1A demonstrated that the molecule can exist in a dimeric state in mammalian cells, but the tightly folded globular state of the molecule makes its oligomeric state difficult to assess (59). This work is consistent with reports that other members of the kinesin-3 family, HsGAKIN/KIF13B, HsKIF1C, DdUNC-104, and NcKin3 (1, 37, 111, 148), exist as homodimeric molecules. Thus, the common classification of kinesin-3 motors as the monomeric family of kinesins does not appear to be widely applicable.

Dimerization now appears to be critical for the fast processive motility of metazoan kinesin-3 motors. While both monomeric and dimeric MmKIF1A motors can undergo one-dimensional diffusion along the microtubule surface, only dimeric motors undergo processive motility at fast velocities (59). Biochemical and structural analyses have shown that dimerization requires not only the neck coil but also several residues of the hinge segment (1, 5, 59, 121). Do dimeric kinesin-3 motors utilize a hand-over-hand mechanism of motility? Analysis of metazoan kinesin-3 motors has not been carried out, but recent work has shown that a dimeric fungal kinesin-3, NcKin3 (*Neurospora crassa*), detaches from the microtubule after one ATPase cycle and is therefore unable to step hand-over-hand to generate processive motility (2, 4). Clearly, further work on the molecular mechanisms that drive kinesin-3 assembly and motility are likely to be informative and interesting.

AUTOINHIBITION IN THE ABSENCE OF CARGO

Motor activity must be precisely regulated in cells to avoid futile ATP consumption and to ensure the proper subcellular localization of motors and their cargoes. In recent years, a general model has emerged that kinesin motors, when not bound to cargo, are kept in an inactive state by an autoinhibition mechanism. Autoinhibition allows motor activation to occur with precise spatial and temporal control.

Autoinhibition as a control mechanism was first described for kinesin-1, and to date, this remains the best-studied motor for autoinhibitory control (3, 135). Kinesin-1 motors exist in two conformations: an extended conformation in the active state and a folded conformation in the inactive state. Folding is due to hinge 2, which interrupts the long coiled-coil stalk of kinesin-1 (**Figure 2**) and allows C-terminal regions to interact with and inhibit the N-terminal motor domain (15, 25, 33, 42). A critical component of the autoinhibitory mechanism is the KHC tail domain, which interacts directly with a key enzymatic helix in the KHC motor domain and blocks intrinsic and microtubule-stimulated release of ADP from the nucleotide binding pocket (15, 25, 33, 42, 53, 55, 56, 127, 136, 139, 144). A second intramolecular mechanism for inhibition involves the KLC subunit. Studies on full-length kinesin-1 molecules expressed in bacteria and mammalian cells have shown that the KLC subunit suppresses motor activity, perhaps by physically separating the two motor domains (15, 25, 42, 136). However, recent studies with recombinant KHC and KLC fragments provide a contradictory view and suggest that the KLC subunits contribute to motor activation by blocking interactions of the KHC tail with both the KHC motor and the microtubule (145), although many of these interactions, in particular the tail-microtubule interaction, are based on ionic interactions and are thus sensitive to buffer conditions (139).

An autoinhibitory control mechanism has also been described for kinesin-3 motors. Full-length MmKIF1A and CeUNC-104 motors are inactive in motility assays (59, 104, 109), likely because of intramolecular interactions within the globular conformation of kinesin-3. The C-terminal half of CeUNC-104 or MmKIF1A is dispensable for autoinhibition as motors that are truncated after coil 2 remain inactive in single-molecule motility assays (59, 109). Rather, autoinhibition involves intramolecular interactions within the N-terminal half of kinesin-3 motors, in particular between the neck coil and coil 1 and between the FHA domain and coil 2 (5, 90, 148), suggesting that, like kinesin-1, kinesin-3 motors utilize dual autoinhibitory mechanisms. A potential mechanism for how the neck coil–coil 1 interaction functions in autoinhibition has been suggested on

MAP: microtubule-associated protein

the basis of biochemical and structural work. Motor constructs containing both the neck coil and coil 1 of MmKIF1A are monomeric and not processive, whereas removal of coil 1 results in dimeric processive motors (59, 90). Structural analysis demonstrated that the neck coil and coil 1 of CeUNC-104 can either self-fold in the monomer state or engage in intermolecular interactions in the dimeric state (5). Thus, neck coil–coil 1 interactions contribute to the regulation of both the autoinhibited state and the monomer-to-dimer transition, although a functional role in the full-length motor remains to be verified.

Homodimeric members of the kinesin-2 family are also regulated by autoinhibition. Full-length CeOSM-3 and HsKIF17 motors are inactive but can be activated by mutation of hinge 2 in the coiled-coil stalk domain that causes a corresponding shift to a more extended conformation (58, 68). The inactive folded conformation suggests that C-terminal regions could interact directly with and inhibit the N-terminal motor domain. Indeed, two intramolecular interactions that contribute to regulation were identified. First, the C-terminal tail interferes with microtubule binding of the motor domain, and second, coil 2 blocks processive motility (58). Thus, like autoinhibition of kinesin-1, regulation of homodimeric kinesin-2 motors utilizes two intramolecular mechanisms that collectively maintain the motor in an inactive state.

It remains largely unclear how the activity of heterotrimeric members of the kinesin-2 family is regulated. Purified native and recombinant forms of kinesin-2 are active in microtubule gliding assays and when attached to beads, yet they show no motility in single-molecule assays (8, 9, 14, 100, 109, 149), suggesting that kinesin-2 motors are regulated by autoinhibition. Recent work showed that autoinhibition is intrinsic to the motor-containing KIF3 subunits, as CeKLP-20/KLP-11 motors undergo few motility events even when attached to beads (14). Autoinhibition may involve the adoption of a folded conformation as purified sea urchin SpKRP85/KRP95/KAP

(*Strongylocentrotus purpuratus*) molecules undergo a salt-dependent shift to a more extended conformation (140).

Given the autoinhibited state of kinesin motors, the simplest model of motor activation posits that cargo binding relieves autoinhibition. Indeed, several motors are activated by binding of cellular cargo proteins. The first demonstration of kinesin activation was for kinesin-1. As two intramolecular interactions contribute to autoinhibition of kinesin-1 (the KHC globular tail and the KLC subunit), binding partners for each of these regions were required for activation of heterotetrameric motors (11). In the absence of KLC, only the KHC binding partner is needed for activation (22, 56). A kinesin-3 motor, GAKIN/KIF13B, is also activated by binding of a cargo protein (148). Although kinesin motors can additionally be regulated by mechanisms such as phosphorylation (135), these results provide a basis for understanding how kinesin activity is coupled to cargo binding.

MOVEMENT ALONG HETEROGENEOUS CELLULAR MICROTUBULES

The elegant biochemical and biophysical approaches that have elucidated the mechanics of kinesin motors have been carried out mostly with recombinant truncated proteins in dilute buffer conditions and along homogeneous populations of microtubules that are taxol stabilized and free of microtubule-associated proteins (MAPs). Such experimental systems lack relevant cargo and associated proteins and operate far from the physical and physiological constraints of the cytoplasm. Thus, how do the motile properties of kinesin motors in vitro compare to those in vivo?

The first step in understanding how kinesin motility is affected by cellular conditions is to compare the motile properties of single motors in vitro and in cells. In this regard, two approaches have been taken. Truncated kinesin motors either have been purified in recombinant form, labeled with quantum dots, and then introduced into cells via osmotic lysis of pinocytic vesicles or lipid-mediated transfection (24, 152), or kinesin motors have been genetically tagged with fluorescent proteins (FPs) and expressed endogenously in the cell of interest (16, 17). Overall, the motile properties of kinesin-1 in cells are similar to those observed previously in vitro, indicating that the basic movement of kinesin-1 along cellular microtubules is neither hindered by the crowded conditions in cells nor upregulated by unknown factors.

Further work used single-molecule imaging to compare the motile properties of kinesin-1 (KIF5C), kinesin-2 (KIF17), and kinesin-3 (KIF1A) motors in mammalian cells (16). In each case, the overall speed and processivity of the motor in cells were similar to its previously determined in vitro properties (16). However, differences between the motors were found in how they responded to the cellular complex of microtubules. The first difference observed was the ability of the motors to navigate the heterogeneous populations of microtubules generated by the cell. Kinesin-1 was observed to be a selective motor, preferentially utilizing stable microtubules marked by specific posttranslational modifications (PTMs). In contrast, kinesin-2 and kinesin-3 motors were not selective and utilized both stable and dynamic microtubules (16). The second difference observed between the motors was that kinesin-2 and kinesin-3 motors appear to tip-track on the plus ends of growing microtubules (16). For KIF17, this is due likely to interactions with plus-end-tracking proteins (+TIPs) (71). These results indicate that there are family-specific differences in how kinesin motors respond to the heterogeneous microtubule tracks that exist in cells.

Which cellular factors on microtubules could differentially influence kinesin motors? At least three differences in microtubule structure and/or composition can be envisaged. The first is tubulin isotype. In most organisms, multiple genes encode for α- and β-tubulins, resulting in a variety of isotypes (96). Most of the differences between isotypes are found in the

C-terminal tails, which provide chemical diversity on the microtubule surface and could thus influence the motor-microtubule interaction, although little work has been done to address this possibility.

The second factor that could influence kinesin motors is the PTMs of α- and β-tubulin subunits that mark subsets of microtubules in cells. To date, the best-studied PTMs of tubulin are acetylation of Lys40 in α-tubulin, detyrosination of α-tubulin, and polyglutamylation and polyglycylation of the C-terminal tails of α- and β-tubulin (**Figure 3**) (67, 134, 143). Kinesin-1 appears to be most sensitive to the

PTM:
posttranslational modification

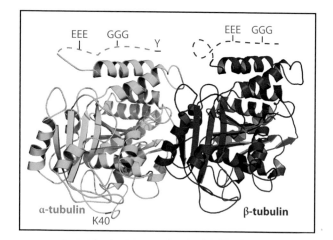

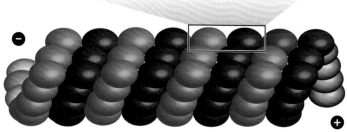

Figure 3

Microtubule posttranslational modifications (PTMs). Within the microtubule lattice, α- and β-tubulin subunits can be modified by various PTMs. The best-studied modifications are illustrated on the crystal structure of a single αβ-tubulin heterodimer (*top*, PDB code 1TUB). Acetylation of Lys40 (K40) of α-tubulin occurs in the lumen of the microtubule, whereas the other modifications occur on the microtubule surface. The C-terminal tyrosine (Y) residue of α-tubulin can be removed to generate detyrosinated tubulin. Glutamate residues in the C-terminal tails of both α- and β-tubulin can be modified by chains of glutamate (E) or glycine (G) residues to generate polyglutamylated and polyglycylated tubulin, respectively.

presence of these tubulin PTMs. Kinesin-1 binds with higher affinity to microtubules marked by acetylation and detyrosination (39, 80, 93, 113), and changing the levels of these modifications can alter kinesin-1 transport events in cells (35, 43, 60, 80, 113). Acetylation and detyrosination may provide two separate cues to kinesin-1. For example, in polarized hippocampal neurons, detyrosination but not acetylation can regulate the axon-specific targeting of kinesin-1 (60, 80). A molecular mechanism for the ability of kinesin-1 to distinguish acetylation and detyrosination cues was provided by Konishi & Setou (80), who showed that a tyrosine-glutamate-arginine-phenylalanine TERF sequence in the kinesin-1 (KIF5C) motor domain is critical for the recognition of detyrosinated but not acetylated microtubules.

Kinesin-2 and kinesin-3 motors appear less sensitive to acetylation and detyrosination marks on microtubules (16, 80). Rather, kinesin-3 motors may be more sensitive to polyglutamylation. In ROSA22 mice that lack functional PGs1, a subunit of α-tubulin-selective polyglutamylase, there was a decrease in the localization of kinesin-3 (KIF1A) motors to neurites and synaptic vesicle cargoes to axon terminals (66). Taken together, there exists significant evidence that PTMs of cellular microtubules can serve as kinesin-specific traffic signals and regulate intracellular trafficking events.

The third factor that could influence kinesin motors in cells is obstacles on microtubule tracks. For example, the MAPs that are believed to organize or stabilize microtubules (31) may pose roadblocks for kinesins. Most studies have focused on the neuronal MAP tau, but several studies have shown that MAP2C negatively influences the interaction of kinesin-1 motors with microtubules, whereas ensconsin promotes the recruitment of kinesin-1 to microtubules (6, 119, 129). In neuronal cells, overexpression of tau impairs kinesin-dependent transport with a specific reduction in attachment rates but not velocity of transport (40, 125, 132). Similar effects were found in single-molecule assays where tau reduced the attach-

ment rate of truncated kinesin-1 motors but had no effect on velocity (34, 119, 137). Whether tau affects the run length of kinesin-1 is less clear as different effects have been found in vivo and in vitro (119, 132, 137). When present at higher concentrations, tau can also regulate the motor off-rate as single kinesin-1 motors detached from the microtubule when they encountered patches of tau protein (34). Using beads coated with varying numbers of kinesin-1 motors, it was found that tau has a greater effect under multiple-motor conditions as tau-dependent decreases were observed in the attachment rate, run length, force production, and detachment rate of kinesin-1 (137). The finding that filamentous tau results in a decrease in the speed of vesicle transport in isolated squid axoplasm is another example of tau effects in a multimotor situation (86). Thus, although the effects of tau on kinesin-1 cannot always be directly compared between studies due to differences in tau isoform, concentration, and ability to form filaments, the general view is that tau negatively affects kinesin-1 motility, with the majority of studies noting an effect on attachment and detachment rates.

Other obstacles on microtubule tracks can also influence kinesin-1 motility. Using single-molecule imaging, Telley et al. (130) found that kinesin-1 motors detach when microtubules are crowded with active or nonmotile mutant kinesin motors, although others (38) have observed kinesin pausing and bypassing such obstacles. In similar assays, Korten & Diez (81) found that kinesin-1 motors stopped and/or paused when encountering streptavidin obstacles on biotinylated microtubules and that about half of the kinesin-1 molecules were able to overcome this obstacle. When faced with larger obstacles such as intersecting microtubules, single kinesin-1 motors were more likely to pass the obstacle (i.e., they stayed on the same track) than to detach (116). It was proposed that the presence of multiple motors on a bead or cargo would enable kinesin-1 to better navigate obstacles, and indeed, kinesin-1-coated beads were more likely to continue on the same track or cross to the intersecting track than to detach (116). Thus, the data available to date indicate that there are a variety of microtubule-associated factors in cells that can influence kinesin motility whether by directing transport to specific microtubule tracks or by altering the motor-microtubule interaction. It appears the effects of tau may be a specific biological function of this MAP, as kinesin-1 motors detach in the presence of tau but are able to navigate around other and larger obstacles. The specific influence of each factor on kinesin motors in the single and multiple motor states is a topic that is likely to occupy investigators for many years to come.

COORDINATION OF MULTIPLE MOTORS

Coordination Between Multiple Motors of the Same Type

The motile properties of cellular cargoes are due largely to the collective action of multiple motors present on the cargo surface. These can be multiple motors of the same type (e.g., multiple plus-end-directed motors or multiple kinesin-1 motors) or multiple motors of different directionality (e.g., kinesin-1 and dynein). Research on understanding how multiple motors of the same type coordinate their motions to work collectively has focused mostly on kinesin-1 and understanding the effects on intrinsic motor properties (e.g., velocity, processivity, and force generation), although there is also evidence that step size can be altered in the multimotor situation (89).

Whether kinesin-1 motors cooperate to increase the speed of motion has been controversial. In now-classic experiments using microtubule gliding assays, the speed of microtubule translocation was independent of the number of kinesin-1 motors at saturating ATP concentration and under no load (64). Even under load conditions, multiple kinesin-1 motors do not increase the speed of movement beyond that of a single motor (65, 79, 97, 138). It is thought that the resistive load applied by an optical trap, or a cargo moving in

the cytoplasm, is distributed among the motors such that each motor experiences a negligible load (83). That cargo velocity is insensitive to motor number is supported by an analysis of lipid droplets in *Drosophila* embryos where the average velocity of lipid droplet motion was largely unaffected by a genetic 50% reduction in kinesin-1 concentration (123) and live-cell imaging of intraflagellar transport particles in *Chlamydomonas* flagella (85). Yet a thorough understanding of how multiple motors coordinate their activities requires precise control over motor number. In this respect, it is exciting to see recent studies using linear DNA duplexes to link exactly two kinesin-1 motors. In this case, a two-motor construct showed no change in velocity at low load but could maintain single-motor velocities at high load (69, 114), again arguing that multiple motors of the same type do not cooperate to increase the speed of their collective motion.

However, several studies tracking vesicles or organelles in live cells have shown that the velocity histograms contain regularly spaced peaks (62, 84, 122, 154). Although the similarity of velocity profiles across transport events, cells, and species argues that these peaks reflect a shared biological phenomenon, a description of the molecular events that contribute to this phenomenon is yet unclear. One possibility is that the stepwise changes between velocity peaks indicate the addition or removal of active motors from the cargo surface (62, 84, 92, 122). This possibility seems incompatible with the in vitro studies described above and subsequent in vivo work (85, 137). A second possibility is that the velocity peaks occur when motors work in the viscous cytoplasmic environment, as similar distinct peaks in the instantaneous velocity histogram were identified when kinesin-1 worked against a load in microtubule gliding assays (44). A third possibility is that motor activity can be regulated by cellular systems such that motors are capable of faster speeds in vivo than in vitro. While single truncated motors display motile properties in vivo similar to those in vitro (16, 17, 24, 152), the motile properties of cargo-attached motors can be modulated by binding partners to produce faster speeds and/or higher forces (95). A fourth possibility is that the faster speeds in vivo result from motors and cargoes piggy-backing on microtubules that are themselves moved as cargoes along other tracks. Strong support for this possibility comes from experiments in *Drosophila* S2 cells where a photoswitchable tubulin construct was used to specifically mark and track microtubules (72). A final possibility is that the velocity peaks indicate transport events carried by different motors. Support for this possibility comes from the fact that genetic ablation of the kinesin-3 motor CeUNC-104 specifically decreased the fast velocities of dense core vesicle transport events in neurons (154). In addition, studies on intraflagellar transport particles in the sensory cilia of *C. elegans* chemosensory neurons have shown that fast (homodimeric CeOSM-3) and slow (heterotrimeric kinesin-2) motors can cooperate to produce intermediate velocities both in vivo in the initial segment of cilia and in vitro in microtubule gliding assays (105, 108). Similar in vitro assays using fast (wild-type kinesin-1) and slow (mutant kinesin-1) motors found that even a small number of fast motors are able to dominate motility and shift the balance in their favor (87).

Cellular cargoes generally move for longer distance along microtubule tracks than single motors do in vitro, most likely due to cooperation between motors that results in an increase in run length. This was first shown by Block et al. (12), who used kinesin-1-coated beads in optical trapping experiments to reveal that beads driven by two kinesin-1 motors moved longer distances than those driven by single motors. Following up on these experiments, the Gross lab determined detailed statistics for motor number and run length for kinesin-1 motors attached to beads (137). In particular, these results showed that beads driven by two motors of the same type can move more than 8 μm, much longer than the average run length of single kinesin-1 motors. Interestingly, recent experiments using two kinesin-1 motors linked by a 50-nm DNA duplex found that the average run length of two linked motors is only ~1.5-fold higher than that of a single kinesin-1 motor (114). These latter experiments are more in line with theoretical work that predicted a relatively small increase in run length for cooperating motors of the same type (78). The differences between these studies can be explained by the arrangement of the motors. It seems that reattachment to the microtubule track is more likely to occur for motors nonspecifically adsorbed to a bead surface, and perhaps also for motors attached to vesicular cargoes in cells, than for a group of clustered motors. Thus, in contrast to the effects of motor coordination on speed, there is good agreement that multiple motors of the same type can work collectively to increase the distance traveled.

There is also good agreement that multiple motors can cooperate for increased force generation, particularly for multiple motors attached to beads and analyzed in vitro (137). Force measurements of lipid droplets in *Drosophila* embryos showed that the forces exerted by multiple kinesin-1 motors are nearly additive (123), in agreement with theoretical work (83). Whether individual motors generate equivalent forces in vitro and in vivo is less clear. Soppina et al. (124) measured the average force exerted by motors on endosomes in *Dictyostelium* and concluded that the total force exerted by kinesin-3 (DdUnc104) in vitro and in vivo is similar. In contrast, force measurements of lipid droplets in *Drosophila* embryos showed that the force exerted by a single motor in vivo was smaller than that determined in vitro (123). Clearly, more work needs to be done to better understand the relationship between motor number and force production on a vesicle surface as well as the influence of parameters intrinsic to cellular systems such as motor-cargo linker stiffness, cellular viscosity and other resistive forces, and stochastic effects such as load sharing (82).

Coordination Between Motors of Different Types

Cooperation between motors of different types (e.g., kinesin-1 and dynein) results in

Tug-of-war

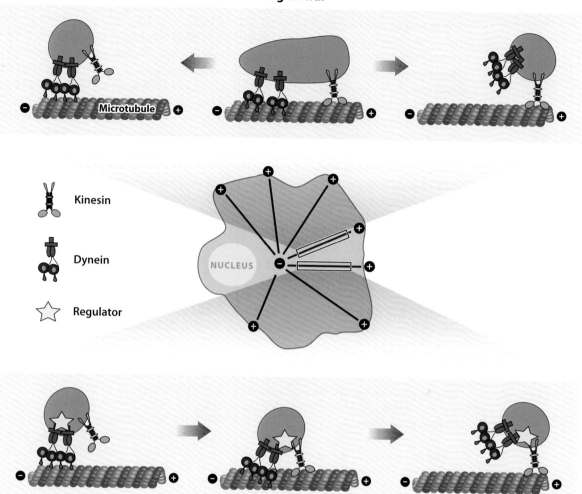

Regulated switching

Figure 4

Models for bidirectional cargo transport. Many cargoes in eukaryotic cells contain both kinesin and dynein motors and thus move in a bidirectional manner where periods of persistent motion are interrupted by reversals in directions. Two models have been put forth to explain bidirectional transport: the regulated switching model and the tug-of-war model. In the regulated switching model, the activity of kinesin and dynein motors is regulated to ensure that only one type of motor is active at any time. In the tug-of-war model, the two types of motors generate force against each other such that one motor wins to ensure a reversal in its favor.

bidirectional motility. In some cases, the cargo moves for long distances in each direction before undergoing a remodeling and/or switching event to generate movement in the opposite direction. For example, intraflagellar transport in cilia and flagella is carried out by kinesin-2 and cytoplasmic dynein motors, with the change in motor activity, and thus cargo direction, occurring at the distal end of the cilium (118). Recent imaging in live *Chlamydomonas* cells showed that these motors are reciprocally coordinated for exclusive transport in one direction (85).

In other cases, bidirectional movement involves short periods of directed motion interrupted by reversals in direction. Two popular models have been put forth to explain reversals: regulated switching of motor activity or tug-of-war between opposite motors (**Figure 4**) (49, 141). A complete understanding of bidirectional transport will require methods to determine the number of active motors in real time. However, as both models have considerable experimental support, it appears that regulated switching and tug-of-war mechanisms are employed in cells in a cargo-specific manner.

In the regulated switching model, minus-end- and plus-end-directed motors alternate their activities due to regulatory mechanisms that rapidly activate and inactivate motors in a mutually exclusive manner. In some cases of bidirectional transport, inhibition or removal of one motor abolishes transport in both directions rather than enhances transport in the opposite direction (142). This indicates, first, that the motors are not competing with each other and, second, that both motors are required for the regulatory systems to operate. The regulatory systems, such as the dynactin complex, have been envisioned to involve specific cargo proteins that regulate individual motors (29, 50, 57, 110).

The tug-of-war model requires that motors of opposite directionality pull against each other and that one class of motor wins to effect a reversal in its favor. Experimental support for this model has come from studies on endosomes purified from *Dictyostelium* cells (124). During motility events, the endosomes were observed to elongate in morphology and slow in speed as they underwent reversals in direction due to a tug-of-war between motors with opposite directionality. Further support has come from recent work using endosomal vesicles purified from mouse brain. Quantitative analysis determined that low numbers of kinesin and dynein motors are stably attached to the vesicles and drive bidirectional motility via force-dependent interactions that exert a tug-of-war on the vesi-

cle (61). Evidence that bidirectional transport is intrinsic to the motors themselves comes from analysis of peroxisome transport in *Drosophila* S2 cells. By knocking down kinesin-1 and cytoplasmic dynein motors and replacing them with various plus-end- or minus-end-directed motor constructs, Ally et al. (7) determined that the specific linkage of the motor to the organelle is irrelevant, that the specific plus-end- or minus-end-directed motor is irrelevant, and that the presence of active motors capable of ATP-dependent motility is required for bidirectional motility. These results demonstrate that bidirectional motility is not due to motor-specific sequences or regulators but rather that mechanical tension between opposite-polarity motors is sufficient for activation of motility in both directions.

CONCLUSIONS

In this review, we have highlighted recent work on how the assembly and motility properties of kinesin motors give rise to intracellular transport events, particularly in metazoan cells. Future work is likely to continue to yield important and exciting information on the motile properties of kinesin motors. First, it is still unclear whether the hand-over-hand mechanism that drives kinesin-1 motility is generally applicable to kinesin motors that drive intracellular trafficking events. Quantitative and qualitative analyses of various kinesin motors in traditional in vitro assays are likely to continue to provide important information on how the kinesin motor domain has been adapted over evolution to provide specific chemomechanical functions. Second, it will be important to understand how multiple motors present on the same cargo coordinate their motilities. Third, the influence of biochemical cues on the microtubule tracks will continue to provide mechanisms for cells to segregate and guide cargo transport. Finally, how these mechanisms are integrated in cells is the ultimate challenge in understanding intracellular trafficking events.

SUMMARY POINTS

1. The kinesin motor domain has been expanded and adapted over evolution to generate families of kinesins with distinct oligomerization states, cargo-carrying capabilities, and regulation mechanisms.

2. The presence of two motor domains in a dimeric kinesin molecule allows for processive motility.

3. In the absence of cargo, kinesin motors are regulated by an autoinhibition mechanism in which intramolecular interactions maintain the motor in an inactive state.

4. Kinesin motors can read and respond to biochemical cues on microtubule tracks, including PTMs and MAPs, in order to segregate and direct trafficking events.

5. Most cargoes are carried by multiple motors that work collectively to increase force and run length but not velocity of movement.

6. Bidirectional cargo transport can result from regulated switching between plus-end- and minus-end-directed motors or from tug-of-war, in which directionality is determined by the dominant motor.

FUTURE ISSUES

1. Mechanisms of motility need to be determined for members of the kinesin-2 and kinesin-3 families. For heterotrimeric kinesin-2 motors, the requirement for two different motor domains is unclear, whereas for kinesin-3 motors, the oligomerization state and mechanisms of processive motility remain to be determined.

2. How kinesin motors navigate the complex cellular environment is an area of active research. The contributions of tubulin isotypes, PTMs, MAPs, and other obstacles to the regulation and direction of motor transport need to be worked out. These factors may work together or may provide conflicting cues to the motor.

3. Future work is also needed to understand how multiple motors cooperate to increase transport fidelity without interfering with each other's motion. The contribution of regulated switching and tug-of-war models to the transport of various cargoes needs to be solved. And other models need to be developed for transport events that are not adequately described by these models.

DISCLOSURE STATEMENT

The authors are not aware of any affiliations, memberships, funding, or financial holdings that might be perceived as affecting the objectivity of this review.

ACKNOWLEDGMENTS

We are grateful to Jay Pieczynski for help with figures and to Edgar Meyhofer for many fruitful discussions. Work in the lab of KJV is supported by the National Institutes of Health and the Human Frontier Science Program.

LITERATURE CITED

1. Adio S, Bloemink M, Hartel M, Leier S, Geeves MA, Woehlke G. 2006. Kinetic and mechanistic basis of the nonprocessive kinesin-3 motor NcKin3. *J. Biol. Chem.* 281:37782–93

2. Adio S, Jaud J, Ebbing B, Rief M, Woehlke G. 2009. Dissection of kinesin's processivity. *PLoS One* 4:e4612

3. Adio S, Reth J, Bathe F, Woehlke G. 2006. Review: regulation mechanisms of kinesin-1. *J. Muscle Res. Cell Motil.* 27:153–60

4. Adio S, Woehlke G. 2009. Properties of the kinesin-3 NcKin3 motor domain and implications for neck function. *FEBS J.* 276:3641–55

5. Al-Bassam J, Cui Y, Klopfenstein D, Carragher BO, Vale RD, Milligan RA. 2003. Distinct conformations of the kinesin Unc104 neck regulate a monomer to dimer motor transition. *J. Cell Biol.* 163:743–53

6. Al-Bassam J, Ozer RS, Safer D, Halpain S, Milligan RA. 2002. MAP2 and tau bind longitudinally along the outer ridges of microtubule protofilaments. *J. Cell Biol.* 157:1187–96

7. Ally S, Larson AG, Barlan K, Rice SE, Gelfand VI. 2009. Opposite-polarity motors activate one another to trigger cargo transport in live cells. *J. Cell Biol.* 187:1071–82

8. Berezuk MA, Schroer TA. 2004. Fractionation and characterization of kinesin II species in vertebrate brain. *Traffic* 5:503–13

9. Berezuk MA, Schroer TA. 2007. Dynactin enhances the processivity of kinesin-2. *Traffic* 8:124–29

10. Bieling P, Telley IA, Piehler J, Surrey T. 2008. Processive kinesins require loose mechanical coupling for efficient collective motility. *EMBO Rep.* 9:1121–27

11. Blasius TL, Cai D, Jih GT, Toret CP, Verhey KJ. 2007. Two binding partners cooperate to activate the molecular motor kinesin-1. *J. Cell Biol.* 176:11–17

12. Block SM, Goldstein LS, Schnapp BJ. 1990. Bead movement by single kinesin molecules studied with optical tweezers. *Nature* 348:348–52

13. Brendza RP, Serbus LR, Duffy JB, Saxton WM. 2000. A function for kinesin I in the posterior transport of oskar mRNA and Staufen protein. *Science* 289:2120–22

14. Brunnbauer M, Mueller-Planitz F, Kosem S, Ho TH, Dombi R, et al. 2010. Regulation of a heterodimeric kinesin-2 through an unprocessive motor domain that is turned processive by its partner. *Proc. Natl. Acad. Sci. USA* 107:10460–65

15. Cai D, Hoppe AD, Swanson JA, Verhey KJ. 2007. Kinesin-1 structural organization and conformational changes revealed by FRET stoichiometry in live cells. *J. Cell Biol.* 176:51–63

16. Cai D, McEwen DP, Martens JR, Meyhofer E, Verhey KJ. 2009. Single molecule imaging reveals differences in microtubule track selection between kinesin motors. *PLoS Biol.* 7:e1000216

17. Cai D, Verhey KJ, Meyhofer E. 2007. Tracking single kinesin molecules in the cytoplasm of mammalian cells. *Biophys. J.* 92:4137–44

18. Cai Y, Singh BB, Aslanukov A, Zhao H, Ferreira PA. 2001. The docking of kinesins, KIF5B and KIF5C, to Ran-binding protein 2 (RanBP2) is mediated via a novel RanBP2 domain. *J. Biol. Chem.* 276:41594–602

19. Chana M, Tripet BP, Mant CT, Hodges RS. 2002. The role of unstructured highly charged regions on the stability and specificity of dimerization of two-stranded alpha-helical coiled-coils: analysis of the neck-hinge region of the kinesin-like motor protein Kif3A. *J. Struct. Biol.* 137:206–19

20. Chana MS, Tripet BP, Mant CT, Hodges R. 2005. Stability and specificity of heterodimer formation for the coiled-coil neck regions of the motor proteins Kif3A and Kif3B: the role of unstructured oppositely charged regions. *J. Pept. Res.* 65:209–20

21. Cho KI, Cai Y, Yi H, Yeh A, Aslanukov A, Ferreira PA. 2007. Association of the kinesin-binding domain of RanBP2 to KIF5B and KIF5C determines mitochondria localization and function. *Traffic* 8:1722–35

22. Cho KI, Yi H, Desai R, Hand AR, Haas AL, Ferreira PA. 2009. RANBP2 is an allosteric activator of the conventional kinesin-1 motor protein, KIF5B, in a minimal cell-free system. *EMBO Rep.* 10:480–86

23. Cole DG, Chinn SW, Wedaman KP, Hall K, Vuong T, Scholey JM. 1993. Novel heterotrimeric kinesin-related protein purified from sea urchin eggs. *Nature* 366:268–70

24. Courty S, Luccardini C, Bellaiche Y, Cappello G, Dahan M. 2006. Tracking individual kinesin motors in living cells using single quantum-dot imaging. *Nano Lett.* 6:1491–95

25. Coy DL, Hancock WO, Wagenbach M, Howard J. 1999. Kinesin's tail domain is an inhibitory regulator of the motor domain. *Nat. Cell Biol.* 1:288–92

26. Crevenna AH, Madathil S, Cohen DN, Wagenbach M, Fahmy K, Howard J. 2008. Secondary structure and compliance of a predicted flexible domain in kinesin-1 necessary for cooperation of motors. *Biophys. J.* 95:5216–27

27. De Marco V, Burkhard P, Le Bot N, Vernos I, Hoenger A. 2001. Analysis of heterodimer formation by Xklp3A/B, a newly cloned kinesin-II from *Xenopus laevis*. *EMBO J.* 20:3370–79

28. De Marco V, De Marco A, Goldie KN, Correia JJ, Hoenger A. 2003. Dimerization properties of a *Xenopus laevis* kinesin-II carboxy-terminal stalk fragment. *EMBO Rep.* 4:717–22

29. Deacon SW, Serpinskaya AS, Vaughan PS, Lopez Fanarraga M, Vernos I, et al. 2003. Dynactin is required for bidirectional organelle transport. *J. Cell Biol.* 160:297–301

30. DeBoer SR, You Y, Szodorai A, Kaminska A, Pigino G, et al. 2008. Conventional kinesin holoenzymes are composed of heavy and light chain homodimers. *Biochemistry* 47:4535–43

31. Dehmelt L, Halpain S. 2005. The MAP2/Tau family of microtubule-associated proteins. *Genome Biol.* 6:204

32. DeLuca JG, Newton CN, Himes RH, Jordan MA, Wilson L. 2001. Purification and characterization of native conventional kinesin, HSET, and CENP-E from mitotic HeLa cells. *J. Biol. Chem.* 276:28014–21

33. Dietrich KA, Sindelar CV, Brewer PD, Downing KH, Cremo CR, Rice SE. 2008. The kinesin-1 motor protein is regulated by a direct interaction of its head and tail. *Proc. Natl. Acad. Sci. USA* 105:8938–43

34. Dixit R, Ross JL, Goldman YE, Holzbaur EL. 2008. Differential regulation of dynein and kinesin motor proteins by tau. *Science* 319:1086–89

35. Dompierre JP, Godin JD, Charrin BC, Cordelieres FP, King SJ, et al. 2007. Histone deacetylase 6 inhibition compensates for the transport deficit in Huntington's disease by increasing tubulin acetylation. *J. Neurosci.* 27:3571–83

36. Doodhi H, Ghosal D, Krishnamurthy M, Jana SC, Shamala D, et al. 2009. KAP, the accessory subunit of kinesin-2, binds the predicted coiled-coil stalk of the motor subunits. *Biochemistry* 48:2248–60

37. Dorner C, Ullrich A, Haring HU, Lammers R. 1999. The kinesin-like motor protein KIF1C occurs in intact cells as a dimer and associates with proteins of the 14-3-3 family. *J. Biol. Chem.* 274:33654–60

38. Dreblow K, Kalchishkova N, Bohm KJ. 2010. Kinesin passing permanent blockages along its protofilament track. *Biochem. Biophys. Res. Commun.* 395:490–95

39. Dunn S, Morrison EE, Liverpool TB, Molina-Paris C, Cross RA, et al. 2008. Differential trafficking of Kif5c on tyrosinated and detyrosinated microtubules in live cells. *J. Cell Sci.* 121:1085–95

40. Ebneth A, Godemann R, Stamer K, Illenberger S, Trinczek B, Mandelkow E. 1998. Overexpression of tau protein inhibits kinesin-dependent trafficking of vesicles, mitochondria, and endoplasmic reticulum: implications for Alzheimer's disease. *J. Cell Biol.* 143:777–94

41. Elluru RG, Bloom GS, Brady ST. 1995. Fast axonal transport of kinesin in the rat visual system: functionality of kinesin heavy chain isoforms. *Mol. Biol. Cell* 6:21–40

42. Friedman DS, Vale RD. 1999. Single-molecule analysis of kinesin motility reveals regulation by the cargo-binding tail domain. *Nat. Cell Biol.* 1:293–97

43. Friedman JR, Webster BM, Mastronarde DN, Verhey KJ, Voeltz GK. 2010. ER sliding dynamics and ER-mitochondrial contacts occur on acetylated microtubules. *J. Cell Biol.* 190:363–75

44. Gagliano J, Walb M, Blaker B, Macosko JC, Holzwarth G. 2010. Kinesin velocity increases with the number of motors pulling against viscoelastic drag. *Eur. Biophys. J.* 39:801–13

45. Gennerich A, Vale RD. 2009. Walking the walk: how kinesin and dynein coordinate their steps. *Curr. Opin. Cell Biol.* 21:59–67

46. Gerdes JM, Davis EE, Katsanis N. 2009. The vertebrate primary cilium in development, homeostasis, and disease. *Cell* 137:32–45

47. Glater EE, Megeath LJ, Stowers RS, Schwarz TL. 2006. Axonal transport of mitochondria requires milton to recruit kinesin heavy chain and is light chain independent. *J. Cell Biol.* 173:545–57

48. Glotzer M. 2009. The 3Ms of central spindle assembly: microtubules, motors and MAPs. *Nat. Rev. Mol. Cell Biol.* 10:9–20

49. Gross SP. 2004. Hither and yon: a review of bi-directional microtubule-based transport. *Phys. Biol.* 1:R1–11

50. Gross SP, Welte MA, Block SM, Wieschaus EF. 2002. Coordination of opposite-polarity microtubule motors. *J. Cell Biol.* 156:715–24

51. Gyoeva FK, Bybikova EM, Minin AA. 2000. An isoform of kinesin light chain specific for the Golgi complex. *J. Cell Sci.* 113:2047–54

52. Gyoeva FK, Sarkisov DV, Khodjakov AL, Minin AA. 2004. The tetrameric molecule of conventional kinesin contains identical light chains. *Biochemistry* 43:13525–31

53. Hackney DD, Baek N, Snyder AC. 2009. Half-site inhibition of dimeric kinesin head domains by monomeric tail domains. *Biochemistry* 48:3448–56

54. Hackney DD, Levitt JD, Wagner DD. 1991. Characterization of alpha 2 beta 2 and alpha 2 forms of kinesin. *Biochem. Biophys. Res. Commun.* 174:810–15

55. Hackney DD, Stock MF. 2000. Kinesin's IAK tail domain inhibits initial microtubule-stimulated ADP release. *Nat. Cell Biol.* 2:257–60

56. Hackney DD, Stock MF. 2008. Kinesin tail domains and Mg2+ directly inhibit release of ADP from head domains in the absence of microtubules. *Biochemistry* 47:7770–78

57. Haghnia M, Cavalli V, Shah SB, Schimmelpfeng K, Brusch R, et al. 2007. Dynactin is required for coordinated bidirectional motility, but not for dynein membrane attachment. *Mol. Biol. Cell* 18:2081–89

58. Hammond JW, Blasius TL, Soppina V, Cai D, Verhey KJ. 2010. Autoinhibition of the kinesin-2 motor KIF17 via dual intramolecular mechanisms. *J. Cell Biol.* 189:1013–25

59. Hammond JW, Cai D, Blasius TL, Li Z, Jiang Y, et al. 2009. Mammalian kinesin-3 motors are dimeric in vivo and move by processive motility upon release of autoinhibition. *PLoS Biol.* 7:e72

60. Hammond JW, Huang CF, Kaech S, Jacobson C, Banker G, Verhey KJ. 2010. Posttranslational modifications of tubulin and the polarized transport of kinesin-1 in neurons. *Mol. Biol. Cell* 21:572–83

61. Hendricks AG, Perlson E, Ross JL, Schroeder HW 3rd, Tokito M, Holzbaur EL. 2010. Motor coordination via a tug-of-war mechanism drives bidirectional vesicle transport. *Curr. Biol.* 20:697–702

62. Hill DB, Plaza MJ, Bonin K, Holzwarth G. 2004. Fast vesicle transport in PC12 neurites: velocities and forces. *Eur. Biophys. J.* 33:623–32

63. Hirokawa N, Noda Y, Tanaka Y, Niwa S. 2009. Kinesin superfamily motor proteins and intracellular transport. *Nat. Rev. Mol. Cell Biol.* 10:682–96

64. Howard J, Hudspeth AJ, Vale RD. 1989. Movement of microtubules by single kinesin molecules. *Nature* 342:154–58

65. Hunt AJ, Gittes F, Howard J. 1994. The force exerted by a single kinesin molecule against a viscous load. *Biophys. J.* 67:766–81

66. Ikegami K, Heier RL, Taruishi M, Takagi H, Mukai M, et al. 2007. Loss of alpha-tubulin polyglutamylation in ROSA22 mice is associated with abnormal targeting of KIF1A and modulated synaptic function. *Proc. Natl. Acad. Sci. USA* 104:3213–18

67. Ikegami K, Setou M. 2010. Unique post-translational modifications in specialized microtubule architecture. *Cell Struct. Funct.* 35:15–22

68. Imanishi M, Endres NF, Gennerich A, Vale RD. 2006. Autoinhibition regulates the motility of the *C. elegans* intraflagellar transport motor OSM-3. *J. Cell Biol.* 174:931–37

69. Jamison DK, Driver J, Rogers A, Constantinou P, Diehl MR. 2010. Two kinesins transport cargo primarily via the action of one motor: implications for intracellular transport. *Biophys. J.* 99:2967–77

70. Jaud J, Bathe F, Schliwa M, Rief M, Woehlke G. 2006. Flexibility of the neck domain enhances kinesin-1 motility under load. *Biophys. J.* 91:1407–12

71. Jaulin F, Kreitzer G. 2010. KIF17 stabilizes microtubules and contributes to epithelial morphogenesis by acting at MT plus ends with EB1 and APC. *J. Cell Biol.* 190:443–60

72. Jolly AL, Kim H, Srinivasan D, Lakonishok M, Larson AG, Gelfand VI. 2010. Kinesin-1 heavy chain mediates microtubule sliding to drive changes in cell shape. *Proc. Natl. Acad. Sci. USA* 107:12151–56

73. Kanai Y, Dohmae N, Hirokawa N. 2004. Kinesin transports RNA: isolation and characterization of an RNA-transporting granule. *Neuron* 43:513–25

74. Kardon JR, Vale RD. 2009. Regulators of the cytoplasmic dynein motor. *Nat. Rev. Mol. Cell Biol.* 10:854–65

75. Kikkawa M, Okada Y, Hirokawa N. 2000. 15 Å resolution model of the monomeric kinesin motor, KIF1A. *Cell* 100:241–52

76. Klopfenstein DR, Tomishige M, Stuurman N, Vale RD. 2002. Role of phosphatidylinositol(4,5)bisphosphate organization in membrane transport by the Unc104 kinesin motor. *Cell* 109:347–58
77. Klopfenstein DR, Vale RD. 2004. The lipid binding pleckstrin homology domain in UNC-104 kinesin is necessary for synaptic vesicle transport in *Caenorhabditis elegans*. *Mol. Biol. Cell* 15:3729–39
78. Klumpp S, Lipowsky R. 2005. Cooperative cargo transport by several molecular motors. *Proc. Natl. Acad. Sci. USA* 102:17284–89
79. Kojima H, Muto E, Higuchi H, Yanagida T. 1997. Mechanics of single kinesin molecules measured by optical trapping nanometry. *Biophys. J.* 73:2012–22
80. Konishi Y, Setou M. 2009. Tubulin tyrosination navigates the kinesin-1 motor domain to axons. *Nat. Neurosci.* 12:559–67
81. Korten T, Diez S. 2008. Setting up roadblocks for kinesin-1: mechanism for the selective speed control of cargo carrying microtubules. *Lab. Chip* 8:1441–47
82. Kunwar A, Mogilner A. 2010. Robust transport by multiple motors with nonlinear force-velocity relations and stochastic load sharing. *Phys. Biol.* 7:16012
83. Kunwar A, Vershinin M, Xu J, Gross SP. 2008. Stepping, strain gating, and an unexpected force-velocity curve for multiple-motor-based transport. *Curr. Biol.* 18:1173–83
84. Kural C, Kim H, Syed S, Goshima G, Gelfand VI, Selvin PR. 2005. Kinesin and dynein move a peroxisome in vivo: a tug-of-war or coordinated movement? *Science* 308:1469–72
85. Laib JA, Marin JA, Bloodgood RA, Guilford WH. 2009. The reciprocal coordination and mechanics of molecular motors in living cells. *Proc. Natl. Acad. Sci. USA* 106:3190–95
86. LaPointe NE, Morfini G, Pigino G, Gaisina IN, Kozikowski AP, et al. 2009. The amino terminus of tau inhibits kinesin-dependent axonal transport: implications for filament toxicity. *J. Neurosci. Res.* 87:440–51
87. Larson AG, Landahl EC, Rice SE. 2009. Mechanism of cooperative behaviour in systems of slow and fast molecular motors. *Phys. Chem. Chem. Phys.* 11:4890–98
88. Lawrence CJ, Dawe RK, Christie KR, Cleveland DW, Dawson SC, et al. 2004. A standardized kinesin nomenclature. *J. Cell Biol.* 167:19–22
89. Leduc C, Ruhnow F, Howard J, Diez S. 2007. Detection of fractional steps in cargo movement by the collective operation of kinesin-1 motors. *Proc. Natl. Acad. Sci. USA* 104:10847–52
90. Lee JR, Shin H, Choi J, Ko J, Kim S, et al. 2004. An intramolecular interaction between the FHA domain and a coiled coil negatively regulates the kinesin motor KIF1A. *EMBO J.* 23:1506–15
91. Lee JR, Shin H, Ko J, Choi J, Lee H, Kim E. 2003. Characterization of the movement of the kinesin motor KIF1A in living cultured neurons. *J. Biol. Chem.* 278:2624–29
92. Levi V, Serpinskaya AS, Gratton E, Gelfand V. 2006. Organelle transport along microtubules in *Xenopus melanophores*: evidence for cooperation between multiple motors. *Biophys. J.* 90:318–27
93. Liao G, Gundersen GG. 1998. Kinesin is a candidate for cross-bridging microtubules and intermediate filaments. Selective binding of kinesin to detyrosinated tubulin and vimentin. *J. Biol. Chem.* 273:9797–803
94. Ling SC, Fahrner PS, Greenough WT, Gelfand VI. 2004. Transport of *Drosophila fragile* X mental retardation protein-containing ribonucleoprotein granules by kinesin-1 and cytoplasmic dynein. *Proc. Natl. Acad. Sci. USA* 101:17428–33
95. Loiseau P, Davies T, Williams LS, Mishima M, Palacios IM. 2010. *Drosophila* PAT1 is required for kinesin-1 to transport cargo and to maximize its motility. *Development* 137:2763–72
96. Luduena RF. 1998. Multiple forms of tubulin: different gene products and covalent modifications. *Int. Rev. Cytol.* 178:207–75
97. Meyhofer E, Howard J. 1995. The force generated by a single kinesin molecule against an elastic load. *Proc. Natl. Acad. Sci. USA* 92:574–78
98. Miki H, Okada Y, Hirokawa N. 2005. Analysis of the kinesin superfamily: insights into structure and function. *Trends Cell Biol.* 15:467–76
99. Muresan V, Abramson T, Lyass A, Winter D, Porro E, et al. 1998. KIF3C and KIF3A form a novel neuronal heteromeric kinesin that associates with membrane vesicles. *Mol. Biol. Cell* 9:637–52
100. Muthukrishnan G, Zhang Y, Shastry S, Hancock WO. 2009. The processivity of kinesin-2 motors suggests diminished front-head gating. *Curr. Biol.* 19:442–47
101. Okada Y, Higuchi H, Hirokawa N. 2003. Processivity of the single-headed kinesin KIF1A through biased binding to tubulin. *Nature* 424:574–77

102. Okada Y, Hirokawa N. 1999. A processive single-headed motor: kinesin superfamily protein KIF1A. *Science* 283:1152–57

103. Okada Y, Hirokawa N. 2000. Mechanism of the single-headed processivity: diffusional anchoring between the K-loop of kinesin and the C terminus of tubulin. *Proc. Natl. Acad. Sci. USA* 97:640–45

104. Okada Y, Yamazaki H, Sekine-Aizawa Y, Hirokawa N. 1995. The neuron-specific kinesin superfamily protein KIF1A is a unique monomeric motor for anterograde axonal transport of synaptic vesicle precursors. *Cell* 81:769–80

105. Ou G, Blacque OE, Snow JJ, Leroux MR, Scholey JM. 2005. Functional coordination of intraflagellar transport motors. *Nature* 436:583–87

106. Palacios IM, Johnston DS. 2002. Kinesin light chain-independent function of the kinesin heavy chain in cytoplasmic streaming and posterior localisation in the *Drosophila* oocyte. *Development* 129:5473–85

107. Pan X, Acar S, Scholey JM. 2010. Torque generation by one of the motor subunits of heterotrimeric kinesin-2. *Biochem. Biophys. Res. Comm.* 401:53–57

108. Pan X, Ou G, Civelekoglu-Scholey G, Blacque OE, Endres NF, et al. 2006. Mechanism of transport of IFT particles in *C. elegans* cilia by the concerted action of kinesin-II and OSM-3 motors. *J. Cell Biol.* 174:1035–45

109. Pierce DW, Hom-Booher N, Otsuka AJ, Vale RD. 1999. Single-molecule behavior of monomeric and heteromeric kinesins. *Biochemistry* 38:5412–21

110. Pilling AD, Horiuchi D, Lively CM, Saxton WM. 2006. Kinesin-1 and dynein are the primary motors for fast transport of mitochondria in *Drosophila* motor axons. *Mol. Biol. Cell* 17:2057–68

111. Pollock N, de Hostos EL, Turck CW, Vale RD. 1999. Reconstitution of membrane transport powered by a novel dimeric kinesin motor of the Unc104/KIF1A family purified from *Dictyostelium*. *J. Cell Biol.* 147:493–506

112. Rashid DJ, Wedaman KP, Scholey JM. 1995. Heterodimerization of the two motor subunits of the heterotrimeric kinesin, KRP85/95. *J. Mol. Biol.* 252:157–62

113. Reed NA, Cai D, Blasius TL, Jih GT, Meyhofer E, et al. 2006. Microtubule acetylation promotes kinesin-1 binding and transport. *Curr. Biol.* 16:2166–72

114. Rogers AR, Driver JW, Constantinou PE, Kenneth Jamison D, Diehl MR. 2009. Negative interference dominates collective transport of kinesin motors in the absence of load. *Phys. Chem. Chem. Phys.* 11:4882–89

115. Rogers KR, Weiss S, Crevel I, Brophy PJ, Geeves M, Cross R. 2001. KIF1D is a fast non-processive kinesin that demonstrates novel K-loop-dependent mechanochemistry. *EMBO J.* 20:5101–13

116. Ross JL, Shuman H, Holzbaur EL, Goldman YE. 2008. Kinesin and dynein-dynactin at intersecting microtubules: Motor density affects dynein function. *Biophys. J.* 94:3115–25

117. Salinas S, Bilsland LG, Schiavo G. 2008. Molecular landmarks along the axonal route: axonal transport in health and disease. *Curr. Opin. Cell Biol.* 20:445–53

118. Scholey JM. 2008. Intraflagellar transport motors in cilia: moving along the cell's antenna. *J. Cell Biol.* 180:23–29

119. Seitz A, Kojima H, Oiwa K, Mandelkow EM, Song YH, Mandelkow E. 2002. Single-molecule investigation of the interference between kinesin, tau and MAP2c. *EMBO J.* 21:4896–905

120. Shastry S, Hancock WO. 2010. Neck linker length determines the degree of processivity in kinesin-1 and kinesin-2 motors. *Curr. Biol.* 20:939–43

121. Shimizu Y, Morii H, Arisaka F, Tanokura M. 2005. Stalk region of kinesin-related protein Unc104 has moderate ability to form coiled-coil dimer. *Biochem. Biophys. Res. Commun.* 337:868–74

122. Shtridelman Y, Cahyuti T, Townsend B, DeWitt D, Macosko JC. 2008. Force-velocity curves of motor proteins cooperating in vivo. *Cell Biochem. Biophys.* 52:19–29

123. Shubeita GT, Tran SL, Xu J, Vershinin M, Cermelli S, et al. 2008. Consequences of motor copy number on the intracellular transport of kinesin-1-driven lipid droplets. *Cell* 135:1098–107

124. Soppina V, Rai AK, Ramaiya AJ, Barak P, Mallik R. 2009. Tug-of-war between dissimilar teams of microtubule motors regulates transport and fission of endosomes. *Proc. Natl. Acad. Sci. USA* 106:19381–86

125. Stamer K, Vogel R, Thies E, Mandelkow E, Mandelkow EM. 2002. Tau blocks traffic of organelles, neurofilaments, and APP vesicles in neurons and enhances oxidative stress. *J. Cell Biol.* 156:1051–63

126. Steinberg G, Schliwa M. 1995. The *Neurospora* organelle motor: a distant relative of conventional kinesin with unconventional properties. *Mol. Biol. Cell* 6:1605–18

127. Stock MF, Guerrero J, Cobb B, Eggers CT, Huang TG, et al. 1999. Formation of the compact conformer of kinesin requires a COOH-terminal heavy chain domain and inhibits microtubule-stimulated ATPase activity. *J. Biol. Chem.* 274:14617–23

128. Straube A, Hause G, Fink G, Steinberg G. 2006. Conventional kinesin mediates microtubule-microtubule interactions in vivo. *Mol. Biol. Cell* 17:907–16

129. Sung HH, Telley IA, Papadaki P, Ephrussi A, Surrey T, Rorth P. 2008. *Drosophila* ensconsin promotes productive recruitment of kinesin-1 to microtubules. *Dev. Cell* 15:866–76

130. Telley IA, Bieling P, Surrey T. 2009. Obstacles on the microtubule reduce the processivity of kinesin-1 in a minimal in vitro system and in cell extract. *Biophys. J.* 96:3341–53

131. Tomishige M, Klopfenstein DR, Vale RD. 2002. Conversion of Unc104/KIF1A kinesin into a processive motor after dimerization. *Science* 297:2263–67

132. Trinczek B, Ebneth A, Mandelkow EM, Mandelkow E. 1999. Tau regulates the attachment/detachment but not the speed of motors in microtubule-dependent transport of single vesicles and organelles. *J. Cell Sci.* 112:2355–67

133. Valentine MT, Gilbert SP. 2007. To step or not to step? How biochemistry and mechanics influence processivity in kinesin and Eg5. *Curr. Opin. Cell Biol.* 19:75–81

134. Verhey KJ, Gaertig J. 2007. The tubulin code. *Cell Cycle* 6:2152–60

135. Verhey KJ, Hammond JW. 2009. Traffic control: regulation of kinesin motors. *Nat. Rev. Mol. Cell Biol.* 10:765–77

136. Verhey KJ, Lizotte DL, Abramson T, Barenboim L, Schnapp BJ, Rapoport TA. 1998. Light chain-dependent regulation of kinesin's interaction with microtubules. *J. Cell Biol.* 143:1053–66

137. Vershinin M, Carter BC, Razafsky DS, King SJ, Gross SP. 2007. Multiple-motor based transport and its regulation by Tau. *Proc. Natl. Acad. Sci. USA* 104:87–92

138. Visscher K, Schnitzer MJ, Block SM. 1999. Single kinesin molecules studied with a molecular force clamp. *Nature* 400:184–89

139. Watanabe TM, Yanagida T, Iwane AH. 2010. Single molecular observation of self-regulated kinesin motility. *Biochemistry* 49:4654–61

140. Wedaman KP, Meyer DW, Rashid DJ, Cole DG, Scholey JM. 1996. Sequence and submolecular localization of the 115-kD accessory subunit of the heterotrimeric kinesin-II (KRP85/95) complex. *J. Cell Biol.* 132:371–80

141. Welte MA. 2004. Bidirectional transport along microtubules. *Curr. Biol.* 14:R525–37

142. Welte MA. 2010. Bidirectional transport: matchmaking for motors. *Curr. Biol.* 20:R410–13

143. Wloga D, Gaertig J. 2010. Post-translational modifications of microtubules. *J. Cell Sci.* 123:3447–55

144. Wong YL, Dietrich KA, Naber N, Cooke R, Rice SE. 2009. The kinesin-1 tail conformationally restricts the nucleotide pocket. *Biophys. J.* 96:2799–807

145. Wong YL, Rice SE. 2010. Kinesin's light chains inhibit the head- and microtubule-binding activity of its tail. *Proc. Natl. Acad. Sci. USA* 107:11781–86

146. Wordeman L. 2010. How kinesin motor proteins drive mitotic spindle function: lessons from molecular assays. *Semin. Cell Dev. Biol.* 21:260–68

147. Wozniak MJ, Allan VJ. 2006. Cargo selection by specific kinesin light chain 1 isoforms. *EMBO J.* 25:5457–68

148. Yamada KH, Hanada T, Chishti AH. 2007. The effector domain of human Dlg tumor suppressor acts as a switch that relieves autoinhibition of kinesin-3 motor GAKIN/KIF13B. *Biochemistry* 46:10039–45

149. Yamazaki H, Nakata T, Okada Y, Hirokawa N. 1995. KIF3A/B: a heterodimeric kinesin superfamily protein that works as a microtubule plus end-directed motor for membrane organelle transport. *J. Cell Biol.* 130:1387–99

150. Yamazaki H, Nakata T, Okada Y, Hirokawa N. 1996. Cloning and characterization of KAP3: a novel kinesin superfamily-associated protein of KIF3A/3B. *Proc. Natl. Acad. Sci. USA* 93:8443–48

151. Yang Z, Goldstein LS. 1998. Characterization of the KIF3C neural kinesin-like motor from mouse. *Mol. Biol. Cell* 9:249–61

152. Yoo J, Kambara T, Gonda K, Higuchi H. 2008. Intracellular imaging of targeted proteins labeled with quantum dots. *Exp. Cell Res.* 314:3563–69

153. Yu Y, Feng YM. 2010. The role of kinesin family proteins in tumorigenesis and progression: potential biomarkers and molecular targets for cancer therapy. *Cancer* 116:5150–60

154. Zahn TR, Angleson JK, MacMorris MA, Domke E, Hutton JF, et al. 2004. Dense core vesicle dynamics in *Caenorhabditis elegans* neurons and the role of kinesin UNC-104. *Traffic* 5:544–59

155. Zhou HM, Brust-Mascher I, Scholey JM. 2001. Direct visualization of the movement of the monomeric axonal transport motor UNC-104 along neuronal processes in living *Caenorhabditis elegans*. *J. Neurosci.* 21:3749–55

Stochastic Conformational Pumping: A Mechanism for Free-Energy Transduction by Molecules

R. Dean Astumian

Department of Physics, University of Maine, Orono, Maine 04469;
email: astumian@maine.edu

Annu. Rev. Biophys. 2011. 40:289–313

First published online as a Review in Advance on February 22, 2011

The *Annual Review of Biophysics* is online at biophys.annualreviews.org

This article's doi:
10.1146/annurev-biophys-042910-155355

Copyright © 2011 by Annual Reviews.
All rights reserved

1936-122X/11/0609-0289$20.00

Keywords

fluctuating enzymes, molecular motors, molecular pumps, microscopic reversibility, fluctuation-dissipation theorem

Abstract

Proteins and other macromolecules can act as molecular machines that convert energy from one form to another through cycles of conformational transitions. In a macroscopically fluctuating environment or at the single-molecule level, the probability for a molecule to be in any state j fluctuates, and the probability current from any other state i to state j is given as the sum of a steady-state current and a pumped current, $I_{ij} = I_{ij}^{ss} + F_{ij}dP_j/dt$, where F_{ij} is the fraction of the fluctuating current into and out of state j coming directly from state i, and dP_j/dt is the rate of change of the probability for the molecule to be in state j. If the fluctuations arise from an equilibrium source, microscopic reversibility guarantees that the time average of the pumped current is zero. If, however, the fluctuations arise due to the action of a nonequilibrium source, the time average of the pumped current is not in general zero and can be opposite in sign to the steady-state current. The pumped current provides a mechanism by which fluctuations, whether generated externally or arising from an internal nonequilibrium chemical reaction, can do electrical, mechanical, or chemical work on a system by coupling into the equilibrium conformational transitions of a protein. In this review I examine work elaborating the mechanism of stochastic pumping and also discuss a thermodynamically consistent approach for modeling the effects of dynamic disorder on enzymes and other proteins.

Contents

INTRODUCTION

The ability to convert free energy from one form to another is essential for life. Our cells store energy in chemical form (often ATP or GTP) and then use the energy released by catalytic breakdown of these molecules to perform various tasks, e.g., pumping ions from low to high electrochemical potential, polymerizing and synthesizing necessary macromolecules, or powering motion of molecular motors to move material from one place to another (**Figure 1**). Despite the fundamental similarity between these different energy conversion processes, molecular motors are typically described very differently than molecular pumps or enzymes such as synthases or polymerases.

Models for molecular motors (43, 69) have focused on an ATP-driven mechanical power stroke—a viscoelastic relaxation process in which the protein starts from a nonequilibrium, strained conformation due to the action of ATP at the active site. The subsequent relaxation following product (ADP or Pi) release can be visualized much like the contraction of a stretched rubber band.

In contrast, molecular pumps are most often modeled in terms of chemical kinetics (51), where ATP energy is used to change the relative affinities of and barrier heights between binding sites by sequentially favoring different conformational states of the protein as ATP is bound and hydrolyzed and the products are released. The conformational relaxation and molecular transport across the membrane are thermally activated steps.

The perspective I develop in this review is that at the single-molecule level molecular machines are mechanically equilibrated systems that serve as conduits for the flow of energy between a source such as an external field or a nonequilibrium chemical reaction and the environment in which work is to be done (9). Consider the process

$$\cdots \rightleftharpoons B \rightleftharpoons \cdots \underset{k_{Ai}}{\overset{}{\rightleftharpoons}} A \underset{k_{BA}}{\overset{k_{AB}}{\rightleftharpoons}} B \overset{k_{Bj}}{\rightleftharpoons} \cdots \rightleftharpoons A \rightleftharpoons \cdots,$$

where A and B are states in the conformational cycle of one of the free-energy transduction processes shown in **Figure 1**. We normally focus on the steady-state probability current $I_{AB}^{ss} = k_{AB} P_A^{ss} - k_{BA} P_B^{ss}$ to assess the average direction of cycling—whether the net flow is to the left or to the right. In a fluctuating environment or at the single-molecule level, however, we must add a term that reflects the

a   Molecular pump

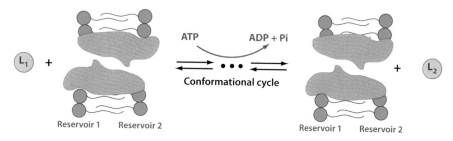

ATP   ADP + Pi

Conformational cycle

Reservoir 1   Reservoir 2   Reservoir 1   Reservoir 2

b   Coupled enzyme

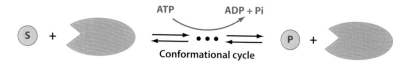

ATP   ADP + Pi

Conformational cycle

c   Molecular motor

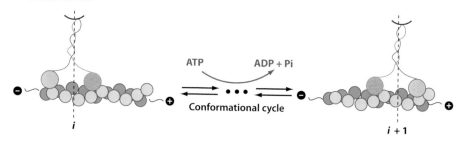

ATP   ADP + Pi

Conformational cycle

i $i + 1$

Figure 1

Schematic illustrations of molecular machines that use energy from ATP hydrolysis to accomplish specific tasks. (*a*) A molecular pump that moves some ligand across a membrane, possibly from low to high electrochemical potential. (*b*) A coupled enzyme that synthesizes some necessary substance. (*c*) A molecular motor that walks along a polymeric track.

correlations between the fluctuating rates and fluctuating probabilities:

$$I_{AB}(t) = I_{AB}^{ss}(t) + F_{AB}(t)\frac{dP_B(t)}{dt}, \qquad 1.$$

where $F_{AB} = \delta I_{AB}(t)/\sum_k \delta I_{kB}(t)$ is the fraction of the transient change in the probability to be in state B, dP_B/dt, coming directly from (to) state A (5, 8, 42), and $\delta I_{ij} = k_{ij}\delta P_i - k_{ji}\delta P_j$ is the fluctuation of the current between any two states i and j around the steady-state level I_{ij}^{ss}. If the fluctuations in F_{AB} and in P_B are uncorrelated, the long time average of the second term, the pumped current, is zero. If, however, the system is pumped, e.g., by simultaneously raising and lowering the free energy of state B and the energy barrier between states B and A

(and hence modulating k_{AB} and k_{Bj}, but with the product $k_{AB}k_{Bj}$ constant), the average pumped flux is positive because $F_{AB}(t)$ is greater than 1/2 when P_B is increasing and $F_{AB}(t)$ is less than 1/2 when P_B is decreasing (18).

The physical motion by which a single molecule of protein in state A is converted to state B is the same irrespective of how different the ensemble probability ratio P_B/P_A is from the equilibrium constant $K_{AB} = k_{AB}/k_{BA} \equiv P_B^{eq}/P_A^{eq}$ or whether the system is pumped. The conformational transitions in the cycles by which these machines carry out their function are intrinsically equilibrium processes between states that are close to thermal equilibrium. A time-dependent external energy source or a nonequilibrium chemical reaction modulates the relative stabilities of the states in the cycle and the rate constants for the conformational transitions between them in a correlated way, thereby driving net flux through the cycle and the performance of work by a mechanism known as stochastic pumping (15, 74) or chemical peristalsis (6).

AN ADIABATICALLY PUMPED MOLECULAR MACHINE

A particularly simple example of stochastic pumping involves a recently synthesized catenane-based molecular motor (52). Catenanes are molecules with two or more interlocked rings. **Figure 2a** illustrates a three-ring catenane and **Figure 2b** illustrates a two-ring catenane. The salient feature of these two molecules is that binding sites or recognition stations (the blue, red, and green boxes labeled 1, 2, and 3, respectively) for the small purple rings can be designed and located on the large orange ring. The stations can be designed such that their interaction energies with the purple rings can be independently externally controlled, e.g., by protonation and deprotonation or by oxidation and reduction (41). A sequence of external cyclical changes to the interaction energies of the stations can squeeze the purple rings to undergo directional rotation by a mechanism similar to peristalsis (6), but where the transitions result from thermal noise so the system operates as a Brownian motor (3, 16).

Three-Ring Catenance

Let us first consider the three-ring catenane shown in **Figure 2a**. The larger yellow ring has three distinct recognition stations, labeled 1, 2, and 3, for the two identical purple rings. The purple rings cannot pass one another, nor can they occupy the same station, as they make thermally activated transitions from one station to another. Thus, there are a total of three distinguishable states, labeled A, B, and C. The interaction between a purple ring and a station is characterized by an interaction energy $E_1, E_2, E_3 \leq 0$. Each transition involves break-

ing the interaction between one station and one ring. For example the transitions from state A to state B require breaking the interaction of the ring on site 1, as does the transition from state C to state B. By using this analysis, the rate constants for the transitions are

$$
\begin{aligned}
k_{AB} &= k_{CB} = \mathcal{A}^{\beta(E_1 - E^\ddagger)} \\
k_{BC} &= k_{AC} = \mathcal{A}e^{\beta(E_3 - E^\ddagger)}, \\
k_{CA} &= k_{BA} = \mathcal{A}e^{\beta(E_2 - E^\ddagger)}
\end{aligned}
\qquad 2.
$$

where $\mathcal{A}$ is a frequency factor, $\beta = (k_B T)^{-1}$ is the inverse of the thermal energy, and $E^\ddagger > 0$ is the energy of the barrier between the stations, which we assume to be the same for all transitions.

Periodic modulation of the energies E_1, E_2, and E_3, even modulation carried out so slowly that the state probabilities are given by an equilibrium relation ($P_i/P_j = k_{ji}/k_{ij}$ for i, j = A, B, C) at every instant, can drive directional rotation of the small rings about the larger ring. Beginning with $E_1, E_3 \ll E_2$, where the system is almost certainly in state A, the interaction energy E_1 is slowly increased while E_2 is decreased to reach the condition $E_2, E_3 \ll E_1$. During this process, the purple ring originally on station 1 is transferred by an equilibrium process to station 2 to reach state B. Because the energy E_3 remains very low, the ring on station 3 does not move; therefore, the transfer could only have occurred in the clockwise direction. The interaction energy E_3 is then slowly increased and simultaneously E_1 is decreased, reaching the condition $E_1, E_2 \ll E_3$. During this process the ring on station 3 is transferred to station 1, also in the clockwise direction, to reach state C. Finally, the interaction energy E_2 is slowly increased and simultaneously E_3 is decreased to regain the original condition $E_1, E_3 \ll E_2$. During this process the ring on station 2 is transferred to station 3 in the clockwise direction, thus returning to the original state A. The cycling A $\rightarrow$ B $\rightarrow$ C $\rightarrow$ A is accompanied by clockwise cycling of each purple ring through the stations in the order $1 \rightarrow 2 \rightarrow 3 \rightarrow 1$, with two cycles through the states necessary to give each ring one clockwise turn.

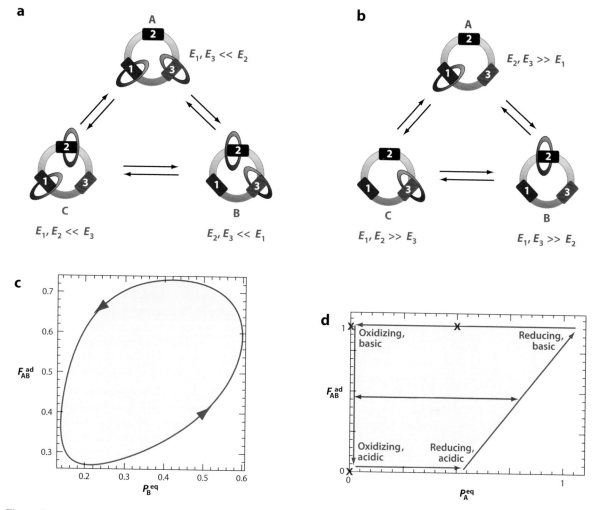

Figure 2

(*a*) A three-ring catenane that can operate as a molecular machine that moves directionally in response to external stimuli (e.g., pH and redox potential modulations). (*b*) A two-ring catenane that can undergo a precise cycle of states in response to an external stimulus, but the motion is not directional. (*c*) Plot of $F_{AB}^{ad} = (1 + e^{\beta(E_3 - E_2)})^{-1}$ and $P_B^{eq} = [1 + e^{\beta(E_1 - E_2)} + e^{\beta(E_1 - E_3)}]^{-1}$ parametrized by time, with $E_1 = [-2 + \cos(\omega t)]\beta^{-1}$, $E_2 = [-2 + \cos(\omega t + \pi/2)]\beta^{-1}$, and $E_3 = -2\beta^{-1}$. (*d*) A parametric plot of the equilibrium probability for state A versus the fraction of the flux into/out of A that comes from/to state B. The red line is based on the rate constants for the two-ring catenane for which F_{AB}^{ad} is constant, and the blue lines are based on the rate constants for the three-ring catenance for which F_{AB}^{ad} is controlled by pH and P_A^{eq} is controlled by the redox potential (8), where for simplicity we follow the cycle $(E_1 = 0, E_2 = -\infty) \rightarrow (E_1 = -\infty, E_2 = -\infty) \rightarrow (E_1 = -\infty, E_2 = 0) \rightarrow (E_1 = 0, E_2 = 0) \rightarrow (E_1 = 0, E_2 = -\infty)$.

Adiabatic Pumping

Directional rotation due to very slow modulation may be surprising given that the instantaneous steady-state current for any transition, e.g., A → B, is zero ($I_{AB}^{ss} = k_{AB} P_A^{eq} - k_{BA} P_B^{eq} = 0$) irrespective of the values of E_1, E_2, and E_3.

However, when the external conditions change, even very slowly, so too do the state probabilities. We must account for the changing state probabilities by adding a term to the steady-state current (5, 8, 42) to get $I_{AB}(t) = I_{AB}^{ss}(t) + F_{AB}(t)dP_B/dt$, the net instantaneous probability current between states A and B. Because

$I_{AB}^{ss}(t) = 0$ at every instant, the net current for cyclic modulation of the energies averaged over a period is

$$\overline{I_{AB}} = \omega \int_0^{\omega^{-1}} F_{AB}(t) \frac{dP_B}{dt} dt = \omega \oint F_{AB}(t) dP_B(t),$$
3.

where ω is the frequency of the modulation. Under adiabatic conditions (very slow modulation) the state probabilities are at equilibrium at every instant and the fraction F_{AB}^{ad} of the change in P_B^{eq} directly to/from state A is independent of whether P_B^{eq} is increasing or decreasing,

$$F_{AB}^{ad} = \frac{k_{BA}}{k_{BA} + k_{BC}} = \frac{k_{AB} P_A^{eq}}{k_{AB} P_A^{eq} + k_{CB} P_C^{eq}}.$$
4.

We find $F_{AB}^{ad} = (1 + e^{\beta(E_3 - E_2)})^{-1}$ and $P_B^{eq} = [1 + e^{\beta(E_1 - E_2)} + e^{\beta(E_1 - E_3)}]^{-1}$ for the rate constants in Equation 2. The adiabatic integrated current per cycle $\oint F_{AB}^{ad} dP_B^{eq}$—the area enclosed in a plot of F_{AB}^{ad} versus P_B^{eq} parametrized by time—is purely geometric and does not depend on frequency. Further, because both F_{AB}^{ad} and P_B^{eq} can vary at most between 0 and 1, the limit is one cycle through the states per cycle of modulation, and one turn for each ring per two cycles of modulation (8, 31). The rate constants in Equation 2 obey the simple relation $k_{AB} k_{BC} k_{CA} = k_{AC} k_{CB} k_{BA}$ irrespective of the values of the energies E_1, E_2, and E_3. This relation is a necessary and sufficient condition for the steady-state component of the current to be zero, $I_{AB}^{ss} = 0$, but, as we have seen, in a fluctuating environment this relation between the rate constants, despite assertions in the literature to the contrary, is not a sufficient condition to assure that the average net cyclic flux $\overline{I_{AB}}$ is zero (20, 66, 85). The stochastic pumping of the three-ring catenane can be implemented experimentally by out-of-phase modulation of the pH and reduction/oxidation (redox) potential as discussed in Reference 8 (**Figure 2c**), where the energy E_1 is controlled by the redox potential and the energy E_2 is controlled by the pH (**Figure 2c**).

The geometric effect by which pumping is achieved for a 3-catenane is termed the geometric phase (25). A similar picture emerges for

a wide variety of physical phenomena including dissipationless pumping of electrons (15, 77, 78), a mechanism for biomolecular ion pumps (5, 75), phase control in oscillating chemical reactions (47), and swimming at low Reynolds number (64, 72).

Preventing Backward Motion: Why a Two-Ring Catenane Doesn't Work

The pumping mechanism for the three-ring-catenane-based molecular motor illustrates an important principle for molecular machines that is very different than the mechanism by which macroscopic machines function. In a macroscopic machine, input energy is used to cause the desired motion, and without the energy there would be no motion at all. For molecular machines, however, the parts of the machine are constantly moving about even at thermal equilibrium. The design of the three-ring catenane motor focuses on restricting or preventing the undesired parts of the thermal motion (4, 11). In each transition, one of the small rings acts as an immobile obstacle, thereby setting the direction of motion of the other ring when, by thermal noise, the mobile ring moves from an unstable station to a stable station. By restriction of the backward motion, only the desired motion remains—a concept often described as biased Brownian motion.

We can better understand this concept by contrasting the case of the three-ring catenane with that of a two-ring catenane, which cannot be induced to undergo directional cycling by adiabatic modulation of the interaction energies between the yellow ring and the stations. For the two-ring catenane in **Figure 2b**, the rate constants out of any state are the same for clockwise and counterclockwise transitions:

$$\begin{aligned} k_{AB} &= k_{AC} = \mathcal{A} e^{\beta(E_1 - E^\ddagger)} \\ k_{BC} &= k_{BA} = \mathcal{A} e^{\beta(E_3 - E^\ddagger)}. \\ k_{CA} &= k_{CB} = \mathcal{A} e^{\beta(E_2 - E^\ddagger)} \end{aligned}$$
5.

With these rate constants we have $F_{AB}^{ad} = 1/2$, a constant, and thus there is no possibility for directional adiabatic pumping. For example, when E_1 is increased, the probability for transition out of state A to state C is exactly the same as the probability for transition to state B. Nonadiabatic pumping (15) at higher frequencies is possible for the two-ring catenane (31, 65), however, because the state probabilities appearing in F_{AB} are no longer given by their equilibrium values and are frequency dependent. A very general and easily implemented computational scheme for calculating the instantaneous state probabilities and currents for a cyclic system with arbitrarily large amplitude and frequency modulation was given by Robertson & Astumian (66). The three-state model for the three-ring catenane is perhaps the simplest example of adiabatic stochastic pumping in which only binding station energies are modulated. The directionality requires the interaction between the two purple rings. If, in addition to a binding station energy, a barrier energy can be directly modulated, a two-state model for pumping is possible (17, 61) in which both adiabatic and nonadiabatic components of the pumping can be analytically evaluated. In the next section we focus on such a two-state model for a membrane pump that is also analogous to the Michaelis-Menten mechanism for enzyme catalysis.

STOCHASTIC PUMPING ACROSS MEMBRANES

ATP-driven pumps are proteins that span a cell or organelle membrane and use energy from ATP hydrolysis to pump ligand (often ions such as Na^+, K^+, H^+, or Ca^{2+}) across the membrane, thus generating and maintaining the ion electrochemical gradients essential for life (51). In a simple picture of a membrane pump (**Figure 1b**), the protein structure presents energy barriers (gates) for ligand permeation at the two entrances, one on either side of the membrane, surrounding an energy well (binding site) in the middle. **Figure 2a** shows an energy diagram for this two-barrier, one-site model of an ion transporter. The differential barrier height u and well energy ε are internal parameters (28) controlled by the conformation of the protein and do not influence the

overall transport equilibrium $\Delta\mu = \mu_2 - \mu_1$. In the absence of input energy, ions flow from high to low electrochemical potential. Conformational fluctuations of the pump protein cause the relative energies of the two gates, as well as the binding energy (well depth) for ligand, to fluctuate—i.e., both $\varepsilon(t)$ and $u(t)$ depend on time through the protein conformation. Undriven fluctuations, however, are not correlated and therefore cannot cause uphill pumping. The absence of correlations in the undriven fluctuations of $\varepsilon(t)$ and $u(t)$ is reflected in the principle of microscopic reversibility (80).

When ATP is bound to the protein, hydrolyzed, and product-released in chemically driven pumping, the protein undergoes shape changes in which the relative gate and binding energies for the ion fluctuate in a correlated way. This correlated fluctuation causes transport of ions across the membrane from low to high electrochemical potential.

In general, different conformations of a protein have different dipole moments. Thus, an external oscillating electric field can also drive structural changes of a pump protein (50) and cause nonequilibrium correlated modulation of the relative barrier height $u(t)$ and the well energy $\varepsilon(t)$, thereby driving uphill pumping. This was shown experimentally by Tsong and colleagues (54, 70, 86, 87), who applied a fluctuating external electric field to suspensions of red blood cells. The zero-average applied fields were able to drive thermodynamically uphill transport via the ion pump Na,K ATPase even under conditions where ATP hydrolysis could not occur. Interpretation of these experiments led to the development of the electroconformational coupling theory (20, 81–83, 85), which explains how, by coupling into intrinsic conformational degrees of freedom of a protein, an external oscillating or fluctuating perturbation can drive pumping of ligand from low to high electrochemical potential (20), catalysis of a chemical reaction away from equilibrium (18), or performance of mechanical work on the environment (12). It seems likely that the same conformational motions are exploited in ATP-hydrolysis-driven pumping.

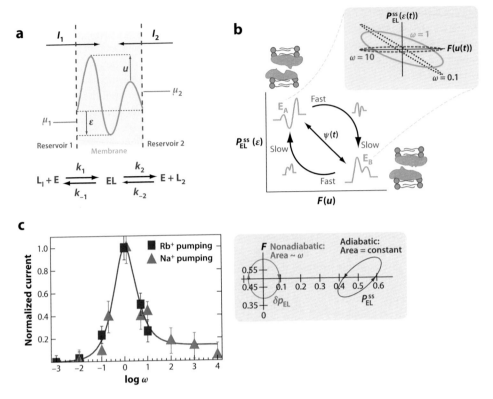

Figure 3

(*a*) Potential energy diagram for a membrane pump. The external parameter $\Delta\mu = \mu_2 - \mu_1$ determines the direction of thermodynamically spontaneous current, while the two internal parameters, u and ε, fluctuate in time due to conformational fluctuations of the protein. If the fluctuations are driven by, e.g., an oscillating field, the correlated fluctuations of u and ε can lead to uphill pumping of ligand. Below the membrane is a kinetic mechanism for the pumping that is analogous to the Michaelis-Menten mechanism for enzyme catalysis. (*b*) Schematic diagram showing how a single external parameter, an oscillating field, can cause the internal parameters u and ε to oscillate out of phase with one another. The inset in the gray box shows how the area enclosed by the parametric plot of F_1 versus P_{EL} is maximized when the external frequency ω matches the system characteristic frequency τ^{-1}. (*c*) Fit of data from Reference 54 to the sum of the adiabatic and nonadiabatic currents from Equation 11. The maximum current was normalized to unity, and the characteristic frequency for Rb^+ pumping (*red squares*) was taken to be 10^3 Hz and the characteristic frequency for Na^+ pumping (*blue triangles*) was taken to be 10^6 Hz, both corresponding to the optimal frequencies for pumping in the experiment. The inset in the gray box shows the parametric plots for the adiabatic (*counterclockwise purple loop on the right*) and the nonadiabatic (*clockwise green loop on the left*) pumping.

Kinetic Mechanism

We can understand conformational pumping in terms of the simple kinetic model in **Figure 3*a***. The instantaneous net current between reservoir 1 and the well is

$$I_1(t) = \frac{I_1(t)}{I_1(t) + I_2(t)} \frac{dP_{EL}(t)}{dt},$$

which, by decomposing the currents $I_i(t) = I_i^{ss}(t) + \delta I_i(t)$ and observing that $I_1^{ss}(t) + I_2^{ss}(t) = 0$ at every instant, can be written

$$I_1 = I_1^{ss} + F_1 \left[\frac{dP_{EL}^{ss}}{dt} + \frac{d\delta p_{EL}}{dt} \right], \qquad 6.$$

where we suppress the explicit denotation of the time dependence of the quantities. The probability for the well to be occupied,

$P_{EL} = P_{EL}^{ss} + \delta p_{EL}$, is split into the instantaneous steady-state value and a deviation from that value, and $F_1 = \delta I_1/(\delta I_1 + \delta I_2)$ is the ratio of the fluctuating current between reservoir 1 and the well to the total fluctuating current into/out of the well. The terms $F_1 dP_{EL}^{ss}/dt$ and $F_1 d\delta p_{EL}/dt$ are the adiabatic and nonadiabatic contributions to the pump current, respectively.

From elementary energetic considerations we have the simple relations between pairs of rate constants and the internal parameters, ε and u, and the external parameters μ_1 and μ_2:

$$\frac{k_1}{k_{-1}} = e^{\beta(\mu_1 - \varepsilon)}; \quad \frac{k_2}{k_{-2}} = e^{\beta(\varepsilon - \mu_2)};$$

$$\frac{k_{-1}}{k_2} = e^{\beta u}; \quad \frac{k_1}{k_{-2}} = e^{\beta(\Delta\mu - u)}. \qquad 7.$$

From Equation 7 we easily derive another relation between all four rate constants in which the internal parameters ε and u disappear,

$$\frac{k_1 k_2}{k_{-1} k_{-2}} = e^{-\beta\Delta\mu}. \qquad 8.$$

The kinetic mechanism for ligand transport shown in **Figure 3a** is identical to the Michaelis-Menten mechanism for catalysis of the chemical reaction $L_1 \rightleftharpoons L_2$, where k_1 and k_{-2} are effective first-order rate constants into which the concentrations $[L_1]$ and $[L_2]$, respectively, have been subsumed. In this model μ_1 and μ_2 are the chemical potentials of substrate, L_1, and product, L_2, respectively. The relative gate height u has the interpretation of the chemical specificity (relative lability) of the enzyme—when $u < 0$ the enzyme is specific for L_1 (i.e., the binding/dissociation of L_1 is faster than the binding/dissociation of L_2) and when $u > 0$ the enzyme is specific for L_2 (i.e., the binding/dissociation of L_2 is faster than the binding/dissociation of L_1). The well depth ε specifies the binding affinity (stability) of the enzyme.

The rate of change of the binding probability is

$$\frac{dP_{EL}}{dt} = -\tau^{-1} P_{EL} + (k_1 + k_{-2}), \qquad 9.$$

where $\tau = (k_1 + k_{-1} + k_2 + k_{-2})^{-1}$ is the relaxation time for ligand binding in the well. The instantaneous steady-state probability for the well to be occupied is obtained by setting $dP_{EL}/dt = 0$ in Equation 9 and solving for P_{EL} to find $P_{EL}^{ss} = (k_1 + k_{-2})\tau$. The fraction of fluctuating well occupancy coming from reservoir 1 is $F_1 = (k_1 + k_{-1})\tau$, and the instantaneous steady-state current $I_1^{ss} = k_1(1 - P_{EL}^{ss}) - k_{-1} P_{EL}^{ss}$ can be written $I_1^{ss} = (k_1 k_2 - k_{-1} k_{-2})\tau = (1 - e^{\beta\Delta\mu})k_2 k_1 \tau$. If the internal parameters u and ε fluctuate, the rate constants and therefore P_{EL}^{ss}, F_1, and I_1^{ss} all vary in time. Nevertheless, irrespective of the instantaneous values of $u(t)$ and $\varepsilon(t)$, the sign of I_1^{ss} is determined solely by $\Delta\mu$. This is not true, however, of the pumped current. If we follow the sequence $(u < 0) \rightarrow (\varepsilon < \mu_1) \rightarrow (u > 0) \rightarrow (\varepsilon > \mu_2) \rightarrow (u < 0)$, the pumped current will be positive (from reservoir 1 to reservoir 2) even though $\mu_2 > \mu_1$. The maximum probability for an ion to be pumped in one cycle is achieved in the limit $|u| \rightarrow \infty$. In this case the steady-state current is nearly zero because one of the gates is very high at every instant. The probability to pump an ion then is the difference in occupancy between the state where $(\varepsilon < \mu_1)$ and the state where $(\varepsilon > \mu_2)$. This probability can be written $\tanh[\beta(\Delta\varepsilon - \Delta\mu)/2]$, since the well equilibrates with the reservoir to which it has finite access. Thus, the maximum average output energy per cycle is $\mathcal{E}_{max, out} = \Delta\mu \tanh[\beta(\Delta\varepsilon - \Delta\mu)/2]$, and the minimum input energy per cycle is $\mathcal{E}_{in, min} = \Delta\varepsilon$, with the limiting thermodynamic efficiency (5, 6)

$$\eta_{max} = \frac{\Delta\mu}{\Delta\varepsilon} \tanh[\beta(\Delta\varepsilon - \Delta\mu)/2]. \qquad 10.$$

For an input energy of $\Delta\varepsilon = 20\beta^{-1}$ (i.e., the energy provided by ATP hydrolysis under physiological conditions), this maximum efficiency is about 75%.

These molecular machines can be interfaced to the external world by arranging them in a macroscopic membrane (35) with electrical coupling to the environment. If the change in dipole moment between the two states is 500 D, the chemical conversion $L_1 \rightarrow L_2$ by a 1 mm^{-2}

surface with 10^8 molecules could supply an output power $P_{out} = 10$ nW with an efficiency greater than 60% and at a current of 15 μA.

Pumped Currents in the Small Perturbation Limit

In order to compare the theory for stochastic pumping with experimental results of Tsong and colleagues (54), consider a situation in which the internal parameters change periodically in time with frequency ω. For small-amplitude oscillations $u(t) = u_0 + \delta u \cos(\omega t)$ and $\varepsilon(t) = \varepsilon_0 + \delta\varepsilon \cos(\omega t - \phi)$, Equation 6 can be evaluated in the small perturbation limit (17) to yield the simple equation for the pumped flux (15)

$$\overline{I_1} - \overline{I_1^{ss}} = K \frac{\sin(\phi)\tilde{\omega} + \cos(\phi)\tilde{\omega}^2}{1 + \tilde{\omega}^2}, \qquad 11.$$

where K is a constant that is proportional to the product $\delta\varepsilon\delta u$. In **Figure 3c** the data of Tsong and colleagues (54) for alternating current (ac)-field-induced pumping of both Rb^+ (an analog of K^+) and Na^+ by the Na,K ATPase as functions of the reduced frequency are fit to Equation 11, with $\tau = 10^{-6}$ s for Na and $\tau = 10^{-3}$ s for Rb based on the optimal pumping frequencies observed in the experiments. The fit parameter $\phi = \pi/2.1$ is very close to $\pi/2$, suggesting that the nonadiabatic contribution is essentially negligible except at very high frequencies $\tilde{\omega} \gg 1$. Thus, we conclude that the Na,K ATPase may work in many respects like an adiabatic pump, where two internal parameters are caused by the applied field to oscillate out of phase with one another.

Figure 3b illustrates a simple two-state mechanism by which a single external parameter, the oscillating external field $\psi(t)$, can cause two internal parameters to oscillate out of phase with one another. There are two major conformational states, E_A and E_B. State E_A has high affinity for ligand ($\varepsilon_A > 0$ and $P_{EL}^{ss,A} > 1/2$) and easy access between the well and the reservoir 1 ($u_A < 0$, and $F_{1,A} > 1/2$), and state E_B has low affinity for ligand ($\varepsilon_B < 0$ and $P_{EL}^{ss,B} < 1/2$) and easy access between the well and the reser-

voir 2 ($u_B > 0$, and $F_{1,B} < 1/2$). If E_A and E_B have different dipole moments, an external ac field will alternately favor one state and then the other state, causing the average values of F_1 and P_{EL}^{ss} to oscillate. Let the conformational transition be governed by two relaxation times: a fast relaxation time that governs u and a slow relaxation time that governs ε. As a result of the different relaxation times F_1, P_{EL}^{ss}, and δp_{EL} oscillate out of phase with one another. The phase lag between F_1 and P_{EL}^{ss} is caused by an internal conformational degree of freedom that is out of equilibrium with the applied modulation. Even at low frequency the system is not in global equilibrium, but only in equilibrium with respect to the degree of freedom corresponding to ion transport. Nonadiabatic flux, in which δp_{EL} fluctuates out of phase with F_1, has also been discussed (17, 19).

In the experiment by Tsong and colleagues (54) the conformational oscillation was driven by an applied oscillating electric field. In chemically driven pumping, where, for example, ATP hydrolysis drives transport, the stochastic binding of reactants and the release of products cause transitions between states of the protein. In this case, after phosphorylation or dephosphorylation, the differential barrier height that controls the parameter F_1 rapidly approaches its final value, followed by a slower relaxation of the well energy (i.e., P_{EL}^{ss}) to its new value. In this way, a stochastic input (ATP hydrolysis) is converted into two on-average phase-shifted outputs. Such hysteretic behavior is very general in proteins or for that matter for any relatively complex molecule (17, 88).

FLUCTUATING PROTEINS AND DYNAMIC DISORDER

Stochastic Pumping and ATP-Driven Pumping

A simple model (19) for stochastic pumping by a single external parameter $\psi(t)$ is shown in **Figure 4a** alongside a kinetic model for ATP-driven pumping involving the same protein states in **Figure 4b**. The model in

External field-driven pumping

ATP hydrolysis-driven pumping

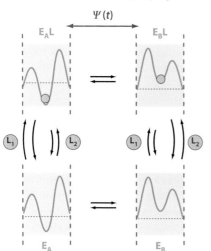

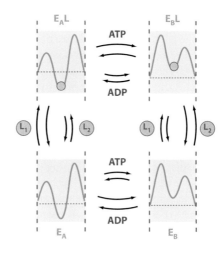

Figure 4

(*a*) Kinetic diagram showing how an external fluctuating field can entrain the equilibrium conformational fluctuations of the pump protein to bind ligand from reservoir 1 and release ligand to reservoir 2 by alternately favoring E_A and E_B. (*b*) Kinetic mechanism showing how, at the single-molecule level, ATP hydrolysis can accomplish the same pumping that the fluctuating electric field causes by alternately phosphorylating the protein (favoring E_B) and dephosphorylating the protein (favoring E_A).

Figure 4*a* can also be written in the form

$$L_1 + E_A \underset{k_{-1A}}{\overset{k_{1A}}{\rightleftharpoons}} E_A L \underset{k_{-2A}}{\overset{k_{2A}}{\rightleftharpoons}} E_A + L_2$$

$$\alpha_{BA} \Big\updownarrow \alpha_{AB} \qquad \gamma_{BA} \Big\updownarrow \gamma_{AB} \qquad \alpha_{BA} \Big\updownarrow \alpha_{AB}$$

$$L_1 + E_B \underset{k_{-1B}}{\overset{k_{1B}}{\rightleftharpoons}} E_B L \underset{k_{-2B}}{\overset{k_{2B}}{\rightleftharpoons}} E_B + L_2 \quad .$$

12.

The sets of rate constants k_{iA} and k_{iB} (i = $\pm 1, \pm 2$) separately satisfy Equations 7 and 8, with u_A and ε_A, and with u_B and ε_B, respectively. Additionally, there is a detailed balance condition (82),

$$\frac{k_{2B}k_{-2A}}{k_{-2B}k_{2A}} = \frac{\alpha_{AB}\gamma_{BA}}{\alpha_{BA}\gamma_{AB}} = \frac{k_{-1B}k_{1A}}{k_{1B}k_{-1A}}, \qquad 13.$$

that constrains how the rate constants for fluctuation between the two states can be assigned. Considering the case in **Figure 4**, neither E_A nor E_B is a good catalyst—there is one large barrier preventing free transport of ligand for each form. The possibility of fluctuations between the two forms can have a significant beneficial effect on the

catalysis. If the α's and γ's are large (i.e., fast fluctuations), the transporter is a much better catalyst than either of its two conformational states alone! By using the pathway

$$\rightleftharpoons \cdots E_A \overset{+L_1}{\rightleftharpoons} E_A L \rightleftharpoons E_B L \overset{-L_2}{\rightleftharpoons} E_B \rightleftharpoons E_A \cdots \rightleftharpoons,$$

ligand can pass between reservoirs 1 and 2 without surmounting a large activation barrier. The flow of ligand, however, is from high to low chemical potential—the protein is just a catalyst, but it is a better catalyst because of the conformational fluctuations. Surprisingly, if we drive the conformational fluctuation with some external forcing $\psi(t)$ such that $\alpha_{AB} = \alpha_{AB}^0 e^{-\beta\psi(t)/2}$, $\alpha_{BA} = \alpha_{BA}^0 e^{\beta\psi(t)/2}$, $\gamma_{AB} = \gamma_{AB}^0 e^{-\beta\psi(t)/2}$, $\gamma_{BA} = \gamma_{BA}^0 e^{\beta\psi(t)/2}$, where $\psi(t)$ is any autonomous function of time (20), there is net flow of ligand from reservoir 1 to reservoir 2 when $\Delta\mu = 0$. Note that ψ drops out of the product $\alpha_{AB}\gamma_{BA}/(\alpha_{BA}\gamma_{AB})$, so the detailed balance conditions are satisfied at every instant.

The protein is a poor catalyst either in state E_A or in state E_B. Equilibrium conformational fluctuation between the two states turns the protein into a better catalyst. These same conformational transitions in the presence of external driving or involved in the catalysis of a nonequilibrium chemical reaction such as ATP hydrolysis allow the protein to function as a free-energy transducer that harvests energy from the external driving or ATP hydrolysis to pump ligand from low to high electrochemical potential.

Xie and colleagues have recently provided compelling experimental evidence that conformational fluctuations at the single-molecule level are important for enzyme function (55). Their results are consistent with much work on the general importance of conformational flexibility in enzyme catalysis reviewed in Reference 40, with a general model for enzyme conformational flexibility given in Reference 24, and with the role of dynamics in protein function (23). How can we theoretically model the effects of internal fluctuations, present even at equilibrium, on enzymes?

Maxwell's and Smoluchowski's Demons: Engineering with Bilability and Bistability

Consider a simple two-state system based on a rotaxane, a mechanically interlinked molecule formed when a long rod-shaped molecule is threaded through a macrocyclic ring compound and then "stopper" groups are added to the ends of the rod to prevent the macrocycle from escaping (see **Figure 5**). As with the catenane discussed in **Figure 2**, separate binding stations for the macrocycle can be chemically incorporated on the rod. We consider two such stations in the model in **Figure 5**, where there is a steric barrier between the two stations hindering but not preventing exchange of the ring between sites 1 and 2. At equilibrium, the occupancy of the macrocycle at the two stations is determined by the relative interaction energies of the stations. There are two ways in which the relative occupancy can be shifted away from the equilibrium value. These are illustrated here by a Maxwell's information demon (**Figure 5a**), an intelligent being that uses information about the location

of the macrocycle to determine when to open and close a gate, and by a Smoluchowski's energy demon (**Figure 5b**), which is drawn as a blindfolded being that randomly raises and lowers the interaction energy between the macrocycle and one of the binding sites (30).

In the Maxwell's information demon case, if the demon, spotting the location of the ring, lowers the barrier when the ring is on station 1 and raises the barrier when ring is on station 2, the ring will obviously spend more time on station 2 than on station 1 despite the fact that the two stations have identical interaction energies at every instant. Exactly this scenario has been implemented experimentally by using both a photo-activated (71) and a chemically activated barrier (1), where the sensitivity of the trigger mechanism depends on the location of the macrocyclic ring. The raising and lowering of the gate illustrates one of the key design principles, bilability (9), for the design of a Brownian motor or stochastic pump. This principle has been investigated experimentally by Chaterjee et al. (30) and by Share et al. (73).

A second important design principle is bistability, illustrated by the Smoluchowski's energy demon, which randomly raises and lowers the interaction energy for one of the stations, thereby switching the relative stabilities of stations 1 and 2 back and forth. When station 1 is less stable, escape to station 2 is rapid, whereas when station 2 is less stable, escape to station 1 is less rapid. As a result of the relative rapidity of escape, the average occupancy of station 2 is greater than the average occupancy of station 1 under the fluctuating conditions, even though on average the interaction energies of station 1 and 2 are the same. Hernandez et al. (41) have used this principle of bistability (coupled with bilability) to design a catenane-based rotary molecular motor.

Developing a Thermodynamically Consistent Model for Molecular Fluctuations and Dynamic Disorder

The two principles of bistability and bilability are the cornerstones of a minimal Brownian motor or stochastic pump (9). Roughly

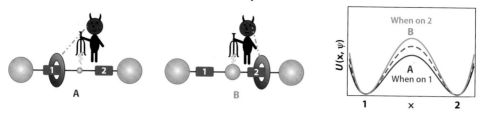

a Maxwell's information demon: correlated lability

b Smoluchowski's energy demon: uncorrelated stability

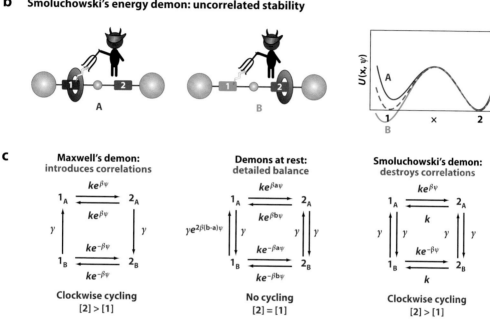

c

Maxwell's demon: introduces correlations	Demons at rest: detailed balance	Smoluchowski's demon: destroys correlations
Clockwise cycling [2] > [1]	No cycling [2] = [1]	Clockwise cycling [2] > [1]

Figure 5

Illustration of (*a*) Maxwell's information demon and (*b*) Smoluchowski's energy demon for controlling a simple two-state rotaxane-based switch. (*a*) Maxwell's demon uses information about the position of the ring to raise a barrier (*green curve*) when the ring is on station 2, and to lower the barrier (*magenta curve*) when the ring is on station 1. Even though the interaction energy between the ring and the two stations is identical at every instant in time, the Maxwell's demon imposes a correlation between the height of the barrier and the position of the ring and thus causes the ring to spend most of the time on station 2. (*b*) Smoluchowski's demon, which is blindfolded, raises (*magenta curve*) and lowers (*green curve*) the interaction energy between the ring and base 1 at random, with equal likelihood to raise the energy when the ring is on station 1 or on station 2. At equilibrium, of course it would be more likely for the interaction energy of station 1 to fluctuate to a high level (*magenta curve*) when the ring is on station 2 than when it is on station 1. Smoluchowski's demon destroys this correlation expected at equilibrium, also causing the ring to spend more than half the time on station 2. When the demons are at rest, i.e., when there is no pumping of the system by an external source, the value of ψ still fluctuates but in a way that is consistent with microscopic reversibility. (*c*) Simple kinetic diagrams illustrating the effects of pumping are shown for a Maxwell's information demon, the demons at rest (no pumping), and Smoluchowski's energy demon.

speaking, the lability of a kinetic pathway is reflected in the term F_{ij} in Equation 1, and the switching of the stability is reflected in the term dP_j/dt in Equation 1. The fact that a single term $\psi(t)$ can drive a system away from equi-

librium even if ψ fluctuates randomly forces us to ask how we can describe, in a single theory, both the assuredly present equilibrium fluctuation in ψ and nonequilibrium driving of ψ (20).

Consider the Michaelis-Menten scheme with rate constants that depend on some control parameter $\psi(t)$ (18):

$$L_1 + E \underset{k_{-1}e^{\beta b\psi(t)}}{\overset{k_{-1}e^{\beta a\psi(t)}}{\rightleftharpoons}} EL \underset{k_{-2}e^{-\beta b\psi(t)}}{\overset{k_2 e^{-\beta a\psi(t)}}{\rightleftharpoons}} E + L_2.$$

In the ratio in Equation 8, ψ cancels in the numerator and denominator. When $a = b$, the system is set up for a Maxwell's information demon that, by increasing ψ when P_{EL} is greater than average and decreasing ψ when P_{EL} is less than average, drives pumped current from L_1 to L_2 even when $\Delta\mu > 0$. When $a \neq 0$ and $b = 0$, the system is set up for a Smoluchowski's energy demon that, by causing random fluctuations, also drives pumped current from L_1 to L_2 (17–20). How can we model both equilibrium and nonequilibrium fluctuations in $\psi(t)$?

The equation for the rate at which the bound state probability changes can be written (see Equation 9) as

$$\frac{dP_{EL}}{dt} = -\tau^{-1}(\psi)[P_{EL} - P_{EL}^{ss}(\psi)]. \qquad 14.$$

It is tempting to consider the model proposed by Zwanzig (89) and subsequently adopted by Wang & Wolynes (84), Schenter et al. (68), and Lerch et al. (53), among others, to describe the effects of dynamic disorder on proteins, including enzymes. Zwanzig's model assumes that it is reasonable to have an internal stochastic control variable [e.g., $\psi(t)$] that influences the dynamics of the rate process dP_{EL}/dt but that is not influenced by the value of P_{EL}, so that the equation of motion for $\psi(t)$ is given by the autonomous Langevin equation $d\psi/dt = -\lambda\psi + \xi(t)$, where $\xi(t)$ is taken to be white noise. This picture, however, is not thermodynamically consistent for endogenous (internal) noise, as had been previously pointed out by Astumian et al. (20). The joint trajectories $[P_{EL}(t), \psi(t)]$ do not obey microscopic reversibility, and when $\Delta\mu = 0$, there is still net pumped flux from reservoir 1 to reservoir 2, in violation of the second law of thermodynamics. For a thermodynamically consistent picture, we must consider the back-reaction of the enzyme state on the likelihood

of the control parameter to adopt some particular value (20) by augmenting the equation of motion for $\psi(t)$ to read (12, 57)

$$\frac{d\psi}{dt} = g(P_{EL}, \psi) + f_{SD}(t) + f_{MD}(P_{EL}) + \xi(t), \qquad 15.$$

where we require

$$\frac{\partial g(P_{EL}, \psi)}{\partial P_{EL}} + \frac{\partial\{\tau^{-1}(\psi)[P_{EL} - P_{EL}^{ss}(\psi)]\}}{\partial\psi} = 0. \qquad 16.$$

The autonomous function $f_{SD}(t)$ depends only on time and describes the action of Smoluchowski's energy demon. The function $f_{MD}(P_{EL})$ depends on the probability P_{EL} for the protein to be in the bound state and describes the action of Maxwell's information demon. When $f_{SD}(t) = f_{MD}(P_{EL}) = 0$ (when the demons are at rest), the trajectories $[P_{EL}(t), \psi(t)]$ obey microscopic reversibility, the direction of the enzyme reaction is given solely by the chemical potential difference $\Delta\mu$, and the net flux is zero when $\Delta\mu = 0$. The choice of $g(P_{EL}, \psi)$ defined in Equation 16 assures that in this case the curl of the vector field of the two reciprocally coupled rates in Equations 14 and 15 is zero $\nabla \times [\frac{dP_{EL}}{dt}, \frac{d\psi}{dt}] = 0$.

With a time-dependent forcing $f_{SD}(t)$, or when there is a mechanism such as allosteric interaction by which $\psi(t)$ changes depending on whether the active site is occupied $f_{MD}(P_{EL})$, the influence of $\psi(t)$ on the protein conformational transitions on which it acts allows free energy to be transduced from the source of the fluctuation $f_{SD}(t)$ or $f_{MD}(P_{EL})$ to do work on the system by breaking the microscopic reversibility present at equilibrium.

MICROSCOPIC REVERSIBILITY AND CONFORMATIONAL TRANSITIONS

Microscopic reversibility and its corollary, detailed balance, are among the most important fundamental principles necessary for understanding free-energy transduction at the single-molecule level, and yet there is great confusion in the literature concerning these principles and their applicability to nonequilibrium systems

(10, 27). In this section I examine how microscopic reversibility constrains possible designs for molecular machines by considering a simple example of ligand binding to a protein.

Myoglobin Binding and Dissociation

Myoglobin is one of the most well-studied proteins (21, 63) and one of the first for which the X-ray crystal structure was determined (49). When oxygen or carbon monoxide binds to the heme group of myoglobin, the heme undergoes a transition from a configuration in which the iron atom is out of the plane of the heme to a configuration in which the iron is in-plane. The local configurational change is followed by a large-scale conformational change of the protein. The mechanism is schematically illustrated in **Figure 6a**.

Thinking about the reverse of the binding process, it is tempting to imagine a scenario in which oxygen dissociates followed by the return of the heme group to its original

---

a **Thermally activated binding**

b **Photodissociation**

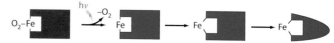

c **Thermally activated dissociation**

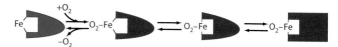

d **Reciprocal cycle**

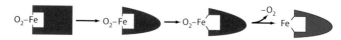

e **Nonreciprocal cycle**

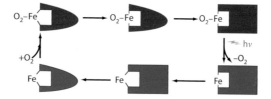

f **Energy levels for reciprocal and nonreciprocal cycles**

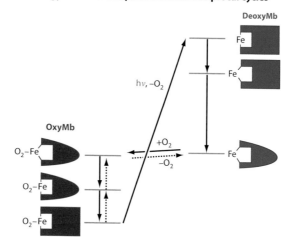

Figure 6

Schematic mechanism for (*a*) thermally activated binding of oxygen to myoglobin contrasted with (*b*) photochemically activated dissociation of oxygen and (*c*) thermally activated dissociation of oxygen. Combination of thermally activated association and dissociation gives a reciprocal cycle (*d*) in which the forward and backward processes are the microscopic reverses of each other and cannot drive directed motion. On the other hand, a combination of thermally activated binding with photochemically activated dissociation gives rise to a nonreciprocal cycle (*e*), which can in principle provide a mechanism for directed motion. (*f*) Energy level diagram for the states involved in the two cycles. The conformational rearrangements following photochemically activated dissociation are patently nonequilibrium processes, involving dissipation of more than 50 $k_B T$ (fifty times the thermal energy) per transition. These types of conformational transitions are called functionally important motions by Frauenfelder and colleagues (2), but they are relevant only for photochemical processes. For thermally activated transitions, the energy changes are much more modest (at most around 20 $k_B T$) and occur at and away from equilibrium.

out-of-plane configuration, with a subsequent global rearrangement of the molecule to restore the initial equilibrium conformation for the protein where the binding site is unoccupied. Indeed, this picture seems to be well supported by experiments on ligand dissociation from myoglobin where at low temperature a ligand is caused to dissociate by a LASER pulse, and the subsequent conformational relaxation is studied. These investigations reveal that after light-induced dissociation the myoglobin molecule undergoes a local rearrangement followed by a global conformational change in what has been termed a protein quake (2). The mechanism for photolytic dissociation is shown in **Figure 6b**. The photolytic mechanism is clearly not the microscopic reverse of the reaction by which binding of oxygen occurs. However, according to microscopic reversibility, we must have for the most probable pathway for nonphotochemically assisted dissociation the microscopic reverse of the binding reaction (i.e., for thermally activated dissociation), as shown in **Figure 6c**. These two different mechanisms for dissociation of ligand from the heme group, depending on whether the reaction occurs by photolysis or by thermal activation, conform to the principle of microscopic reversibility as defined in the International Union of Pure and Applied Chemists (IUPAC) Compendium of Chemical Terminology (**http://goldbook.iupac.org/**), informally known as the Gold Book:

> Microscopic Reversibility—In a reversible reaction, the mechanism in one direction is exactly the reverse of the mechanism in the other direction. This does not apply to reactions that begin with a photochemical excitation.

The idea that after either binding or dissociation of ligand at the heme the conformational rearrangement of the protein starts locally and propagates through the protein until the global change to the new equilibrium conformational state has occurred follows very naturally from macroscopic analogy. When a finger is inserted into water, ripples propagate from the finger outward. When the finger is

removed from the water, ripples once again propagate outward from where the finger was. This picture was explicitly suggested by Ansari et al. (2) for ligand association/dissociation to myoglobin. These authors stated that "binding or dissociation of a ligand at the heme iron causes a protein-quake," in which the heme is the focus of the quake. Such a picture is not consistent with microscopic reversibility and is possible only in the case of photochemically induced dissociation. If thermally activated binding causes a quake propagating outward from the focus, then, counterintuitive though it may be, thermally activated dissociation must arise by an inward propagating unquake that triggers release of the ligand.

Conformational transitions (i.e., shape changes) of a protein [or any other deformable body (64, 72)] cause the center of mass of the protein to move relative to the fluid in which the protein is immersed. The combination of any set of transitions constrained by microscopic reversibility such that the backward reaction (e.g., mechanism; **Figure 6c**) is the microscopic reverse of the forward reaction (e.g., **Figure 6a**) gives rise to a reciprocal process (cycle) (**Figure 6d**) that, according to Purcell's scallop theorem (64), cannot in the absence of inertia cause net directed motion in a cycle of the forward and backward transitions. Whatever is done in the forward process is undone in the backward process. Thus, the thermally activated binding and release of oxygen or carbon monoxide to myoglobin does not provide, even in principle, a mechanism for propulsion of the protein through solution.

On the other hand, although there are doubtless many practical reasons that net motion induced by shape changes is not biologically relevant for myoglobin, there is no fundamental reason that a cycle of thermally activated binding and photochemically induced dissociation of a ligand could not provide an effective mechanism for self-propulsion under the right circumstances. The combination of panels *a* and *b* in **Figure 6** is shown in panel *e*, where it is apparent that the conformational relaxation following photoassisted dissociation is not the microscopic reverse of the conformational relaxation following thermally activated binding. As a matter of principle, any nonreciprocal conformational cycle of a protein or polymer (or anything else) in viscous solution can, and in general will, lead to directed motion (72), whether it be of an ion across a membrane, of a molecular motor stepping along a polymeric track, or of a bacterium through the aqueous solvent (67). The example called pushmepullyou has been proposed and discussed by Avron et al. (22). Let us now consider how such nonreciprocal cyclical processes can be driven without photochemical activation.

Cycles of Molecular Machines

Togashi & Mikhailov (79) proposed that a polymer, described as an elastic network, could be constructed to operate as a cyclic machine powered by ligand binding. The binding was modeled by forming elastic links between the ligand and nearby nodes of the elastic network and allowing the network to relax to its new conformational energy minimum. The ligand was then removed (the elastic links were deleted) and the system again was allowed to undergo conformational relaxation. The overall process resulting from adding ligand, relaxation, removing ligand, relaxation, adding ligand, etc. was described by a simple cycle shown in **Figure 7a**. Thermal noise was not included in the computational study, and the transitions $E_A L \rightarrow E_B L$ and $E_B \rightarrow E_A$ were deterministic overdamped elastic relaxation processes. The mechanism is robust. Trajectories begun off the relaxation pathway feed into the pathway.

The proposed mechanism for autonomous generation of nonreciprocal cyclic motion is not consistent with microscopic reversibility if the ligand that binds to state E_A is the same molecule as the ligand that dissociates from state $E_B L$. Clearly, the energy of state E_A is less than that of state E_B, and the energy of state $E_B L$ is less than that of state $E_A L$. No matter the fixed arrangement of the energies of the bound states relative to the energies of the nonbound states, the overall cycle of binding ligand to E_A,

relaxing to $E_B L$, releasing ligand from E_B, and relaxing back to E_A needs energy; energy cannot be provided by the binding and release of the same ligand under the same conditions.

In Togashi & Mikhailov's work, the ligand dissociating from the polymer was implicitly different than the ligand that had associated. The details of the ligand binding $E_A \rightarrow E_A L$ and dissociation $E_B L \rightarrow E_B$ were not explicitly discussed, nor was the role of the chemical potential of ligand. The cycling of $E_A \rightarrow E_A L \rightarrow E_B L \rightarrow E_B \rightarrow E_A$ can be used to do work, W, on the environment in a manner similar to a single-molecule optomechanical cycle (44) so long as the motion on each of the two potentials is downhill, $aW < \Delta U_{\text{free}}$ and $(1 - a)W < \Delta U_{\text{bound}}$, where $\Delta U_{\text{free}} = U_{E_B} - U_{E_A}$ and $\Delta U_{\text{bound}} = U_{E_A L} - U_{E_B L}$.

Molecular Machines in a Thermal Environment

Our goal is to understand how molecular motors convert chemical energy into nonreciprocal conformational cycling, and hence into directed motion and mechanical work, in solution at room temperature where thermal noise is very strong and where there is a continual, reversible exchange of energy between each polymer molecule and its environment. The presence of thermal has important ramifications for how we should describe and think about molecular motors.

When we look at the mechanism in **Figure 7a**, it is tempting to term the elastic relaxation processes $E_A L \rightarrow E_B L$ and $E_B \rightarrow E_A$ to be power strokes, and indeed they are— power is dissipated as the system undergoes elastic relaxation. In a thermal environment, though, we can compare the power dissipated during the power stroke with the power that is continually and reversibly exchanged between the polymer and the environment to gauge the relative importance of mechanical versus thermal effects. If at some point on its energy profile the polymer experiences a very large force of 100 pN that at that instant moves the center of mass of the polymer with the very large velocity of 1 m s^{-1}, the power instantaneously dissipated by the power stroke is 10^{-10} J s^{-1}. In contrast, the power reversibly and continuously exchanged with the environment at room temperature is $k_B T = 4 \times 10^{-21}$ J every thermal relaxation time $\approx 10^{-12}$ s or 4×10^{-9} J s^{-1}, 40 times greater than the maximum power dissipated during even a very powerful

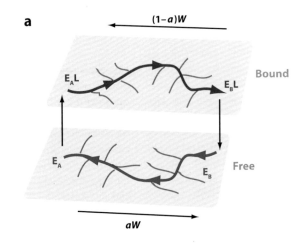

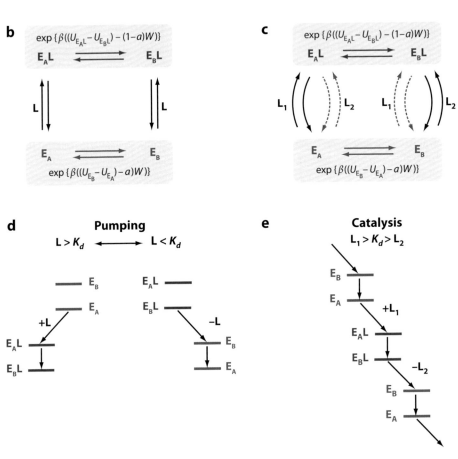

molecular power stroke (9)! Further, as recognized by Huxley (45), the unpower strokes $E_BL \rightarrow E_AL$ and $E_A \rightarrow E_B$ also occur with appreciable rates in a thermal environment. This was shown experimentally (52) for a simple rotaxane molecule. The ratio of the probability for an uphill fluctuation to a downhill relaxation by the microscopic reverse process is given by the simple relation (7, 26)

$$\frac{P(E_B \rightarrow E_A)}{P(E_A \rightarrow E_B)} = e^{\beta(\Delta U_{\text{free}} - aW)};$$

$$\frac{P(E_AL \rightarrow E_BL)}{P(E_BL \rightarrow E_AL)} = e^{\beta(\Delta U_{\text{bound}} - (1-a)W)}. \quad 17.$$

In contrast to the patently nonequilibrium "functionally important motions" following photodissociation of oxygen or carbon monoxide from myoglobin (2), the conformational changes by which chemically driven molecular motors move are equilibrium processes. The only difference between equilibrium and nonequilibrium for a thermally activated mechanism is that, away from equilibrium, the probability to bind ligand when the polymer is in state E_A is different than the probability to bind ligand when the molecule is in state E_B. The physical motions of the molecule that follow binding of ligand are exactly the same at and away from equilibrium. That the energy difference between the bound states and the nonbound states depends on the chemical potential of ligand immediately suggests two approaches for how to use ligand binding and dissociation to drive nonreciprocal cycling of the polymer conformational states—pumping and catalysis.

Pumping

Nonreciprocal conformational cycling (58) can be pumped by externally driven oscillations or fluctuations between large and small concentrations of the ligand. The large concentration $[L]_{\text{large}} > K_d$ favors binding of ligand to E_A followed by elastic relaxation $E_AL \rightarrow E_BL$. The small concentration $[L]_{\text{small}} < K_d$ favors dissociation of ligand from E_BL followed by elastic relaxation $E_B \rightarrow E_A$. The dissociation constant K_d is the concentration of ligand at which half the polymer is bound and half is free.

The oscillation (or fluctuation) of the concentration of L can be repeated, resulting in continual cycling. This pumping mechanism requires external oscillation or fluctuation of the ligand concentration, although in principle, if the ligand were some intermediate in an oscillating chemical reaction such as the Belousov-Zhabotinsky reaction (37), the process could be driven without direct experimental manipulation of the concentrations.

Figure 7

(*a*) Illustration of cycling induced in an elastic network by binding ligand and allowing viscous relaxation on the bound energy surface, followed by removal of ligand and relaxation on the free-energy surface. Whenever ligand is added or removed, energy is deposited into the system and dissipated during relaxation. If the system is set up to harness this energy by attaching the relaxing network to a load, some of the energy deposited upon addition and removal of ligand can by harnessed to do work (*W*) in the environment. (*b*) In a thermal environment both forward and backward transitions are possible. The ratio of the probability for an uphill fluctuation to a downhill dissipation is a state function (26) (Equation 17). (*c*) If the polymer can bind some substrate L_1 and catalytically convert it to a different molecule L_2, then the energy in the chemical potential difference of L_1 and L_2 can drive nonreciprocal cycling of the polymer and do work on the environment. (*d*) Illustration of a pumping mechanism for driving nonreciprocal cycling by oscillation of the ligand concentration between a low level, where the polymer is most likely free, and a high level, where the polymer is most likely bound. (*e*) Illustration of a catalysis mechanism where the concentration of L_1 is greater than the dissociation constant, and the concentration of L_2 is less than the dissociation constant; so on average the polymer will bind L_1, undergo conformational relaxation, release L_2, undergo conformational relaxation, bind L_1, etc. For both the pumping and the catalysis mechanisms only unidirectional arrows have been shown. In a thermal environment, however, the molecule occasionally carries out a cycle in reverse.

The pumped energy flux into the system that allows work to be done on the environment comes from the fact that, on average, ligand is bound while the chemical potential of ligand is high and dissociates when the chemical potential of ligand is low. Through a cycle of oscillation of the ligand concentration, the energy available to drive flux through the conformational cycle is, at most, equal to the amplitude of the oscillation of the ligand's chemical potential, but for very large oscillations between very small ([L]$_{small}$ ≪ K_d) and very large ([L]$_{large}$ ≫ K_d) concentrations, the ratio of the probability to complete a clockwise versus counterclockwise cycle is

$$\frac{P(E_A \to E_A L \to E_B L \to E_B \to E_A)}{P(E_A \to E_B \to E_B L \to E_A L \to E_A)} = e^{\beta(\Delta U_{free} + \Delta U_{bound} - W)}. \qquad 18.$$

It is necessary to have at least one of the conformational changes involve a viscoelastic relaxation (i.e., either $\Delta U_{free} > 0$ and/or $\Delta U_{bound} > 0$) for conformational pumping by an external source. The essential mechanism is that of an energy ratchet (14) or Smoluchowski's energy demon. As soon as L binds when [L] ≫ K_d, the polymer relaxes from state $E_A L$ to state $E_B L$ because $\Delta U_{bound} > 0$, and as soon as L dissociates when [L] ≪ K_d, the polymer relaxes from state E_B to state E_A because $\Delta U_{free} > 0$.

Catalysis

The second approach for driving directional cycling is catalysis. If a polymer can be designed to catalyze a reaction $L_1 \rightleftharpoons L_2$, then, when the chemical potentials of L_1 and L_2 are not equal, $\mu_{L_1} \neq \mu_{L_2}$, the polymer will most likely bind whichever of L_1 and L_2 has the higher chemical potential and release whichever of L_1 and L_2 has the lower chemical potential. Thus, catalysis autonomously achieves the bind high/release low by which pumping drives directional cycling. Because L_1 and L_2 must be related chemically (i.e., they are interconvertible), either L_1 or L_2 can bind to either state E_A or state E_B (**Figure 7c**) but with possibly different rates. We can express the ratio of the probabilities for clockwise and counterclockwise conforma-

tional cycling in terms of only the ratios of off rate constants for L_1 and L_2 from E_A and E_B (the selectivities or specificities), the chemical potential difference $\Delta \mu = \mu_{L_1} - \mu_{L_2}$, and the work, W, on the environment required for nonreciprocal conformational cycling as (14)

$$\frac{P(E_A \to E_A L \to E_B L \to E_B \to E_A)}{P(E_A \to E_B \to E_B L \to E_A L \to E_A)}$$
$$= \frac{(s_A + 1)(s_B e^{\beta \Delta \mu} + 1)}{(s_B + 1)(s_A e^{\beta \Delta \mu} + 1)} e^{-\beta W}, \qquad 19.$$

where $s_i = k_{off,i}^{L_1} / k_{off,i}^{L_2}$, i = A, B. In contrast to the case of external pumping, for catalytically driven conformational cycling this ratio is independent of the elastic energy differences ΔU_{free} and ΔU_{bound}. We can think of the mechanism as a chemically driven information ratchet (1, 14) where the active site functions as a Maxwell's information demon (30), selecting for L_1 in state E_A and for L_2 in state E_B through an allosteric specificity switch (36, 46). Note, however, that irrespective of how strongly asymmetric the selectivity of the active site is, the mechanism fails to drive directed transport or to do work in the environment if $\Delta \mu = 0$, in consistency with the second law of thermodynamics. The nonreciprocal conformational cycling induced by catalysis in general may well be the mechanism for enhanced diffusion during active catalysis by an enzyme (60), where the rapid rotational rearrangement prevents the appearance of net directed motion.

CONCLUSIONS

The transitions within the conformational cycles by which molecular machines function are equilibrium processes. The physical motions of the individual molecules are exactly the same at and away from statistical equilibrium. Conformational pumping occurs when an external source or an energy-releasing chemical reaction entrains these equilibrium motions to occur preferentially in a particular sequence, thereby breaking microscopic reversibility and allowing for a nonreciprocal conformational cycle by which directed motion occurs. Unsatisfying as it may be, the best description of the mechanism by which, e.g., ATP hydrolysis drives a molecular motor to move in one direction along a biopolymeric track is mass action. The fact that binding ATP, converting it to ADP and Pi at the active site, and releasing ADP and Pi are more likely than the reverse when the ATP hydrolysis reaction is away from equilibrium imposes a temporal ordering on the equilibrium fluctuations of the protein, giving rise to correlations that drive directed motion, pumping ligand across a membrane, or synthesis of important biopolymers. Because in a kinetic cycle all states can equilibrate with one another even if one transition is kinetically blocked, a molecular machine can do work against significant loads by designing mechanisms that sequentially move kinetic blockades through the cycle, thereby preventing slip even under the influence of load (8).

Great progress has been made in the design and synthesis of artificial molecular machines (38, 48, 59, 76) and DNA (39) and small-molecule (33, 34) walkers. It is becoming clear that these are first and foremost molecules, governed by the laws of chemistry rather than mechanics. The dynamical behavior of machines based on chemical principles can be described as a random walk on a network of states. In contrast to macroscopic machines, whose function is determined predominately by the connections between the elements of the machine, the function of a Brownian machine in response to an external stimulus is completely specified by the equilibrium energies of the states and by the heights of the barriers between the states. Chemists have much experience with approaches for controlling stabilities and labilities of molecules, and for designing systems with sterically or energetically hindered pathways allowing for kinetic rather than thermodynamic control of mechanisms. This experience will be crucial in the next steps of interfacing synthetic molecular machines with the macroscopic world.

SUMMARY POINTS

1. Irrespective of how far a system is from statistical equilibrium, the instantaneous probability flux between any two conformational states of a macromolecule is the sum of a steady-state current and a pumped current, $I_{ij} = I_{ij}^{ss} + F_{ij}\frac{dP_j}{dt}$. In the absence of driving, the time average of the pumped current is guaranteed to be zero by microscopic reversibility. In the presence of time-dependent driving, however, the average of the pumped current in general is not zero and can be opposite in sign and larger in magnitude than the steady-state current.

2. Conformational fluctuations present at equilibrium can be exploited as a mechanism by which an external modulation can do work on the environment. The same equilibrium conformational fluctuations are important in free-energy transduction from a nonequilibrium chemical reaction such as ATP hydrolysis.

3. A thermodynamically consistent model for the effects of fluctuations on proteins requires inclusion of a reciprocal reaction between the source of the fluctuation and the protein. A key test for any model is that, in the absence of external driving, the joint fluctuations of the protein and the noise must obey microscopic reversibility.

4. Far from being predominately mechanical devices slightly perturbed by thermal noise, molecular motors are overwhelmingly dominated by thermal effects and operate based on the laws of chemistry rather than the laws of macroscopic mechanics. A key design principle is that molecular machines take advantage of omnipresent thermal noise and function by using input energy in part to prevent backward motion rather than to cause forward motion.

FUTURE ISSUES

1. Specificity switching is essential for molecular machines powered by catalysis of a nonequilibrium chemical reaction. Development of an understanding of how to incorporate allosteric switches into macromolecular and supramolecular structures will be necessary for future progress in designing artificial molecular motors and pumps.

2. The conformational transitions by which molecular machines carry out their function are equilibrium processes. Can this fact be used to design a machine that optimizes both power output and thermodynamic efficiency under the same conditions?

3. How do the microscopic concepts developed for behavior of individual molecular machines scale when the molecules are linked together to perform a macroscopic task?

4. Is direct mechanical coupling the best way to interface molecular machines to the macroscopic environment, or can this be better accomplished through electrical coupling?

DISCLOSURE STATEMENT

The author is not aware of any affiliations, memberships, funding, or financial holdings that might be perceived as affecting the objectivity of this review.

ACKNOWLEDGMENTS

I am grateful to the German Humboldt Foundation for facilitating this work through conferral of a Humboldt Research Award.

LITERATURE CITED

1. Alvarez-Perez M, Goldup SM, Leigh DA, Slawin AMZ. 2008. A chemically driven molecular information ratchet. *J. Am. Chem. Soc.* 130:1836–38
2. Ansari A, Berendzen J, Bowne SF, Frauenfelder H, Iben IET, et al. 1985. Protein states and proteinquakes. *Proc. Natl. Acad. Sci. USA* 82:5000–4
3. Astumian RD. 1997. Thermodynamics and kinetics of a Brownian motor. *Science* 276:917–22
4. Astumian RD. 2001. Making molecules into motors. *Sci. Am.* 285:45–51
5. Astumian RD. 2003. Adiabatic pumping mechanism for ion motive ATPases. *Phys. Rev. Lett.* 91:118102
6. Astumian RD. 2005. Chemical peristalsis. *Proc. Natl. Acad. Sci. USA* 102:1843–47
7. Astumian RD. 2006. The unreasonable effectiveness of equilibrium theory in describing non-equilibrium experiments. *Am. J. Phys.* 74:683–88
8. Astumian RD. 2007. Adiabatic operation of a molecular machine. *Proc. Natl. Acad. Sci. USA* 104:19715–18
9. Astumian RD. 2007. Design principles for Brownian molecular machines: how to swim in molasses and walk in a hurricane. *Phys. Chem. Chem. Phys.* 9:5067–83
10. Astumian RD. 2009. Comment: detailed balance revisited. *Phys. Chem. Chem. Phys.* 11:9592
11. Astumian RD. 2010. Kinetics and thermodynamics of molecular motors. *Biophys. J.* 98:2401–9
12. Astumian RD, Bier M. 1994. Fluctuation driven ratchets: molecular motors. *Phys. Rev. Lett.* 72:1766–69
13. Astumian RD, Bier M. 1996. Mechanochemical coupling of the motion of molecular motors to ATP hydrolysis. *Biophys. J.* 70:637–53
14. Astumian RD, Derenyi I. 1998. Fluctuation driven transport and models of molecular motors and pumps. *Eur. Biophys. J.* 27:474–89
15. Astumian RD, Derenyi I. 2001. Towards a chemically driven molecular electron pump. *Phys. Rev. Lett.* 86:3859–62

16. Astumian RD, Hanggi P. 2002. Brownian motors. *Phys. Today* 55(11):33–39
17. Astumian RD, Robertson B. 1989. Nonlinear effect of an oscillating electric field on membrane proteins. *J. Chem. Phys.* 91:4891
18. Astumian RD, Robertson B. 1993. Imposed oscillations of kinetic barriers can cause an enzyme to drive a chemical reaction away from equilibrium. *J. Am. Chem. Soc.* 115:11063–68
19. Astumian RD, Chock PB, Tsong TY, Westerhoff HV. 1989. Effects of oscillations and energy-driven fluctuations on the dynamics of enzyme catalysis and free-energy transduction. *Phys. Rev. A* 39:6416–35
20. Astumian RD, Tsong TY, Chock PB, Chen YD, Westerhoff HV. 1987. Can free energy be transduced from electric noise? *Proc. Natl. Acad. Sci. USA* 84:434–38
21. Austin RH, Beeson KW, Eisenstein L, Frauenfelder H, Gunsalus IC. 1975. Dynamics of ligand binding to myoglobin. *Biochemistry* 14:5355–73
22. Avron JE, Kenneth O, Oaknin DH. 2005. Pushmepullyou: an efficient microswimmer. *New J. Phys.* 7:234–42
23. Bahar I, Lezon TR, Yan LW, Eran E. 2010. Global dynamics of proteins: bridging between structure and function. *Annu. Rev. Biophys.* 39:23–42
24. Benkovic SJ, Hammes GG, Hammes-Schiffer S. 2008. Free-energy landscape of enzyme catalysis. *Biochemistry* 47:3317–21
25. Berry MV. 1990. Anticipations of the geometric phase. *Phys. Today* 43(12):34–40
26. Bier M, Derenyi I, Kostur M, Astumian RD. 1999. Intrawell relaxation of over-damped particles. *Phys. Rev. E* 59:6422–32
27. Blackmond DG. 2009. "If Pigs Could Fly" chemistry: a tutorial on the principle of microscopic reversibility. *Ang. Chem. Int. Ed.* 48:2648–54
28. Burbaum JJ, Raines RT, Albery WJ, Knowles JK. 1989. Evolutionary optimization of the catalytic effectiveness of an enzymes. *Biochemistry* 28:9293–303
29. Carter NJ, Cross RA. 2005. Mechanics of the kinesin step. *Nature* 435:308–12
30. Chatterjee MN, Kay ER, Leigh DA. 2006. Beyond switches: ratcheting a particle energetically uphill with a compartmentalized molecular machine. *J. Am. Chem. Soc.* 128:4058–73
31. Chernyak VY, Sinitsyn NA. 2009. Robust quantization of a molecular motor motion in a stochastic environment. *J. Chem. Phys.* 131:181101
32. Cressman A, Togashi Y, Mikhailov AS, Kapral R. 2008. Mesoscale modeling of molecular machines: cyclic dynamics and hydrodynamical fluctuations. *Phys. Rev. E* 77:050901
33. Delius M, Geertsema EM, Leigh DA. 2010. A synthetic small molecule that can walk down a track. *Nat. Chem.* 2:96–101
34. Delius M, Geertsema EM, Leigh DA, Tang DD. 2010. Design, synthesis and operation of small molecules that walk along tracks. *J. Am. Chem. Soc.* 132:1613–45
35. Derenyi I, Astumian RD. 1998. Spontaneous onset of coherence and energy storage by membrane transporters in an RLC electric circuit. *Phys. Rev. Lett.* 180:4602–5
36. Eisenberg E, Hill TL. 1985. Muscle contraction and free energy transduction in biological systems. *Science* 227:999–1006
37. Epstein IR, Showalter K. 1996. Nonlinear chemical dynamics: oscillations, patterns, and chaos. *J. Phys. Chem.* 100:13132–47
38. Feringa BL. 2007. The art of building small: from molecular switches to molecular motors. *J. Org. Chem.* 72:6635–52
39. Green SJ, Bath J, Turberfield AJ. 2008. Coordinated chemomechanical cycles: a mechanism for autonomous molecular motion. *Phys. Rev. Letts.* 101:238104
40. Hammes-Schiffer S, Benkovic SJ. 2006. Relating protein motion to catalysis. *Annu. Rev. Biochem.* 75:519–41
41. Hernandez JV, Kay ER, Leigh DA. 2004. A reversible synthetic rotary molecular motor. *Science* 306:1532–37
42. Horowitz JM, Jarzynski C. 2009. Exact formula for currents in strongly pumped diffusive systems. *J. Stat. Phys.* 136:917–25
43. Howard J. 2001. *Mechanics of Motor Proteins and the Cytoskeleton*. Sunderland, MA: Sinauer

44. Hugel T, Holland NB, Cattani A, Moroder L, Seitz M, Gaub HE. 2002. Single-molecule optomechanical cycle. *Science* 296:1103–6

45. Huxley AF. 1957. Muscle structure and theories of contraction. *Prog. Biophys. Biophys. Chem.* 7:255–318

46. Jencks WP. 1995. The mechanism of coupling chemical and physical reactions by the calcium ATPase of sarcoplasmic reticulum and other coupled vectorial systems. *Biosci. Rep.* 15:283–87

47. Kagan ML, Kepler TB, Epstein IR. 1991. Geometric phase shifts in chemical oscillators. *Nature* 349:506–8

48. Kay ER, Leigh DA, Zerbetto F. 2007. Synthetic molecular motors and mechanical machines. *Ang. Chem. Int. Ed.* 46:72–191

49. Kendrew JC, Dickerson RE, Strandberg BE, Hart RG, Davies DR, et al. 1960. Structure of myoglobin: a three-dimensional Fourier synthesis at 2 Å resolution. *Nature* 185:422–27

50. Kim YC, Furchtgott LA, Hummer G. 2009. Biological proton pumping in an oscillating electric field. *Phys. Rev. Lett.* 103:268102

51. Lauger P. 1991. *Electrogenic Ion Pumps*. Sunderland, MA: Sinauer

52. Leigh DA, Wong JKY, Dehez F, Zerbetto F. 2003. Unidirectional rotation in a mechanically interlocked molecular rotor. *Nature* 424:174–79

53. Lerch HP, Rigler R, Mikhailov AS. 2005. Functional conformational motions in the turnover cycle of cholesterol oxidase. *Proc. Natl. Acad. Sci. USA* 102:10807–12

54. Liu DS, Astumian RD, Tsong TY. 1990. Activation of Na^+ and K^+ pumping mode of (Na,K)-ATPase by an oscillating electric field. *J. Biol. Chem.* 265:7260–67

55. Lu HP, Xun L, Xie XS. 1998. Single-molecule enzymatic dynamics. *Science* 282:1877–81

56. Maes C, Netocny K, Thomas SR. 2010. General no-go condition for stochastic pumping. *J. Chem. Phys.* 132:234116

57. Magnasco MO. 1994. Molecular combustion motors. *Phys. Rev. Lett.* 72:2656–59

58. Markin VS, Tsong TY, Astumian RD, Robertson B. 1990. Energy transduction between a concentration gradient and an alternating electric field. *J. Chem. Phys.* 93:5062–66

59. Michl J, Sykes EC. 2009. Molecular motors and rotors: recent advances and future challenges. *ACS Nano* 3:1042–48

60. Muddana HS, Sengupta S, Mallouk TE, Sen A, Butler PJ. 2010. Substrate catalysis enhances single-enzyme diffusion. *J. Am. Chem. Soc.* 132:2110–11

61. Ohkubo J. 2008. Current and fluctuation in a two-state stochastic system under nonadiabatic periodic perturbation. *J. Chem. Phys.* 129:205102

62. Panman MR, Bodis P, Shaw DJ, Bakker BH, Newton AC, et al. 2010. Operation mechanism of a molecular machine revealed using time-resolved vibrational spectroscopy. *Science* 328:1255–58

63. Parak FG, Nienhaus GU. 2002. Myoglobin, a paradigm in the study of protein dynamics. *ChemPhysChem* 3:249–54

64. Purcell E. 1977. Life at low Reynolds number. *Am. J. Phys.* 45:3–11

65. Rahav S, Horowitz J, Jarzynski C. 2008. Directed flow in nonadiabatic stochastic pumps. *Phys. Rev. Lett.* 101:140602

66. Robertson B, Astumian RD. 1990. Kinetics of a multistate enzyme in a large oscillating field. *Biophys. J.* 57:689–96

67. Sakaue T, Kapral R, Mikhailov AS. 2010. Nanoscale swimmers: hydrodynamic interactions and propulsion of molecular machines. *Eur. Phys. J. B* 75:381–87

68. Schenter GK, Lu HP, Xie XS. 1999. Statistical analysis and theoretical models of single-molecule enzymatic dynamics. *J. Phys. Chem. A* 103:10477–88

69. Schliwa M, Woehlke G. 2003. Molecular motors. *Nature* 422:759–65

70. Serpersu EH, Tsong TY. 1984. Activation of electrogenic Rb+ transport of (Na,K)-ATPase by an electric field. *J. Biol. Chem.* 259:7155–62

71. Serreli V, Lee CF, Kay ER, Leigh DA. 2007. A molecular information ratchet. *Nature* 445:523–27

72. Shapere A, Wilczek F. 1987. Self-propulsion at low Reynolds number. *Phys. Rev. Lett.* 58:2051–54

73. Share AI, Parimal K, Flood AH. 2010. Bilability is defined when one electron is used to switch between concerted and step-wise pathways in Cu(I)-based bistable [2/3]pseudorotaxanes. *J. Am. Chem. Soc.* 132:1665–75

74. Sinitsyn NA. 2009. The stochastic pump effect and geometric phases in dissipative and stochastic systems. *J. Phys. A Math. Theor.* 42:193001

75. Sinitsyn NA, Nemenman I. 2007. The Berry phase and the pump flux in stochastic chemical kinetics. *Europhys. Lett.* 77:58001

76. Siwy ZS, Powell MR, Kalman E, Astumian RD, Eisenberg RS. 2006. Negative incremental resistance induced by calcium in asymmetric nanopores. *Nano Lett.* 6:473–77

77. Switkes M, Marcus CM, Campman K, Gossard AC. 1999. An adiabatic quantum electron pump. *Science* 283:1905–8

78. Thouless DJ. 1983. Quantization of particle transport. *Phys. Rev. B* 27:6083–87

79. Togashi Y, Mikhailov AS. 2007. Nonlinear relaxation dynamics in elastic networks and design principles of molecular machines. *Proc. Natl. Acad. Sci. USA* 104:8697–702

80. Tolman RC. 1938. *The Principles of Statistical Mechanics*. Oxford: Clarendon Press

81. Tsong TY, Astumian RD. 1986. Absorption and conversion of electric field energy by membrane bound ATPases. *Bioelectrochem. Bioenerg.* 15:457–76

82. Tsong TY, Astumian RD. 1987. Electroconformational coupling and membrane protein function. *Prog. Biophys. Mol. Biol.* 50:1–45

83. Tsong TY, Astumian RD. 1988. Electroconformational coupling: how membrane bound ATPase transduces energy from dynamic electric fields. *Annu. Rev. Physiol.* 50:273–90

84. Wang J, Wolynes P. 1999. Intermittency of activated events in single molecules: the reaction diffusion description. *J. Chem. Phys.* 110:4812–19

85. Westerhoff HV, Tsong TY, Chock PB, Chen YD, Astumian RD. 1986. How enzymes can capture and transmit free energy from an oscillating field. *Proc. Natl. Acad. Sci. USA* 83:4734–38

86. Xie TD, Chen YD, Marszalek P, Tsong TY. 1997. Fluctuation-driven directional flow in biochemical cycles: further study of electric activation of Na,K pumps. *Biophys. J.* 72:2496–502

87. Xie TD, Marszalek P, Chen YD, Tsong TY. 1994. Recognition and processing of randomly fluctuating electric signals by Na,K-ATPase. *Biophys. J.* 67:1247–51

88. Xu D, Phillips JC, Schulten K. 1996. Protein response to external electric fields: relaxation, hysteresis, and echo. *J. Phys. Chem.* 100:12108–21

89. Zwanzig RJ. 1990. Rate processes with dynamic disorder. *Acc. Chem. Res.* 23:148–52

Protein Self-Organization: Lessons from the Min System

Martin Loose,[1,2] Karsten Kruse,[3] and Petra Schwille[1,2]

[1] Biophysics, BIOTEC, Dresden University of Technology, 01307 Dresden, Germany; email: petra.schwille@biotec.tu-dresden.de

[2] Max-Planck-Institute for Molecular Cell Biology and Genetics, 01307 Dresden, Germany

[3] Theoretical Physics, Saarland University, 66041 Saarbrücken, Germany

Annu. Rev. Biophys. 2011. 40:315–36

The *Annual Review of Biophysics* is online at biophys.annualreviews.org

This article's doi: 10.1146/annurev-biophys-042910-155332

Copyright © 2011 by Annual Reviews. All rights reserved

1936-122X/11/0609-0315$20.00

Keywords

Min oscillations, pattern formation

Abstract

One of the most fundamental features of biological systems is probably their ability to self-organize in space and time on different scales. Despite many elaborate theoretical models of how molecular self-organization can come about, only a few experimental systems of biological origin have so far been rigorously described, due mostly to their inherent complexity. The most promising strategy of modern biophysics is thus to identify minimal biological systems showing self-organized emergent behavior. One of the best-understood examples of protein self-organization, which has recently been successfully reconstituted in vitro, is represented by the oscillations of the Min proteins in *Escherichia coli*. In this review, we summarize the current understanding of the mechanism of Min protein self-organization in vivo and in vitro. We discuss the potential of the Min oscillations to sense the geometry of the cell and suggest that spontaneous protein waves could be a general means of intracellular organization. We hypothesize that cooperative membrane binding and unbinding, e.g., as an energy-dependent switch, may act as an important regulatory mechanism for protein oscillations and pattern formation in the cell.

Contents

INTRODUCTION

The ability to self-organize and spontaneously form dynamic and spatially variable structures is among the most intriguing features of living systems. Classic examples of biological pattern formation are provided by animal fur and sea shells, by the collective dynamic behavior of organisms in ant colonies, during termite nest building, in schools of fish or flocks of birds, and during the formation of the fruiting bodies in *Dictyostelium discoideum* and *Myxococcus*

Collective dynamic behavior: behavior of a system that emerges from the interactions of a large number of components in a spontaneous way

xanthus (15). However, the ability to form temporal and spatial architectures and patterns, as cells and particularly organisms continuously do with high fidelity, also has its origin in the specific properties of the basic elements, i.e., proteins. This means that self-organization not only plays an important role for pattern formation on the level of whole organisms and tissues, but it is also important for the spatial and temporal organization of molecules inside cells (58, 61, 83).

In 1944, Schrödinger (103) formulated in his visionary view on biology that self-organization is only possible if energy is continuously flowing through the system, with the additional requirement that this energy is of low entropy, i.e., significantly more ordered than thermal energy. During the past century, chemists and biochemists together with physicists and mathematicians developed an impressive framework of concepts and models of how self-organization of biological molecules may emerge from the interplay between their structural features and the specific types of interactions between them (15, 58).

Importantly, in self-organized systems, order can emerge even from a situation lacking any prepatterning or structured cues if spontaneous local fluctuations, e.g., variations in the concentrations of the players involved, have the tendency to get amplified (91, 92), e.g., due to cooperative or autocatalytic effects. Thus, spatial and temporal pattern formation results from the nonlinear properties of the reaction network. For some of the most prominent of these systems, such as glycolytic oscillations, allosteric regulation of the enzymes responsible for this behavior could be identified (44). These nonlinear effects allow the system to switch between different states, when a specific parameter assumes a critical value, leading to the breaking of symmetry in time and/or space, comprehensively described as bifurcations (30, 31), and possibly giving rise to oscillations in the biochemical reaction system.

In 1952, Alan Turing published a landmark paper (114) extrapolating the same principles of nonlinear amplification of spontaneous fluctuations to the central question of developmental biology: how organisms assume their structures during growth from embryos to adults. In his model, the instability driving the nonlinear dynamic system into patterns is also based on diffusion, specifically, on two interacting molecular species differing significantly in their diffusion characteristics. This paper was remarkable in many ways, coining the term morphogen as one of the most influential concepts of tissue differentiation and organism development, and boldly subjecting one of the most complex biological processes one could possibly think of (and that Turing had no practical knowledge about) to rigorous reductionism, followed by a mathematical description. Due to the abrupt end of his work by Turing's untimely death, this paper was not followed up on for quite some time, until the concept of morphogenesis based on nonlinear dynamics in a reaction-diffusion system was revived and further specified by the work of Gierer & Meinhardt (39). One of their most prominent additions to Turing's original concept was to show that only a restricted class of reaction-diffusion systems is capable of generating patterns. Indeed, it is necessary to have an antagonistic pair of molecular species, one of which (the activator) is self-enhancing and of short range and coupled to the other one (the inhibitor) of long range. This model comprises Turing's original suggestion and causes a homogeneous distribution to become unstable (80). Gierer & Meinhardt applied this mechanism to a number of specific biological systems, ranging from the freshwater polyp hydra to the shells of molluscs, and for the first time tried to identify specific molecules that play the role of the short-range activator and the long-range inhibitor.

These models of biological pattern formation are in contrast to the findings showing that morphogenesis is often guided by mechanical or parental tissue-specific cues, i.e., the specific localization of distinct proteins or RNA (123), a mechanism that has been favored by many experimental developmental biologists. Another reason for the initially low acceptance

of Turing's model was certainly the impossibility to prove it in a simple, reproducible experiment, as usually required for a successful reductionist approach. The only direct evidence for the potential of molecules to form large-scale patterns came in the form of the oscillating, Belousov-Zhabotinsky reaction originally described in the 1950s (29, 125). Although composed solely from inorganic molecules, this reaction served as a paradigm of nonequilibrium biological phenomena for a long time. For cells and organisms, however, studies on reaction-diffusion patterning were almost exclusively theoretical, until Kondo & Asai (62, 63) could indeed demonstrate its impact in studies on live animals. During the past years, following the release of genetically encoded fluorescent probes, a true revolution in imaging of live cells and organisms with high spatial and temporal resolution has been triggered, enabling researchers to study pattern formation and tissue differentiation with much more molecular detail and quantitative rigor. Therefore, it is finally possible to test many of the predictions from theoretical/computational studies, thus potentially leading to deep new insights into this field.

However, many of the mechanisms leading to biological pattern formation may not be conclusively testable in a living system, simply because it is impossible to truly reduce the number of controllable functional elements without seriously interfering with essential live functions. In light of this, cell-free minimal systems and in vitro reconstitution approaches have become increasingly important in past years. The idea is to reduce the system under investigation to the smallest possible combination of molecules that is still able to reproduce a specific process or function, and subject this system to highly quantitative analytics that are impossible to apply with the same rigor to living systems. Strictly speaking, this could be termed a bottom-up or synthetic approach toward biology (6, 70, 105). Particularly attractive are systems with the intrinsic potential to oscillate, as the experimental realization of an oscillating system already has a considerable persuasive power on its own (66).

Self-organization: emergence of a spatiotemporal pattern as a collective dynamic behavior

Nonlinear dynamic system: refers to a dynamic system in which the sum of two solutions to the equations is not solution; it usually reflects collective dynamics

Reaction-diffusion systems: theoretical models that explain how the concentrations of interacting substances in a dynamic system can change due to local chemical reactions and diffusion

Although several examples of biological oscillators have been rigorously investigated, many of them rely on the regulation of gene expression and are thus not simple enough to be easily reconstituted in a cell-free system. Recently, two systems were identified that displayed oscillatory behavior with supposedly stunning simplicity, based solely on biochemical reactions between a few proteins. The first system is the circadian clock from cyanobacteria based on a set of three proteins, KaiA, KaiB, and KaiC. In contrast to other organisms, which depend on a transcription-translation feedback loop, KaiC phosphorylation is the sole pacemaker for the cyanobacterial circadian system (see Reference 77 for a review). By simply incubating KaiC with KaiA and KaiB in a test tube, the KaiC phosphorylation cycle shows circadian periods comparable to those in vivo (85), demonstrating the robustness of fundamental biological oscillations in a minimal reconstituted system.

The second system is the oscillation of the Min system, which regulates cell division in *Escherichia coli*. A number of mechanisms underlying the oscillations have been proposed for the Min system. Our group has successfully reconstituted the Min system in a cell-free assay consisting of only two proteins, a membrane, and ATP as an energy source (72). In contrast to other oscillating systems that are solution based or coupled to gene transcription, an important regulatory feature of the Min protein system is the reversible binding of proteins to the membrane. Thus, in addition to the structural possibilities that membranes provide to biological systems, including the separation of a variety of reactants, ions, and compounds in 2D and 3D, and the multitude of morphological transformations and self-assembly of the amphiphilic lipid molecules, one of the membranes' most remarkable features may actually be the regulatory power that reversible attachment to and detachment from membranes constitutes for a system of proteins. In this sense, the Min proteins are an exemplary system of how the role of membrane binding motifs can be crucial for the emergence of biological self-organization.

INTRACELLULAR WAVES OF Min PROTEINS CONTROL CYTOKINESIS IN SPACE

Before cells can divide, a key challenge is to accurately determine the site of division. Bacterial cell division is initiated by the polymerization of the tubulin homolog FtsZ into the so-called Z-ring, which then recruits other proteins required for cell division (1). In *E. coli* and other rod-shaped bacteria, Z-ring assembly is usually restricted to the center of the cell. Targeting to midcell is a highly precise process, with a deviation of less than 3%, giving rise to two daughter cells of about equal size (40, 113). Two complementary spatial regulators achieve this high precision of cell division: The first, nucleoid occlusion, describes the observation that Z-ring formation is inhibited at positions close to the chromosome of the bacterium (122). This mechanism allows chromosome segregation to couple to cell division. Although the proteins required for this inhibitory effect have been identified for *E. coli* (SlmA) (8) and *B. subtilis* (Noc) (124), the exact mechanism of how it affects Z-ring assembly is still unclear. The second, better-characterized determinant is the Min system, which not only inhibits cell division close to the cell poles (22), but also further increases the precision of Z-ring positioning at midcell (40).

The activity of the Min system relies on three proteins, MinC, MinD, and MinE, all of which are encoded on the *minB* operon. Initial phenotypic analysis by electron microscopy suggested that MinD and MinC form a stable cap at the cell poles blocking FtsZ assembly and confining cell division to midcell (21, 22). MinE was originally described to form a static ring close to the cell center, keeping the midcell free of the cell division inhibitor MinCD (97). Using GFP fusion proteins and live-cell imaging, Raskin & de Boer (99) revealed that this static picture is wrong by describing the fascinating dynamics of the Min proteins in *E. coli*: MinC and MinD oscillate rapidly between the two poles of the cell in the form of a standing wave. During one half of the oscillation cycle, which

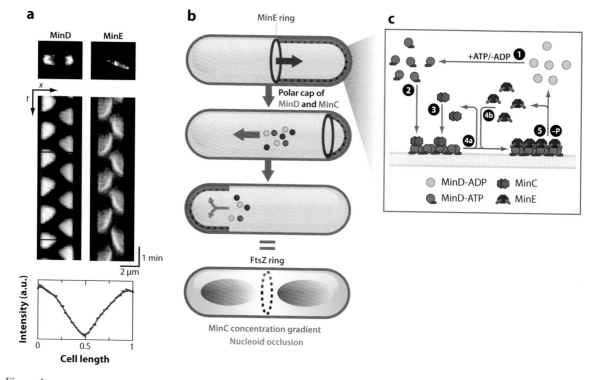

Figure 1

Oscillatory patterns of Min proteins in vivo. (*a*) Top: Micrographs of GFP-MinD and MinE-GFP in vivo. Middle: Kymograph of pole-to-pole oscillations of MinD and MinE in cells of normal length (shorter than 5 μm). Bottom: Time-averaged intensity profile along the red rectangle shown in the kymograph. (*b*) Illustration of Min protein dynamics in *Escherichia coli*: MinD-ATP (*green*) binds to the membrane and recruits MinC (*blue*). MinE (*red*) displaces MinC and then stimulates MinD ATPase activity, causing the release of proteins from the membrane. All three proteins diffuse through the cytoplasm and, after nucleotide exchange by MinD, rebind to the plasma membrane. Bottom: Illustration of the inhibitory gradient of MinC and nucleoid occlusion (*gray ellipsoids*), which together restrict FtsZ polymerization (*purple*) to the center of the cell. (*c*) Biochemical reactions underlying the oscillations. After binding to ATP [1], MinD binds to the membrane as dimers [2]. MinC [3] and MinE [4b] can bind to membrane-bound MinD, with MinE able to displace MinC from MinD [4a]. MinE stimulates the ATPase activity of MinD, whereupon both proteins detach from the membrane [5]. Finally, MinD exchanges its bound nucleotide [1].

takes approximately 1–2 min to complete, a cap of membrane-bound MinD usually starts to grow from one pole toward midcell and then shrinks again toward the same pole (**Figure 1***a*). As this polar zone disappears, a new one is established at the opposite cell end. In contrast to the early notion that MinE forms a static ring close to midcell, which prevents the polar cap of the MinCD complex to grow toward midcell, the intracellular dynamics of MinE are tightly coupled to the behavior of MinD (35, 41). Also, MinE oscillates between the two cell poles, with the E-ring becoming most apparent when it is progressing toward the cell pole (**Figure 1***a*).

Because the oscillation occurs on a timescale much shorter than the cell cycle, it leads to a spatial gradient of Min proteins on time average, with the maximal protein concentration at the cell poles and the minimal concentration at midcell (see **Figure 1***a*). However, this minimum concentration at midcell would occur only when the cell exceeds a certain length. Because the cells grow longer during their life cycle, the Min oscillations could thereby also define the time of cell division.

THE BIOCHEMICAL BASIS OF Min OSCILLATIONS

What is the molecular mechanism of this striking spatiotemporal pattern? Biochemical studies provided a detailed picture of the molecular interactions between the Min proteins and the membrane (**Figure 1c**): MinD was identified to be an ATPase of the ParA family, whose members share the common feature of a deviant Walker motif for ATP binding (21, 74). In contrast to other members of the family, MinD contains a short C-terminal nascent helix, the so-called membrane-targeting sequence (MTS), through which it can interact with phospholipids upon binding of ATP (51, 111). Although only the structure of soluble, monomeric MinD has been solved (18, 42), several lines of evidence suggest that MinD targeting to the cell membrane is controlled by ATP-dependent dimerization of MinD. First, a closely related protein involved in plasmid segregation, Soj, binds to DNA after ATP-dependent dimerization (69). Second, by attaching the MTS peptide to a GFP molecule, Szeto et al. (110) showed that at least two MTS peptides are required to allow for membrane binding. Third, MinD monomers lacking the MTS are not able to bind to the membrane, which is required for MinD-MinD interactions (112). Fourth, FRET studies demonstrated that MinD monomers interact with each other only in the presence of ATP and a phospholipid membrane (82).

For protein concentrations approximately six times higher than those in vivo, MinD assembles into membrane-bound filaments in vitro that tubulate vesicles (47). This indicates that membrane-bound MinD dimers have the potential to further interact with each other to form higher-order structures. In vitro experiments also showed that membrane binding of MinD did not conform to a Langmuir adsorption reaction, but that it can be described by a Hill function with a Hill coefficient of 2 (67). This cooperative binding of MinD to the membrane could arise from dimerization of cytoplasmic MinD, interactions between membrane-bound MinD dimers, or both.

MinD has a low basal ATPase activity, which is stimulated ~10-fold in the presence of phospholipids and MinE (47, 76). Recent FRET experiments indicate that MinE is a dimer when it binds to membrane-bound MinD (71), where it stimulates the ATPase activity of MinD via an N-terminal nascent helix. After ATP hydrolysis, MinD detaches from the membrane. Genetic and biochemical studies have shown that MinE consists of two domains, each of which is supposed to have a separate function. The N-terminal domain (amino acids 1–33), called anti-MinCD domain, is required and sufficient to interact with MinD, to counteract MinCD-mediated division inhibition, and to stimulate the ATPase activity of MinD (75, 76, 90, 97). Topological specificity, which is the ability of MinE to counteract cell division inhibition by MinCD only at midcell, is conferred by a structurally autonomous C-terminal domain (MinE amino acids 33–88) called topological specificity domain (TSD). Originally, this domain was thought to infer binding to a specific site close to midcell, which was assumed to explain the formation of the E-ring in vivo (97), but this has not been verified. The main characteristic of the TSD is that it allows for dimerization of MinE and possibly has some regulatory function on the activity of the anti-MinCD domain (38, 96, 127). NMR spectroscopy has shown that the TSD forms a homodimeric sandwich structure (60). The N-terminal anti-MinCD domain was not resolved in this structure but has been proposed to exist as an α-helix (59, 60). Additional NMR studies corroborated this proposal but also showed that a C-terminal portion of the anti-MinCD domain assumes a β-conformation (96). Newer crystallography studies on MinE from *Helicobacter pylori* refined the published structure and revealed that part of the N-terminal region previously thought to belong to the anti-MinCD domain indeed forms a β-strand, which contributes to the folding of the TSD (57). Moreover, this study also found evidence for extensive dimer-dimer interactions, which could result in the formation of MinE multimers. Although this kind of dimer-dimer interaction could be involved in the formation of the E-ring, evidence for a plausible mechanism underlying its formation and how exactly the TSD contributes to the homogenous oscillation of Min proteins remained elusive.

MinC, the third Min protein, is not required for the oscillations (48, 98), but it is the actual inhibitor of Z-ring formation (9). Recruitment of MinC to membrane-bound MinD increases its efficiency to inhibit Z-ring formation about 25-fold (52). MinC forms homodimers and consists of two domains. Its C-terminal domain is required for binding to membrane-bound MinD and prevents lateral interactions between FtsZ filaments, and its N-terminal domain is responsible for the inhibition of Z-ring assembly by weakening the longitudinal bonds between FtsZ molecules (20, 49). Importantly, MinE and MinC share the same binding site on MinD, with MinE able to displace MinC from MinD (51, 67, 128).

Biochemical characterization of Min proteins revealed the molecular basis for their membrane-bound or cytoplasmic localization (**Figure 1c**). However, it failed to show whether and, if so, how oscillations emerge solely from these molecular events, how directionality of cap retraction is achieved, and how the E-ring is formed.

Min PROTEINS IN INTERPLAY WITH THE BACTERIAL CELL

During the oscillation, the MinCD complex alternately covers the membrane of each of the two cell poles. Most commonly, proteins in the bacterial cell are thought to find their position in the cell by randomly diffusing through the cytoplasm until they are recruited to a specific site in the cell, a mechanism termed diffusion-and-capture (101, 102). A straightforward explanation for the intracellular distribution of Min proteins could therefore be that an additional component located at the cell poles specifically recruits MinD (10, 19, 26). MinD does not require other proteins to bind to the lipid membrane, but it binds stronger to negatively charged lipids (82, 126). Because the highly negatively charged lipid cardiolipin is enriched at the cell poles, one possible spatial cue is polar membrane domains of cardiolipin (54). An alternative possibility is that MinD can sense the high negative curvature of the cell poles directly as do other cell division proteins (55, 68, 94, 95).

If such topological factors were responsible for recruitment of MinD to the cell poles, the distance between the poles would define the length scale of the oscillation. However, a strong argument against this mechanism came from studies on FtsZ-depleted cells, which grow much longer than wild-type cells. As soon as the cells exceeded a certain length, the Min proteins did not only bind to the two cell poles, but showed an additional density maximum at midcell (99). In other words, the oscillation switched to a standing wave pattern, displaying several nodes with a characteristic distance of about 7–8 μm (**Figure 2**). Moreover, the oscillations still occur in spherical (17) and even branched cells (115); thus, the formation of coherent protein patterns does not require the

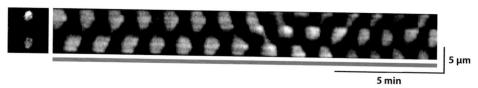

5 μm

5 min

Figure 2
_____
Oscillatory pattern of Min proteins in vivo. Kymograph of the in vivo pattern of MinD-GFP. The pattern of MinD-GFP in longer cells can switch from a pole-to-pole (*green line*) to a pole-midcell-pole oscillatory pattern (*orange line*) (taken from Reference 32).

rod-like shape of wild-type cells. These observations strongly argued against spatial markers, such as lipids or proteins, that facilitate binding of MinD to specific regions of the cytoplasmic membrane. Instead, they suggest that the dynamic pattern of the Min system has an intrinsic length scale defined by the rates of the biochemical reactions involved and their diffusion constants. In other words, Min oscillations emerge from MinD and MinE self-organizing in the presence of a uniform membrane.

However, the proteins are still confined to the finite volume of the cell, and cell size can have an influence on the dynamic pattern (33). If the cell radius is increased by the use of specific drugs, Min proteins do not form homogenous pole-to-pole oscillations but show movements in different directions inside of the cell (32). Furthermore, because of the characteristic length scale of the oscillation, one might expect a minimal length above which oscillations occur. Moreover, the number of nodes of the standing protein wave increases with cell length (99).

How the Min system structures the intracellular space differs from other systems. For instance, Min oscillations are in stark contrast to mechanisms, such as kinase-phosphatase or small G-protein systems, that need one of the players in the interaction network to be locally confined and another one free to diffuse in the cytoplasm (2, 7, 116). In these cases, a steady-state gradient on the length scale of the cell can be formed, which encodes for spatial information. Another mechanism is employed by DivIVa, the spatial regulator of cell division in the rod-shaped cell of *Bacillus subtilis*. This protein is thought to diffuse through the cytoplasm and specifically attach to the cell poles using a diffusion-and-capture mechanism. The spatial cue in this case is the negative curvature of the membrane at the cell pole to which the protein preferentially binds (68, 95). Once it has reached the cell poles, DivIVa recruits MinJ, which in turn binds to MinCD, giving rise to a gradient directed toward the cell center inhibiting cell division (12, 88). In contrast to the Min oscillation or G-protein systems, which are based on the hydrolysis of nucleotides, the intracellular positioning of DivIVa by diffusion-and-capture represents a self-assembly process that does not require the constant dissipation of energy.

The discovery that inhibitory gradients regulate cell division not only in *E. coli* but also in *B. subtilis*, *Caulobacter crescentus* (11, 28), and *Schizosaccharomyces pombe* (78, 84) suggests that such gradients represent a generic mechanism for rod-shaped cells to link cell size and shape to cell division. The evolutionary advantage of protein oscillations employed by *E. coli* to regulate the localization of cell division is unknown. Recently, live-cell imaging experiments of Min protein dynamics addressed how the oscillations are affected by the phase of the cell cycle; i.e., these studies focused on differences in oscillatory behavior of the proteins just before, during, and right after cell division. A recent study by Juarez & Margolin (56) focused on the problem of how formation of the division septum, representing a local constriction of the cell volume, can perturb protein dynamics. Surprisingly, they found that GFP-MinD often paused at midcell and that the frequency of midcell pausing increased as cells grew longer and cell division approached. They argued that this change in the protein pattern could support the equal distribution of Min proteins among the daughter cells at division. Fischer-Friedrich et al. (33) discovered that right after *E. coli* cells divide, the Min proteins often do not oscillate regularly as in normal-sized cells, but display stochastic switching between the two poles; that is, the time the proteins spend in one of the two cell halves varied randomly. When the cells grew longer, this stochastic switching changed back to regular oscillatory behavior. Furthermore, cell-to-cell variability was reduced with increasing cell length. This observation shows that cell length can act as a nontrivial control parameter, as it qualitatively influences the Min protein pattern. Protein self-organization thus offers a good way to link cell behavior to cellular conditions.

THEORETICAL ANALYSIS SUGGESTS BASIC REQUIREMENTS FOR Min OSCILLATIONS

Important information has been gathered about the biochemical interactions and intracellular dynamics of Min proteins. Taken together, they indicated that the Min oscillations arise by self-organization. However, individual MinD molecules alone, which bind site nonspecifically to the membrane in combination with MinE-induced release, cannot explain the observed regular oscillatory pattern of Min proteins. A spontaneous collective pattern of nanometer-sized molecules that extends over several microns necessarily requires interparticle interactions. As a result of such interactions, the homogenous distribution of Min proteins on the membrane can become unstable; i.e., perturbations of the distribution do not dissipate, but instead grow. Because the parameters of the system dynamically change, different states become stable at different times, eventually leading to the oscillatory pattern (65). Only by implementing conceivable interaction mechanisms into a mathematical framework can one hope to identify minimal requirements for self-organized Min oscillations. Using theoretical approaches, several studies have established that the initial cooperative binding of MinD to the membrane can generate a dynamic instability of the homogenous state (45, 53, 81). Cooperative binding could, at least in part, be due to the requirement of MinD to dimerize before binding to the membrane (67, 82). On the basis of experimental evidence, which emphasized the importance of interactions between different MinD dimers bound to the membrane (47, 112), other early theoretical works (64, 79) assumed a so-called aggregation current. In this process, MinD diffusing on the membrane gets trapped by aggregates of membrane-bound MinD, which effectively results in a current of membrane-bound MinD toward regions of higher MinD concentrations. These works showed that the nonlinearity required for dynamic instability could also be self-association and aggregation of already membrane-bound MinD and not cooperativity during membrane binding of MinD.

The emergence of these patterns depends crucially on the difference of diffusion constants of the proteins either bound to the membrane or in solution. If the reacting species has similar diffusion constants, the system can only lead to oscillation in time. An example for this kind of oscillation is the Kai system, which is the circadian clock from cyanobacteria.

All theoretical works show that the E-ring per se is not required for collective Min protein dynamics. Nevertheless, the E-ring is a prominent feature of the Min protein pattern, and retraction of the Min protein cap to a cell pole starting from the distal end is generally attributed to the presence of the E-ring. The mechanism of E-ring formation is still unclear, but a number of different proposals have been made. In one possible scenario, MinE is binding cooperatively to membrane-bound MinDE complexes. For example, the presence of MinDE on the membrane could accelerate binding of new MinE (72, 81). Another possibility relies on augmented binding of MinE to the rim of the MinD cap via slow cytoplasmic diffusion toward the opposite pole combined with a high membrane-binding rate (53). Still other works suggested that the E-ring is generated by persistent binding of MinE to the MinD cap, due either to rapid rebinding after membrane detachment (23, 79) or to transient interaction with the membrane of MinE itself after stimulation of the MinD ATPase (3). It has also been put forward that the E-ring represents inactive dimerized MinE, whereas active monomeric MinE would be present outside the ring (41). Cooperative effects during MinE-induced ATP hydrolysis by MinD have also been suggested (47, 50). However, in vivo and biochemical bulk experiments were not able to provide decisive arguments for or against either possibility, leaving the mechanism of MinE-ring formation uncertain.

Dynamic instability: a dynamic state that was initially stable with respect to perturbations can suddenly lose stability if the parameters of the system are changed

Min POLYMERS

Further challenges for identifying the mechanism underlying Min oscillations are posed by indications of filament-like MinD structures in vivo and in vitro. Incidentally, all homologues of MinD and members of the ParA family are indeed able to polymerize into filaments, either by themselves or when they are bound to a surface (73). The discovery of MinD polymers assembled on lipid vesicles in vitro (47, 108) and the observation of membrane-associated patches of MinD on deconvoluted micrographs in fixed and live cells (5, 106) suggest that 1D polymers of MinD might play an important functional role for Min oscillations (see Reference 117 for a review of filamentous protein assemblies in bacteria). According to this view, MinCD forms a membrane-bound polymer coil in one half of the cell; MinE is associated mainly with this structure toward the cell center, where it would block extension of the MinCD helix. MinE stimulates depolymerization and membrane detachment of the MinCD polymer. Because the number of Min proteins in the cell is too small to densely cover the cytoplasmic membrane of one cell half, MinD polymers could explain how MinD is able to influence FtsZ assembly in the entire cell. However, it remains uncertain whether these helical structures are important for generating Min oscillations. Even their existence in different strains is not fully established and might also depend on different growth and imaging conditions (107). Nevertheless, the idea of dynamic Min polymers in the cell was considered by a number of theoretical models (19, 23, 26, 89). For some of these models to work, sites for initiating MinD polymerization located at the cell poles had to be assumed, which conflicts with the pattern observed in filamentous cells.

IN VITRO RECONSTITUTION

Although based on known biochemical properties of the Min proteins and able to reproduce the oscillations found in vivo, the proposed computational models differed in crucial aspects and needed to be tested experimentally.

However, because of the small size of bacteria and because bulk biochemical experiments were not able to elucidate the spatiotemporal nature of the events during pattern formation, many assumptions could not be verified with the experimental approaches described so far. In a theoretical study, it had been predicted that if MinD, MinE, ATP, and a membrane were sufficient to generate the Min oscillations, then MinD and MinE would form dynamic patterns also in an open geometry on a flat membrane (34). Confirming this hypothesis, the in vitro reconstitution of Min protein dynamics on a supported lipid bilayer mimicking the cytoplasmic membrane of the bacterial cell presented a major breakthrough that allowed the behavior of the Min proteins to be studied in great detail and under controlled conditions (72).

When the purified and fluorescently labeled proteins were incubated on a flat supported lipid bilayer, MinD formed a homogenous protein carpet in the presence of ATP, which is consistent with the homogenous distribution of MinD in the absence of MinE in vivo. Upon addition of MinE to the system, this symmetric situation became dynamically unstable and MinD detachment was initiated from little foci of increased MinE density. After several rounds of protein detachment and reattachment, and after a transient phase during which small protein waves were moving in different directions, the proteins organized themselves into a steady state of parallel protein waves traveling on the membrane (**Figure 3a**). In addition to parallel waves the proteins could self-organize into spiral waves (**Figure 3d**). These patterns exist only when energy is dissipated, that is, when sufficient chemical energy is present in the form of ATP. These reconstitution experiments demonstrate the potential of the Min system to self-organize into large-scale patterns, due solely to the interactions among two proteins, MinD and MinE, a membrane, and ATP.

The Min protein waves in vitro share important features with the wave patterns formed in vivo. First, the waves travel at similar speeds as MinD caps retract in vivo, which is determined by the ratio of MinE to MinD. Second,

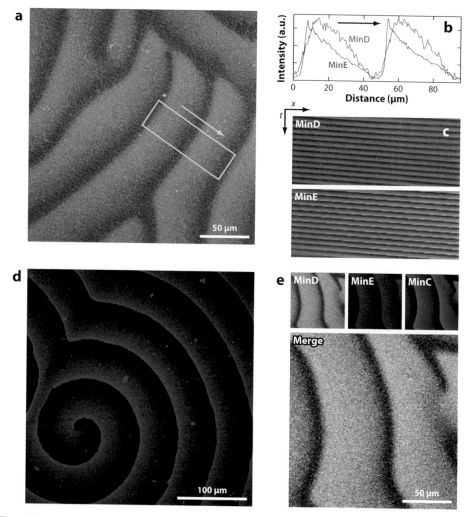

Figure 3

Wave patterns of Min protein in vitro. (*a*) Parallel surface waves formed by MinE and MinD in the presence of ATP on a supported lipid bilayer in vitro. (*b*) Intensity profiles corresponding to the yellow rectangle shown in panel *a*. The maximum of MinE is located behind the maximum of MinD. (*c*) The protein waves travel at a constant speed, giving rise to parallel lines in the kymograph corresponding to the same rectangle. (*d*) Min proteins can also form spiral patterns in vitro. (*e*) When MinC is added to the system, it is incorporated into the traveling waves of MinE and MinD (panels *a–d* taken from Reference 72, panel *e* from Reference 71).

the in vitro patterns show a protein distribution similar to that of in vivo patterns, with the maximum of MinE located behind the maximum of MinD. Third, as later studies showed, MinC is the passenger of the pattern formed by MinD and MinE (**Figure 3*d***) (71). However, this initial in vitro study was not able to reveal exactly

how the characteristic density distributions of MinD and MinE within the traveling protein band were achieved. A comparison of the experimental observation with different computational models favored cooperative attachment of MinD over an aggregation current as the principal source of the dynamic instability.

Furthermore, a theoretical analysis suggested additional cooperative effects during MinE binding to membrane-bound MinDE are important for generating the observed in vitro patterns (72). However, cooperativity during MinE binding could not be confirmed by more recent experimental studies (71), raising the question about not yet identified alternative nonlinear reaction steps present in the Min system.

Single-molecule experiments have helped to elucidate the mechanism of Min protein pattern formation. By using total internal reflection fluorescence (TIRF) microscopy and small amounts of labeled proteins, it was possible to resolve and analyze the dynamics of single proteins within traveling protein bands. During wave propagation, single proteins attached to the membrane, where they diffused isotropically until they detached again. Toward the rear of the wave, the proteins diffused more slowly and bound longer to the membrane, suggesting that with rising density membrane-bound MinD dimers increasingly self-associated, giving rise to stronger attachment to the membrane. This diffusivity gradient could support the formation of a focused, rather than blurred, band of proteins. Although the pronounced decrease in diffusivity suggests strong interactions between membrane-bound MinD dimers, the proteins diffused in an isotropic fashion, which argues against the presence of rigid filament-like MinD polymers. From these studies a dynamic protein carpet appears to be more likely.

Analysis of the single-molecule data also showed that MinE bound longer to the MinD carpet than MinD bound to the membrane. Although the difference in residence times observed in these experiments was only 40%, it has some important functional consequences. It can explain how MinE piles up at the rear of the traveling wave, giving rise to a density distribution different than the distribution of MinD, i.e., a peak in the MinE/MinD ratio found at the end of the wave (**Figure 4a,b**). It also supports the theoretical models, which assumed either transient binding of MinE to the membrane (3, 46) or fast reattachment to membrane-bound MinD (23, 79) as the basis of E-ring formation. Another conceivable possibility, which could allow for persistent binding of MinE, is the formation of higher-order complexes as proposed based on recent structural studies (57). All three different scenarios could lead to the observed difference in residence times.

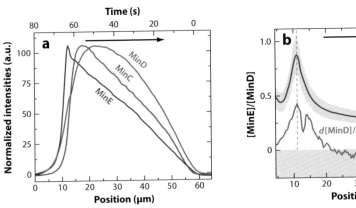

Figure 4
Intensity profiles reveal mutual dependence of Min proteins during wave propagation. (*a*) Averaged intensity profiles of MinC, MinD, and MinE obtained using total internal reflection fluorescence microscopy. MinC is displaced from MinD just before MinE accumulates to a peak. (*b*) The change in MinD density on the membrane (*green line*) agrees with a change in the [MinE]/[MinD] ratio (*gray line*) and peaks at the same position within the wave. Black arrows indicate the traveling direction of the protein waves (from Reference 71).

As a result of persistent binding of MinE, MinE can occupy more and more membrane-bound MinD dimers, which, first, efficiently displaces MinC from MinD (**Figure 4a** and **Figure 5**) and, second, leads to the stimulation and rapid detachment of all membrane-bound MinD (**Figure 4b** and **Figure 5**). Because MinE remains incorporated in the wave while MinD is detaching, the MinE/MinD ratio further increases, leading to even faster activation and detachment of MinD. The net result of this mechanism is a positive-feedback loop that accelerates MinD detachment at the end of the traveling wave (**Figure 5**). This positive feedback between the MinE/MinD ratio and the detachment rate of MinD explains how distribution symmetry is broken and pattern formation is initiated in a homogenous protein: Owing to positive feedback generated during protein detachment, small fluctuations in the MinE/MinD ratio are amplified, giving rise to the accumulation of MinE in small foci, where dominant protein detachment is initiated. Similar persistent binding of an activator to its corresponding ATPase has been proposed to be the basis of ParA-driven segregation of chromosomes in *C. crescentus* (93) and of low-copy-number plasmids in *E. coli* (100), although direct evidence for this mechanism is still missing.

Although many aspects of the in vivo oscillations are well captured by the in vitro protein waves, they do not comprehensively reflect the behavior seen in vivo. Most importantly, the in vitro pattern is about 10 times larger than in an *E. coli* cell. Generally, the length scales of reaction-diffusion patterns are governed by the diffusion constant of the reacting species. Whereas the bacterial membrane consists of 60% protein (129), the membrane used in vitro is composed of lipids only. Therefore, molecular crowding could be a reason for the larger length scale observed in vitro. Also, judged from the number of Min proteins measured in vivo, we expect that membrane concentrations reached in vitro are larger than those in the cell. This can also give rise to different membrane binding and detachment rates.

CYTOSKELETAL WAVES

The Min system illustrates that spontaneous protein waves can provide structural information in living cells. What might at first seem to be rather unique to a rod-shaped prokaryotic cell is in fact also employed by higher organisms, i.e., eukaryotic cells. Intracellular organization in a eukaryotic cell is usually achieved by its cytoskeleton, which can organize itself into different structures to fulfill a variety of cellular functions such as cell division, motility, and intracellular transport. Self-organization of cytoskeletal filaments and molecular motors, which can bundle, transport, and align filaments, has long been known to play an important role in the formation of these structures (58, 86). Indeed, purified filament-motor systems can spontaneously form steady-state patterns in vitro (4, 43, 87, 109). More recently, spontaneous intracellular polymerization waves of actin, which arise independently of directed transport by molecular motors, have gained some attention. Most studies have been performed in *Dictyostelium discoideum*, in which actin waves form during cell locomotion (118, 119) and during recovery from drug-induced disassembly of the actin network (**Figure 6a**) (13, 14, 37, 104). Similarly, spontaneous actin waves have been detected in naive human neutrophilic granulocytes (**Figure 6b**) (120).

Similar to the Min system, these wave-like patterns related to actin polymerization do not emerge from a simple linear pathway, but require sufficient nonlinear reaction dynamics. The molecular scale mechanisms underlying actin polymerization waves are, in contrast to the Min system, not well understood. In all cases of actin waves, nucleation of actin polymerization by nucleation-promoting factors (NPFs) and negative feedback from the actin filaments inactivating the NPFs seem to be involved. Just as for the Min system, computational modeling helped to test possible functional circuits that give rise to actin waves. Most models have implemented actin-filament-induced inhibition of the NPF as a core feedback, and by emphasizing different aspects of this

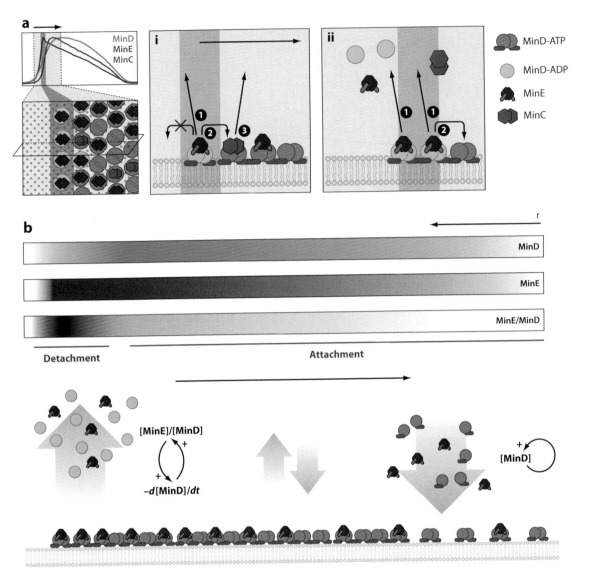

Figure 5

Model of Min wave propagation. (*a*) Illustration of the order of events at the rear of the protein wave. Top and side views of Min proteins bound to the membrane. After stimulation of ATP hydrolysis and detachment of MinD (*i, ii step 1*), MinE can rebind to a neighboring membrane-bound MinD (*i, ii step 2*). Because the density of membrane-bound MinD is higher toward the front of the wave, rebinding MinE is biased in this direction. Owing to MinE rebinding, MinE occupies every MinD on the membrane, forcing it to detach (*i, ii step 3*). (*b*) Starting from the front of the protein wave (or at the beginning of an oscillation cycle) (*right*), MinD-ATP begins to bind to the membrane. The higher the MinD dimer density becomes, the longer they bind to the membrane and the slower the proteins diffuse. MinE dimers bind to membrane-bound MinD, but the concentration of MinE is at first too low to lead to dominant membrane detachment. At a sufficiently high MinE/MinD ratio, more and more MinD dimers are in complex with MinE and protein detachment begins to dominate. Owing to rapid MinE rebinding to MinD, the [MinE]/[MinD] ratio continuously increases toward the rear of the wave. This behavior guarantees that eventually all membrane-bound MinD dimers are in complex with MinE and rapidly leave the membrane (taken from Reference 71).

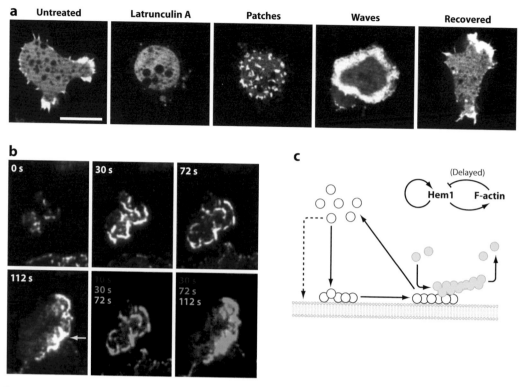

Figure 6

Actin waves in vivo. (*a*) Waves of actin polymerization observed after latrunculin A–induced depolymerization of actin filaments. First, static spots of actin filaments are formed on the membrane, which then organize themselves to waves of actin polymerization. Eventually, the original pattern of actin filaments recovers (from Reference 37). (*b*) Waves of Hem1 in crawling neutrophil cells (from Reference 120). (*c*) Model for the generation of Hem1 waves. Hem1 (*open circles*) preferentially binds to regions already covered with Hem1 (see solid arrows compared to dashed arrow). There, it activates the Scar-WAVE complex (not shown). This leads to nucleation of new actin filaments (*orange circles*), which in turn inactivate Scar-WAVE by inducing the release of Hem1 from the membrane. Abbreviations: Scar-WAVE, suppressor of cAMP receptor/WASP-family verprolin homologous protein.

problem, spontaneous polymerization waves were obtained (16, 25, 120, 121). In all cases the feedback was based on the assumption that the NPF is active while bound to the membrane, whereas actin filaments disengaged the NPF from the membrane. In the case of actin waves in human neutrophils, cooperativity during membrane binding of the NPF Hem1, a component of the Scar/WAVE (suppressor of cAMP receptor/WASP-family verprolin homologous protein) complex, has also been considered in theoretical models (see **Figure 6c**).

The physiological relevance of spontaneous actin waves might be to spatially organize the cytoskeleton during cell locomotion, inducing cell shape changes. Indeed, waves of actin polymerization are associated with protrusions of the cell membrane (13, 120).

If one wanted to draw an analogy to the Min system, then the NPF would play the role of MinD and actin that of MinE. Note, however, that the actin system requires a whole set of auxiliary proteins to form waves and that the actual mechanism is probably much more complex.

CONCLUSIONS

In this review, we illustrated how intracellular protein waves organize the intracellular space. Whereas the molecular order of events for actin polymerization waves still remains uncertain, the mechanism of Min protein waves is studied in great detail. Especially, the in vitro reconstitution of the Min protein dynamics offered unprecedented access to features such as residence times and molecular motility of the players involved. Taken together, in vitro studies on the Min system helped to clarify the order of events in space and time and the generic features required for pattern formation: the interplay between cooperative binding of MinD to the membrane and positive feedback during protein detachment.

In light of recent discussions about the role of membranes in biological systems, an interesting aspect resulting from our studies on Min protein self-organization is that the reversible, cooperative, and particularly energy-dependent membrane attachment of proteins allows for one source of nonlinearity in the system, which is crucial to constitute self-organization and emergent behavior. Although in the case of the Min system this mechanism is due most likely to cooperative protein-protein interactions, proteins can also facilitate their own recruitment by indirect mechanisms, for example by modulating either the chemical composition of the membrane, such as its phosphoinositide content (see Reference 24 for a review), or its physical properties, such as membrane curvature or fluidity (27, 36). Both mechanisms are important for cell signaling and vesicle trafficking and represent analogous ways to allow for switching between two different physical states of the proteins, either soluble in the cytoplasm or bound to the membrane.

We believe that the minimal systems approach, as successfully applied to the Min system, in combination with protein engineering and synthetic biology will provide a wealth of insight into similarly beautiful protein self-organization phenomena in the near future.

OPEN QUESTIONS

What Is the Structural Basis of Min Oscillations?

In order to fully understand the biochemical basis of Min protein self-organization, we need to know the interactions between the players involved at the molecular level. So far, however, only the structures of monomeric, soluble MinD and MinE have been solved. The observation of isotropic diffusion of membrane-bound MinD suggests that MinD does not polymerize into membrane-bound filaments. Instead, a two-dimensional polymer, similar to a protein carpet, seems to be more likely. Structural studies on membrane-bound MinD and the membrane-bound MinDE complex could help to clarify the molecular basis of Min oscillations; that is, it could provide information about the MinD dimer-dimer interactions and how exactly MinE stimulates MinD ATPase activity. Furthermore, it could give a molecular explanation for persistent binding of MinE.

Why Is the In Vitro Pattern Much Larger than the In Vivo Pattern?

A theoretical description suggests that the different length scales of the patterns observed in vitro and in vivo can be explained by lower diffusion constants of the proteins in the cell. However, so far, this hypothesis has not been verified experimentally. Because the length scale of the evolving pattern in reaction-diffusion systems is given by the ratio of the diffusion constants to the reaction rates of the reacting components, an alternative explanation for the smaller length scale of the in vivo pattern could be higher reaction rates, for example, due to an enhanced ATPase activity.

How Can Persistent Binding of MinE Explain the Formation of the E-Ring In Vivo?

Why is collective protein detachment in vivo always induced close to the cell center and not at the cell poles? A possible explanation is given by the notion of positive feedback between the MinE/MinD ratio and the MinD detachment rate. Because the local density of MinD on the membrane is lower at midcell than at the pole, collective MinD detachment would be initiated at midcell first. During protein detachment, MinE would further accumulate, continuously inducing protein detachment toward the cell pole. A theoretical description including persistent binding of MinE could test this hypothesis.

Can Oscillations be Reconstituted In Vitro?

In the reconstitution experiment, the Min proteins were incubated on a supported lipid bilayer in an open, unrestricted geometry where the proteins form waves that travel in a distinct direction. In wild-type cells, however, the wave changes its direction of propagation; i.e., the proteins oscillate because they are confined to the small volume of the cell. In cells that grow much longer than normal wild-type cells, a switch from oscillations to traveling waves was observed, confirming the idea that it is the geometry of the cell and not any additional constraint that results in oscillatory behavior.

Therefore, encapsulating the proteins, for instance, in a vesicle of limited size, should lead to oscillations similar to those in vivo. This remains to be verified.

What Is the Physiological Relevance of Cytoskeletal Waves and Can They be Reconstituted In Vitro?

A number of observations suggest that cytoskeletal waves are present in cells under physiological conditions. The functional importance of the waves for cell processes and/or cell behavior is, however, unclear. It was suggested that waves might organize the cytoskeleton during cell locomotion. This and other possible roles of cytoskeletal waves need to be explored. Although experimentally challenging, in vitro reconstitution could help to investigate the mechanism of actin waves.

SUMMARY POINTS

1. A self-organized oscillatory wave of Min proteins restricts cell division to the center of the cell.

2. The self-organized pattern is in interplay with the geometry of the cell, thereby partially defining the time and space of cell division.

3. Self-organization of the Min protein is based on a bistability emerging from cooperative attachment of MinD and persistent binding of MinE.

4. The intracellular pattern can change depending on the length of the cell.

5. The properties that lead to dynamic patterns in the Min system can also give rise to dynamic patterns in other systems, e.g., waves of actin polymerization.

DISCLOSURE STATEMENT

The authors are not aware of any affiliations, memberships, funding, or financial holdings that might be perceived as affecting the objectivity of this review.

LITERATURE CITED

1. Adams DW, Errington J. 2009. Bacterial cell division: assembly, maintenance and disassembly of the Z ring. *Nat. Rev. Microbiol.* 7:642–53

2. Altschuler SJ, Angenent SB, Wang Y, Wu LF. 2008. On the spontaneous emergence of cell polarity. *Nature* 454:886–89

3. Arjunan SN, Tomita M. 2010. A new multicompartmental reaction-diffusion modeling method links transient membrane attachment of *E. coli* MinE to E-ring formation. *Syst. Synth. Biol.* 4:35–53

4. Backouche F, Haviv L, Groswasser D, Bernheim-Groswasser A. 2006. Active gels: dynamics of patterning and self-organization. *Phys. Biol.* 3:264–73

5. Barák I, Muchová K, Wilkinson AJ, O'Toole PJ, Pavlendová N. 2008. Lipid spirals in *Bacillus subtilis* and their role in cell division. *Mol. Microbiol.* 68:1315–27

6. Bashor CJ, Horwitz AA, Peisajovich SG, Lim WA. 2010. Rewiring cells: synthetic biology as a tool to interrogate the organizational principles of living systems. *Annu. Rev. Biophys.* 39:515–37

7. Bastiaens P, Caudron M, Niethammer P, Karsenti E. 2006. Gradients in the self-organization of the mitotic spindle. *Trends Cell Biol.* 16:125–34

8. Bernhardt TG, de Boer PA. 2005. SlmA, a nucleoid-associated, FtsZ binding protein required for blocking septal ring assembly over chromosomes in *E. coli. Mol. Cell.* 18:555–64

9. Bi EF, Lutkenhaus J. 1991. FtsZ ring structure associated with division in *Escherichia coli. Nature* 354:161–64

10. Borowski P, Cytrynbaum EN. 2009. Predictions from a stochastic polymer model for the MinDE protein dynamics in *Escherichia coli. Phys. Rev. E* 80:041916

11. Bowman GR, Comolli LR, Zhu J, Eckart M, Koenig M, et al. 2008. A polymeric protein anchors the chromosomal origin/ParB complex at a bacterial cell pole. *Cell* 134:945–55

12. Bramkamp M, van Baarle S. 2009. Division site selection in rod-shaped bacteria. *Curr. Opin. Microbiol.* 12:683–88

13. Bretschneider T, Anderson K, Ecke M, Muller-Taubenberger A, Schroth-Diez B, et al. 2009. The three-dimensional dynamics of actin waves, a model of cytoskeletal self-organization. *Biophys. J.* 96:2888–900

14. Bretschneider T, Diez S, Anderson K, Heuser J, Clarke M, et al. 2004. Dynamic actin patterns and Arp2/3 assembly at the substrate-attached surface of motile cells. *Curr. Biol.* 14:1–10

15. Camazine S. 2001. *Self-Organization in Biological Systems.* Princeton, NJ/Oxford: Princeton Univ. Press. 538 pp.

16. Carlsson AE. 2010. Dendritic actin filament nucleation causes traveling waves and patches. *Phys. Rev. Lett.* 104:228102

17. Corbin BD, Yu XC, Margolin W. 2002. Exploring intracellular space: function of the Min system in round-shaped *Escherichia coli. EMBO J.* 21:1998–2008

18. Cordell SC, Löwe J. 2001. Crystal structure of the bacterial cell division regulator MinD. *FEBS Lett.* 492:160–65

19. Cytrynbaum EN, Marshall BD. 2007. A multi-stranded polymer model explains MinDE dynamics in *E. coli* cell division. *Biophys. J.* 93:1134–50

20. Dajkovic A, Lan G, Sun SX, Wirtz D, Lutkenhaus J. 2008. MinC spatially controls bacterial cytokinesis by antagonizing the scaffolding function of FtsZ. *Curr. Biol.* 18:235–44

21. de Boer PA, Crossley RE, Hand AR, Rothfield LI. 1991. The MinD protein is a membrane ATPase required for the correct placement of the *Escherichia coli* division site. *EMBO J.* 10:4371–80

22. de Boer PA, Crossley RE, Rothfield LI. 1989. A division inhibitor and a topological specificity factor coded for by the minicell locus determine proper placement of the division septum in *E. coli. Cell* 56:641–49

23. Derr J, Hopper JT, Sain A, Rutenberg AD. 2009. Self-organization of the MinE protein ring in subcellular Min oscillations. *Phys. Rev. E* 80:011922

24. Di Paolo G, De Camilli P. 2006. Phosphoinositides in cell regulation and membrane dynamics. *Nature* 443:651–57

25. Doubrovinski K, Kruse K. 2008. Cytoskeletal waves in the absence of molecular motors. *Europhys. Lett.* 83:18003

26. Drew DA, Osborn MJ, Rothfield LI. 2005. A polymerization-depolymerization model that accurately generates the self-sustained oscillatory system involved in bacterial division site placement. *Proc. Natl. Acad. Sci. USA* 102:6114–18

27. Drin G, Casella J-F, Gautier R, Boehmer T, Schwartz TU, Antonny B. 2007. A general amphipathic α-helical motif for sensing membrane curvature. *Nat. Struct. Mol. Biol.* 14:138–46

28. Ebersbach G, Briegel A, Jensen GJ, Jacobs-Wagner C. 2008. A self-associating protein critical for chromosome attachment, division, and polar organization in *Caulobacter. Cell* 134:956–68

16. Describes a stochastic particle-based simulation of actin in the vicinity of a membrane and reports the appearance of spontaneous actin waves due to negative feedback of actin on an NPF.

29. Epstein IR, Pojman JA. 1998. *An Introduction to Nonlinear Chemical Dynamics: Oscillations, Waves, Patterns, and Chaos*. New York/Oxford: Oxford Univ. Press. 392 pp.

30. Ferrell JE. 1996. Tripping the switch fantastic: how a protein kinase cascade can convert graded inputs into switch-like outputs. *Trends Biochem. Sci.* 21(12):460–66

31. Ferrell JE, Xiong W. 2001. Bistability in cell signaling: how to make continuous processes discontinuous, and reversible processes irreversible. *Chaos* 11:227–36

32. Fischer-Friedrich E. 2008. *Pattern Formation by the Min System of* Escherichia coli. PhD thesis. Univ. Saarlandes, Saarbrücken, Ger.

33. Fischer-Friedrich E, Meacci G, Lutkenhaus J, Chate H, Kruse K. 2010. Intra- and intercellular fluctuations in Min-protein dynamics decrease with cell length. *Proc. Natl. Acad. Sci. USA* 107:6134–39

34. Fischer-Friedrich E, van Yen RN, Kruse K. 2007. Surface waves of Min proteins. *Phys. Biol.* 4:38–47

35. Fu X, Shih YL, Zhang Y, Rothfield LI. 2001. The MinE ring required for proper placement of the division site is a mobile structure that changes its cellular location during the *Escherichia coli* division cycle. *Proc. Natl. Acad. Sci. USA* 98:980–85

36. Gallop JL, Jao CC, Kent HM, Butler PJ, Evans PR, et al. 2006. Mechanism of endophilin N-BAR domain-mediated membrane curvature. *EMBO J.* 25:2898–910

37. Gerisch G, Bretschneider T, Muller-Taubenberger A, Simmeth E, Ecke M, et al. 2004. Mobile actin clusters and traveling waves in cells recovering from actin depolymerization. *Biophys. J.* 87:3493–503

38. Ghasriani H, Ducat T, Hart CT, Hafizi F, Chang N, Al-Baldawi A, et al. 2010. Appropriation of the MinD protein-interaction motif by the dimeric interface of the bacterial cell division regulator MinE. *Proc. Natl. Acad. Sci. USA* 107:18416–21

39. Gierer A, Meinhardt H. 1972. A theory of biological pattern formation. *Kybernetik* 12:30–39

40. Guberman JM, Fay A, Dworkin J, Wingreen NS, Gitai Z. 2008. PSICIC: noise and asymmetry in bacterial division revealed by computational image analysis at sub-pixel resolution. *PLoS Comput. Biol.* 4:e1000233

41. Hale CA, Meinhardt H, de Boer PA. 2001. Dynamic localization cycle of the cell division regulator MinE in *Escherichia coli*. *EMBO J.* 20:1563–72

42. Hayashi I, Oyama T, Morikawa K. 2001. Structural and functional studies of MinD ATPase: implications for the molecular recognition of the bacterial cell division apparatus. *EMBO J.* 20:1819–28

43. Hentrich C, Surrey T. 2010. Microtubule organization by the antagonistic mitotic motors kinesin-5 and kinesin-14. *J. Cell Biol.* 189:465–80

44. Hess B. 1997. Periodic patterns in biochemical reactions. *Q. Rev. Biophys.* 30:121–76

45. Howard M, Rutenberg AD, de Vet S. 2001. Dynamic compartmentalization of bacteria: accurate division in *E. coli*. *Phys. Rev. Lett.* 87:278102

46. Hsieh C-W, Lin T-Y, Lai H-M, Lin C-C, Hsieh T-S, Shih Y-L. 2010. Direct MinE-membrane interaction contributes to the proper localization of MinDE in *E. coli*. *Mol. Microbiol.* 499–512

47. Hu Z, Gogol EP, Lutkenhaus J. 2002. Dynamic assembly of MinD on phospholipid vesicles regulated by ATP and MinE. *Proc. Natl. Acad. Sci. USA* 99:6761–66

48. Hu Z, Lutkenhaus J. 1999. Topological regulation of cell division in *Escherichia coli* involves rapid pole to pole oscillation of the division inhibitor MinC under the control of MinD and MinE. *Mol. Microbiol.* 34:82–90

49. Hu Z, Lutkenhaus J. 2000. Analysis of MinC reveals two independent domains involved in interaction with MinD and FtsZ. *J. Bacteriol.* 182:3965–71

50. Hu Z, Lutkenhaus J. 2001. Topological regulation of cell division in *E. coli*. Spatiotemporal oscillation of MinD requires stimulation of its ATPase by MinE and phospholipid. *Mol. Cell.* 7:1337–43

51. Hu Z, Lutkenhaus J. 2003. A conserved sequence at the C-terminus of MinD is required for binding to the membrane and targeting MinC to the septum. *Mol. Microbiol.* 47:345–55

52. Hu Z, Mukherjee A, Pichoff S, Lutkenhaus J. 1999. The MinC component of the division site selection system in *Escherichia coli* interacts with FtsZ to prevent polymerization. *Proc. Natl. Acad. Sci. USA* 96:14819–24

53. Huang KC, Meir Y, Wingreen NS. 2003. Dynamic structures in *Escherichia coli*: spontaneous formation of MinE rings and MinD polar zones. *Proc. Natl. Acad. Sci. USA* 100:12724–28

39. Establishes the theoretical concepts of how patterns can form starting from an initial homogenous distribution of interacting species (see also Reference 114).

54. Huang KC, Mukhopadhyay R, Wingreen NS. 2006. A curvature-mediated mechanism for localization of lipids to bacterial poles. *PLoS Comput. Biol.* 2:e151

55. Huang KC, Ramamurthi KS. 2010. Macromolecules that prefer their membranes curvy. *Mol. Microbiol.* 76:822–32

56. Juarez JR, Margolin W. 2010. Changes in the Min oscillation pattern before and after cell birth. *J. Bacteriol.* 192:4134–42

57. Kang GB, Song HE, Kim MK, Youn HS, Lee JG, et al. 2010. Crystal structure of *Helicobacter pylori* MinE, a cell division topological specificity factor. *Mol. Microbiol.* 76:1222–31

58. Karsenti E. 2008. Self-organization in cell biology: a brief history. *Nat. Rev. Mol. Cell Biol.* 9:255–62

59. King GF, Rowland SL, Pan B, Mackay JP, Mullen GP, Rothfield LI. 1999. The dimerization and topological specificity functions of MinE reside in a structurally autonomous C-terminal domain. *Mol. Microbiol.* 31:1161–69

60. King GF, Shih YL, Maciejewski MW, Bains NP, Pan B, et al. 2000. Structural basis for the topological specificity function of MinE. *Nat. Struct. Biol.* 7:1013–17

61. Kirschner M, Mitchison T. 1986. Beyond self-assembly: from microtubules to morphogenesis. *Cell* 45:329–42

62. Kondo S. 2002. The reaction-diffusion system: a mechanism for autonomous pattern formation in the animal skin. *Genes Cells* 7:535–41

63. Kondo S, Asai R. 2002. A reaction-diffusion wave on the skin of the marine angelfish *Pomacanthus*. *Nature* 376:765–68

64. Kruse K. 2002. A dynamic model for determining the middle of *Escherichia coli*. *Biophys. J.* 82:618–27

65. Kruse K, Howard M, Margolin W. 2007. An experimentalist's guide to computational modelling of the Min system. *Mol. Microbiol.* 63:1279–84

66. Kruse K, Jülicher F. 2005. Oscillations in cell biology. *Curr. Opin. Cell. Biol.* 17:20–26

67. Lackner LL, Raskin DM, de Boer PAJ. 2003. ATP-dependent interactions between *Escherichia coli* Min proteins and the phospholipid membrane in vitro. *J. Bacteriol.* 185:735–49

68. Lenarcic R, Halbedel S, Visser L, Shaw M, Wu L, et al. 2009. Localisation of DivIVA by targeting to negatively curved membranes. *EMBO J.* 28:2272–82

69. Leonard TA, Butler PJ, Löwe J. 2005. Bacterial chromosome segregation: structure and DNA binding of the Soj dimer—a conserved biological switch. *EMBO J.* 24:270–82

70. Liu AP, Fletcher DA. 2009. Biology under construction: in vitro reconstitution of cellular function. *Nat. Rev. Mol. Cell Biol.* 10:644–50

71. Loose M, Fischer-Friedrich E, Herold C, Kruse K, Schwille P. 2010. Single-molecule imaging of Min protein dynamics reveals the general properties of pattern formation. *Nat. Struct. Mol. Biol.* In press

72. Demonstrates the potential of the MinD and MinE proteins to form a highly dynamic pattern in a minimal reconstituted system.

72. Loose M, Fischer-Friedrich E, Ries J, Kruse K, Schwille P. 2008. Spatial regulators for bacterial cell division self-organize into surface waves in vitro. *Science* 320:789–92

73. Löwe J, Amos LA. 2009. Evolution of cytomotive filaments: the cytoskeleton from prokaryotes to eukaryotes. *Int. J. Biochem. Cell Biol.* 41:323–29

74. Lutkenhaus J, Sundaramoorthy M. 2003. MinD and role of the deviant Walker A motif, dimerization and membrane binding in oscillation. *Mol. Microbiol.* 187:629–63

75. Ma L, King GF, Rothfield L. 2004. Positioning of the MinE binding site on the MinD surface suggests a plausible mechanism for activation of the *Escherichia coli* MinD ATPase during division site selection. *Mol. Microbiol.* 54:99–108

76. Ma LY, King G, Rothfield L. 2003. Mapping the MinE site involved in interaction with the MinD division site selection protein of *Escherichia coli*. *J. Bacteriol.* 185:4948–55

77. Markson JS, O'Shea EK. 2009. The molecular clockwork of a protein-based circadian oscillator. *FEBS Lett.* 583:3938–47

78. Martin SG, Berthelot-Grosjean M. 2009. Polar gradients of the DYRK-family kinase Pom1 couple cell length with the cell cycle. *Nature* 459:852–56

79. Meacci G, Kruse K. 2005. Min-oscillations in *Escherichia coli* induced by interactions of membrane-bound proteins. *Phys. Biol.* 2:89–97

80. Meinhardt H. 2008. Models of biological pattern formation: from elementary steps to the organization of embryonic axes. *Curr. Top. Dev. Biol.* 81:1–63

81. Meinhardt H, de Boer PA. 2001. Pattern formation in *Escherichia coli*: a model for the pole-to-pole oscillations of Min proteins and the localization of the division site. *Proc. Natl. Acad. Sci. USA* 98:14202–7

82. Mileykovskaya E, Fishov I, Fu X, Corbin BD, Margolin W, Dowhan W. 2003. Effects of phospholipid composition on MinD-membrane interactions in vitro and in vivo. *J. Biol. Chem.* 278:22193–98

83. Mitchison TJ. 1992. Self-organization of polymer-motor systems in the cytoskeleton. *Philos. Trans. R. Soc. Lond. B Biol. Sci.* 336:99–106

84. Moseley JB, Mayeux A, Paoletti A, Nurse P. 2009. A spatial gradient coordinates cell size and mitotic entry in fission yeast. *Nature* 459:857–60

85. Nakajima M, Imai K, Ito H, Nishiwaki T, Murayama Y, et al. 2005. Reconstitution of circadian oscillation of cyanobacterial KaiC phosphorylation in vitro. *Science* 308:414–15

86. Nedelec F, Surrey T, Karsenti E. 2003. Self-organisation and forces in the microtubule cytoskeleton. *Curr. Opin. Cell Biol.* 15:118–24

87. Nedelec FJ, Surrey T, Maggs AC, Leibler S. 1997. Self-organization of microtubules and motors. *Nature* 389:305–8

88. Patrick JE, Kearns DB. 2008. MinJ (YvjD) is a topological determinant of cell division in *Bacillus subtilis*. *Mol. Microbiol.* 70:1166–79

89. Pavin N, Paljetak HC, Krstić V. 2006. Min-protein oscillations in *Escherichia coli* with spontaneous formation of two-stranded filaments in a three-dimensional stochastic reaction-diffusion model. *Phys. Rev. E* 73:021904

90. Pichoff S, Vollrath B, Touriol C, Bouché JP. 1995. Deletion analysis of gene *minE* which encodes the topological specificity factor of cell division in *Escherichia coli*. *Mol. Microbiol.* 18:321–29

91. Prigogine I, Lefever R, Goldbeter A, Herschkowitz-Kaufman M. 1969. Symmetry breaking instabilities in biological systems. *Nature* 223:913–16

92. Prigogine I, Nicolis G, Babloyantz A. 1974. Nonequilibrium problems in biological phenomena. *Ann. N. Y. Acad. Sci.* 231:99–105

93. Ptacin JL, Lee SF, Garner EC, Toro E, Eckart M, et al. 2010. A spindle-like apparatus guides bacterial chromosome segregation. *Nat. Cell. Biol.* 12:791–98

94. Ramamurthi KS, Lecuyer S, Stone HA, Losick R. 2009. Geometric cue for protein localization in a bacterium. *Science* 323:1354–57

95. Ramamurthi KS, Losick R. 2009. Negative membrane curvature as a cue for subcellular localization of a bacterial protein. *Proc. Natl. Acad. Sci. USA* 106:13541–45

96. Ramos D, Ducat T, Cheng J, Eng NF, Dillon J-AR, Goto NK. 2006. Conformation of the cell division regulator MinE: evidence for interactions between the topological specificity and anti-MinCD domains. *Biochemistry* 45:4593–601

97. Raskin DM, de Boer PA. 1997. The MinE ring: an FtsZ-independent cell structure required for selection of the correct division site in *E. coli*. *Cell* 91:685–94

98. Raskin DM, de Boer PA. 1999. MinDE-dependent pole-to-pole oscillation of division inhibitor MinC in *Escherichia coli*. *J. Bacteriol.* 181:6419–24

99. Raskin DM, de Boer PA. 1999. Rapid pole-to-pole oscillation of a protein required for directing division to the middle of *Escherichia coli*. *Proc. Natl. Acad. Sci. USA* 96:4971–76

100. Ringgaard S, van Zon J, Howard M, Gerdes K. 2009. Movement and equipositioning of plasmids by ParA filament disassembly. *Proc. Natl. Acad. Sci. USA* 106:19369–74

101. Rudner DZ, Losick R. 2010. Protein subcellular localization in bacteria. *Cold Spring Harb. Perspect. Biol.* 2:a000307

102. Rudner DZ, Pan Q, Losick RM. 2002. Evidence that subcellular localization of a bacterial membrane protein is achieved by diffusion and capture. *Proc. Natl. Acad. Sci. USA* 99:8701–6

103. Schrödinger E. 1944. *What Is Life? The Physical Aspect of the Living Cell*. Cambridge, UK: Cambridge Univ. Press

104. Schroth-Diez B, Gerwig S, Ecke M, Hegerl R, Diez S, Gerisch G. 2009. Propagating waves separate two states of actin organization in living cells. *HFSP J.* 3:412–27

105. Schwille P, Diez S. 2009. Synthetic biology of minimal systems. *Crit. Rev. Biochem. Mol. Biol.* 44:223–42

94. Gives the first compelling experimental evidence for protein localization induced by membrane curvature (see also Reference 95).

95. Gives the first compel'ing experimental evidence for protein localization induced by membrane curvature (see also Reference 94).

99. The first observation of the spectacular dynamic pattern formed by the Min proteins in vivo.

100. Identifies an intriguing mechanism for the segregation of low-copy-number plasmids during cell division that is based on filaments pulling on the plasmids and releasing them in a length-dependent manner.

106. Shih Y-L, Fu X, King GF, Le T, Rothfield L. 2002. Division site placement in *E. coli*: Mutations that prevent formation of the MinE ring lead to loss of the normal midcell arrest of growth of polar MinD membrane domains. *EMBO J.* 21:3347–57

107. Strahl H, Hamoen LW. 2010. Membrane potential is important for bacterial cell division. *Proc. Natl. Acad. Sci. USA* 107:12281–86

108. Suefuji K, Valluzz R, RayChaudhuri D. 2002. Dynamic assembly of MinD into filament bundles modulated by ATP, phospholipids, and MinE. *Proc. Natl. Acad. Sci. USA* 99:16776–81

109. Surrey T, Nedelec F, Leibler S, Karsenti E. 2001. Physical properties determining self-organization of motors and microtubules. *Science* 292:1167–71

110. Szeto TH, Rowland SL, Habrukowich CL, King GF. 2003. The MinD membrane targeting sequence is a transplantable lipid-binding helix. *J. Biol. Chem.* 278:40050–56

111. Szeto TH, Rowland SL, Rothfield LI, King GF. 2002. Membrane localization of MinD is mediated by a C-terminal motif that is conserved across eubacteria, archaea, and chloroplasts. *Proc. Natl. Acad. Sci. USA* 99:15693–98

112. Taghbalout A, Ma L, Rothfield L. 2006. Role of MinD-membrane association in Min protein interactions. *J. Bacteriol.* 188:2993–3001

113. Trueba FJ. 1982. On the precision and accuracy achieved by *Escherichia coli* cells at fission about their middle. *Arch. Microbiol.* 131:55–59

114. Establishes the theoretical concepts of how patterns can form starting from a initial homogenous distribution of interacting species (see also Reference 39).

114. **Turing AM. 1952. The chemical basis of morphogenesis. *Philos. Trans. R. Soc. Lond. B* 237:37–72**

115. Varma A, Huang KC, Young KD. 2008. The Min system as a general cell geometry detection mechanism: Branch lengths in Y-shaped *Escherichia coli* cells affect Min oscillation patterns and division dynamics. *J. Bacteriol.* 190:2106–17

116. Vartak N, Bastiaens P. 2010. Spatial cycles in G-protein crowd control. *EMBO J.* 29:2689

117. Vats P, Yu J, Rothfield L. 2009. The dynamic nature of the bacterial cytoskeleton. *Cell. Mol. Life Sci.* 66:3353–62

118. Vicker MG. 2000. Reaction-diffusion waves of actin filament polymerization/depolymerization in *Dictyostelium pseudopodium* extension and cell locomotion. *Biophys. Chem.* 84:87–98

119. Vicker MG. 2002. Eukaryotic cell locomotion depends on the propagation of self-organized reaction-diffusion waves and oscillations of actin filament assembly. *Exp. Cell. Res.* 275:54–66

120. Weiner OD, Marganski WA, Wu LF, Altschuler SJ, Kirschner MW. 2007. An actin-based wave generator organizes cell motility. *PLoS Biol.* 5:e221

121. Whitelam S, Bretschneider T, Burroughs NJ. 2009. Transformation from spots to waves in a model of actin pattern formation. *Phys. Rev. Lett.* 102:198103

122. Woldringh CL, Mulder E, Huls PG, Vischer N. 1991. Toporegulation of bacterial division according to the nucleoid occlusion model. *Res. Microbiol.* 142:309–20

123. Wolpert L. 2006. *Principles of Development*. Oxford, UK: Oxford Univ. Press. 576 pp.

124. Wu LJ, Errington J. 2004. Coordination of cell division and chromosome segregation by a nucleoid occlusion protein in *Bacillus subtilis*. *Cell* 117:915–25

125. Zaikin AN, Zhabotinsky AM. 1970. Concentration wave propagation in two-dimensional liquid-phase self-oscillating system. *Nature* 225:535–37

126. Zhou H, Lutkenhaus J. 2003. Membrane binding by MinD involves insertion of hydrophobic residues within the C-terminal amphipathic helix into the bilayer. *J. Bacteriol.* 185:4326–35

127. Ghasriani H, Ducat T, Hart CT, Hafizi F, Chang N, et al. 2010. Appropriation of the MinD protein-interaction motif by the dimeric interface of the bacterial cell division regulator MinE. *Proc. Natl. Acad. Sci. USA* 107:18416–21

128. Wu W, Park KT, Holyoak T, Lutkenhaus J. 2011. Determination of the structure of the MinD-ATP complex reveals the orientation of MinD on the membrane and the relative location of the binding sites for MinE and MinC. *Mol. Microbiol.* 79:1515–28

129. Morris DM, Jensen GJ. 2008. Toward a biomechanical understanding of whole bacterial cells. *Annu. Rev. Biochem.* 77:583–613

Protein Folding at the Exit Tunnel

Daria V. Fedyukina and Silvia Cavagnero

Department of Chemistry, University of Wisconsin-Madison, Madison, Wisconsin 53706;
email: cavagnero@chem.wisc.edu

Annu. Rev. Biophys. 2011. 40:337–59

First published online as a Review in Advance on March 1, 2011

The *Annual Review of Biophysics* is online at biophys.annualreviews.org

This article's doi:
10.1146/annurev-biophys-042910-155338

Copyright © 2011 by Annual Reviews.
All rights reserved

1936-122X/11/0609-0337$20.00

Keywords

ribosome, molecular chaperones, ribosomal exit tunnel, nascent protein, molecular crowding, ribosome-bound nascent chain

Abstract

Over five decades of research have yielded a large body of information on how purified proteins attain their native state when refolded in the test tube, starting from chemically or temperature-denatured states. Nevertheless, we still know little about how proteins fold and unfold in their natural biological playground: the living cell. Indeed, a variety of cellular components, including molecular chaperones, the ribosome, and crowding of the intracellular medium, modulate folding mechanisms in physiologically relevant environments. This review focuses on the current state of knowledge in protein folding in the cell with emphasis on the early stage of a protein's life, as the nascent polypeptide traverses and emerges from the ribosomal tunnel. Given the vectorial nature of ribosome-assisted translation, the transient degree of chain elongation becomes a relevant variable expected to affect nascent protein foldability, aggregation propensity and extent of interaction with chaperones and the ribosome.

Contents

PRINCIPLES OF IN VITRO PROTEIN FOLDING

Since Christian Anfinsen's pioneering article on the relation between protein sequence and structure in 1954 (1) and his formulation of the thermodynamic hypothesis of protein folding in 1962 (35), thousands of articles have been written on how proteins travel through energy landscapes and reach their native state. The large majority of this body of work considers the in vitro refolding mechanisms of pure proteins, starting from a thermally or chemically denatured state diluted into a buffer at physiologically relevant pH. Most experimental and computational studies have so far been carried out on small single-domain proteins. Multidomain proteins are still largely unexplored and have started to receive attention only recently (4).

Over five decades of research on the mechanisms of protein folding in vitro have revealed that there is a wide variability in the way different proteins fold in the test tube (9, 15, 77). Nonetheless, a few important trends of general significance have emerged. The main concepts are worth a summary here because they can be considered the basis for understanding fundamental aspects and mechanistic differences once proteins are allowed to fold and unfold in the complex cellular environment.

First, protein folding does not proceed via a random search (51), and protein energy landscapes are highly funneled (8, 99). The above facts greatly contribute to optimize the efficiency of the conformational search to reach the native state. As a result, a variety of parallel paths are typically present as proteins fold, each generally comprising the formation of numerous transiently populated species, i.e., kinetic intermediates (some experimentally undetectable) separated by energy barriers in the case of rugged landscapes, or progressively evolving conformations undergoing barrierless diffusion toward the native state. The latter scenario typically applies only to very small (<60 residues) proteins. In experimental studies, single-exponential kinetics is often observed. It is important to keep in mind that single-exponential folding is fully compatible with the concept of parallel folding pathways, and it does not necessarily imply a truly two-state folding, which is rarely observed. Indeed, multiple unfolded or partially folded conformations often interconvert faster than the rate-determining steps; hence they do not give rise to distinct kinetic phases (22). In addition, computer simulations suggest that kinetic intermediates are usually present, yet they may be poorly populated and therefore experimentally undetectable (16). Several proteins fold via experimentally detectable folding intermediates, which in some cases are en route to the native state.

Second, individual elements of secondary structure may form very fast (18), as in the case of α-helices (typically <1 μs), but are usually not stable in the absence of long-range tertiary contacts. Therefore, protein folding is generally not a rigorously hierarchical process, and it is extremely rare that high populations of secondary structure (e.g., helices) fold first, followed by collapse and tertiary structure formation. This idea is schematically illustrated in **Figure 1** as the class of paths denoted by dashed gray lines, comprising type 1, 2, and 5 species. Studies on isolated polypeptides representing

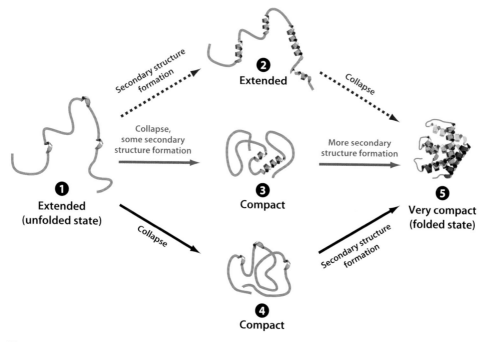

Figure 1

Scheme illustrating limiting in vitro protein folding mechanisms denoted by dashed gray (rarely observed), dark blue, and black arrows. The experiments leading to the formulation of these models are typically performed in purified protein solutions and involve the refolding of unfolded states generated chemically or by temperature jumps. Note that the species other than the unfolded and folded states (denoted 2, 3, and 4) may be either intermediates, transition states, or transient species populated along diffusive downhill routes.

apoMb:
apomyoglobin

Molten globule:
a highly dynamic
nonnative compact
state lacking a
considerable fraction
of a protein's
secondary structure

Transition state
ensemble:
a collection of
conformations that lie
at the maximum of
energy barriers in
protein folding energy
landscapes

IDP: intrinsically
disordered protein

NMR: nuclear
magnetic resonance

portions of primary structure of entire proteins show that individual helices and sheets are usually unstructured in the absence of surrounding tertiary contacts (23). Investigations on the early stages of protein folding showed that only small populations of secondary structure are detectable before chain collapse [exceptions are some members of the engrailed homeodomain family and protein A (15)]. Furthermore, protein variants containing destabilized versions of highly intrinsically helical regions of the chain are folding competent (12).

Third, the timescale for protein chain collapse is highly variable (nanoseconds to seconds) and sequence dependent (77). Collapse may (*a*) occur after most of the secondary structure is formed, as rarely observed experimentally (gray path in **Figure 1**); (*b*) be concurrent with most secondary structure formation, as seen in a number of apparently two-state folders (blue path in **Figure 1**) giving rise to relatively slow collapse with topology-dependent rates; (*c*) be concurrent with some secondary structure formation followed by slower acquisition of additional secondary structure, as in proteins with detectable folding intermediates such as apomyoglobin (apoMb) (42) (blue path in **Figure 1**), or (*d*) precede most secondary structure formation (black path in **Figure 1**).

The sequence determinants for the above options are not entirely clear yet and represent an outstanding challenge in in vitro protein folding. On the other hand, there are two apparent emerging trends. Collapse is slower when it occurs concomitantly with secondary structure formation, pointing to the kinetic difficulties in assembling secondary and tertiary structure together. In addition, secondary structure formation starting from a collapsed intermediate is also typically slow, pointing to the kinetic challenges in sampling conformational space from collapsed species (especially in large proteins). The above is true even if these species have significant internal dynamics, for instance, in the case of molten globules, and may bear a solvated nonpolar core.

Fourth, the starting species of in vitro folding experiments, the so-called unfolded

state, is sometimes far from lacking a structure; therefore, it is not truly unfolded (60, 80). Only expanded highly dynamic unfolded state ensembles follow the three criteria outlined above. Unfolded states bearing significant secondary structure and/or compaction are clearly posed to apply biases to the conformational search, sometimes making it more efficient. The presence of secondary/tertiary structure in proteins under strongly denaturing conditions is particularly interesting in the context of this review, given that the unfolded state populated under physiologically relevant conditions sometimes behaves differently from a self-avoiding Gaussian chain (71).

Fifth, a significant fraction (~40% in Eukarya) of the proteins expressed in the cell is actually natively unfolded (102). Representatives of this class are known as intrinsically disordered proteins (IDPs) and lack a well-defined independent structure at physiologically relevant pH and ionic strength. IDPs often fold upon interaction with their biological counterparts: Their folding mechanisms, still poorly explored, are beyond the scope of this work.

PROTEIN FOLDING IN THE CELL

In Vitro and In Vivo Protein Folding

Based on the results of pioneering nuclear magnetic resonance (NMR) experiments in live cells, the native structure of medium-size proteins in the intracellular environment is believed to be similar to the one populated in vitro in buffered solution. However, folding mechanisms in the cell are bound to be different from in vitro folding (**Figure 2**) due to the presence of a different unfolded state (see below); molecular chaperones; the ribosome; a highly crowded medium (200–300 mg ml^{-1} total protein concentration); cofactors such as heme, NADH, and others; intracellular processes such as posttranslational modifications; and quality-control processes such as protein degradation. In addition, some proteins are also subject to translocation into and out of different cell

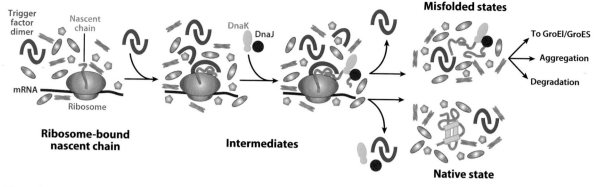

Figure 2

Schematic representation of key aspects of cotranslational protein folding in the crowded milieu of the cellular cytosol.

compartments, secretion, and cotranslational insertion into membranes. The latter processes are neglected in this review, which focuses on the folding of cytosolic soluble proteins.

The Unfolded State

Protein folding and unfolding in the cell can occur either during or after protein biosynthesis, i.e., co- or posttranslationally. In both cases, the nature of the unfolded state is poorly understood, yet likely profoundly different from the nonphysiological unfolded state ensemble of in vitro experiments in denaturants. For instance, in the case of full-length proteins in aqueous media at pH 7, the unfolded state is more compact than in the presence of denaturants (101). The effect of molecular crowding on the unfolded state under native conditions has yet to be studied in depth.

Molecular Chaperones

Molecular chaperones are key components of the cellular environment in bacteria, eukarya, and archaea. Their identity and roles have been reviewed elsewhere (17, 30, 37). Chaperones assist protein folding in the cell by preventing protein misfolding and aggregation and possibly also promoting folding. Interestingly, many chaperones in bacteria have overlapping specificities and their roles can sometimes be swapped (32), except for the bacterial

GroEL/ES, the lack of which is lethal to the cell.

FOLDING ON THE RIBOSOME: WHAT IS SPECIAL ABOUT IT?

Incomplete Protein Chains from Single-Domain Proteins Do not Generally Assume a Native-Like Conformation

The presence of ribosome-bound incomplete protein chains is one of the unique features of cotranslational events. In 1967, i.e., soon after the discovery that the biosynthesis of most proteins is catalyzed by the ribosome and proceeds vectorially from the N terminus to the C terminus, Phillips (65) formulated the hypothesis that the N-terminal portion of nascent proteins may start folding during translation. Two years later, Taniuchi & Anfinsen (81) responded by showing that cotranslational folding is unlikely for small- and medium-size single-domain proteins because individual purified N-terminal fragments of staphylococcal nuclease (SNase) of increasing length do not achieve any stable fold until their length closely approaches that of the complete protein. Since then, additional experimental model studies on SNase showed that the C-terminally truncated protein can indeed become compact yet partially disordered with only some of its secondary structure if very few residues are removed from its C terminus (29).

SNase: staphylococcal nuclease

TF: trigger factor

This finding suggests that the thermodynamic driving force for native-like tertiary structure formation develops during the very latest stages of chain elongation. Analogous studies on chymotrypsin inhibitor 2 and barnase are in agreement with the above ideas (61). Computational studies based on the burial of nonpolar surface as a function of chain elongation further support this concept (49, 50).

Incomplete Protein Chains can be Prone to Aggregation

Chain elongation model studies on purified model polypeptides from the medium-size (17 kDa) all-α-helical protein sperm whale apoMb (13) provide additional support to the idea that the native fold can be achieved only at lengths close to that of the complete primary structure. In addition, this study shows that incomplete N-terminal chains (from 36 to 119 residues of the 153-residue full-length protein), rich in nonpolar residues, exhibit a strong tendency to aggregate and form nonnative β-strands. The above misfolding/aggregation progressively decreases in magnitude as chain length approaches the full-length protein. This model system study highlights a unique feature of incomplete protein chains bearing a high nonpolar content: their tendency to aggregate. Aggregation of incomplete nascent chains is not tolerable in the cellular environment. The ability of the ribosome to keep chains maximally segregated during translation was demonstrated by a recent cryo-electron tomography study of *Escherichia coli* polysomes (7). Individual ribosome components of the polysome adopt a staggered or pseudohelical mutual arrangement, with nascent chains maximally spaced and pointing toward the cytosol. This 3D arrangement is naturally posed to minimize self-association of nascent proteins. Another investigation showed that the ribosome's ability to keep chains segregated prevents the aggregation of incomplete chains of the tailspike protein from the *Salmonella* phage P22 even in the absence of the cotranslationally

active trigger factor (TF) chaperone (25). As soon as ribosome release of the tailspike nascent chains is induced, the incomplete-length chains undergo self-association. The intrinsic ability of the ribosome to prevent the aggregation of rhodanese and lysozyme was also shown (33).

The Ribosome and Its Exit Tunnel Provide a Unique Environment for Nascent Chain Conformational Sampling

The tethering of all nascent polypeptides to the ribosome leads to expanding the function of this amazing machine from that of an mRNA-decoding center and catalyst for peptide bond formation to an obligatory scaffold, and possibly interaction counterpart, during the cotranslational conformational sampling of nascent polypeptides and proteins. The last few years have witnessed enormous progress in the elucidation of the archaeal and bacterial ribosome structure, structure-function relations, and assembly (66, 70, 79, 88, 94). The structures of the 50S large subunit and the entire ribosome solved at high resolution by X-ray crystallography (2, 36, 74, 91) provide ideal support to all studies of protein folding in and out of the exit tunnel. **Figure 3a** shows the high-resolution 3D structure of the *E. coli* ribosome, including the small and large subunits and the ribosomal proteins. **Figure 3b,c** provide schematics of a section of both the bacterial and archaeal ribosomes, respectively, highlighting the ribosomal exit tunnel and the proteins that directly face the tunnel's interior (L4, L22, L24) or are in close proximity to the tunnel (L23 and L29 in bacteria, and L23, L29, and L39e in archaea). Nascent proteins traverse the tunnel from the ribosome active site (i.e., the peptidyl transferase center, which houses the nascent protein C terminus) up until the tunnel's exit (31, 85). The tunnel is not completely straight and has a bend. Its length spans 80 to 100 Å, depending on where the exit-side end of the tunnel is defined (**Figure 3d**) (85).

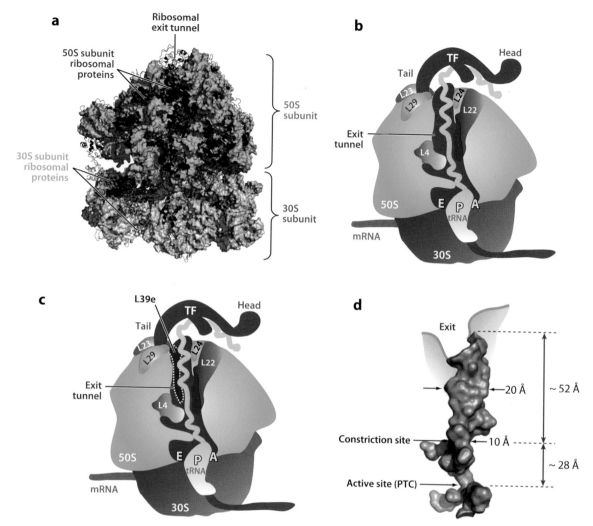

Figure 3

(*a*) Crystal structure of the *Escherichia coli* ribosome at 3.5 Å resolution (PDB IDs: 2AVY and 2AW4) (72). The ribosomal RNA is represented as surfaces (23S and 5S RNAs, *turquoise*; 16S RNA, *beige*). Ribosomal proteins are shown as ribbons (proteins in 50S subunit, *purple*; proteins in 30S subunit, *green*). Schematic representation of a vertical section of the 70S (*b*) prokaryotic and (*c*) archaeal ribosomes highlighting the ribosomal proteins facing or near the exit tunnel and the ribosome-associated TF chaperone. A representative hypothetical nascent polypeptide is drawn in yellow. (*d*) Structure of the ribosomal exit tunnel (PDB file kindly provided by N.R. Voss and P.B. Moore) (85). Abbreviations: PTC, peptidyl transferase center; TF, trigger factor.

Cotranslationally Active Molecular Chaperones Assist the Earliest Stages of a Protein's Life

The identity of cotranslationally active molecular chaperones varies depending on the kingdom of life and specific organism (30). For instance, in prokaryotes, the ribosome-associated dragon-shaped TF chaperone welcomes a large fraction of all nascent proteins emerging from the ribosomal tunnel, due to its high local concentration (89), and forms an arch above the tunnel (**Figure 3*b***). The resulting constrained environment encompasses sufficient space to host the folding of a small protein domain (3, 28) and serves as a protective shield (38)

Spheroplast: a
bacterium that has
been deprived of the
cell wall

for nascent chains capable of interacting with it (82). TF latches onto the ribosome via the L23 and L29 proteins (**Figure 3b**). TF does not exist in eukarya, and it is replaced by a number of other ribosome-associated chaperones (86).

The cotranslationally active chaperone DnaK (i.e., bacterial Hsp70) plays a role complementary to that of TF. The mechanism of action of DnaK and its co-chaperones DnaJ and GrpE has been reviewed (58). More than one Hsp70 are found in eukarya (30).

Finally, the ribosome itself may be far more than a spectator in co- and posttranslational protein folding. For instance, earlier studies by Dasgupta and colleagues (97, 98) and Hardesty (48) showed that ribosomes can promote the folding of denatured proteins. This finding prompted the Hardesty group to suggest that, among its many other activities, the ribosome also plays the role of a chaperone. Future studies hold promise to shed additional light on this interesting proposal.

The Kinetics of TF Binding/ Unbinding is Coordinated with Chain Elongation Rates and Nascent Protein Folding

The lifetime of TF-ribosome complexes is much longer (multiple seconds) (55, 68) than the average lifetime of nascent chain-TF complexes ($\geq$ms) in the absence of the ribosome (56). However, the latter lifetime can be modulated by the extent of nascent chain interaction with TF; i.e., it can increase significantly with the size of the nascent protein's nonpolar-region-binding TF and with ribosome-induced proximity. Furthermore, the presence of the ribosome enhances the association rates between the nascent protein chain and TF. Upon measuring the apparent association and dissociation rates of ribosome-nascent chain complexes with fluorescently labeled TF, Rutkowska et al. (68) proposed a kinetic model for the interplay of chaperone binding/release, cotranslational chain elongation, and protein folding, as shown in **Figure 4**. This interesting scheme shows how fast association and release of nascent chain to TF (within the ribosomal

complex) is compatible with polypeptide chain elongation. However, when the emerged nonpolar region is sufficiently large to slow down release of TF from the nascent chain, the TF chaperone may stay bound to the nascent chain even if released from the ribosome. Additional kinetic studies will certainly clarify how the above events are coordinated with co- and post-translational folding, so that nascent and newly synthesized proteins have a kinetic (and thermodynamic) opportunity to sample conformational space during translation and upon release from the ribosome.

Kinetic Considerations on Cotranslational Protein Folding

The best way to study folding at the exit tunnel is undoubtedly to watch the development of nascent protein structure and dynamics concurrently with translation. As shown in the next section, following up on this opportunity is especially desirable for large proteins, given that their translation rates approach intrinsic folding rates (62) and that it is likely that codon usage and ribosomal pausing are posed to affect the actual mechanism of folding.

Indeed, studying the cotranslational folding of fairly small single-domain proteins would also be extremely useful to verify that translation rates are slower than conformational sampling on the ribosome. However, to the best of our knowledge no such studies have been performed, although there are excellent prospects for progress in this area in the near future.

Cotranslational protein folding studies need to preserve the natural translation rates (so that they can be compared with folding rates) and are therefore best performed in vivo. However, working in an in vivo environment is challenging due to (a) the difficulties in selectively detecting folding in the complex cellular environment, and (b) the inability to synchronize translation given the stochastic nature of the process. Biological approaches pioneered by A. Helenius and F.U. Hartl have solved challenge a by monitoring protein activity cotranslationally, and challenge b by pulse-chase experiments often performed in bulk spheroplasts.

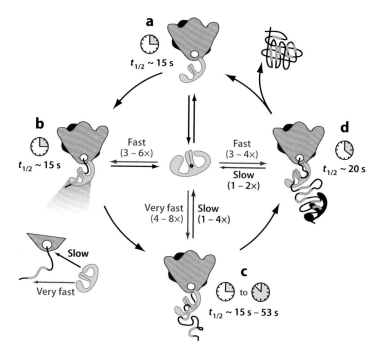

Figure 4

Model for the dynamic interaction of the trigger factor (TF) chaperone with ribosomes. The symbol $t_{1/2}$ denotes the half-life for the dissociation of the TF-ribosome binary complex or the apparent half-life for the dissociation of the TF-ribosome-nascent chain ternary complex. (*a,b*) The apparent association rate constant of TF (*green*) to ribosomes increases when a peptide chain emerges from the ribosomal exit tunnel. (*c*) Some longer nascent chains can increase the half-life $t_{1/2}$ for complex dissociation up to ~53 s. (*d*) The association rate of TF for ribosomes eventually decreases when a large nascent polypeptide is exposed to the ribosomal surface. TF may remain associated with some nascent chains even after dissociation of TF from its ribosome-binding site. Nonpolar stretches serving as TF binding sites are in blue. Adapted from Reference 68.

WHAT A DIFFERENCE TRANSLATION MAKES

Biosynthesis Rates Affect the Extent of Cotranslational Folding in Multidomain Proteins and Can Be Ad Hoc Modulated

Protein synthesis proceeds at variable rates in different environments and organisms (see Reference 93 and table 1 in Reference 13). For instance, translation rates are faster in vivo than under cell-free conditions. In addition, translation proceeds faster in prokarya (15–20 amino acids s^{-1}) than in eukarya (3–4 amino acids s^{-1}), leading to an average timescale for the production of a small-/medium-size protein of ~10 s and ~65 s in prokarya and eukarya,

respectively. These fairly long timescales are similar to chaperone binding/unbinding times and longer than the folding/unfolding timescales of small proteins. The above suggests that, in vivo, nascent chains encoding small proteins may have sufficient time to adopt preferred conformations as they are synthesized. On the other hand, very large proteins take multiple seconds to fold and may or may not attain stable conformations cotranslationally. Accordingly, the absence (62) or presence (63) of in vivo cotranslational folding in *E. coli* seems to be highly protein and codon dependent. Rare codon clusters (14), sometimes localized at interdomain junctions in large proteins, are emerging as important sites for an orchestrated pausing. This pausing is responsible for

RNC: ribosome-
bound nascent chain

facilitating cotranslational domain folding
before synthesis of the following domain
is initiated (92). A proper balance between
translation rates and co- and posttranslational
folding is important for the production of active
ribosome-released multidomain proteins. For
instance, mutant ribosomes displaying slower
translation than wild-type *E. coli* ribosomes
enhance the production of active multidomain
proteins (of eukaryotic origin) in bacteria (76).
A detailed review of this topic was recently
published by Zhang & Ignatova (93).

Preparation and Analysis of RNCs for Model Studies on the Conformation of Nascent Proteins at Equilibrium

The highest resolution information on protein folding at the exit tunnel has so far been
achieved via studies on purified arrested ribosomes bearing nascent proteins, sometimes labeled with fluorophores or NMR-active tags.
These studies implicitly assume that nascent
protein chains have the opportunity to conformationally equilibrate faster than the rate of
translation. This assumption is likely acceptable in many cases, especially for small proteins.
However, in general, caution should be exer-
cised and it is desirable to assess the validity of
this approximation in each case.

The most common methods to prepare
ribosome-bound nascent chains (RNCs) for
model studies at equilibrium are outlined in
Figure 5a. An exhaustive overview of these
methodologies is beyond the scope of this review. Therefore, we simply provide general
guidelines here.

RNCs can be prepared via genetic
approaches exploiting addition or in situ production of truncated mRNAs. These methodologies have been employed in vitro, either in
cell-free systems (e.g., from *E. coli*, wheat germ,
or rabbit reticulocyte) (20, 96, 103) or via
reconstituted systems containing all the necessary components for translation (e.g., PURE,
protein synthesis using recombinant elements,
based on prokaryotic components) (64, 75).

Reconstituted in vitro expression systems
have recently emerged as a convenient option
because of their complete lack of nucleases,
proteases, tmRNA, and other undesired
components.

Alternatively, stalled ribosomes can be
generated via protein-based approaches by
controlling translation arrest via special gene
products, i.e., short amino acid sequences (typically ~15–25 residues) that interact strongly
with specific portions of the ribosomal tunnel
and cause translation to stop (**Figure 5a**). The
most popular of these approaches is based on
generating very stable RNCs via the 17-residue
SecM arrest sequence (26, 67, 69), as shown in
Figure 5c. The SecM approach is particularly
convenient when RNCs are generated in vivo,
given the affordability of the method and its
high yields.

On the other hand, the SecM sequence is by
far not the only available method to generate arrested ribosomes by protein-based approaches.
Other strategies, based on both intrinsic and in-
ducible classes of ribosome-stalling sequences
(e.g., MifM, TnaC, and the Arg attenuator
peptide), have been reviewed recently (41)
(**Figure 5a**).

Generation of RNCs at equilibrium enabled analysis of the structure and dynamics of
nascent proteins at a level of detail presently
unattainable by in vivo studies. This analysis
has allowed addressing several questions regarding the stepwise generation of 3D protein
structures in nature. **Figure 6** provides a global
overview of the RNC structural and dynamic
aspects that have been addressed thus far. This
figure also links these specific folding-related
aspects to the technique used to gain the desired
information. As shown in **Figure 6**, a large variety of spectroscopy- and microscopy-based and
biological techniques have been used synergis-
tically. Perhaps even more importantly, the diagram shows that the same structural/dynamic
question can often be addressed by both
biological and spectroscopy/microscopy biophysical approaches. This synergism has been
particularly valuable in the field of RNC

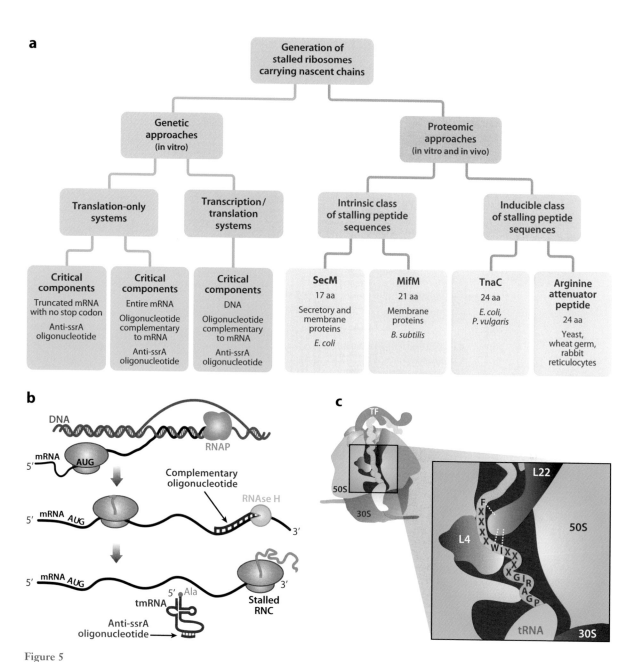

Figure 5

(*a*) Overview of currently available methods to generate RNCs of well-defined chain length. Step-by-step procedures based on (*b*) in vitro (cell-free) coupled transcription-translation and (*c*) SecM stalling. For simplicity, cotranslationally active chaperones are omitted. The 17-residue SecM peptide-stalling sequence (FXXXXWIXXXXGIRAGP) is shown inside the ribosomal tunnel. The underlined amino acids (*in red*) experience critical interactions with the ribosomal tunnel (*white dashed lines*) with L22. Abbreviations: RNAP, RNA polymerase; RNC, ribosome-bound nascent chain; X, any residue.

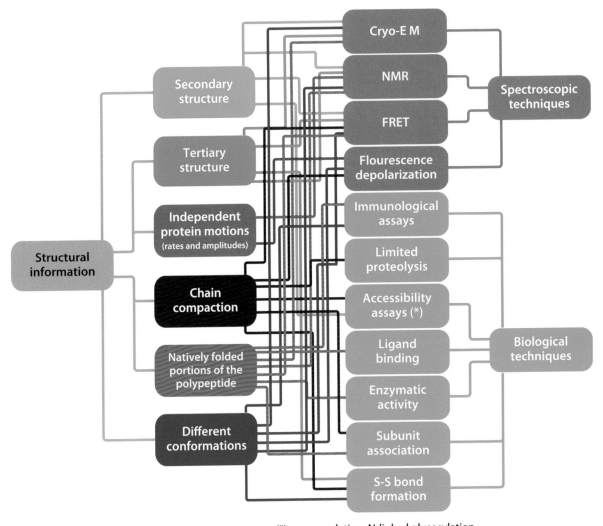

Figure 6

Relationship between specific RNC structural features and biological or spectroscopic techniques employed to elucidate them. Abbreviations: cryo-EM, cryo-electron microscopy; FRET, Förster resonance energy transfer; NMR, nuclear magnetic resonance; RNC, ribosome-bound nascent chain.

Ribosome exit tunnel: a narrow (10–20 Å) tunnel in the interior of the ribosome large subunit that nascent proteins need to traverse as they are being synthesized, before reaching the ribosome's surface

folding, given the challenging features of RNC for direct biophysical analysis (e.g., large internal dynamics, conformational heterogeneity, and large size of the RNC complexes). Despite the potential of the biophysical methods to provide higher resolution insights, biological approaches taking advantage of properties such as protein activity and antibody response are often efficient and highly informative.

Investigations on Nascent Polypeptides Inside the Ribosomal Tunnel

The ribosome exit tunnel (**Figure 3d**) has a width that ranges from 10 Å (constriction site) to 20 Å (widest region). These dimensions are incompatible with tertiary structure formation within the tunnel's interior. Major tunnel dynamics, presumably accompanied by extensive ribosome rearrangements, would be required for the tunnel to host even a simple tertiary fold such as a helical hairpin. Therefore, although it is recognized that the ribosome is a dynamic entity, it is likely that no tertiary structure, and only secondary structure, can be populated in nascent chains inside the ribosomal tunnel (85). This argument is supported by the finding of a highly spatially confined environment inside the tunnel, revealed by recent investigations showing that the N terminus of nascent polypeptides buried in the tunnel experiences narrow local motions (21).

The lower limit of the tunnel length is 80 Å, as proposed by Voss et al. (85) for the archaeal ribosome. Given this value, a fully α-helical polypeptide would bury approximately 53 residues (assuming an effective length of 1.5 Å per residue for the α-helix), and a fully extended polypeptide would bury approximately 23 residues (assuming an effective length of 3.5 Å per residue for an extended chain). These geometrical considerations prompt the question of whether any specific secondary structure is supported by the tunnel. Pioneering experiments by Malkin & Rich in 1967 (57), using in vivo pulse-chase techniques followed by cell lysis and proteolysis, showed that approximately 30 to 35 residues of nascent globin are protected from proteolysis in eukaryotic polysomes, implying that those residues are buried inside the ribosomal exit tunnel. These results are consistent with later investigations (reviewed in Reference 47) showing that there are 30 to 40 protected residues in nascent proteins. The above finding supports the presence of a partially helical conformation inside the tunnel. Computational studies by Ziv et al. (95) showed that a helical conformation can be entropically favored in a cylinder that models the ribosomal tunnel's dimensions. This result suggests that even a Teflon-like noninteracting tunnel (2) may be capable of inducing helical structure, especially in the case of nascent polypeptides whose coil state is highly disordered. Some recent high-resolution experiments further clarify this matter.

Förster resonance energy transfer (FRET) investigations on peptide sequences from a soluble secretory protein and a membrane protein showed that inside the eukaryotic tunnel the former is less helical than the latter (43). As the polypeptide chain elongates, helices can persist beyond the tunnel if they are stable in that environment. For instance, a peptide sequence from an integral membrane protein stays helical outside the tunnel as it is inserted into the membrane. However, the same sequence loses its helicity if the ribosomal surface faces bulk solution and is not bound to the membrane. Moreover, peptide sequences from soluble proteins have negligible helicity both inside and outside the tunnel (43). In summary, the ribosomal tunnel is capable of inducing helicity in nascent polypeptides, and this phenomenon is highly sequence dependent.

Additional investigations from Deutsch and coworkers (52–54) support the above conclusion by exploiting ingenuous accessibility assays, which enabled the detection of distinct tunnel zones characterized by different (highly negative) electrostatic potential. The authors also showed that some of these tunnel regions promote polypeptide chain compaction, suggestive of helix formation (46, 83, 84).

Recent cryo-electron microscopy (cryo-EM) work by Beckman and coworkers (5, 6) provides to date the highest-resolution insights on nascent secondary structure within the exit tunnel. As shown in **Figure 7**, the authors detected helical structure for sequences with high helical propensity in distinct regions of the tunnel (5). However, sequences with lower intrinsic helical propensity are disordered (5, 73). This work effectively complements and supports the findings by Johnson and Deutsch (46, 52, 84, 90, 103).

Whether the secondary structure formation in distinct regions of the tunnel results from specific polypeptide-tunnel interactions or whether it is driven (or at least contributed by) by entropic effects is still unclear and in need of further investigation. In the specific case of the SecM and TnaC ribosome-stalling sequences, convincing evidence for

FRET: Förster resonance energy transfer

cryo-EM: cryo-electron microscopy

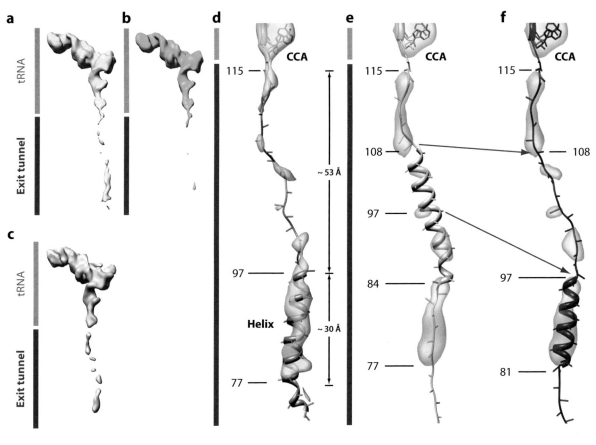

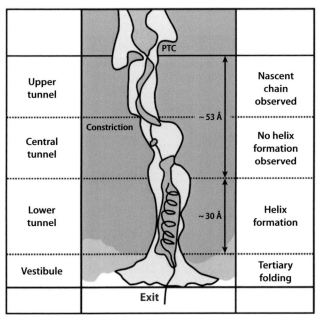

tunnel-polypeptide interactions was presented (73, 90).

Investigations on Nascent Proteins Emerging from the Ribosomal Tunnel

Our understanding of polypeptide conformation and dynamics as nascent proteins emerges from the ribosomal tunnel is not as advanced as our knowledge on nascent peptide structure inside the tunnel. The many experimental challenges presented by out-of-tunnel RNCs include the high conformational heterogeneity of the nascent chain and the variable effects introduced by cotranslationally active chaperones. Nevertheless, considerable progress has been made, and recent technical advances hold promise for additional exciting future progress. Comprehensive reviews of earlier work are available (24, 27, 47). Here, we focus on recent findings. In short, several examples of independent nascent structure and dynamics were discovered in RNCs emerging from the tunnel, defying the earlier proposal (81) that proteins acquire an independent conformation only after departing from the ribosome.

Investigations by Merz et al. (59) based on chemical cross-linking showed that nascent proteins with a significant nonpolar content emerging out the tunnel have a tendency to interact with the TF chaperone via its elongated binding surface (**Figure 8**). It is plausible that, while interacting with TF, the nascent protein also binds/unbinds TF and, possibly, other chaperones as it gets elongated, therefore maintaining its ability to sample conformational space while transiently non-TF-bound.

Cryo-EM images of polypeptides emerging from the ribosomal tunnel (34, 59) provided somewhat moderate structural detail. On the other hand, these studies were important to establish the possibility of tertiary structure formation outside the exit tunnel, in small single domain proteins. Additional evidence on 3D structure development comes from nascent chains from the ion channels, where tertiary structure was detected close to the ribosomal tunnel exit, via accessibility experiments based on side chain pegylation (45, 46). These results are supported by recent computational investigations (100).

Analysis of fluorescence depolarization decays of RNCs and ribosome-released fluorophore-labeled apoMb in the frequency domain (87) enabled Ellis et al. (20) to study the dynamics of nascent apoMb's N terminus on the subnanosecond timescale and follow the formation of an independent protein domain on the nanosecond timescale, as shown in **Figure 9a,b**. ApoMb RNCs acquire independent dynamics, indicative of compact or semicompact species, only when a significant portion of the sequence emerges from the ribosomal tunnel. The rotational correlation time reporting on the protein's nanosecond local motions increases significantly upon nascent protein release from the ribosome, showing that the structure of the full-length RNC differs from that of the ribosome-released native apoMb. RNCs encoding the natively unfolded protein PIR (phosphorylated insulin receptor interaction region) experience no motions on the nanosecond timescale, suggesting that PIR does not fold on the ribosome. The spatial amplitude of the nascent chain local motions is very narrow inside and, surprisingly, even outside the ribosomal tunnel (**Figure 9c,d**) (21). This is true even when RNCs are depleted of bound chaperones (TF and Hsp70). This

---

Figure 7

Cryo-EM maps of different peptidyl tRNAs inside the eukaryotic ribosome's P-site and exit tunnel. (*a*) 80S–helix 1 RNC, (*b*) 80S–DPAP RNC, (*c*) 80S–helix2 RNC, and (*d*) enlarged view of transparent density of panel *a* with fitted ribbon model for tRNA and nascent chain. (*e*, *f*) Enlarged view of panel *c* with alternative models for helix 2 nascent chain. Red arrows indicate corresponding region (residues 97–108) modeled as helical (*e*) or extended (*f*). (*g*) Schematic cross-section of 80S–helix 1 RNC representing helix formation within the exit tunnel. Abbreviations: cryo-EM, cyro-electron microscopy; RNC, ribosome-bound nascent chain; DPAP, dipeptidylaminopeptidase; PTC, peptidyl transferase center. Adapted by permission from Macmillan Publishers Ltd: *Nature Structural & Molecular Biology* (Reference 5), copyright (2010).

Figure 8

Mode for TF chaperone binding to nascent polypeptides based on cross-linking experiments by Merz et al. (59). TF directs the nascent chains through its interior in a sequence- and length-dependent manner. Interactions with TF are (*a*) moderate for nascent chains 40 to 60 residues long, and (*b*) considerable for nascent chains up to 90 residues, where the nascent chain's N terminus reaches up to the TF PPIase domain (head). (*c*) Upon further elongation, the nascent chain may leave TF or it may accumulate in the interior of the TF chaperone. Abbreviations: PDF, protein deformylase; MAP, methionine aminopeptidase; SRP, signal recognition particle; TF, trigger factor. Adapted by permission from Macmillan Publishers Ltd: *EMBO Journal* (Reference 59), copyright (2008).

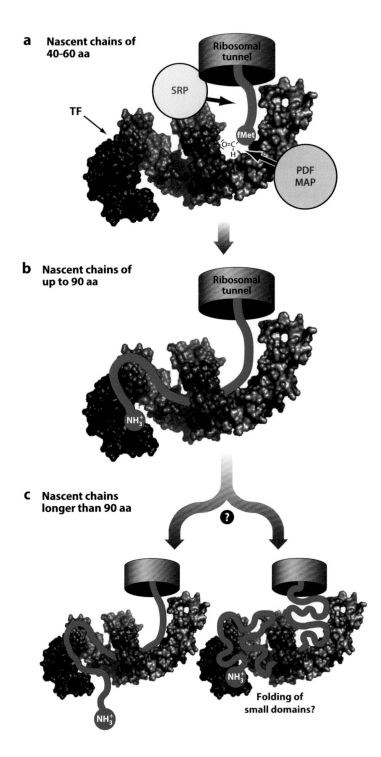

a Nascent chains of 40-60 aa

b Nascent chains of up to 90 aa

c Nascent chains longer than 90 aa

Folding of small domains?

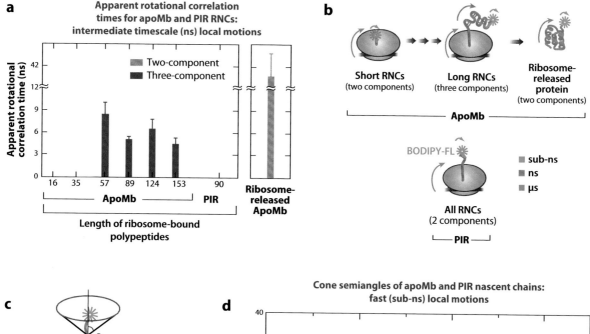

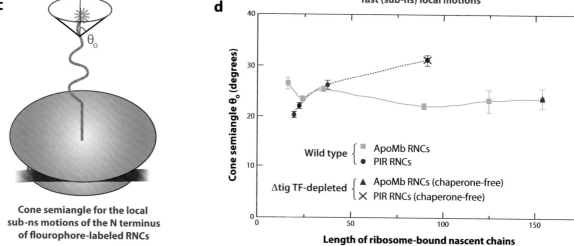

Figure 9

(*a*) Frequency domain dynamic fluorescence depolarization of ribosome-bound apoMb and PIR nascent chains generated in an *Escherichia coli* cell-free system. Data are shown only for the nanosecond local motions that reveal the presence of a small compact or semicompact species. (*b*) Scheme highlighting the motions associated with each fluorescence phase with each associated component of the motion. (*c*) Scheme illustrating the spatial amplitude of the subnanosecond local motion of the N terminus of the fluorophore-labeled RNC. The symbol θ_o represents the cone semiangle (*in red*) assessed in panel *d*. (*d*) Amplitude of the fast (subnanosecond) motions experienced by the N termini of nascent apoMb and natively unfolded PIR nascent polypeptides of increasing length under different conditions. Data were collected for samples prepared from either wild-type or Δtig TF-depleted cell strains. Panels *a* and *b* adapted with permission from References 20 and 21, respectively. Copyright 2008 and 2009, respectively, American Chemical Society and John Wiley and Sons. Abbreviations: apoMb, apomyoglobin; PIR, phosphorylated insulin receptor interaction region; RNC, ribosome-bound nascent chain; TF, trigger factor chaperone.

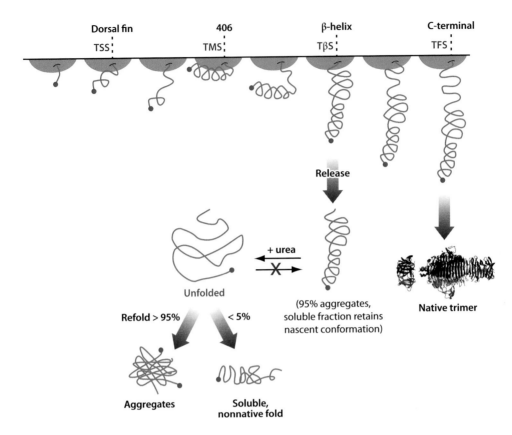

Figure 10

Model for the cotranslational folding of P22 tailspike nascent protein chains. Abbreviations: TSS, tailspike short stalled nascent chain; TMS, tailspike mid-length stalled nascent chain; TβS, tailspike stalled nascent chain with the entire β-helix exposed; TFS, tailspike full stalled nascent chain. Reprinted from the *Journal of Molecular Biology*, Vol. 383, Evans MS, Sander IM, Clark PL. "Cotranslational folding promotes beta-helix formation and avoids aggregation in vivo" pp. 683–92, Copyright (2008), with permission from Elsevier.

result suggests that both the tunnel and the outer surface of the ribosome exert a severe local confinement on nascent apoMb and PIR.

The limits of NMR spectroscopy have been pushed by recent studies on RNCs at atomic resolution (10, 11, 19, 39, 40). These investigations revealed that nascent single-domain proteins are not fully structured before they have entirely emerged from the ribosomal tunnel, consistent with the expectation that the C-terminal portion of the chain plays an important role in folding (49, 50).

Taken together, the above findings suggest that relatively small, full-length single-domain nascent proteins may adopt compact conforma-

tions outside the ribosomal tunnel. However, the nascent chains whose buried C-terminal residues are not available for folding may retain a considerable degree of disorder. Additional future studies are needed to provide more extensive evidence for these emerging trends.

The influence of the ribosome on protein folding is striking particularly for very large proteins unable to fold in vitro in the absence or presence of molecular chaperones. For such systems (e.g., the trimeric phage P2 tailspike protein), cotranslational folding is an irreplaceable requirement to attain the folded state and exploit biological activity (25). This important concept is illustrated in **Figure 10**.

SUMMARY POINTS

1. Chain compaction preceding or concurrent with secondary structure formation is a dominant class of mechanisms for the in vitro folding of small- and medium-size proteins, starting from largely unstructured unfolded ensembles.

2. The unfolded state of full-length proteins is believed to be rather compact in aqueous solution and physiological pH. Hence, secondary structure formation from compact states may be an important motif in posttranslational protein folding in the cell. Landscapes corresponding to this process may be rather rugged.

3. Incomplete N-terminal protein fragments (lacking the C terminus) often lack much of the native structure and may aggregate in aqueous solution and physiological pH, in the absence of the ribosome and molecular chaperones.

4. What do incomplete protein chains look like before translation is complete? The answer to this question is still largely unknown but great progress has been made over the past few years. There is a lot of activity in this exciting area.

5. The ribosomal tunnel is narrow and it provides an extremely spatially constrained environment for nascent polypeptides. The tunnel is capable of inducing helical structure, even in nascent polypeptides (derived from soluble proteins) that lack independent structure in solution. However, this process is highly sequence dependent.

6. The ribosomal tunnel consists of zones that differ in chemical potential and may promote secondary structure formation to a different degree.

7. Folding-competent proteins emerging from the ribosomal exit tunnel can assume a compact or semicompact conformation. Small single-domain proteins experience variations in their chain dynamics (and possibly folding) as they are released from the ribosome.

8. Very large proteins such as P22 tailspike are incapable of reaching their native state unless they are allowed to fold vectorially on the ribosome.

DISCLOSURE STATEMENT

The authors are not aware of any affiliations, memberships, funding, or financial holdings that might be perceived as affecting the objectivity of this review.

ACKNOWLEDGMENTS

We are grateful to all the past and present members of the Cavagnero group for their invaluable contributions. We thank Bernd Bukau, Anna Rutkowska, Elke Deuerling, Frieder Merz, Patricia Clark, Roland Beckmann, and Daniel Wilson for providing figures. The protein folding research in the Cavagnero group was funded by the National Science Foundation (grants MCB-0951209 and MCB-0544182), the National Institutes of Health (grants R21AI079656, R01GM068535, and R21GM071012), the Research Corporation Research Innovation Award, and the Shaw and Vilas Associates Awards.

LITERATURE CITED

1. Anfinsen CB, Redfield RR, Choate WI, Page J, Carroll WR. 1954. Studies on the gross structure, cross-linkages, and terminal sequences in ribonuclease. *J. Biol. Chem.* 207:201–10

2. Ban N, Nissen P, Hansen J, Moore PB, Steitz TA. 2000. The complete atomic structure of the large ribosomal subunit at 2.4 angstrom resolution. *Science* 289:905–20

3. Baram D, Pyetan E, Sittner A, Auerbach-Nevo T, Bashan A, Yonath A. 2005. Structure of trigger factor binding domain in biologically homologous complex with eubacterial ribosome reveals its chaperone action. *Proc. Natl. Acad. Sci. USA* 102:12017–22

4. Batey S, Nickson AA, Clarke J. 2008. Studying the folding of multidomain proteins. *HFSP J.* 2:365–77

5. Bhushan S, Gartmann M, Halic M, Armache J-P, Jarasch A, et al. 2010. Alpha-helical nascent polypeptide chains visualized within distinct regions of the ribosomal exit tunnel. *Nat. Struct. Mol. Biol.* 17:313–17

6. Bhushan S, Meyer H, Starosta AL, Becker T, Mielke T, et al. 2010. Structural basis for translational stalling by human cytomegalovirus and fungal arginine attenuator peptide. *Mol. Cell* 40:138–46

7. Brandt F, Etchells SA, Ortiz JO, Elcock AH, Hartl FU, Baumeister W. 2009. The native 3D organization of bacterial polysomes. *Cell* 136:261–71

8. Bryngelson JD, Onuchic JN, Socci ND, Wolynes PG. 1995. Funnels, pathways, and the energy landscape of protein folding: a synthesis. *Proteins Struct. Funct. Bioinform.* 21:167–95

9. Brockwell DJ, Smith DA, Radford SE. 2000. Protein folding mechanisms: new methods and emerging ideas. *Curr. Opin. Struct. Biol.* 10:16–25

10. Cabrita LD, Dobson CM, Christodoulou J. 2010. Early nascent chain folding events on the ribosome. *Isr. J. Chem. USA* 50:99–108

11. Cabrita LD, Hsu STD, Launay H, Dobson CM, Christodoulou J. 2009. Probing ribosome-nascent chain complexes produced in vivo by NMR spectroscopy. *Proc. Natl. Acad. Sci. USA* 106:22239–44

12. Cavagnero S, Dyson HJ, Wright PE. 1999. Effect of H helix destabilizing mutations on the kinetic and equilibrium folding of apomyoglobin. *J. Mol. Biol.* 285:269–82

13. Chow CC, Chow C, Raghunathan V, Huppert TJ, Kimball EB, Cavagnero S. 2003. Chain length dependence of apomyoglobin folding: structural evolution from misfolded sheets to native helices. *Biochemistry* 42:7090–99

14. Clarke TF 4th, Clark PL. 2008. Rare codons cluster. *PLOS One* 3:e3412

15. Daggett V, Fersht AR. 2003. Is there a unifying mechanism for protein folding? *Trends Biochem. Sci.* 28:18–25

16. Daggett V, Fersht A. 2003. The present view of the mechanism of protein folding. *Nat. Rev. Mol. Cell Biol.* 4:497–502

17. Deuerling E, Bukau B. 2004. Chaperone-assisted folding of newly synthesized proteins in the cytosol. *Crit. Rev. Biochem. Mol. Biol.* 39:261–77

18. Eaton WA, Munoz V, Thompson PA, Henry ER, Hofrichter J. 1998. Kinetics and dynamics of loops, alpha-helices, beta-hairpins, and fast-folding proteins. *Acc. Chem. Res.* 31:745–53

19. Eichmann C, Preissler S, Riek R, Deuerling E. 2010. Cotranslational structure acquisition of nascent polypeptides monitored by NMR spectroscopy. *Proc. Natl. Acad. Sci. USA* 107:9111–16

20. Ellis JP, Bakke CK, Kirchdoerfer RN, Jungbauer LM, Cavagnero S. 2008. Chain dynamics of nascent polypeptides emerging from the ribosome. *ACS Chem. Biol.* 3:555–66

21. Ellis JP, Culviner PH, Cavagnero S. 2009. Confined dynamics of a ribosome-bound nascent globin: Cone angle analysis of fluorescence depolarization decays in the presence of two local motions. *Protein Sci.* 18:2003–15

22. Ellison PA, Cavagnero S. 2006. Role of unfolded state heterogeneity and en-route ruggedness in protein folding kinetics. *Protein Sci.* 15:564–82

23. Epand RM, Scheraga A. 1968. Influence of long-range interactions on structure of myoglobin. *Biochemistry* 7:2864–72

24. Evans MS, Clark TF, Clark PL. 2005. Conformations of co-translational folding intermediates. *Protein Pept. Lett.* 12:189–95

25. Evans MS, Sander IM, Clark PL. 2008. Cotranslational folding promotes beta-helix formation and avoids aggregation in vivo. *J. Mol. Biol.* 383:683–92

26. Evans MS, Ugrinov KG, Frese M-A, Clark PL. 2005. Homogeneous stalled ribosome nascent chain complexes produced in vivo or in vitro. *Nat. Methods* 2:757–62

27. Fedorov AN, Baldwin TO. 1997. Cotranslational protein folding. *J. Biol. Chem.* 272:32715–18

28. Ferbitz L, Maier T, Patzelt H, Bukau B, Deuerling E, Ban N. 2004. Trigger factor in complex with the ribosome forms a molecular cradle for nascent proteins. *Nature* 431:590–96

29. Flanagan JM, Kataoka M, Shortle D, Engelman DM. 1992. Truncated staphylococcal nuclease is compact but disordered. *Proc. Natl. Acad. Sci. USA* 89:748–52

30. Frydman J. 2001. Folding of newly translated proteins in vivo: the role of molecular chaperones. *Annu. Rev. Biochem.* 70:603–47

31. Fulle S, Gohlke H. 2009. Statics of the ribosomal exit tunnel: implications for cotranslational peptide folding, elongation regulation, and antibiotics binding. *J. Mol. Biol.* 387:502–17

32. Genevaux P, Keppel F, Schwager F, Langendijk-Genevaux PS, Hartl FU, Georgopoulos C. 2004. In vivo analysis of the overlapping functions of DnaK and trigger factor. *EMBO Rep.* 5:195–200

33. Ghosh N, Hazra K, Sarkar SN. 1996. Ribosome facilitates refolding of rhodanese and lysozyme by suppressing aggregation. *Prog. Biophys. Mol. Biol.* 65:85

34. Gilbert RJC, Fucini P, Connell S, Fuller SD, Nierhaus KH, et al. 2004. Three-dimensional structures of translating ribosomes by cryo-EM. *Mol. Cell* 14:57–66

35. Haber E, Anfinsen CB. 1962. Side-chain interactions governing the pairing of half-cystine residues in ribonuclease. *J. Biol. Chem.* 237:1839–44

36. Harms J, Schluenzen F, Zarivach R, Bashan A, Gat S, et al. 2001. High-resolution structure of the large ribosomal subunit from a mesophilic eubacterium. *Cell* 107:679–88

37. Hartl FU, Hayer-Hartl M. 2009. Converging concepts of protein folding in vitro and in vivo. *Nat. Struct. Mol. Biol.* 16:574–81

38. Hoffmann A, Merz F, Rutkowska A, Zachmann-Brand B, Deuerling E, Bukau B. 2006. Trigger factor forms a protective shield for nascent polypeptides at the ribosome. *J. Biol. Chem.* 281:6539–45

39. Hsu ST, Cabrita LD, Fucini P, Christodoulou J, Dobson CM. 2009. Probing side-chain dynamics of a ribosome-bound nascent chain using methyl NMR spectroscopy. *J. Am. Chem. Soc.* 131:8366–67

40. Hsu ST, Fucini P, Cabrita LD, Launay H, Dobson CM, Christodoulou J. 2007. Structure and dynamics of a ribosome-bound nascent chain by NMR spectroscopy. *Proc. Natl. Acad. Sci. USA* 104:16516–21

41. Ito K, Chiba S, Pogliano K. 2010. Divergent stalling sequences sense and control cellular physiology. *Biochem. Biophys. Res. Commun.* 393:1–5

42. Jennings PA, Wright PE. 1993. Formation of a molten globule intermediate early in the kinetic folding pathway of apomyoglobin. *Science* 262:892–96

43. Johnson AE. 2004. Functional ramifications of FRET-detected nascent chain folding far inside the membrane-bound ribosome. *Biochem. Soc. Trans.* 32:668–72

44. Katzen F, Chang G, Kudlicki W. 2005. The past, present and future of cell-free protein synthesis. *Trends Biotechnol.* 23:150–56

45. Kosolapov A, Deutsch C. 2009. Tertiary interactions within the ribosomal exit tunnel. *Nat. Struct. Mol. Biol.* 16:405–11

46. Kosolapov A, Tu L, Wang J, Deutsch C. 2004. Structure acquisition of the T1 domain of Kv1.3 during biogenesis. *Neuron* 44:295–307

47. Kramer G, Ramachandiran V, Hardesty B. 2001. Cotranslational folding—omnia mea mecum porto? *Int. J. Biochem. Cell Biol.* 33:541–53

48. Kudlicki W, Coffman A, Kramer G, Hardesty B. 1997. Ribosomes and ribosomal RNA as chaperones for folding of proteins. *Fold. Des.* 2:101–8

49. Kurt N, Cavagnero S. 2005. The burial of solvent-accessible surface area is a predictor of polypeptide folding and misfolding as a function of chain elongation. *J. Am. Chem. Soc.* 127:15690–91

50. Kurt N, Mounce BC, Ellison PA, Cavagnero S. 2008. Residue-specific contact order and contact breadth in single-domain proteins: implications for folding as a function of chain elongation. *Biotechnol. Progr.* 24:570–75

51. Levinthal C. 1968. Are there pathways for protein folding? *J. Chim. Phys. Phys.-Chim. Biol.* 65:44–45

52. Lu J, Deutsch C. 2005. Folding zones inside the ribosomal exit tunnel. *Nat. Struct. Mol. Biol.* 12:1123–29

53. Lu J, Deutsch C. 2008. Electrostatics in the ribosomal tunnel modulate chain elongation rates. *J. Mol. Biol.* 384:73–86

54. Lu J, Kobertz WR, Deutsch C. 2007. Mapping the electrostatic potential within the ribosomal exit tunnel. *J. Mol. Biol.* 371:1378–91

55. Maier R, Eckert B, Scholz C, Lilie H, Schmid FX. 2003. Interaction of trigger factor with the ribosome. *J. Mol. Biol.* 326:585–92

56. Maier R, Scholz C, Schmid FX. 2001. Dynamic association of trigger factor with protein substrates. *J. Mol. Biol.* 314:1181–90

57. Malkin LI, Rich A. 1967. Partial resistance of nascent polypeptide chains to proteolytic digestion due to ribosomal shielding. *J. Mol. Biol.* 26:329–46

58. Mayer MP, Bukau B. 2005. Hsp70 chaperones: cellular functions and molecular mechanism. *Cell. Mol. Life Sci.* 62:670–84

59. Merz F, Boehringer D, Schaffitzel C, Preissler S, Hoffmann A, et al. 2008. Molecular mechanism and structure of trigger factor bound to the translating ribosome. *EMBO J.* 27:1622–32

60. Millett IS, Doniach S, Plaxco KW. 2002. Toward a taxonomy of the denatured state: small angle scattering studies of unfolded proteins. *Adv. Protein Chem.* 62:241–62

61. Neira JL, Fersht AR. 1999. Exploring the folding funnel of a polypeptide chain by biophysical studies on protein fragments. *J. Mol. Biol.* 285:1309–33

62. Netzer WJ, Hartl FU. 1997. Recombination of protein domains facilitated by co-translational folding in eukaryotes. *Nature* 388:343–49

63. Nicola AV, Chen W, Helenius A. 1999. Co-translational folding of an alphavirus capsid protein in the cytosol of living cells. *Nat. Cell Biol.* 1:341–45

64. Ohashi H, Kanamori T, Shimizu Y, Ueda T. 2010. A highly controllable reconstituted cell-free system—a breakthrough in protein synthesis research. *Curr. Pharm. Biotechnol.* 11:267–71

65. Phillips DC. 1967. The hen egg-white lysozyme molecule. *Proc. Natl. Acad. Sci. USA* 57:483–95

66. Ramakrishnan V. 2008. What we have learned from ribosome structures. *Biochem. Soc. Trans.* 036:567–74

67. Rutkowska A, Beerbaum M, Rajagopalan N, Fiaux J, Schmieder P, et al. 2009. Large-scale purification of ribosome-nascent chain complexes for biochemical and structural studies. *FEBS Lett.* 583:2407–13

68. Rutkowska A, Mayer MP, Hoffmann A, Merz F, Zachmann-Brand B, et al. 2008. Dynamics of trigger factor interaction with translating ribosomes. *J. Biol. Chem.* 283:4124–32

69. Schaffitzel C, Ban N. 2007. Generation of ribosome nascent chain complexes for structural and functional studies. *J. Struct. Biol.* 159:302–10

70. Schmeing TM, Ramakrishnan V. 2009. What recent ribosome structures have revealed about the mechanism of translation. *Nature* 461:1234–42

71. Schuler B, Eaton WA. 2008. Protein folding studied by single-molecule FRET. *Curr. Opin. Struct. Biol.* 18:16–26

72. Schuwirth BS, Borovinskaya MA, Hau CW, Zhang W, Vila-Sanjurjo A, et al. 2005. Structures of the bacterial ribosome at 3.5 Å resolution. *Science* 310:827–34

73. Seidelt B, Innis CA, Wilson DN, Gartmann M, Armache J-P, et al. 2009. Structural insight into nascent polypeptide chain-mediated translational stalling. *Science* 326:1412–15

74. Selmer M, Dunham CM, Murphy FV 4th, Weixlbaumer A, Petry S, et al. 2006. Structure of the 70S ribosome complexed with mRNA and tRNA. *Science* 313:1935–42

75. Shimizu Y, Kanamori T, Ueda T. 2005. Protein synthesis by pure translation systems. *Methods* 36:299–304

76. Siller E, DeZwaan DC, Anderson JF, Freeman BC, Barral JM. 2010. Slowing bacterial translation speed enhances eukaryotic protein folding efficiency. *J. Mol. Biol.* 396:1310–18

77. Sinha KK, Udgaonkar JB. 2009. Early events in protein folding. *Curr. Sci.* 96:1053–70

78. Deleted in proof

79. Steitz TA. 2008. A structural understanding of the dynamic ribosome machine. *Nat. Rev. Mol. Cell Biol.* 9:242–53

80. Tang YF, Rigotti DJ, Fairman R, Raleigh DP. 2004. Peptide models provide evidence for significant structure in the denatured state of a rapidly folding protein: the villin headpiece subdomain. *Biochemistry* 43:3264–72

81. Taniuchi H, Anfinsen CB. 1969. An experimental approach to the study of the folding of staphylococcal nuclease. *J. Biol. Chem.* 244:3864–75

82. Tomic S, Johnson AE, Hartl FU, Etchells SA. 2006. Exploring the capacity of trigger factor to function as a shield for ribosome bound polypeptide chains. *FEBS Lett.* 580:72–76

83. Tu L, Wang J, Deutsch C. 2007. Biogenesis of the T1 S1 linker of voltage-gated K+ channels. *Biochemistry* 46:8075–84

84. Tu LW, Deutsch C. 2010. A folding zone in the ribosomal exit tunnel for Kv1.3 helix formation. *J. Mol. Biol.* 396:1346–60

85. Voss NR, Gerstein M, Steitz TA, Moore PB. 2006. The geometry of the ribosomal polypeptide exit tunnel. *J. Mol. Biol.* 360:893–906

86. Wegrzyn RD, Deuerling E. 2005. Molecular guardians for newborn proteins: ribosome-associated chaperones and their role in protein folding. *Cell. Mol. Life Sci.* 62:2727–38

87. Weinreis SA, Ellis JP, Cavagnero S. 2010. Dynamic fluorescence depolarization: a powerful tool to explore protein folding on the ribosome. *Methods* 52:57–73

88. Williamson JR. 2008. Biophysical studies of bacterial ribosome assembly. *Curr. Opin. Struct. Biol.* 18:299–304

89. Witt SN. 2009. Tethering creates unusual kinetics for ribosome-associated chaperones with nascent chains. *Protein Pept. Lett.* 16:631–34

90. Woolhead CA, Johnson AE, Bernstein HD. 2006. Translation arrest requires two-way communication between a nascent polypeptide and the ribosome. *Mol. Cell* 22:587–98

91. Yusupov MM, Yusupova GZ, Albion B, Lieberman K, Earnest TN, et al. 2001. Crystal structure of the ribosome at 5.5 angstrom resolution. *Science* 292:883–96

92. Zhang G, Hubalewska M, Ignatova Z. 2009. Transient ribosomal attenuation coordinates protein synthesis and co-translational folding. *Nat. Struct. Mol. Biol.* 16:274–80

93. Zhang G, Ignatova Z. 2010. Folding at the birth of the nascent chain: coordinating translation with co-translational folding. *Curr. Opin. Struct. Biol.* 21:1–7

94. Zimmerman E, Yonath A. 2009. Biological implications of the ribosome's stunning stereochemistry. *ChemBioChem* 10:63–72

95. Ziv G, Haran G, Thirumalai D. 2005. Ribosome exit tunnel can entropically stabilize alpha-helices. *Proc. Natl. Acad. Sci. USA* 102:18956–61

96. Behrmann M, Koch HG, Hengelage T, Wieseler B, Hoffschulte HK, Muller M. 1998. Requirements for the translocation of elongation-arrested, ribosome-associated OmpA across the plasma membrane of *Escherichia coli*. *J. Biol. Chem.* 273:13898–904

97. Chattopadhyay S, Das B, Dasgupta C. 1996. Reactivation of denatured proteins by 23S ribosomal RNA: Role of domain V. *Proc. Natl. Acad. Sci. USA* 93:8284–87

98. Das B, Chattopadhyay S, Das Gupta C. 1992. Reactivation of denatured fungal glucose 6-phosphate dehydrogenase and alkaline phosphatase with ribosome. *Biochem. Biophys. Res. Commun.* 183:774–80

99. Dill KA, Bromberg S, Yue K, Fiebig KM, Yee DP, Thomas PD, Chan HS. 1995. Principles of protein folding: a perspective from simple exact models. *Protein Sci.* 4:561–602

100. O'Brien EP, Christodoulou J, Vendruscolo M, Dobson CM. 2011. New scenarios of protein folding can occur on the ribosome. *J. Am. Chem. Soc.* 133:513–26

101. Schuler B, Lipman EA, Eaton WA. 2002. Probing the free-energy surface for protein folding with single-molecule fluorescence spectroscopy. *Nature* 419:743–47

102. Uversky VN, Dunker AK. 2010. Understanding protein non-folding. *Biochim. Biophys. Acta* 1804:1231–64

103. Woolhead CA, McCormick PJ, Johnson AE. 2004. Nascent membrane and secretory proteins differ in FRET-detected folding far inside the ribosome and in their exposure to ribosomal proteins. *Cell* 116:725–36

Mechanosignaling to the Cell Nucleus and Gene Regulation

G.V. Shivashankar

Mechanobiology Institute & Department of Biological Sciences, National University of Singapore, 117543 Singapore; email: shiva.gvs@gmail.com

Annu. Rev. Biophys. 2011. 40:361–78

First published online as a Review in Advance on March 8, 2011

The *Annual Review of Biophysics* is online at biophys.annualreviews.org

This article's doi:
10.1146/annurev-biophys-042910-155319

Copyright © 2011 by Annual Reviews.
All rights reserved

1936-122X/11/0609-0361$20.00

Keywords

physicochemical signals, cytoplasmic-nuclear links, nuclear architecture, chromosome assembly, transcription compartments

Abstract

Cells integrate physicochemical signals on the nanoscale from the local microenvironment, resulting in altered functional nuclear landscape and gene expression. These alterations regulate diverse biological processes including stem cell differentiation, establishing robust developmental genetic programs and cellular homeostatic control systems. The mechanisms by which these signals are integrated into the 3D spatiotemporal organization of the cell nucleus to elicit differential gene expression programs are poorly understood. In this review I analyze our current understanding of mechanosignal transduction mechanisms to the cell nucleus to induce differential gene regulation. A description of both physical and chemical coupling, resulting in a prestressed nuclear organization, is emphasized. I also highlight the importance of spatial dimension in chromosome assembly, as well as the temporal filtering and stochastic processes at gene promoters that may be important in understanding the biophysical design principles underlying mechanoregulation of gene transcription.

Contents

INTRODUCTION

Biological cells are mechanical in origin and sense their local microenvironment using cell surface receptors (31). In addition to soluble signals, cells are sensitive to extracellular matrix rigidity, to the geometry or local curvature of the matrix, and to the applied stress or strain (119). These physical signals are an integral component of cellular behavior during differentiation (24), during development (1, 11), or in disease phenotypes (45). A variety of molecular intermediates describing cell matrix interactions and the remodeling of cytoskeletal networks in response to extracellular matrix signals have now been elucidated. Although the cytoskeleton is a well-appreciated critical component of cellular morphology (28), emerging evidence suggests that it may also have important consequences for maintenance of functional nuclear architecture and its mechanical properties (23, 121). Work from a number of laboratories, including mine, is beginning to suggest an elaborate physicochemical network of protein assemblies coupling the cytoskeleton to the nucleus (71, 110). This coupling results in a prestressed nuclear organization in living cells that balances contractile forces of the cytoskeleton and condensation forces of the chromatin (75–77). In addition, the prestressed nuclear architecture could perhaps serve as a substrate for transducing mechanical signals to the nucleus.

The regulation of gene expression in response to cellular matrix and geometric cues requires mechanisms that act at a distance (15, 120, 121). A number of canonical signaling pathways are activated in response to such mechanical signals converging on transcription factors—such as NFκB—that translocate to the nucleus upon activation (79, 126). Although these soluble factors translocate to the nucleus via diffusive processes, recent evidence also highlights the physical transmission of active stresses via cytoplasmic-nuclear connections to remodel chromatin assembly (121). The physicochemical signals targeted to the nucleus have to be further sorted to appropriate promoter sequences within the 3D architecture of the cell nucleus to bring about changes in gene expression programs (22, 61). Although the location of promoter sequences along the 1D contour of the DNA polymer is known from genome sequencing, their 3D location when folded into chromatin via histone and nonhistone proteins within the nucleus is largely unknown (83). Further, a number of essential posttranslational modifications of histone proteins (50) result in specificity and accessibility of promoter sequences to initiate gene transcription.

The regulation of gene expression programs depends on genome organization, posttranslational modifications of histone proteins, and assembly of transcription apparatus at gene promoters. Recent work in the literature has shown that genome organization is nonrandom (21) and perhaps specific to cell type (93). Work from several laboratories, including mine, is beginning to show that chromosomes are highly plastic in stem cells (4, 74, 81, 92) and that gene-rich chromosomes in differentiated cells may be spatially clustered (29) and are correlated with the transcriptome network of the cell (33). In addition, gene transcription requires transcription machinery enriched in dynamic compartments within the eukaryotic cell nucleus (113). The spatiotemporal assembly of transcription apparatus onto promoter sites, bound with the appropriate initiation signals derived through cellular mechanical cues, would finally result in the desired genetic outputs. However, it remains to be seen how such spatially coded maps of chromosome positioning in the nucleus and dynamic transcription machines enable efficient sorting of mechanical signals and initiate gene transcription in a highly regulated manner. The following sections elaborate on the modularity in mechanoregulation of gene transcription, as outlined above in eukaryotic cells.

MECHANOSIGNAL GENERATION FOR GENE REGULATION

Cells have evolved a plethora of sensing mechanisms in their local microenvironment. Such sensing could be physical and/or chemical in nature (24, 66, 116, 122). For example, when soluble factors such as EGF bind to EGF receptors, downstream signaling cascades impinge on gene expression. More importantly, the fact that cells can detect their local mechanical environment physically suggests that there must be distinct modularity in cellular sensing mechanisms to sample local geometry or substrate rigidity to which cells are adhered (31, 119). In addition, cells discriminate shear from compressive forces (62) and the direction they are applied from. Physical sensing modes are elicited, for example, through either stretch-activated receptors or integrin-like transmembrane receptors that attach to extracellular matrix proteins (53, 85), thereby sensing local rigidity of the matrix at the focal adhesion sites to transduce signals to the nucleus. Further, a

number of other molecular sensing mechanisms via receptor clustering are also being uncovered. These physicochemical signals integrate to alter cell behavior and function by modulating the cytoskeleton and gene expression—the theme of this review. Mechanisms that regulate the assembly of focal adhesion complexes mediated via the actomyosin machinery and the subsequent activation of mechanosensing proteins that impinge on cytoskeleton remodeling (48, 66, 116) and alter diverse cellular behaviors, including transmigration (18), are beyond the scope of this review. The reader is directed to recent comprehensive reviews of this aspect of mechanosignaling (31, 49, 94, 119).

Stem cells sample both physical and chemical signals, leading them to differentiate into multiple cell types, suggesting that modules of mechanosensing mechanisms used to alter gene expression programs are functional in these cells (17, 26, 55). Interestingly, lineage-specific transcription programs stabilize these differentiating cells. The cells are optimized or to sense matrix in the context of epithelial cells or sense soluble signals in the context of blood lineages. In the context of dedifferentiation (70, 128), reprogramming fibroblasts, for example, to a pluripotent cell or to a blood cell lineage suggests that modules of mechanosensing via feedback from the local microenvironment must be conserved, whereas specialized (differentiated) cells have the ability to use either one or the other mechanosensing module in a context-dependent manner.

Cellular mechanosensing at the plasma membrane has to be transduced to the nucleus, a few microns away from sensing sites in mammalian systems, to elicit differential gene expression programs. Typical diffusion constants of membrane-bound proteins are in the range of less than 1 μm^{-2} s^{-1}—perhaps determining the timescales in early events in mechanosensing. Transduction of mechanosignals from the inner leaflet of the plasma membrane via a number of protein conformational switches to phosphorylate buried amino acid residues and thereby transduce second-messenger calcium flux transport occurs in milliseconds to seconds. Although these signals are generated within seconds to minutes, the early gene transcriptional events are detected only after a few minutes, providing a temporal window on gene regulatory processes. Matrix-related genes are upregulated in tens of minutes (16, 120), whereas stem-cell-specific genes are modulated with matrix or geometric cues (26, 55) in a few hours to days, suggesting the need for integrating sustained versus short-time input signals required for mechanoregulation of gene expression. In the next sections I explore how input mechanical signals result in both physical remodeling of cytoskeleton and nucleus and transfer of signal intermediates via distinct canonical signaling pathways, such as MAPK/ERK or JAK-STAT (10).

CHEMICAL TRANSDUCTION: TRANSCRIPTION FACTORS AS MOLECULAR INTERMEDIATES

The transduction of mechanical signals eventually converges on transcription factors and cofactors that bind to regulatory DNA sites to determine which genes are upregulated or downregulated. For this, proteins activated, for example, via phosphorylation by kinases at the inner leaflet of the plasma membrane are transmitted either directly to the nucleus or by secondary mechanisms. Upon application of mechanical force, proteins such as paxillin (an essential linker between integrins and actin) and zyxin, which are usually resident at focal adhesion complexes, shuttle between the cytoplasm and nucleus (123). Although these proteins contain the Lim domain (52), which is involved in diverse cellular processes including cell adhesion and transcription control, the functional roles of these focal adhesion–resident proteins in the nucleus remain to be elucidated. Similarly, mechanosignal-induced phosphorylation of β-catenin bound to E-cadherins, present at the cell-cell junctions, results in β-catenin translocation into the nucleus (124).

Whereas some proteins can be translocated directly from focal adhesion or cell-cell junction to the nucleus, very often downstream signaling processes activate transcription-relevant factors either in the cytoplasm or in the nucleus. For example, activation of Notch receptors, which are transmembrane proteins, results in proteolytic cleavage of the intracellular Notch domain, which is then targeted to the nucleus (42). In contrast, transcription factors such as NFκB (114), STAT3 (10), and NFAT (40) that are sequestered in the cytoplasm upon a phosphorylation step translocate to the nucleus. Other examples such as MAL (118), which is bound to actin in the cytoplasm, are transferred from F-actin to G-actin, resulting in their translocation to the nucleus. Formins (82), nucleators for actin polymerization, exhibit cytoplasmic to nuclear shuttling. In general, although transcription factors and cofactors are activated in the cytoplasm upon stimulation and then translocated to the nucleus, there is a class of transcription-relevant factors that reside in the nucleus. These factors are present in an inactive form, but when stimulated, they target specific regulatory sites on the DNA. For example, EGR1 (67), a transcription factor stimulated by serum-responsive factors, is activated in the nucleus, resulting in binding to DNA target sites to regulate gene expression, including that of matrix-related genes.

Most signals targeted to the nucleus should contain nuclear localization signals (NLS) that are then recognized by the nuclear importin machinery. The sequence of events that follows transduction of chemical signals to the nucleus to regulate gene expression is highly variable. For example, transcription factors and cofactors bind regulatory sites and then further recruit a complex assembly of transcription machines to regulate gene expression (56). In the cytoplasm, whether some proteins translocate directly to the nucleus or whether others require secondary and tertiary relay mechanisms upon mechanical stimulation is currently unclear. However, early-response genes use a modularity in which transcriptional cues are targeted directly to the nucleus, whereas genes that encode for transcription factors and cofactors and thereby regulate a whole set of additional genes often are connected with sequential relay mechanisms for signal transduction. In addition, most proteins that shuttle to the nucleus are short-lived and are exported via the nuclear export signal (NES) that is recognized by the nuclear export machinery (19) or they are degraded by ubiquitination. Proteins composed of NLS or NES sequences are part of the protein family that shuttles to and from the nucleus (39), with a typical turnover timescale of a few minutes, suggesting kinetic mechanisms to actively access genetic information.

PHYSICAL TRANSDUCTION OF MECHANOSIGNALS VIA CYTOPLASMIC-NUCLEAR LINKS

Although chemical signals are targeted to the nucleus via soluble transcription intermediates, recent evidence reveals transduction of active stresses between the cytoplasm and nucleus that are inherent in biological cells (44, 76, 77, 105). The cytoskeleton, composed primarily of actin, microtubules, and intermediate filaments, has dynamic links to the nuclear membrane and chromatin assembly (117). The actin filaments that originate at the focal adhesion points also polymerize into long filaments (stress fibers) that have been implicated in transmitting stress signals throughout the cytoskeleton (28, 89, 94) and perhaps to the nucleus. In addition, microtubules that originate from the microtubule-organizing centers are highly filamentous structures that provide structural integrity to the cellular architecture (9) and may act as cables to transmit stresses from a distance. Further, the intermediate filaments (38) that form a dense structure in the cytoplasm provide a scaffold through which physical cues could be transmitted. These cytoskeletal filaments are highly cross-linked, providing a gel-like architecture to the cytoskeleton. In addition, the cytoskeletal structures act as substrates to sequester both soluble signaling intermediates and tracks for transporting signal-enriched vesicular cargos via the motors that run on them.

Recent work has revealed that the cytoskeletal structures bind to a number of nuclear membrane proteins (104, 110). In particular, the Sun and Kash domain proteins, residing on the outer and inner nuclear membrane, respectively, connect to cytoskeleton on one side and to nuclear lamina and chromatin assembly on the other side, bridging the internuclear membrane space via physical links (36, 37). Outer nuclear membrane proteins, such as Nesprin (132), interact with actin, while Klaroid binds to microtubules and plectin with intermediate filaments (117). The nuclear morphology is therefore determined as a result of these physical links (8, 100). For example, the nuclear volume in an adherent cell is much larger than the nucleus devoid of the cytoskeleton (76). Experiments have revealed that inhibiting myosin motors by blebbistatin or actin polymerization by cytochalasin D results in a dramatic decrease in nuclear size of adherent cells, whereas inhibiting microtubule polymerization by nocodazole results in an increase in nuclear size (77). These results suggest that the eukaryotic cell nucleus is held under tensile load by the actomyosin links, whereas the microtubules offer a compressive loading on the nucleus, suggesting a dynamic prestressed nuclear organization. The dynamic loading is further balanced by the intermediate filaments, and the degree of nuclear prestress is fine-tuned depending on cellular adhesion, with increased nuclear prestress in adherent cells. The degree of nuclear prestress has a number of important biological consequences. For example, it determines (*a*) the plasticity of the stem cell nucleus, (*b*) the ability of the nucleus to rotate, (*c*) the cytoplasmic-nuclear links mediating cell migration, and (*d*) the degree of mechanical unfolding of individual proteins or proteinaceous scaffolds including chromatin assembly and its remodeling.

In stem cells, the reduced cytoplasmic-to-nuclear ratio results in the lack of nuclear prestress and thus a soft, highly plastic, and mechanically pliable cell nucleus (4, 92). The emergence of cell differentiation programs leads to the dynamic reorganization of cytoskeleton to nuclear membranes, progressively leading to a stiffer nucleus as lineage-committed cells emerge (77, 92). On the other hand, the proximity of microtubule filaments to the nuclear membrane causes the rotation of the nucleus by dynein motors (64). These nuclear rotations are influenced by geometric constraints imposed by cytoplasmic-nuclear links. While in interphase cells, dynein motors bind to nuclear membrane proteins; their binding is controlled by the organization of microtubules, such that the nuclear rotations are highly regulated. Although these geometric constraints are far from clear, evidence of nuclear rotation suggests dynamic links between the cytoskeleton and the nucleus. These links are essential mechanical intermediates for cell migration (69) and for defining nuclear mechanical integrity via the perinuclear actin cap (54). In support of these links, a number of molecules currently have been revealed by monitoring the alterations in nuclear morphology upon ablation of specific genes in imaging-based RNAi screens.

In the context of this review, the cytoskeletal filaments that provide nuclear prestress might transmit active stresses from the plasma membrane to the nuclear membrane. This transduction mechanism has two functions. First, it acts as a mechanosensor with which to unfold proteins (51) linked to the cytoskeleton for phosphorylation events. Second, it may provide a means of unfolding chromatin assembly from a distance. The timescales of these force-induced unfolding processes depend directly on the direct transmission of stresses or stress waves limited by the acoustic transmission rates (121). The ability to alter chromatin structure from a distance may influence chromatin-remodeling enzymes (e.g., histone acetylase transferases and histone deacetylase transferases) by shifting their dynamic equilibrium toward the bound or unbound forms, which eventually remodels chromatin structure to regulate gene expression. In addition, mechanical links to the nucleus could modulate the opening and closing of nuclear pores (111). This could provide mechanisms to enhance the transduction of soluble chemical intermediates via the nuclear pores in a ligand-dependent nuclear translocation that is influenced by secondary forces exerted by the cytoplasm on the prestressed nucleus. These processes, however, would require highly regulated control of cytoskeletal structure and dynamics within living cells via Rho GTPases (27) and other signaling intermediates. In the next section I describe how these physiochemical signals may be integrated into the 3D architecture of the nucleus and transcription processes.

SPATIAL DIMENSION TO NUCLEAR ORGANIZATION

Mechanochemical signals that integrate into the nucleus have to be sorted within nuclear 3D architecture to elicit gene expression. The eukaryotic nucleus is a highly crowded environment, and its 3D architecture is determined by compaction of DNA (2), which has a typical contour length of ~1 m. Double-stranded DNA is a good example of a flexible polymer chain with a persistence length of ~50 nm, as revealed in DNA micromanipulation experiments (13). For this polymer chain, the typical radius of gyration is ~200 μm—the size of the entropic polymer configuration. This entropic coil is compacted within the cell nucleus (72), about 10 to 50 μm in diameter, depending on the cell type. For compaction, histone and nonhistone proteins bind to the DNA, stabilized by electrostatic interactions, forming a nucleosomal array 11 nm in diameter, and are folded into a dynamic 30-nm chromatin fiber by the interaction of linker histone proteins (3, 30, 68). Further packaging into higher-order chromatin structures involves forming condensed DNA structures called the heterochromatin structure (35), which is stabilized by heterochromatin binding proteins and noncoding RNA (131) and loosely packed euchromatin structure (99, 125). This differential packaging is highly regulated by a number of epigenetic modifications on the amino acid residues of histone and nonhistone proteins and their variants (50, 101). For example, increased acetylation of core histone amino acid tail residues results in more open and flexible chromatin structure; alternatively, methylation of amino acid residues of core histone proteins facilitates the binding of heterochromatin binding proteins to form condensed heterochromatin structure. A 1-m-long 1D sequence of DNA in any given chromosome is thus condensed into a chromosome ~1–2 μm in diameter in an interphase nucleus.

In addition to packaging DNA into a chromosome, the next level of packaging involves positioning chromosomes (for example, 23 pairs of chromosomes in human cells) within the nucleus (5). Because the contour length and folding fraction of each chromosome are different, their eventual volumes are variable. In effect, if each chromosome is treated as a flexible particle (sizes vary within a few microns), 40 such particles must be assembled within the cell nucleus. A number of scaffold proteins, comprising a family of intermediate filaments (lamin proteins) residing in the nucleus, form a structural scaffold to anchor chromosomes impinging on gene expression (36, 73, 104, 107). In particular, lamin proteins are organized as a sheet-like structure anchored to the inner nuclear membrane proteins and to condensed heterochromatin structures via the lamin receptor proteins. However, during mitosis, with the breakdown of the structural scaffolds, chromosomes are further condensed into metaphase chromosomes that regain their excluded volume and perhaps spatial positioning in the two daughter nuclei (7, 41). The nuclei of interphase cells (**Figure 1**) therefore have to balance the outward entropic forces of the DNA fiber and the inward condensation forces of a metaphase chromosome, imposing mechanical constraints on the nucleus (75–77).

Recent experiments have revealed a nonrandom positioning of chromosomes within the nucleus (5, 25). Therefore, the arrangement of 40 chromosomes in mouse cells or 46 chromosomes in human cells requires spatial positioning mechanisms within the 3D architecture of the nucleus. These mechanisms are further augmented by the fact that interphase chromosomes, depending on their size, appear to have preferred 3D radial positions when mapped from the center of the nucleus (87, 112). For

example, gene-inactive chromosomes appear to be located in the nuclear periphery, whereas gene-active chromosomes are located in the nuclear interior in the same cell type (33, 96). Evidence also suggests that cells from different tissues may comprise distinct cell-type-specific positional ordering of chromosomes (93). Such positional coding requires physical anchoring of chromosomes by chromosome-chromosome interactions, anchoring to nuclear scaffold proteins, or both. Cytoplasmic-nuclear links and nuclear scaffold proteins may act as a substrate to position chromosomes to their respective 3D positional coordinates in differentiated cells (86). However, such structural integrity is lacking in the stem cell nuclei, resulting in their plasticity and enhanced collisions between chromosome interfaces, whereas well-defined chromosome interfaces (**Figure 2**) are observed in differentiated cells (78, 98). How such cell-type-specific chromosome positioning and precision in their positional inheritance through mitosis are derived is still far from clear.

TEMPORAL DYNAMICS AND FILTERING BY TRANSCRIPTION FACTORIES

The spatial organization of chromosomes within the 3D architecture of the cell nucleus imposes physical constraints to accessing of genetic information. A typical gene (~2 kbp in linear length) folded around histone and nonhistone proteins is ~50 nm in diameter. Whereas the position of genes on the linear genome sequence is known, their location when folded into a 3D chromosome structure is still unknown (83, 88). However, chromosome surfaces have a largely euchromatin structure—composed of gene sequences—although there are a number of exceptions (32, 57). Thus, spatial positioning of chromosomes within the cell nucleus results in chromosome interfaces rich in coding sequences. Studies using chromosome capture assays have revealed that the interchromosome spacing or territories may contain intermingled flexible euchromatin loops, allowing for easy access to

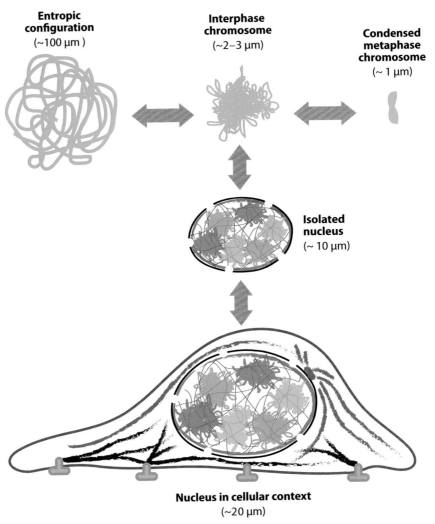

Entropic configuration (~100 μm)

Interphase chromosome (~2–3 μm)

Condensed metaphase chromosome (~ 1 μm)

Isolated nucleus (~ 10 μm)

Nucleus in cellular context (~20 μm)

Figure 1

Schematic illustrating the packaging of chromosomes into a prestressed eukaryotic cell nucleus balanced by cytoplasmic-nuclear links.

genetic information (65). The machinery to transcribe genes therefore must be assembled within these interchromosome territories to regulate gene transcription (6, 109).

A number of studies have experimentally identified colocalization between gene loops and transcription machinery (20). In addition, there is growing evidence for spatial clustering of gene loops within the interchromosome territories for their coregulation (90, 91). Depending on the cellular context, genes could

be accessed by the transcription machinery in their local 3D position or genes might loop out into interchromosome territories for their regulation in *cis* or in *trans* (127, 133). Electron microscopy studies of labeled transcription machinery in the nucleus revealed the compartmentalization of genes into small aggregates ranging from ∼40 to ∼500 nm in size (46, 47, 84). The density of these aggregates is smaller than the number of genes, suggesting shared transcription foci to regulate gene expression.

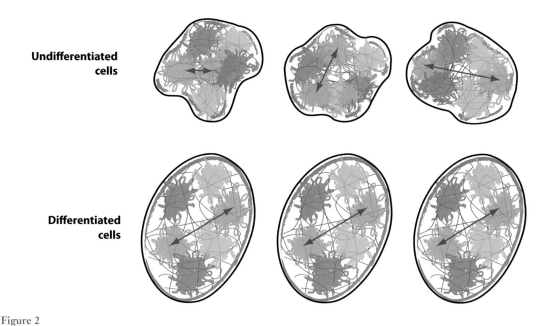

Undifferentiated cells

Differentiated cells

Figure 2

Time-lapse visualization of the cell nucleus. Stem cell nuclei are highly plastic, whereas nuclei in differentiated cells have fixed 3D chromosome organization and nuclear architecture.

This finding is also consistent with the fact that ∼10% of genes code for transcription factors; therefore, compartmentalization of transcription control facilitates sharing multiple transcription factors for combinatorial control of gene expression (83).

Immobile transcription compartments (or factories) require that genes loop out to these foci for their regulation or that these foci are organized in physical proximity to coding genes. However, live-cell imaging experiments have revealed a highly dynamic organization of transcription compartments (106) actively diffusing as nuclear bodies (34, 95) within the interchromosome territories. These compartments perhaps are composed of RNA polymerases (Pol II machinery), various subunits, and NTPs to initiate transcription upon the recruitment of transcription factors (84). Whereas some transcription compartments can be fixed in their 3D location, some are mobile—perhaps switching

between immobile and mobile fractions (106). Single-particle tracking analysis of these compartment dynamics has revealed that they undergo confined to directed movements within interchromosome territories. In addition, their dynamics depend on ATP, chromatin decompaction, and prestressed nuclear architecture.

Live-cell imaging studies have shown that heat-shock genes appear to recruit de novo transcription apparatus with no observed spatial repositioning of gene loci (130). However, fixed-cell immunofluorescence studies of immunoglobin genes show gene looping and spatial repositioning of gene clusters with respect to transcription machinery for their coregulation (12, 90). In this context, the dynamic organization of transcription compartments and their switching between mobile and immobile fractions (106) may reconcile these two disparate views about transcription control. Importantly, the possibility of a dynamic conversion between immobile and mobile fractions of transcription foci could indeed provide a temporal mechanical filtering to access gene regulatory sites within the 3D architecture of the nucleus.

MECHANICAL ACCESSIBILITY AND CHEMICAL SPECIFICITY FOR GENE TRANSCRIPTION IN LIVING CELLS: PUTTING THE PIECES TOGETHER

Biochemical and structural evidence for transcription initiation at the gene promoter has revealed important elements of eukaryotic transcription control (56). In summary, context-dependent transcription factors are targeted to promoter sites, resulting in remodeling of local chromatin structure at the regulatory sites. This follows the recruitment of transcription machinery, dynamic removal of the core histones by the chromatin-remodeling machinery in the coding region, and an efficient repacking of the local DNA structure incorporating histones and histone variants marking the transcription state. However, these processes are highly dynamic, and our understanding of how this is initiated in the context of a living cell nucleus is just beginning. In addition, we know very little about the spatiotemporal ordering of these events. However, the posttranslational modifications of core histones, primarily histone acetylation induced via histone acetyl transferases to loosen chromatin structure at the promoter sites and the recruitment and corecruitment of transcription factors, chromatin remodeling, and transcription machinery, are a necessary step to regulate gene transcription. There are a number of examples in which this recruitment and initiation are regulated via an enhancer sequence and its binding protein, resulting in a distal chromatin looping of a few hundred bases to a few kilobase pairs between the promoter region and the enhancer region to regulate gene transcription (63). The geometric scale of this looping ranges from a few tens to hundreds of nanometers at any given spatial position of the gene within the 3D architecture of the nucleus.

How then do mechanical signals that are exerted on the cell membrane, a few microns away, influence gene transcription within the highly crowded nuclear environment? How does the transcriptional apparatus recruit at a given gene in a context-dependent manner? Although the chemical signals generated at the plasma membrane are translocated via the transcription factor intermediates, diffusing either passively or actively in search of their regulatory sites within the 3D nuclear architecture, the physical integration via the cytoplasmic-nuclear links perhaps modulates exposure of buried regulatory sequences to facilitate the search process of these promoter sites for dynamic recruitment of the transcriptional apparatus. The modularity in gene transcription and transcription-dependent compartmentalization within the nucleus perhaps also provides a mechanical regulation of the dynamic movement of transcription machinery through the interchromosome separation via the modulation of prestressed cytoplasmic-nuclear links. In contrast, the plasticity of chromatin organization and nuclear architecture in stem cells (58, 80, 108) may provide a means to enhance interactions between distinct chromosome locations for gene clustering and their coregulation for lineage commitment.

There is a need to reconcile a number of facts about gene transcription regulation. For example, while all the regulatory machinery is in place, the positioning of the gene within the nucleus also appears to play a critical role in regulating its expression (60). When all the regulatory signaling modules are similar, and when they are positioned at the nuclear periphery or in the interior, genes elicit distinct expression states. On different cell types, identical mechanical forces express different genes, whereas on the same cell type, altering the amplitude of force or geometry results in differential gene expression patterns (15, 43, 102, 115, 120, 129). In addition, cells experiencing compressive or shear forces exhibit distinct gene expression programs. Soft nuclei such as those in stem cells express distinct lineage-specific gene expression patterns based on the substrate stiffness or the

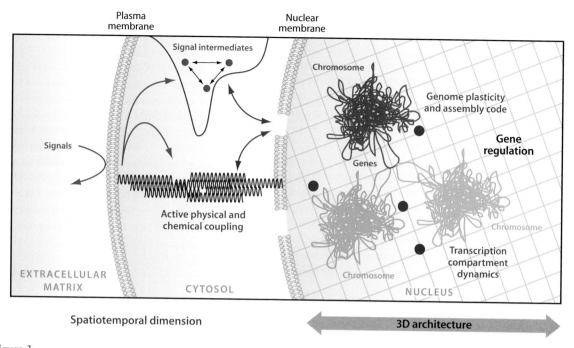

Plasma membrane

Nuclear membrane

Signal intermediates

Chromosome

Genome plasticity and assembly code

Gene regulation

Signals

Genes

Active physical and chemical coupling

Chromosome

Transcription compartment dynamics

EXTRACELLULAR MATRIX

CYTOSOL

NUCLEUS

Chromosome

Spatiotemporal dimension

3D architecture

Figure 3

Schematic of our current understanding of mechanosignaling to the cell nucleus and gene regulation.

geometric pattern to which they are adhered (26, 55). Although we understand the integration of soluble signals (i.e., transcription factors that regulate gene transcription), how spatial and temporal aspects of gene transcription are regulated by cell mechanics cues is unclear. The answers lie in understanding the physicochemical integration between the cytoskeletal filaments and the nuclear scaffold, the spatial organization of chromosomes and their relative positions, and the temporal dynamics of the compartmentalization of transcription apparatus. It is also critical to integrate the stochastic aspects of transcription control (14, 59, 97), given the small number of regulatory molecules and the dynamic self-organization principles of network motifs at the scale of genes and gene clusters, which may serve as substrates for their coregulation.

Figure 3 illustrates a possible hypothesis regarding the integration of mechanosignals to regulate gene expression. Regulatory molecules bound to focal adhesion or cytoskeletal junctions could be released either upon tun-

ing mechanical stretch or compression due to dissociation of the molecules based on their binding affinity, or via phosphorylation events. These regulatory molecules can then dynamically shuttle to the nucleus, self-organizing as network motifs at gene promoters or transcription hubs. Accessibility to regulatory sites is also facilitated by the physical modulation of cytoplasmic-nuclear links, which also impinge on the recruitment of dynamic transcription apparatus to sites of transcription initiation. These transient dynamic aggregates could then be disassembled rapidly and the regulatory proteins could be exported to the cytoplasm or degraded. Such cycles of cytoplasm-to-nucleus shuttling upon cell mechanics cues could enable kinetic assembly of regulatory molecules within the 3D nuclear architecture, for which relative spatial chromosome positions may have been optimized for a given cell type. Tuning the amplitude of mechanical signals could release distinct regulatory molecules bound with different binding affinities into the cytoplasm, resulting in their combinatorial recruitment and induction of differential gene expression programs in differentiated cell types. However, in stem cells, given the reduced cytoplasmic-to-nuclear ratio and the hyperdynamic nature of cytoskeletal filaments and chromatin assembly, an increased flux of regulatory molecules and enhanced collisions between chromosome territories could provide a mechanism to keep an active transcriptome. Matrix rigidity or geometric cues that stiffen the cytoplasmic-nuclear links could then provide cytoskeletal substrates to spatially localize regulatory molecules based on their binding affinities, resulting in lineage-specific gene expression programs. While this is purely speculative, the assumptions on which this hypothesis stands are beginning to garner strong support from a number of scattered experiments in the field.

IMPLICATIONS

In conclusion, we are just beginning to understand the effect of the cellular microenvironment (matrix rigidity or its geometry) on nuclear mechanics and genome regulation (103). In addition, the mechanical integrity of the cell nucleus and nuclear mechanical signaling profoundly influence cellular homeostatic controls by driving cells toward differentiation, proliferation, or apoptosis. Further, diseases such as cancer are hypothesized to originate at the single-cell level in its local mechanical environment. Therefore, understanding the mechanical control of gene function in living cells and the modularity of their control mechanisms will have important implications in understanding cellular behavior in diverse functional contexts.

SUMMARY POINTS

1. Mechanical signals are transduced to the nucleus both by soluble regulatory factors via nuclear pores and by active stresses via the prestressed cytoplasmic-nuclear links.

2. Integration of physicochemical signals from the cytoplasm and dynamic recruitment of compartmentalized transcription machinery at regulatory sites are important for chromatin remodeling and mechanotranscription of genes.

3. Having a stable 3D organization of genes, gene clusters, and chromosome position is critical for regulating the initiation of gene transcription.

4. There exists a spatial and temporal window in length and timescales between mechanosignal generation at the plasma membrane and gene transcription in the nucleus, giving rise to modularity in mechanosignaling to the nucleus via regulatory molecules and active stresses.

5. Chromosomes are positioned in distinct spatial locations and are mechanically interconnected with the nuclear lamina in differentiated cells. In addition, the eukaryotic transcription apparatus is compartmentalized, switching between immobile and mobile fractions.

FUTURE ISSUES

1. Do the physicochemical links to the prestressed nucleus imposed by cellular geometry impinge on the dynamic integration of transcription factors/compartments and active stresses to determine the accessibility and specificity of gene transcription?

2. Is there a coupling between functional genetic networks and the 3D nonrandom organization of chromosomes to facilitate mechanical regulation of gene transcription programs?

3. Does the chromatin and nuclear plasticity in undifferentiated cells (stem cells) play an essential role in increasing collisions between interchromosome territories and thereby regulate mean variability in gene transcription programs?

4. Can one formulate an input (geometry, matrix rigidity)—output (gene transcription programs) function based on modularity in mechanosignaling to functional nuclear architecture?

DISCLOSURE STATEMENT

The author is not aware of any affiliations, memberships, funding, or financial holdings that might be perceived as affecting the objectivity of this review.

ACKNOWLEDGMENTS

The list of references is not exhaustive and I apologize for omitting many key contributions from various laboratories in this emerging area of research. I thank members of my laboratory and a number of collaborators who have contributed to our ongoing research program on nuclear mechanics and genome regulation. I also thank the following agencies for generously supporting our research program: the Mechanobiology Institute and the Department of Biological Sciences, National University of Singapore; the National Center for Biological Sciences, TIFR-Bangalore, India; and the Swarnajayanti Fellowship and the National Nanoscience Initiative of the Department of Science and Technology, India.

LITERATURE CITED

1. Adamo L, Naveiras O, Wenzel PL, McKinney-Freeman S, Mack PJ, et al. 2009. Biomechanical forces promote embryonic haematopoiesis. *Nature* 459:1131–35
2. Banerjee B, Bhattacharya D, Shivashankar GV. 2006. Chromatin structure exhibits spatio-temporal heterogeneity within the cell nucleus. *Biophys. J.* 91:2297–303
3. Bhattacharya D, Mazumder A, Miriam SA, Shivashankar GV. 2006. EGFP-tagged core and linker histones diffuse via distinct mechanisms within living cells. *Biophys. J.* 91:2326–36
4. Bhattacharya D, Talwar S, Mazumder A, Shivashankar GV. 2009. Spatio-temporal plasticity in chromatin organization in mouse cell differentiation and during *Drosophila* embryogenesis. *Biophys. J.* 96:3832–39
5. Bolzer A, Kreth G, Solovei I, Koehler D, Saracoglu K, et al. 2005. Three-dimensional maps of all chromosomes in human male fibroblast nuclei and prometaphase rosettes. *PLoS Biol.* 3:e157
6. Branco MR, Pombo A. 2006. Intermingling of chromosome territories in interphase suggests role in translocations and transcription-dependent associations. *PLoS Biol.* 4:e138
7. Branco MR, Pombo A. 2007. Chromosome organization: new facts, new models. *Trends Cell Biol.* 17:127–34
8. Brandt A, Papagiannouli F, Wagner N, Wilsch-Brauninger M, Braun M, et al. 2006. Developmental control of nuclear size and shape by Kugelkern and Kurzkern. *Curr. Biol.* 16:543–52
9. Brangwynne CP, MacKintosh FC, Kumar S, Geisse NA, Talbot J, et al. 2006. Microtubules can bear enhanced compressive loads in living cells because of lateral reinforcement. *J. Cell Biol.* 173:733–41
10. Brivanlou AH, Darnell JE Jr. 2002. Signal transduction and the control of gene expression. *Science* 295:813–18
11. Brouzes E, Farge E. 2004. Interplay of mechanical deformation and patterned gene expression in developing embryos. *Curr. Opin. Genet. Dev.* 14:367–74
12. Brown JM, Leach J, Reittie JE, Atzberger A, Lee-Prudhoe J, et al. 2006. Coregulated human globin genes are frequently in spatial proximity when active. *J. Cell Biol.* 172:177–87

13. Bustamante C, Bryant Z, Smith SB. 2003. Ten years of tension: single-molecule DNA mechanics. *Nature* 421:423–27

14. Chang HH, Hemberg M, Barahona M, Ingber DE, Huang S. 2008. Transcriptome-wide noise controls lineage choice in mammalian progenitor cells. *Nature* 453:544–47

15. Chiquet M, Gelman L, Lutz R, Maier S. 2009. From mechanotransduction to extracellular matrix gene expression in fibroblasts. *Biochim. Biophys. Acta* 1793:911–20

16. Chiquet-Ehrismann R, Tannheimer M, Koch M, Brunner A, Spring J, et al. 1994. Tenascin-C expression by fibroblasts is elevated in stressed collagen gels. *J. Cell Biol.* 127:2093–101

17. Chowdhury F, Na S, Li D, Poh YC, Tanaka TS, et al. 2010. Material properties of the cell dictate stress-induced spreading and differentiation in embryonic stem cells. *Nat. Mater.* 9:82–88

18. Cinamon G, Shinder V, Alon R. 2001. Shear forces promote lymphocyte migration across vascular endothelium bearing apical chemokines. *Nat. Immunol.* 2:515–22

19. Cook AG, Conti E. 2010. Nuclear export complexes in the frame. *Curr. Opin. Struct. Biol.* 20:247–52

20. Cook PR. 2010. A model for all genomes: the role of transcription factories. *J. Mol. Biol.* 395:1–10

21. Cremer M, von Hase J, Volm T, Brero A, Kreth G, et al. 2001. Non-random radial higher-order chromatin arrangements in nuclei of diploid human cells. *Chromosome Res.* 9:541–67

22. Cremer T, Cremer C. 2001. Chromosome territories, nuclear architecture and gene regulation in mammalian cells. *Nat. Rev. Genet.* 2:292–301

23. Dahl KN, Ribeiro AJ, Lammerding J. 2008. Nuclear shape, mechanics, and mechanotransduction. *Circ. Res.* 102:1307–18

24. Discher DE, Mooney DJ, Zandstra PW. 2009. Growth factors, matrices, and forces combine and control stem cells. *Science* 324:1673–77

25. Duan Z, Andronescu M, Schutz K, McIlwain S, Kim YJ, et al. 2010. A three-dimensional model of the yeast genome. *Nature* 465:363–67

26. Engler AJ, Sen S, Sweeney HL, Discher DE. 2006. Matrix elasticity directs stem cell lineage specification. *Cell* 126:677–89

27. Etienne-Manneville S, Hall A. 2002. Rho GTPases in cell biology. *Nature* 420:629–35

28. Fletcher DA, Mullins RD. 2010. Cell mechanics and the cytoskeleton. *Nature* 463:485–92

29. Fraser P, Bickmore W. 2007. Nuclear organization of the genome and the potential for gene regulation. *Nature* 447:413–17

30. Gasser SM. 2002. Visualizing chromatin dynamics in interphase nuclei. *Science* 296:1412–16

31. Geiger B, Spatz JP, Bershadsky AD. 2009. Environmental sensing through focal adhesions. *Nat. Rev. Mol. Cell Biol.* 10:21–33

32. Gilbert N, Boyle S, Fiegler H, Woodfine K, Carter NP, Bickmore WA. 2004. Chromatin architecture of the human genome: Gene-rich domains are enriched in open chromatin fibers. *Cell* 118:555–66

33. Goetze S, Mateos-Langerak J, Gierman HJ, de Leeuw W, Giromus O, et al. 2007. The three-dimensional structure of human interphase chromosomes is related to the transcriptome map. *Mol. Cell Biol.* 27:4475–87

34. Gorisch SM, Wachsmuth M, Ittrich C, Bacher CP, Rippe K, Lichter P. 2004. Nuclear body movement is determined by chromatin accessibility and dynamics. *Proc. Natl. Acad. Sci. USA* 101:13221–26

35. Grewal SI, Jia S. 2007. Heterochromatin revisited. *Nat. Rev. Genet.* 8:35–46

36. Gruenbaum Y, Margalit A, Goldman RD, Shumaker DK, Wilson KL. 2005. The nuclear lamina comes of age. *Nat. Rev. Mol. Cell Biol.* 6:21–31

37. Haque F, Lloyd DJ, Smallwood DT, Dent CL, Shanahan CM, et al. 2006. SUN1 interacts with nuclear lamin A and cytoplasmic nesprins to provide a physical connection between the nuclear lamina and the cytoskeleton. *Mol. Cell Biol.* 26:3738–51

38. Herrmann H, Bar H, Kreplak L, Strelkov SV, Aebi U. 2007. Intermediate filaments: from cell architecture to nanomechanics. *Nat. Rev. Mol. Cell Biol.* 8:562–73

39. Hervy M, Hoffman L, Beckerle MC. 2006. From the membrane to the nucleus and back again: bifunctional focal adhesion proteins. *Curr. Opin. Cell Biol.* 18:524–32

40. Hogan PG, Chen L, Nardone J, Rao A. 2003. Transcriptional regulation by calcium, calcineurin, and NFAT. *Genes Dev.* 17:2205–32

41. Hubner MR, Spector DL. 2010. Chromatin dynamics. *Annu. Rev. Biophys.* 39:471–89

42. Hurlbut GD, Kankel MW, Lake RJ, Artavanis-Tsakonas S. 2007. Crossing paths with Notch in the hyper-network. *Curr. Opin. Cell Biol.* 19:166–75

43. Illi B, Nanni S, Scopece A, Farsetti A, Biglioli P, et al. 2003. Shear stress-mediated chromatin remodeling provides molecular basis for flow-dependent regulation of gene expression. *Circ. Res.* 93:155–61

44. Ingber DE. 1993. Cellular tensegrity: defining new rules of biological design that govern the cytoskeleton. *J. Cell Sci.* 104(Pt. 3):613–27

45. Jaalouk DE, Lammerding J. 2009. Mechanotransduction gone awry. *Nat. Rev. Mol. Cell Biol.* 10:63–73

46. Jackson DA, Hassan AB, Errington RJ, Cook PR. 1993. Visualization of focal sites of transcription within human nuclei. *EMBO J.* 12:1059–65

47. Jackson DA, Iborra FJ, Manders EM, Cook PR. 1998. Numbers and organization of RNA polymerases, nascent transcripts, and transcription units in HeLa nuclei. *Mol. Biol. Cell* 9:1523–36

48. Jamora C, Fuchs E. 2002. Intercellular adhesion, signalling and the cytoskeleton. *Nat. Cell Biol.* 4:E101–8

49. Janmey PA. 1998. The cytoskeleton and cell signaling: component localization and mechanical coupling. *Physiol. Rev.* 78:763–81

50. Jenuwein T, Allis CD. 2001. Translating the histone code. *Science* 293:1074–80

51. Johnson CP, Tang HY, Carag C, Speicher DW, Discher DE. 2007. Forced unfolding of proteins within cells. *Science* 317:663–66

52. Kadrmas JL, Beckerle MC. 2004. The LIM domain: from the cytoskeleton to the nucleus. *Nat. Rev. Mol. Cell Biol.* 5:920–31

53. Katsumi A, Orr AW, Tzima E, Schwartz MA. 2004. Integrins in mechanotransduction. *J. Biol. Chem.* 279:12001–4

54. Khatau SB, Hale CM, Stewart-Hutchinson PJ, Patel MS, Stewart CL, et al. 2009. A perinuclear actin cap regulates nuclear shape. *Proc. Natl. Acad. Sci. USA* 106:19017–22

55. Kilian KA, Bugarija B, Lahn BT, Mrksich M. 2010. Geometric cues for directing the differentiation of mesenchymal stem cells. *Proc. Natl. Acad. Sci. USA* 107:4872–77

56. Kornberg RD. 2007. The molecular basis of eukaryotic transcription. *Proc. Natl. Acad. Sci. USA* 104:12955–61

57. Kosak ST, Groudine M. 2004. Gene order and dynamic domains. *Science* 306:644–47

58. Kosak ST, Scalzo D, Alworth SV, Li F, Palmer S, et al. 2007. Coordinate gene regulation during hematopoiesis is related to genomic organization. *PLoS Biol.* 5:e309

59. Krishna S, Banerjee B, Ramakrishnan TV, Shivashankar GV. 2005. Stochastic simulations of the origins and implications of long-tailed distributions in gene expression. *Proc. Natl. Acad. Sci. USA* 102:4771–76

60. Kumaran RI, Thakar R, Spector DL. 2008. Chromatin dynamics and gene positioning. *Cell* 132:929–34

61. Lanctot C, Cheutin T, Cremer M, Cavalli G, Cremer T. 2007. Dynamic genome architecture in the nuclear space: regulation of gene expression in three dimensions. *Nat. Rev. Genet.* 8:104–15

62. Leipzig ND, Athanasiou KA. 2008. Static compression of single chondrocytes catabolically modifies single-cell gene expression. *Biophys. J.* 94:2412–22

63. Levine M. 2010. Transcriptional enhancers in animal development and evolution. *Curr. Biol.* 20:R754–63

64. Levy JR, Holzbaur EL. 2008. Dynein drives nuclear rotation during forward progression of motile fibroblasts. *J. Cell Sci.* 121:3187–95

65. Lieberman-Aiden E, van Berkum NL, Williams L, Imakaev M, Ragoczy T, et al. 2009. Comprehensive mapping of long-range interactions reveals folding principles of the human genome. *Science* 326:289–93

66. Liu B, Kim TJ, Wang Y. 2010. Live cell imaging of mechanotransduction. *J. R. Soc. Interface* 7(Suppl. 3):S365–75

67. Liu C, Yao J, Mercola D, Adamson E. 2000. The transcription factor EGR-1 directly transactivates the fibronectin gene and enhances attachment of human glioblastoma cell line U251. *J. Biol. Chem.* 275:20315–23

68. Luger K, Hansen JC. 2005. Nucleosome and chromatin fiber dynamics. *Curr. Opin. Struct. Biol.* 15:188–96

69. Luxton GW, Gomes ER, Folker ES, Vintinner E, Gundersen GG. 2010. Linear arrays of nuclear envelope proteins harness retrograde actin flow for nuclear movement. *Science* 329:956–59

70. MacArthur BD, Ma'ayan A, Lemischka IR. 2009. Systems biology of stem cell fate and cellular reprogramming. *Nat. Rev. Mol. Cell Biol.* 10:672–81

71. Maniotis AJ, Chen CS, Ingber DE. 1997. Demonstration of mechanical connections between integrins, cytoskeletal filaments, and nucleoplasm that stabilize nuclear structure. *Proc. Natl. Acad. Sci. USA* 94:849–54

72. Marenduzzo D, Micheletti C, Cook PR. 2006. Entropy-driven genome organization. *Biophys. J.* 90:3712–21

73. Mattout A, Dechat T, Adam SA, Goldman RD, Gruenbaum Y. 2006. Nuclear lamins, diseases and aging. *Curr. Opin. Cell Biol.* 18:335–41

74. Mattout A, Meshorer E. 2010. Chromatin plasticity and genome organization in pluripotent embryonic stem cells. *Curr. Opin. Cell Biol.* 22:334–41

75. Mazumder A, Roopa T, Basu A, Mahadevan L, Shivashankar GV. 2008. Dynamics of chromatin decondensation reveals the structural integrity of a mechanically prestressed nucleus. *Biophys. J.* 95:3028–35

76. Mazumder A, Shivashankar GV. 2007. Gold-nanoparticle-assisted laser perturbation of chromatin assembly reveals unusual aspects of nuclear architecture within living cells. *Biophys. J.* 93:2209–16

77. Mazumder A, Shivashankar GV. 2010. Emergence of a prestressed eukaryotic nucleus during cellular differentiation and development. *J. R. Soc. Interface* 7(Suppl. 3):S321–30

78. Meaburn KJ, Misteli T. 2007. Cell biology: chromosome territories. *Nature* 445:379–81

79. Mercurio F, Manning AM. 1999. NF-κB as a primary regulator of the stress response. *Oncogene* 18:6163–71

80. Meshorer E, Misteli T. 2006. Chromatin in pluripotent embryonic stem cells and differentiation. *Nat. Rev. Mol. Cell Biol.* 7:540–46

81. Meshorer E, Yellajoshula D, George E, Scambler PJ, Brown DT, Misteli T. 2006. Hyperdynamic plasticity of chromatin proteins in pluripotent embryonic stem cells. *Dev. Cell* 10:105–16

82. Miki T, Okawa K, Sekimoto T, Yoneda Y, Watanabe S, et al. 2009. mDia2 shuttles between the nucleus and the cytoplasm through the importin-α/β- and CRM1-mediated nuclear transport mechanism. *J. Biol. Chem.* 284:5753–62

83. Misteli T. 2007. Beyond the sequence: cellular organization of genome function. *Cell* 128:787–800

84. Mitchell JA, Fraser P. 2008. Transcription factories are nuclear subcompartments that remain in the absence of transcription. *Genes Dev.* 22:20–25

85. Moore SW, Roca-Cusachs P, Sheetz MP. 2010. Stretchy proteins on stretchy substrates: the important elements of integrin-mediated rigidity sensing. *Dev. Cell* 19:194–206

86. Muller I, Boyle S, Singer RH, Bickmore WA, Chubb JR. 2010. Stable morphology, but dynamic internal reorganisation, of interphase human chromosomes in living cells. *PLoS One* 5:e11560

87. Nagele RG, Freeman T, McMorrow L, Thomson Z, Kitson-Wind K, Lee H. 1999. Chromosomes exhibit preferential positioning in nuclei of quiescent human cells. *J. Cell Sci.* 112(Pt. 4):525–35

88. Naumova N, Dekker J. 2010. Integrating one-dimensional and three-dimensional maps of genomes. *J. Cell Sci.* 123:1979–88

89. Orr AW, Helmke BP, Blackman BR, Schwartz MA. 2006. Mechanisms of mechanotransduction. *Dev. Cell* 10:11–20

90. Osborne CS, Chakalova L, Brown KE, Carter D, Horton A, et al. 2004. Active genes dynamically colocalize to shared sites of ongoing transcription. *Nat. Genet.* 36:1065–71

91. Osborne CS, Chakalova L, Mitchell JA, Horton A, Wood AL, et al. 2007. *Myc* dynamically and preferentially relocates to a transcription factory occupied by *Igh*. *PLoS Biol.* 5:e192

92. Pajerowski JD, Dahl KN, Zhong FL, Sammak PJ, Discher DE. 2007. Physical plasticity of the nucleus in stem cell differentiation. *Proc. Natl. Acad. Sci. USA* 104:15619–24

93. Parada LA, McQueen PG, Misteli T. 2004. Tissue-specific spatial organization of genomes. *Genome Biol.* 5:R44

94. Parsons JT, Horwitz AR, Schwartz MA. 2010. Cell adhesion: integrating cytoskeletal dynamics and cellular tension. *Nat. Rev. Mol. Cell Biol.* 11:633–43

95. Platani M, Goldberg I, Lamond AI, Swedlow JR. 2002. Cajal body dynamics and association with chromatin are ATP-dependent. *Nat. Cell Biol.* 4:502–8

96. Pombo A, Branco MR. 2007. Functional organisation of the genome during interphase. *Curr. Opin. Genet. Dev.* 17:451–55

97. Raj A, van Oudernaarden A. 2009. Single-molecule approaches to stochastic gene expression. *Annu. Rev. Biophys.* 38:255–70

98. Rajapakse I, Perlman MD, Scalzo D, Kooperberg C, Groudine M, Kosak ST. 2009. The emergence of lineage-specific chromosomal topologies from coordinate gene regulation. *Proc. Natl. Acad. Sci. USA* 106:6679–84

99. Roopa T, Shivashankar GV. 2006. Direct measurement of local chromatin fluidity using optical trap modulation force spectroscopy. *Biophys. J.* 91:4632–37

100. Rowat AC, Lammerding J, Ipsen JH. 2006. Mechanical properties of the cell nucleus and the effect of emerin deficiency. *Biophys. J.* 91:4649–64

101. Sarma K, Reinberg D. 2005. Histone variants meet their match. *Nat. Rev. Mol. Cell Biol.* 6:139–49

102. Shaik SS, Soltau TD, Chaturvedi G, Totapally B, Hagood JS, et al. 2009. Low intensity shear stress increases endothelial ELR+ CXC chemokine production via a focal adhesion kinase-p38-MAPK-NFkB pathway. *J. Biol. Chem.* 284:5945–55

103. Shivashankar GV, ed. 2010. *Methods in Cell Biology*. Vol. 98: *Nuclear Mechanics and Genome Regulation*, p. xiii. New York: Elsevier

104. Shumaker DK, Kuczmarski ER, Goldman RD. 2003. The nucleoskeleton: Lamins and actin are major players in essential nuclear functions. *Curr. Opin. Cell Biol.* 15:358–66

105. Sims JR, Karp S, Ingber DE. 1992. Altering the cellular mechanical force balance results in integrated changes in cell, cytoskeletal and nuclear shape. *J. Cell Sci.* 103(Pt. 4):1215–22

106. Sinha DK, Banerjee B, Maharana S, Shivashankar GV. 2008. Probing the dynamic organization of transcription compartments and gene loci within the nucleus of living cells. *Biophys. J.* 95:5432–38

107. Spann TP, Goldman AE, Wang C, Huang S, Goldman RD. 2002. Alteration of nuclear lamin organization inhibits RNA polymerase II-dependent transcription. *J. Cell Biol.* 156:603–8

108. Spivakov M, Fisher AG. 2007. Epigenetic signatures of stem cell identity. *Nat. Rev. Genet.* 8:263–71

109. Sproul D, Gilbert N, Bickmore WA. 2005. The role of chromatin structure in regulating the expression of clustered genes. *Nat. Rev. Genet.* 6:775–81

110. Stewart CL, Roux KJ, Burke B. 2007. Blurring the boundary: The nuclear envelope extends its reach. *Science* 318:1408–12

111. Strambio-De-Castillia C, Niepel M, Rout MP. 2010. The nuclear pore complex: bridging nuclear transport and gene regulation. *Nat. Rev. Mol. Cell Biol.* 11:490–501

112. Sun HB, Shen J, Yokota H. 2000. Size-dependent positioning of human chromosomes in interphase nuclei. *Biophys. J.* 79:184–90

113. Sutherland H, Bickmore WA. 2009. Transcription factories: gene expression in unions? *Nat. Rev. Genet.* 10:457–66

114. Tay S, Hughey JJ, Lee TK, Lipniacki T, Quake SR, Covert MW. 2010. Single-cell NFkB dynamics reveal digital activation and analogue information processing. *Nature* 466:267–71

115. Thomas CH, Collier JH, Sfeir CS, Healy KE. 2002. Engineering gene expression and protein synthesis by modulation of nuclear shape. *Proc. Natl. Acad. Sci. USA* 99:1972–77

116. Tzima E, Irani-Tehrani M, Kiosses WB, Dejana E, Schultz DA, et al. 2005. A mechanosensory complex that mediates the endothelial cell response to fluid shear stress. *Nature* 437:426–31

117. Tzur YB, Wilson KL, Gruenbaum Y. 2006. SUN-domain proteins: 'Velcro' that links the nucleoskeleton to the cytoskeleton. *Nat. Rev. Mol. Cell Biol.* 7:782–88

118. Vartiainen MK, Guettler S, Larijani B, Treisman R. 2007. Nuclear actin regulates dynamic subcellular localization and activity of the SRF cofactor MAL. *Science* 316:1749–52

119. Vogel V, Sheetz M. 2006. Local force and geometry sensing regulate cell functions. *Nat. Rev. Mol. Cell Biol.* 7:265–75

120. Wang JH, Thampatty BP, Lin JS, Im HJ. 2007. Mechanoregulation of gene expression in fibroblasts. *Gene* 391:1–15

121. Wang N, Tytell JD, Ingber DE. 2009. Mechanotransduction at a distance: mechanically coupling the extracellular matrix with the nucleus. *Nat. Rev. Mol. Cell Biol.* 10:75–82

122. Wang Y, Botvinick EL, Zhao Y, Berns MW, Usami S, et al. 2005. Visualizing the mechanical activation of Src. *Nature* 434:1040–45

123. Wang Y, Gilmore TD. 2003. Zyxin and paxillin proteins: Focal adhesion plaque LIM domain proteins go nuclear. *Biochim. Biophys. Acta* 1593:115–20

124. Whitehead J, Vignjevic D, Futterer C, Beaurepaire E, Robine S, Farge E. 2008. Mechanical factors activate beta-catenin-dependent oncogene expression in APC mouse colon. *HFSP J.* 2:286–94

125. Woodcock CL. 2006. Chromatin architecture. *Curr. Opin. Struct. Biol.* 16:213–20

126. Xu J, Zutter MM, Santoro SA, Clark RA. 1998. A three-dimensional collagen lattice activates NF-kappaB in human fibroblasts: role in integrin alpha2 gene expression and tissue remodeling. *J. Cell Biol.* 140:709–19

127. Xu M, Cook PR. 2008. Similar active genes cluster in specialized transcription factories. *J. Cell Biol.* 181:615–23

128. Yamanaka S, Blau HM. 2010. Nuclear reprogramming to a pluripotent state by three approaches. *Nature* 465:704–12

129. Yang G, Im HJ, Wang JH. 2005. Repetitive mechanical stretching modulates IL-1beta induced COX-2, MMP-1 expression, and PGE2 production in human patellar tendon fibroblasts. *Gene* 363:166–72

130. Yao J, Munson KM, Webb WW, Lis JT. 2006. Dynamics of heat shock factor association with native gene loci in living cells. *Nature* 442:1050–53

131. Zaratiegui M, Irvine DV, Martienssen RA. 2007. Noncoding RNAs and gene silencing. *Cell* 128:763–76

132. Zhang Q, Ragnauth CD, Skepper JN, Worth NF, Warren DT, et al. 2005. Nesprin-2 is a multi-isomeric protein that binds lamin and emerin at the nuclear envelope and forms a subcellular network in skeletal muscle. *J. Cell Sci.* 118:673–87

133. Zink D, Amaral MD, Englmann A, Lang S, Clarke LA, et al. 2004. Transcription-dependent spatial arrangements of CFTR and adjacent genes in human cell nuclei. *J. Cell Biol.* 166:815–25

Amphipols From A to Z*

J.-L. Popot,[1] T. Althoff, D. Bagnard, J.-L. Banères,
P. Bazzacco, E. Billon-Denis, L.J. Catoire,
P. Champeil, D. Charvolin, M.J. Cocco, G. Crémel,
T. Dahmane, L.M. de la Maza, C. Ebel, F. Gabel,
F. Giusti, Y. Gohon, E. Goormaghtigh, E. Guittet,
J.H. Kleinschmidt, W. Kühlbrandt, C. Le Bon,
K.L. Martinez, M. Picard, B. Pucci, J.N. Sachs,
C. Tribet, C. van Heijenoort, F. Wien, F. Zito,
and M. Zoonens

[1] Institut de Biologie Physico-Chimique, CNRS/Université Paris-7 UMR 7099,
F-75005 Paris, France; email: Jean-Luc.Popot@ibpc.fr

Annu. Rev. Biophys. 2011. 40:379–408

The *Annual Review of Biophysics* is online at
biophys.annualreviews.org

This article's doi:
10.1146/annurev-biophys-042910-155219

Copyright © 2011 by Annual Reviews.
All rights reserved

1936-122X/11/0609-0379$20.00

*Affiliations for all coauthors can be found in the
Acknowledgments section.

Keywords

membrane proteins, amphipathic polymers, membrane biophysics,
membrane biochemistry

Abstract

Amphipols (APols) are short amphipathic polymers that can substitute
for detergents to keep integral membrane proteins (MPs) water soluble. In this review, we discuss their structure and solution behavior; the
way they associate with MPs; and the structure, dynamics, and solution properties of the resulting complexes. All MPs tested to date form
water-soluble complexes with APols, and their biochemical stability is
in general greatly improved compared with MPs in detergent solutions. The functionality and ligand-binding properties of APol-trapped
MPs are reviewed, and the mechanisms by which APols stabilize MPs
are discussed. Applications of APols include MP folding and cell-free
synthesis, structural studies by NMR, electron microscopy and X-ray
diffraction, APol-mediated immobilization of MPs onto solid supports,
proteomics, delivery of MPs to preexisting membranes, and vaccine
formulation.

Contents

INTRODUCTION

Membrane proteins (MPs) are notoriously more difficult to study in vitro than their soluble counterparts, mainly because of two factors: (*a*) the relative scarcity of most MPs and the difficulties encountered to overexpress them under a functional form, and (*b*) their frequent instability outside biological membranes. The complexity of a natural membrane is such that this environment is essentially impossible to mimic, even with artificial lipid bilayers, and much less so in aqueous solutions. Detergents are traditionally used to screen the hydrophobic transmembrane (TM) surface of MPs and make them water soluble, but they tend to inactivate them. There are several reasons why detergents destabilize MPs, possibly including the loss of physical constraints provided by the membrane environment. It is, however, our contention that a major factor is the competition of detergents with the protein-protein and protein-lipid interactions that stabilize the native structure of MPs (for a discussion, see Reference 67). This effect is compounded by the fact that MP/detergent complexes must be handled in the presence of free detergent micelles, because below the critical micellar concentration (CMC) of the detergent, entropic effects drive the aggregation of transmembrane surfaces (for a recent discussion, see Reference 89). Free micelles act as a hydrophobic sink into which lipids, cofactors, and subunits can disperse, which is a leading cause of MP inactivation. This view has been at the inception of the work summarized in the present review.

There are several possible ways to reduce or suppress the hydrophobic sink (for a discussion,

MP: membrane protein

Integral membrane protein: a protein that is in contact with the hydrophobic interior of a biological membrane to which it exposes an extensive transmembrane hydrophobic surface

Detergent: a surfactant with the ability to solubilize fats

see Reference 67). One of them is to replace detergents with compounds that would have such a high affinity for the transmembrane surface of MPs that they would remain associated to it even in the presence of vanishingly low concentrations of free micelles. This can theoretically be achieved by endowing a hydrophilic polymer with a multitude of hydrophobic chains that would act as many anchoring points. This consideration has led to the development of amphipols, a class of amphipathic polymers specially designed to complex MPs and keep them water soluble. Here the term amphipol (APol) refers to "an amphipathic polymer that is able to keep individual MPs soluble under the form of small complexes" (67).

Fifteen years have passed since the feasibility of this approach was first demonstrated (87). In the interim, much work has been devoted to learning to control the synthesis of these molecules and to understanding their physical chemical properties, to investigating the nature and properties of the complexes they form with MPs, and to developing their applications. As will be shown, the latter have extended well beyond the initial idea of enabling the biochemist and biophysicist to carry out basically the same studies as in detergent solution, but under improved conditions. Some APols are by now commercial. The thirty-odd publications that have appeared to date about the properties and uses of APols provide their potential users with a solid body of data to help them to exploit applications that have been validated and to develop novel ones. Previous reviews have discussed the advantages and disadvantages of APols versus other novel approaches to stabilizing MPs and studying them in vitro (12, 35, 67). Two other reviews have focused on APols and cover the field up to 2003 (68, 79).

Over the years, several types of APols have been designed, synthesized, and tested. Their study is at various stages of development. In the present review, we refer mainly to the following four types of APols: (*a*) A8–35 (**Figure 1*a***), the first APol to be validated, whose solubility is due to the presence of carboxylates (87); (*b*) phosphorylcholine-based APols (PC-APols) (**Figure 1*b***) (25, 26); (*c*) glucose-based, nonionic APols (NAPols) (**Figure 1*c***) (5, 6, 71, 83); and (*d*) sulfonated APols (SAPols) (**Figure 1*d***) (23, 64). Other types of APols are described in earlier reviews (67, 68, 79).

CHEMICAL STRUCTURE AND SOLUTION PROPERTIES OF AMPHIPOLS: A SPECIAL CLASS OF AMPHIPATHIC POLYMERS

Using APols efficiently depends on a good understanding of the way these polymers are designed and how they behave in aqueous solutions. In the present section, we summarize the chemical structure and solution behavior of the best field-tested APols and we describe the labeled or functionalized APols that have been developed and tested thus far.

Design, Synthesis, and Solution Properties of Amphipols

APols are amphipathic polymers, carrying both many hydrophilic and many hydrophobic groups (**Figure 1**). Being amphipathic, they belong, along with detergents, to the general class of surfactants. APols present a distinctive feature compared to most other amphipathic polymers: They self-assemble into well-defined particles comprising a few macromolecules [e.g., four molecules, on average, for 9- to 10-kDa A8–35 (32), three for 25-kDa NAPols (6)]. This results from the favorable combination of a short backbone chain length and an appropriate number and size of hydrophobic segments (so that the equilibrium size of the hydrophobic core can be reached with the amount of hydrophobic groups carried by just a few chains, while avoiding the otherwise typical formation of large networks of interconnected polymers due to hydrophobic self-association), flexibility (allowing a small radius of curvature of the interface with the solution), and a limited polydispersity in composition (to avoid large populations of chains with too few or too many hydrophobic groups compared to the optimum). In practice, macromolecules

Critical micellar concentration (CMC): the concentration of a small surfactant at which its molecules start to organize into micelles

Amphipathic polymer: a polymer comprising both hydrophobic and hydrophilic moieties

Amphipol (APol): an amphipathic polymer that can keep membrane proteins water soluble in detergent-free solutions as small individual entities by adsorbing onto their transmembrane surface

A8–35: a particular type of anionic amphipol

PC-APol: phosphorylcholine-based APol

NAPol: nonionic amphipol

SAPol: sulfonated APol

Surfactant: a compound that adsorbs at the air/water interface and thereby lowers the surface tension of water

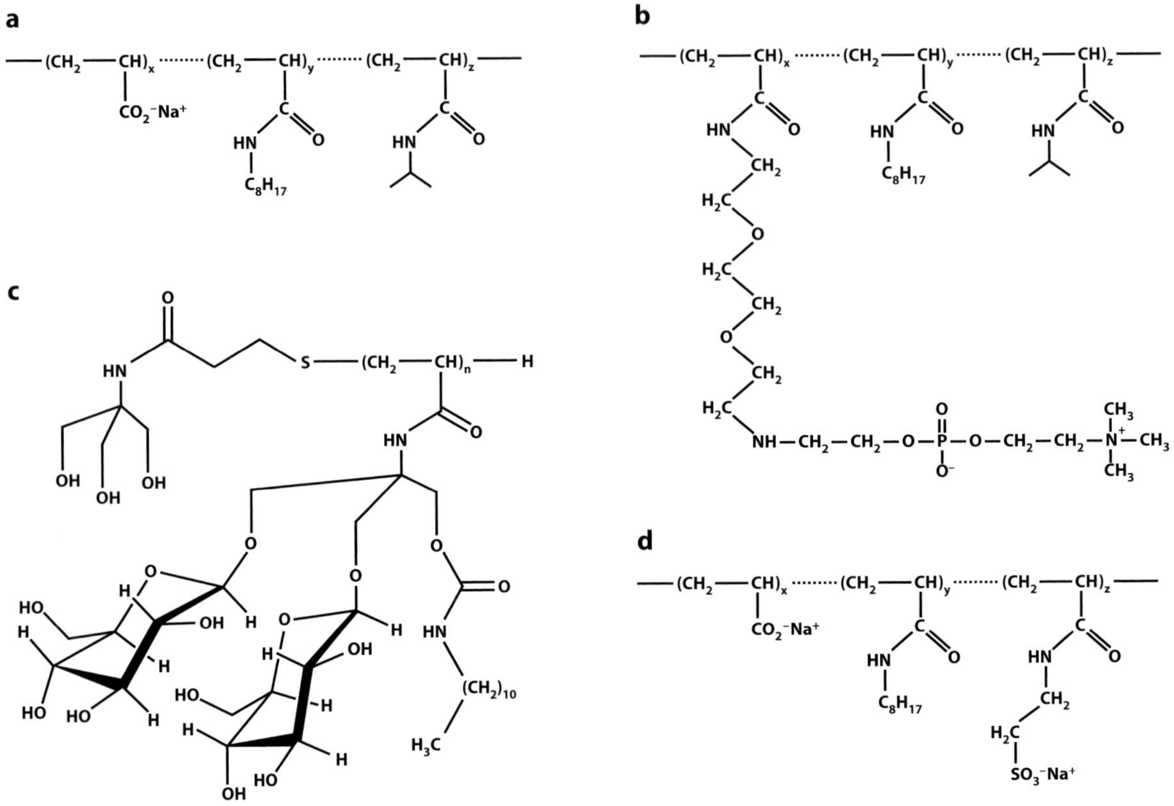

Figure 1

Chemical structures of four types of amphipols. (*a*) A polyacrylate-based APol, A8–35 (87); (*b*) a phosphorylcholine-based APol, C22–43 (26); (*c*) a nonionic, glucose-based APol (71); and (*d*) a sulfonated APol (23).

reaching or approaching such properties belong to one of the two following types: polysoaps, i.e., homopolymers made from amphipathic monomers bearing both alkyl and hydrophilic moieties, as is the case of homopolymeric NAPols (**Figure 1c**), or copolymers featuring a high density of hydrophobic monomers with short alkyl side groups (C_8-C_{12}) separated by one or a few hydrophilic monomers (all other APols described to date). Both types of APols typically contain a few tens of alkyl groups per chain.

When APols were initially devised, their expected mode of interaction with MPs was the primary concern (68, 87). In order to efficiently trap and keep water-soluble individual MPs, it was thought that polymers ought to be small

and highly flexible and to remain soluble despite the presence of multiple hydrophobic chains. The hydrophobic chains should be densely spaced (they are statistically ∼1 nm apart in A8–35) to avoid the formation of extended hydrophilic loops, which would interfere with many types of biophysical approaches. Chemically, this implied that the hydrophobic chains be relatively short (C_{18} chains were considered unsuitable) and that they be interspersed with highly polar groups. Polar groups are charged groups in most of the APols validated to date (23, 25, 26, 36, 59, 64, 87), hydroxyl groups in early NAPols (70), and glucose moieties in the most recently developed NAPols (5, 6, 71, 83). All the polymers that are efficient as APols self-assemble into small, compact, well-defined globular assemblies resembling detergent micelles. This behavior is a result of the constraints initially imposed on their chemical structure.

To prepare APols, various synthetic routes can be followed, which include (a) radical copolymerization of hydrophilic and hydrophobic monomers (25, 70, 83); (b) radical polymerization of a suitable monomer (e.g., an acrylic monomer) followed by random chemical modification of the homopolymer thus obtained (6, 23, 59, 87); and (c) homopolymerization of an amphipathic monomer (71).

A8–35 (**Figure 1a**) is by far the most extensively studied APol to date. Its behavior in aqueous buffers has been examined by size exclusion chromatography (SEC), dynamic and static light scattering, equilibrium and sedimentation velocity analytical ultracentrifugation (AUC), small-angle neutron scattering (SANS), inelastic neutron scattering (INS), and molecular dynamics (MD) (33, 34; F. Giusti, C. Tribet, M. Tehei, G. Zaccaï & J.N. Sachs, unpublished data). From a practical point of view, the most important solution properties of A8–35 can be summarized as follows. At and above pH 7, most carboxylates are ionized in aqueous solutions, conferring A8–35 a high solubility in water (>200 g·liter^{-1}) (34). A8–35 molecules, whose average molecular weight varies between 9 and 10 kDa depending on the average length of the polyacrylic acid used for the synthesis, self-assemble into particles averaging ~40 kDa (33). The particles are monodisperse, much more so than the molecules that form them. Individual particles, therefore, comprise a variable number of molecules depending on the latter's size. On average, A8–35 particles comprise approximately four molecules and 75–80 octyl chains (33).

Investigations of A8–35 and PC-APol C22–43 in water by neutron and light scattering (33) and by isothermal titration calorimetry (26) failed to detect dissociation of the particles down to 0.1 g·liter^{-1}. Down to this concentration, the hydrodynamic radius and molar mass of A8–35 particles are independent of both polymer concentration (33) and chain length polydispersity (C. Tribet &

F. Giusti, unpublished data). Mixtures of fluorescently labeled A8–35 samples have been studied in an attempt to determine the concentration at which individual chains start associating, which gives rise to a Förster resonance energy transfer (FRET) signal. The critical aggregation concentration thus evaluated is extremely low: ~0.002 g·liter^{-1} in 100 mM NaCl, pH 8.0 (F. Giusti & C. Tribet, unpublished data).

A8–35 particles bind a maximum of ~1.2 g water per g polymer; their radius of gyration is ~2.4 nm, and their Stokes radius is ~3.15 nm (33). Their operational specific volume (φ') in 100 mM NaCl is 0.87 mL·g^{-1} and their contrast match point (the percentage of D$_2$O at which their neutron scattering length density is the same as that of the solvent and the particles become invisible to neutrons) is 23.5% D$_2$O (34). The shape of SANS curves indicates that the particles are globular and present a sharp interface with water (33), consistent with the results of recent MD simulations (J.N. Sachs, unpublished data). The latter approach has revealed a few interesting features that had remained inaccessible experimentally, such as the intermingling of the APols chains that constitute the particles or the presence of water molecules in their center, probably a result of steric constraints that oppose the formation of a compact hydrophobic core (67; J.N. Sachs, unpublished data). The comparison of INS data collected on unlabeled versus partially deuterated A8–35 reveals backbone dynamics similar to that of CH$_2$ groups in lipids or free polymers (M. Tehei & G. Zaccaï, personal communication).

Because the aqueous solubility of A8–35 depends on the presence of ionized carboxylates, it is highly sensitive to their protonation, as well as to the presence of multivalent cations, in particular Ca^{2+} ions; MP/A8–35 complexes are similarly affected (19, 25, 33, 34, 64). A8–35 therefore should not be used below pH 7 (preferentially at pH ≥7.5) nor in the presence of millimolar concentrations of Ca^{2+} (Mg^{2+} is better tolerated; 64). Depending on the experiments at hand, a minor degree of aggregation may or may not be acceptable.

Critical aggregation concentration: the concentration of a polymer at which its molecules start to self-assemble

FAPol$_{NBD}$: A8–35
grafted with NBD

Stability: refers either
to physical (colloidal)
stability, i.e., particles
remain water-soluble
and do not aggregate
nor precipitate, or to
biochemical stability,
i.e., MPs remain in
their native state and,
when measurable,
functional

It can, however, become critical in radiation scattering experiments, where the presence of even minute amounts of small oligomers is highly detrimental (32). In NMR studies, working above pH 7 accelerates the exchange of solvent-exposed amide protons.

These experimental constraints have stimulated the development of alternative APols, such as PC-APols, NAPols, and SAPols (**Figure 1b–d**), which are insensitive to low pH and to the presence of Ca^{2+} ions (23, 25, 64). In aqueous buffers, SAPols and glucosylated NAPols form small, compact, globular, well-defined particles whose size and dispersity are comparable to those of the particles formed by A8–35 (6, 23, 83). As for A8–35, the size of NAPol particles seems insensitive to the length of the constitutive polymers (6).

Labeled and Functionalized Amphipols

The chemistry of APols offers rich opportunities for labeling and functionalization. To date, these modifications have been applied only to A8–35 or its close congener A8–75 (which differs from A8–35 by the absence of isopropyl grafts; 87), but they could be readily extended to other APols. Isotopically labeled APols include ^{14}C-labeled A8–75 and ^{3}H-labeled A8–35, which have been used mainly to follow the distribution of polymers during fractionation experiments and to quantify their binding to MPs (32, 68, 88). A form of A8–35 in which the polyacrylic acid chain is hydrogenated but the isopropyl and octyl groups are perdeuterated (34) has been heavily used for SANS, AUC, and NMR experiments (16, 18, 32–34, 99). Four APols functionalized with a fluorophore have been synthesized, namely a form of A8–75 carrying a naphtyl group (94) and three forms of A8–35 carrying a 7-nitrobenz-2-oxa-1,3-diazol-4-yl (NBD) group (FAPol$_{NBD}$) (100), a rhodamine group, or a fluorescein one (F. Giusti, unpublished data), respectively. They have been used particularly in FRET experiments (100) and for observing the distribution of APols in lipid vesicles (94), in

MP crystals (**Figure 6**), and in cells and whole organisms (**Figure 8**). A biotinylated version of A8–35 has been used to mediate the immobilization of APol-trapped MPs onto solid supports (21).

FORMATION, STRUCTURE, AND PROPERTIES OF MEMBRANE PROTEIN/ AMPHIPOL COMPLEXES

A vast body of information about MP/APol complexes is available. In this section, we examine how these complexes are formed; what is known of their composition, structure, and solution behavior; what effects APols have on the stability and functional properties of the MPs they complex; which methodological constraints they impart; which opportunities they offer; and how APol-trapped MPs can be transferred to other environments, be they detergent micelles or lipid bilayers.

Trapping Membrane Proteins with Amphipols

Because APols are poorly dissociating surfactants, they are not efficient at solubilizing biological membranes, even though they partition into them (19, 65). As a result, APols are generally unable to directly extract MPs from cell membranes, even though there are a few exceptions (68).

Usually, MPs are first extracted and purified in detergent by standard methods and then transferred to APols (87). In a typical trapping experiment, MPs in detergent solution are supplemented with aliquots from a concentrated stock solution of APols in water. The sample is briefly incubated, during which time APols mix with the detergent both in solution and at the hydrophobic transmembrane surface of MPs (89, 100). The MP/detergent/APol complexes thus formed can also incorporate endogenous lipids that had been solubilized along with MPs, or lipids added afterward, to increase the protein's stability and/or activity (see below). It is possible that, at this stage, lipids that had been displaced by the detergent rebind (see below). The detergent concentration is then lowered below the CMC, usually by one of the following two methods:

- The solution is diluted under the CMC of the detergent. In this case, the detergent is not physically removed from the solution, but most of it disperses in the aqueous phase as monomers and its concentration in the surfactant layer surrounding the protein drops. This method is sufficient, for instance, to restore native-like allosteric properties to the nicotinic acetylcholine receptor (nAChR), possibly due to lipid rebinding (54). It is also useful to screen trapping conditions, by modulating the MP/APol ratio used and determining by ultracentrifugation, for each condition, which fraction of the protein remains soluble after dilution of the detergent under its CMC (see below).

- Polystyrene beads (Bio-Beads SM2) are added to the sample. Detergent molecules adsorb to the beads, which are removed by low-speed centrifugation (99). The use of Bio-Beads has the advantage of not diluting the protein, and it is usually preferred for preparative protocols. APols (at least A8–35, but this is likely to hold for all APols) do not adsorb significantly to the beads, presumably because their particles do not enter the beads' narrow anfractuosities (98).

Whichever procedure is used, residual detergent molecules, if present, can be removed by dilution/concentration cycles, dialysis, or preparative SEC using surfactant-free buffer (32, 99).

The amount of APols required for efficient trapping is an important parameter that must be established early in any study. The optimal MP/APol ratio depends on the protein of interest and can be determined by carrying out quick trapping tests using a concentration range of APols, dilution below the CMC of the detergent, and ultracentrifugation (25, 83, 87). If the concentration of APols is too low, the protein

aggregates and pellets upon ultracentrifugation [note that with low-CMC detergents such as dodecylmaltoside (DDM), precipitation can be slow and incomplete even in the absence of APols and well below the detergent's CMC]. Upon increasing the APol/protein ratio, a first threshold is reached where most of the protein is kept soluble but MP/APol complexes appear heterogeneous upon SEC or AUC. It is only above a second threshold that MP/APols complexes become homogeneous. Further increasing the APol/MP ratio is useless and can be detrimental: It may increase the viscosity of the sample after concentration (e.g., in solution NMR studies) and compromise the stability of the protein (because it recreates the hydrophobic sink that the use of APols aims to abolish). The minimal mass ratio of A8–35 to MP for a small, deeply membrane-embedded protein such as bacteriorhodopsin (BR) is ~3 g APol per g BR (32). Trapping experiments are usually carried out at a somewhat higher ratio, 5 g APol per g BR. This ratio exceeds by a factor of ~2.5 the amount of APol that actually binds to BR, which is ~2 g APol per g BR (32). The same requirement for a moderate excess of APols is observed with MPs that, like the cytochrome bc_1 complex, feature a much higher ratio of mass to transmembrane surface and therefore bind much less APols per g protein (D. Charvolin, unpublished data).

There are at least two rationales to explain this need for an excess of APols over that amount that actually binds to the protein. The first one, hypothetical, is that the protein may select, in a population of heterogeneous APol molecules, those that contribute to form MP/APol complexes with the lowest free-energy level. The second rationale is well documented: Because APols are not strongly dissociating, they do not prevent the formation of small MP oligomers unless present in some excess over that amount actually bound by the proteins (32, 100).

All the ~30 polytopic and bitopic MPs tested thus far, which differ widely in their origin, secondary structure, topology, size, function, and oligomeric state, form water-soluble complexes with APols (67, 68). This reflects the fact that the process of association is a physical phenomenon that does not depend on details of the protein's transmembrane surface, but simply on it exposing a large enough hydrophobic area for APols to adsorb onto. To our knowledge, no attempts have been reported yet to stabilize monotopic or lipid-anchored MPs with APols.

Composition, Structure, Dynamics, and Solution Properties of Membrane Protein/Amphipols Complexes: Methodological Opportunities and Constraints

MP/A8–35 complexes have been studied by SEC, AUC, small-angle X-ray scattering (SAXS), SANS, FRET, solution NMR, and INS. Their properties have been reviewed recently (67), and only the most salient facts are recalled here. Small MPs with limited extramembrane regions such as BR (27 kDa; seven transmembrane helices plus, in most experiments, bound lipids) bind proportionately more A8–35 ($\sim$2 g per g protein) (32) than do large complexes with extended extramembrane domains such as cytochrome bc_1 (490 kDa; 22 transmembrane helices; $\sim$0.11 g A8–35 per g protein) (D. Charvolin, unpublished data). The amount of A8–35 bound to these two proteins roughly corresponds to half a dozen molecules of average molecular weight. Small transmembrane domains such as single α-helices can be expected to bind approximately one particle of APols, i.e., in the case of A8–35, $\sim$40 kDa. The A8–35 layer is compact (1.5 to 2 nm thick) and does not form an extended corona (32). In SEC, SAXS, SANS, AUC, or NMR experiments, MP/A8–35 complexes indeed behave like compact, globular particles (67). The only MP/APol contacts detected to date are with the transmembrane surface of the protein (18, 99).

Preparing homogeneous MP/A8–35 complexes requires some care (32, 100), and it is difficult to totally avoid the presence of minor fractions of small oligomers, a nuisance in radiation scattering experiments (32). Even oligomer-free preparations do not appear as narrowly distributed as MP/detergent complexes, unless some detergent is added (100). A tentative interpretation is that the large size and small numbers of adsorbed APol molecules make it impossible for the system to finely tune the volume of the adsorbed layer to that corresponding to the lowest free energy, generating differences in the exact amount of APol bound from one MP to the next. Detergent molecules would provide the small change needed to optimize the adjustment. Polydispersity is generated when the complexes are separated from the extra, free polymer present at the end of a trapping experiment (32, 100). The latter oligomerization is reversible upon adding back APols (100).

Conditions that provoke the aggregation of the polymer, such as, for A8–35, working at pH $\leq$ 7 or adding Ca^{2+} ions, also cause MP/A8–35 complexes to aggregate (17, 25, 32, 64, 99). If either of these conditions is essential to the experiment considered and monodispersity is required, other APols must be used.

Upon trapping with APols a mixture of MPs, e.g., the supernatant of solubilized membranes, MPs are trapped under the association state (e.g., monomers, oligomers, and supercomplexes) that was theirs in the detergent solution, and they retain bound lipids (1, 5, 31, 32, 54, 90). APols do not interfere with most purification techniques, such as SEC (6, 19, 32, 100), sucrose gradient fractionation (54, 87), or immobilized ligand (24) or metal (100) affinity chromatography. In the last case, however, it seems that polyhistidine tags fused too close to the transmembrane region may interact less efficiently with the column than in detergent solution, probably because of steric and/or electrostatic effects (100). Ion-exchange chromatography should be restricted to MP/NAPol complexes, as is isoelectrofocusing.

APols do not interfere with most light spectroscopy techniques (for a UV-visible absorbance spectrum of A8–35, see Reference 100), including fluorescence spectroscopy (100; M. Opačić, unpublished data), circular dichroism (CD) (30, 66, 96), and synchrotron radiation circular dichroism (23; T. Dahmane,

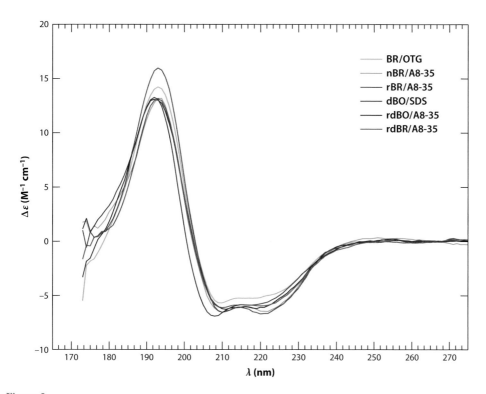

Figure 2

Synchrotron radiation circular dichroism (SR-CD) spectra of denatured and A8–35-refolded bacteriorhodopsin (BR). The spectra correspond to the following samples: BR/OTG (octylthioglucoside), native BR (nBR) solubilized from purple membrane (PM) in OTG (along with PM lipids); nBR/A8–35, nBR transferred from OTG to A8–35 (1:5 BR/APol mass ratio); rBR/A8–35, BR refolded in A8–35 from sodium dodecylsulfate (SDS)-solubilized PM; dBO/SDS, bacterio-opsin (BO) delipidated and separated from retinal in organic solvents (dBO) and transferred to SDS; rdBO/A8–35, dBO refolded in A8–35 in the absence of retinal; rdBR/A8–35, dBO refolded in A8–35 in the presence of retinal. SR-CD spectra were recorded at 25°C on the 3m-nim Bessy and DISCO SOLEIL beamlines (29, 74). They extend down to ~175 nm, clearly resolving the 190 nm π-π* transition of the exciton split originating from peptide bond electrons. Data from Reference 23 and T. Dahmane, M. Zoonens & F. Wien, unpublished data.

M. Zoonens & F. Wien, unpublished data) (**Figure 2**). Attempts have been made to apply Fourier transform infrared spectroscopy to measuring the rate of $^1H/^2H$ exchange at the amide bonds of A8–35-trapped BR. However, all current APols contain amide bonds, and the amide II region of protein spectra (1590–1510 cm^{-1}) is obliterated by APol signals (Y. Gohon & E. Goormaghtigh, unpublished data).

Functionality of Amphipol-Trapped Membrane Proteins

Functional and/or ligand-binding studies have been carried out with a dozen different APol-trapped MPs. These include the calcium ATPase (SERCA1a) from the sarcoplasmic reticulum (SR) (19, 64), the nAChR (21, 54), the BR (5, 23, 32, 66), the transporter enzyme II mannitol (EII^{mtl}) (M. Opačić, J.-L. Popot & J. Broos, unpublished data) and OmpF (68), both from *Escherichia coli*, as well as half a dozen G-protein-coupled receptors (GPCRs) including rhodopsin (68), the leukotriene receptors 1 and 2 (BLT1 and BLT2), the 5HT$_{4(a)}$ serotonin receptor, the CB1 cannabinoid receptor 1, and the GHSR-1a ghrelin receptor (16, 24 and J.-L. Banères, unpublished data). The data indicate that most proteins are still

SERCA1a:
fast-twitch skeletal muscle sarcoplasmic reticulum Ca^{2+}-ATPase

functional after transfer to APols and that the binding of small ligands, toxins, and antibodies is unimpaired compared to that observed in biological membranes.

BR, an archaebacterial light-driven proton pump that accumulates in the so-called purple membrane of *Halobacterium salinarium*, accomplishes its entire photocycle after trapping in A8–35 (32) or in NAPols (5). Compared with the kinetics observed in purple membrane, the first steps of the photocycle are accelerated whether BR is solubilized in detergent or trapped in APols. The last steps feature similar kinetics in APol and in purple membrane, whereas in detergent solution the return to the fundamental state is slower. The kinetics of the photocycle is the same whether BR has been trapped in its native state or refolded in A8-35 from sodium dodecylsulfate (SDS)-solubilized purple membrane, i.e., in the presence of purple membrane lipids (66). On the contrary, when BR is refolded in A8-35 in the absence of lipids, late kinetics is similar to that in detergent solution (23). These observations suggest that the restoration of native-like protein-lipid interactions upon transferring BR from detergent to APol might be responsible for the recovery of purple membrane-like kinetics (23).

The allosteric equilibrium between the resting and desensitized states of the nAChR is strongly perturbed when receptor-rich postsynaptic membranes from the electric organ of *Torpedo marmorata* are solubilized in detergent (20). Membrane-like equilibrium and transition kinetics are recovered when the solubilized preparation is supplemented with A8-35 and diluted below the CMC of the detergent (54). As for BR, one possible explanation is that transfer to APols allows lipids to rebind to critical allosteric sites at the transmembrane surface of the receptor.

Recent studies of five GPCRs folded in APols indicate that their affinity for their ligands is similar in the membrane and in APols. However, as previously observed with rhodopsin (68), G protein activation by A8–35-trapped BLT1 is significantly slowed down compared to that observed in lipid/detergent

mixtures. In contrast, when folded in NAPols, BLT1 catalyzes GDP→GTP exchange on the $G_{\alpha i}$ subunit with kinetics similar to those in lipid/detergent mixtures, suggesting that electrostatic repulsion may account for the slowing down observed in the presence of A8-35 (J.-L. Banères, unpublished data).

Diacylglycerolkinase (DAGK) was fully active when transferred to APol PMAL B-100 (59). Other, less extensive functional studies with such MPs as the maltose transporter from *E. coli* or photosystems I and II are reviewed in Reference 68. The binding of antibodies by APol-trapped MPs is discussed below.

A particularly interesting case is that of SERCA1a, which has been studied in some detail (19, 64) and has been reviewed previously (67, 68). In brief, several APols (A8–35, PMAL C-12, PMAL*A* C-12, and SAPols) both protected SERCA1a against inactivation and slowed down the overall turnover (ATP hydrolysis) and individual steps in the catalytic cycle (release of trapped Ca^{2+}) (19, 64). Intermediate effects are observed in mixtures of APols and detergent. Trivial mechanisms, such as sequestering of Ca^{2+} or interference with ATP binding, do not account for the inhibition (19). Our current hypothesis about the origin of this phenomenon relies on the kinetics of reorganization of APols around membrane proteins when the latter undergo conformational changes that affect their transmembrane surface. Small (subnanometric) conformational changes, such as those that affect or may affect BR (40) and the nAChR (22, 39), can probably be accommodated by displacements of the APol's alkyl chains, which are expected to be rapid (submicrosecond) and to not involve a high free energy of activation. Larger interfacial movements (nanometer), such as those undergone by the transmembrane helix bundle of SERCA1a upon transiting between the E1 and E2 states (61), may cause a reorganization of the polymer's backbone, which could entail a higher free-energy penalty in APol than in detergent and thereby slow down the enzymatic cycle. As discussed below, this mechanism (called the Gulliver effect; 19, 68) may well contribute to the protective effect of APols against MP denaturation.

Biochemical Stability of Amphipol-Trapped Membrane Proteins

As a rule, transferring a MP from a detergent solution to APols improves its stability, often dramatically (67, 68). In order to understand the underlying mechanism(s), one should pay attention to the volume of nonmonomeric surfactant to which the protein is exposed to distinguish effects that merely result from a decrease of the hydrophobic sink from those that reflect the interactions of the protein with the surfactant layer that surrounds its transmembrane region.

The effects of trapping a fixed quantity of BR with increasing amounts of APols have been examined in detail (23). Native BR was extracted from purple membrane along with purple membrane lipids (32). Upon trapping it with A8–35 at BR/APol mass ratios ranging from 1:5 to 1:50, ternary BR/lipids/APol complexes formed. Control samples were stored in either 18-mM or 25-mM octylthioglucoside. Under the experimental conditions used, the volume of the hydrophobic sink was roughly comparable in 18- or 25-mM octylthioglucoside and at BR/A8–35 ratios of 1:10 or 1:20, respectively. Even at 40°C, BR is highly stabilized by A8–35, denaturing by <10% over a week (**Figure 3**). Under the same conditions, the protein in octylthioglucoside is totally inactivated in less than a day. Whereas a large excess of APols (ratio 1:50) is well tolerated at 4°C and at room temperature, at 40°C it negatively affects the stability of BR, approximately one-third of which is denatured after 6 days (**Figure 3**). This effect is most likely a consequence of delipidation, which is favored by increasing the volume of the hydrophobic sink (23). Inactivation is nevertheless slow compared

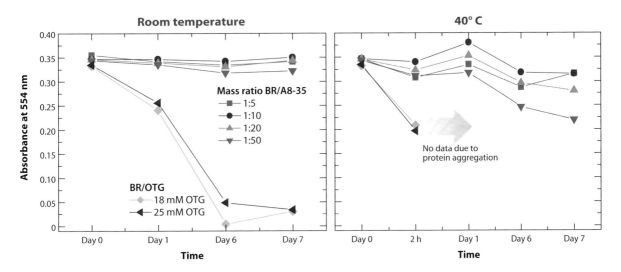

Figure 3

Time stability of bacteriorhodopsin (BR) in amphipol A8–35 versus octylthioglucoside (OTG). BR was extracted with OTG from *Halobacterium halobium* purple membrane (PM), along with PM lipids, trapped in A8–35 (32) at various BR/A8–35 mass ratios, and stored in the dark either at room temperature or at 40°C in a buffer containing 100 mM NaCl and 20 mM sodium phosphate, pH 7.0 ([BR] = 0.22 g·liter^{-1}). Its absorbance at 554 nm, which is proportional to the concentration of the holoprotein, was followed as a function of time. Control samples were kept in 18- or 25-mM OTG (total OTG concentration, including bound detergent). These two concentrations correspond to roughly the same mass concentration of free detergent as that of free A8–35 particles in the samples trapped respectively at 1:10 and 1:20 BR/A8–35 mass ratios. The absence of data points in OTG at 40°C past 2 h is due to the aggregation of the protein, accompanied by complete bleaching. Adapted from Reference 23.

LTB$_4$: leukotriene B$_4$

MOMP: major outer membrane protein from *Trachomatis pneumoniae*

to that in octylthioglucoside. Thermostability is also improved: A8–35-trapped BR stands well being exposed at 60°C for 20 min, whereas BR in octylthioglucoside denatures at 40°C (23). BR/A8–35 preparations at a mass ratio of 1:5 (i.e., with ~3 g free APol per g BR) do not exhibit any denaturation even after six months of storage at 4°C (32). BR/A8–35 complexes can be frozen and thawed without denaturing the protein (32), but they cannot be lyophilized (Y. Gohon, unpublished data).

BLT1, one of the receptors of leukotriene B$_4$ (LTB$_4$), is significantly stabilized by A8–35. After 25 days at 4°C, no loss of activity is observed whether or not lipids are present. Under the same conditions, BLT1 kept in fos-choline-16/asolectin solution loses about half of its activity over the same period. In detergent/lipid mixed micelles, BLT1 denatures at ~27°C, whereas it is stable up to ~35°C when trapped in pure A8–35 and up to ~39°C in the presence of A8–35 plus lipids (24).

Data about the relative ability of various APols to stabilize MPs remain scattered, but they seem to point in at least one direction: The less charges APols bear, the more stabilizing they are. Thus, BR is more stable when trapped in A8–35 than in A8–75, which is a similar polyacrylic acid–derived APol that carries ~75% free carboxylate groups rather than ~35% (87; C. Tribet, unpublished data). Similarly, SERCA1a is more stable in A8–35 than in SAPols, which also carry ~75% of charged groups (64). Cytochrome b_6f, a highly detergent-sensitive complex (13), is not very stable in A8–35 (5, 87), but much more so in NAPols (5, 70). The stability of BR in SAPols is much greater in the presence of 100 mM NaCl than in its absence (23). A simple rationale for these effects is that, as may be the case for detergents, electrostatic repulsion between net charges carried by the surfactant bound to a MP tends to favor the formation of particles with a small radius of curvature. This would drive the opening of the protein's structure or, in the case of a multisubunit assembly such as cytochrome b_6f, fragmentation.

Little information is available on the stability of APol-trapped β-barrel MPs. On the one hand, A8–35-trapped OmpA, a monomeric protein, denatures at lower urea concentrations than in the detergent lauryldimethylaminoxide, perhaps a consequence of electrostatic repulsion between the protein and the APol at pH 10 (J.H. Kleinschmidt, unpublished observations). On the other hand, the major outer membrane protein (MOMP) from *Chlamydia trachomatis*, a trimeric porin, is considerably thermostabilized (by >30°C) when transferred from Zwittergent 3–14 to A8–35 (M.J. Cocco, unpublished data). Thus, whether APols stabilize β-barrel MPs seems to depend on the protein, the method of denaturation, or both.

The study of SERCA1a has yielded highly valuable information on the mechanisms underlying the protective effects of APols (19, 64). Three points are particularly worth noting. First, the mere addition of A8–35 to a solution of sarcoplasmic reticulum in $C_{12}E_8$ or DDM markedly improves the stability of the ATPase, even though it increases the volume of the hydrophobic sink, which should drive delipidation (19). This is yet another indication that an APol environment is intrinsically less denaturing than a detergent environment. Second, as discussed above, trapping with APols reversibly inhibits the activity of the ATPase, to an extent that depends on the nature of the APol and on the presence or absence of detergent (19, 64). Third, there is a correlation between the degree of inhibition and the degree of protection against denaturation: The more stabilizing the environment is, the more inhibitory. Thus, SERCA1a surrounded by a mixture of A8–35 and $C_{12}E_8$ is both partially inhibited and partially stabilized (19), as is SERCA1a trapped in SAPols rather than in A8–35 (64). These observations suggest that the same mechanism may underlie the two phenomena. The high sensitivity of calcium-free SERCA1a to detergents (19 and references therein) can probably be attributed to the detergent prying apart the transmembrane helix bundle (57), which is stabilized by Ca^{2+} ions (86). One may speculate that the Gulliver effect suggested above slows down both the natural conformational changes that take place during the enzymatic cycle and the nonnatural conformational excursions that open the structure and initiate denaturation, which could explain the correlation between the two effects.

Transferring MPs from Amphipols to Detergent Solutions or Membranes

APols bind to the hydrophobic surface of MPs in a noncovalent way, but, thanks to their multiple contact points, with an extremely slow dissociation rate. This makes their association with MPs permanent in the absence of a competing surfactant (68, 88, 89, 100), even at extreme dilutions (100). Nevertheless, APols do desorb from MPs in the presence of an excess of competing surfactant, be it free APols (88, 100), detergents (88, 89, 100), or lipids (59, 66).

The displacement of MP-bound APols by free APols was first evidenced upon centrifuging MPs trapped with $[^{14}C]A8–75$ through a sucrose gradient containing unlabeled A8–75 (88). Subsequently, the exchange rate was determined by measuring the decrease of the FRET signal between tOmpA (the transmembrane domain of E. coli OmpA) and $FAPol_{NBD}$ upon addition of an excess of unlabeled A8–35 (100). At low ionic strength, the kinetics of exchange extends over tens of hours. In the presence of 100 mM NaCl, the exchange is essentially over after 10 min. These observations are consistent with a mechanism involving collisions between MP/APol complexes and free APol particles, probably followed by fusion, mixing, and fission. Such a mechanism is also suggested by many other observations: (a) The concentration of nonassembled A8–35 molecules is extremely low (see above), making it highly improbable that they can mediate the exchange between free and bound APol. (b) Retinal, a hydrophobic molecule that does not partition significantly in water, can be delivered from A8–35 particles to bacterio-opsin (the apoprotein) refolded beforehand in the same APol (23). (c) A8–35-trapped tOmpA reversibly oligomerizes upon elimination of free APol particles (100).

APols and detergents mix rapidly and efficiently (25, 26). Isothermal titration calorimetry measurements indicate that neutral detergents (DDM, octylthioglucoside, C_8E_4) partition in anionic (A8–35) or cationic (C22–43) APols, and their mixing is quasi-ideal (26). This result can seem surprising given that neutral and charged detergents do not mix ideally. Charged detergents are typically prone to form smaller assemblies than neutral detergents (and they facilitate the dispersion and denaturation of proteins) because of the coulombic repulsion between their ionized head groups. In ionic APols, however, the charged groups are covalently linked to one another. Covalent association balances the repulsion between vicinal ionic groups, which cancels the coulombic effects at short length scales.

When $MP/FAPol_{NBD}$ complexes are mixed with a nonionic detergent (DDM, $C_{12}E_8$) in a stopped-flow instrument, the kinetics of displacement of MP-bound APols, as followed by FRET, is extremely rapid (<1 s) (100). The thermodynamics of the exchange of A8–35 and C22–43 for octylthioglucoside or C_8E_4 has been studied in detail by isothermal titration calorimetry (89). The exchange is isoenthalpic and entropy-driven. Fluorescence and isothermal titration calorimetry data converge toward the conclusion that APols mix with MP/detergent complexes and MP-free detergent micelles nearly ideally, so that at equilibrium the layer of MP-bound surfactant and the free mixed micelles have almost the same composition (89, 100). This is an important observation because it implies that APols have no special affinity for MP transmembrane surfaces. Isothermal titration calorimetry studies also provide insights into the origin of the difference of behavior of MP/APol versus MP/detergent complexes at low surfactant concentration. Upon dilution of a solution of MP/detergent complexes below the CMC of the detergent, entropy favors the aggregation of the protein because it frees hundreds of detergent molecules as monomers. APols, on the contrary, stick to MPs even at extreme dilutions (100), because their desorption entails no

tOmpA:
transmembrane domain of OmpA from Escherichia coli

Folding: refers to a MP acquiring the 3D structure of the native protein, without having ever been correctly folded previously

significant entropy gain (for a discussion, see Reference 89).

APols can also be displaced by lipids. Two studies have described the spontaneous delivery of APol-trapped MPs to preformed lipid vesicles and black lipid films. In the first study, DAGK was delivered from complexes with the APol OAPA-20 to palmitoyloleoylphosphatidylcholine vesicles while retaining its ability to phosphorylate dibutyrylglycerol (59). The second study reported the transfer of two A8–35-trapped β-barrel MPs, OmpA and FomA, to black lipid membranes (66). After transfer, the two proteins formed channels with properties similar to those observed after reconstitution from a classical detergent solution.

At concentrations sufficient to deliver MPs, APols are neither cytolytic nor toxic when they are applied to living cells or injected into mice (68, 82; D. Bagnard, G. Crémel, L.M. de la Maza, unpublished observations). This opens onto extremely interesting basic and biomedical applications, where APols are used to deliver MPs, MP fragments, or other hydrophobic molecules to cell membranes.

There are two caveats to APol-mediated delivery of MPs to preexisting membranes. First, this type of transfer can be expected to be a high-risk event for the protein, as some of its hydrophilic extramembrane regions have to cross the membrane, which necessarily exerts strong distorting forces on the structure. DAGK, FomA, and OmpA are robust proteins, and in the experiments mentioned above (59, 66), it is not known which fraction of them actually adopt a transmembrane orientation without being denatured. This somewhat brutal approach is unlikely to result in high yields of functional insertion with more fragile proteins, unless they refold easily. Second, the APols with which the protein was initially trapped can be expected to remain associated with the target membrane while most likely diffusing away from the protein. When A8–35-refolded OmpA and FomA were applied to black lipid films, the properties of the channels they formed indeed appeared perturbed unless A8–35 was also added on the *trans* side

of the film (66). This suggests that adsorbed APol molecules affect the electrostatics and/or the internal pressure gradient of the lipid bilayer (66). Other studies have shown that APols by themselves can form pores in natural (19, 82) or artificial (51, 82, 92, 93) membranes. As shown below, when FAPol$_{NBD}$-trapped transmembrane peptides are delivered to COS cells, the polymer integrates the plasma membrane along with the peptides (see **Figure 8**). Vigilance regarding possible artefactual effects is therefore in order.

APPLICATIONS

The original idea behind the design of APols was that, by trapping MPs under a water-soluble form more stable than that in detergent solutions, APols would make it easier to purify them and to explore in vitro their structural and functional properties. As the study of APols progressed, however, it became apparent that their applications extended beyond this original concept. In the present section, we examine those applications of APols that have been tested and either validated or found problematic, along with an assessment of their prospects and/or limitations.

Folding Membrane Proteins to their Native State

Overexpressing MPs as inclusion bodies allows production of large amounts of protein, which is hard to achieve when directing them to cell membranes. However, folding to their native state the inactive MPs obtained in this way are difficult. To date, in vitro folding of MPs in detergents, detergent/lipid mixtures, or lipid vesicles has been achieved for a dozen of β-barrel MPs (14) and about as many α-helical MPs. The latter include BR (42, 69); the major light-harvesting complex II of higher plants (75); DAGK (37); the small multidrug transporter from *E. coli*, EmrE (58); and a few GPCRs (2–4, 47). Conditions for efficient folding tend to be highly idiosyncratic.

The feasibility of using APols to assist MP folding was first established using as models two β-barrel MPs, OmpA and FomA, from the eubacteria *E. coli* and *Fusobacterium nucleatum*, respectively, and an α-helical MP, BR (66). Following dilution of urea-unfolded OmpA or FomA into urea-free buffer in the presence of A8–35, both proteins developed β-sheet secondary structure, became protected from proteolysis by trypsin, and recovered the electrophoretic mobility during sodium dodecyl sulfate-polyacrylamide gel electrophoresis (SDS-PAGE) that is characteristic of the folded β-barrel (66). The folding yields were near 100%. Upon application of the refolded MPs to black lipid films, single-channel recordings showed the formation of functional pores (66). OmpA has also been refolded in NAPols, albeit with much slower kinetics (J.H. Kleinschmidt, unpublished data). tOmpA has been renatured in SAPols, and the recovery of the native structure has been demonstrated by solution NMR (23).

When a solution of denatured BR in SDS solution is supplemented with A8–35 in various mass ratios and dodecyl sulfate precipitated as its potassium salt (PDS), the purple color of native BR starts to develop within minutes. Following removal of PDS crystals by centrifugation and overnight dialysis, UV-visible absorption spectra indicate that BR renaturation is complete at BR/A8–35 wt/wt ratios of either 1:5 or 1:10. The synchrotron radiation circular dichroism spectrum of refolded BR is indistinguishable from that of the native protein (**Figure 2**) (23). Upon illumination, A8–35-refolded BR undergoes its full photocycle (23, 66). Similar results have been obtained in NAPols (5).

The ability of APols to induce and stabilize the native fold of MPs has been further explored using a series of GPCRs from *E. coli* inclusion bodies purified under denaturing conditions (in SDS solutions). Conditions initially established to refold BR were applied virtually without changes to folding the LTB$_4$ receptors BLT1 and BLT2, serotonin receptor 5-HT$_{4a}$, cannabinoid receptor CB1 (24), and

ghrelin receptor GHSR-1a (J.-L. Banères, unpublished data). Folding yields of 40%–70% were systematically achieved, based on ligand-binding experiments. For all five receptors tested, addition of exogenous lipids (in 1:5 mass ratio to the APol) significantly increased the refolding yields (e.g., from ~50% to ~70% for BLT1). The pharmacological properties of APol-folded receptors, including the K_d values for agonists, are close to those observed in either detergent solutions or membrane fractions. As observed for most APol-trapped MPs, GPCRs folded in A8–35 are significantly more stable than those kept in lipid/detergent mixtures (24). Folding and stabilization of the BLT2 receptor in A8–35 have made it possible to determine by solution NMR the structure of BLT2-bound LTB$_4$. The BLT1 and GHSR-1a receptors have also been folded in NAPols, with yields similar to those achieved in A8–35 (J.-L. Banères, unpublished data).

Cell-Free Synthesis of Membrane Proteins

In vivo overexpression of MPs has to navigate the Charybdis of limited expression under a functional form and the Scylla of poor folding yields of mass-produced but inactive protein. An alternative approach that has attracted much attention over the past few years is to express target MPs in vitro, using a cell lysate that contains the machinery for transcribing and translating genes carried by appropriate plasmids (84, 101). Cell-free synthesis presents the advantages of doing away with toxicity issues and of lending itself to straightforward labeling using limited amounts of labeled amino acids (48). MPs, however, will precipitate if synthesized in the absence of any surfactant. Three main strategies have been described: (*a*) expression in the absence of surfactant, followed by solubilization of the precipitate with a detergent; (*b*) expression in the presence of a detergent; and (*c*) expression in the presence of a lipid bilayer, in the form of either lipid vesicles or nanodiscs (15, 44, 46). The third approach

Refolding: refers to MPs that were initially obtained under their native state and have been denatured using urea, SDS, or organic solvents, prior to being brought back to their initial 3D structure

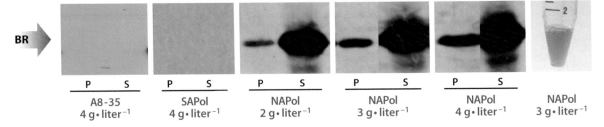

BR

P	S	P	S	P	S	P	S	P	S		
A8-35		SAPol		NAPol		NAPol		NAPol		NAPol	
4 g·liter⁻¹		4 g·liter⁻¹		2 g·liter⁻¹		3 g·liter⁻¹		4 g·liter⁻¹		3 g·liter⁻¹	

Figure 4

Amphipol-assisted cell-free expression of bacteriorhodopsin (BR). In vitro synthesis of polyhistidine-tagged BR in the presence of retinal and of A8–35, SAPols, or NAPols at the indicated concentrations. At the end of the synthesis, the samples were centrifuged at $16,000 \times g$ for 20 min. The proteins present in the pellet (P) and supernatant (S) were separated by SDS-PAGE on a 12% polyacrylamide gel and detected using an anti-His-tag antibody. Far right panel: BR after purification on a nickel column. The purple color of the solution indicates the presence of native BR (representing at least two-thirds of the protein present in the purified sample). Adapted from Reference 5.

suffers from two disadvantages: The amount of protein that can be thus produced tends to be limited by the small volume of lipids that can be made available, and not all MPs insert correctly into preformed bilayers in the absence of the translocon machinery (50). The first two approaches involve the use of detergents. Resuspending precipitated MPs does not always yield a functional protein (49). Expression in the presence of detergent is a priori a hazardous, and certainly not a universal approach: Given that many, if not most, fully folded MPs become unstable upon being exposed to detergents, it is to be expected that many of them will not reach their native state when synthesized in their presence. Indeed, while many MPs can be expressed in this way (81), demonstrations of their functionality are rare (7, 45).

It is therefore of great interest to examine whether less aggressive surfactants are compatible with cell-free synthesis (11, 62, 63, 72). Both A8–35 and SAPols, however, inhibit the synthesis of MPs (**Figure 4**), without affecting that of a control soluble protein, the green fluorescent protein (63). It may be that these APols, both of which are polyanions, inhibit translation by simultaneously binding to MP hydrophobic segments as they appear out of the ribosome tunnel and interacting with positively charged patches at the surface of the ribosome. This hypothesis is supported by the observation that excellent results are obtained when BR is expressed in

vitro in the presence of NAPols: Not only is the yield of synthesis (∼0.4 g·liter⁻¹ in the presence of 3 g·liter⁻¹ NAPol) higher and more reproducible (5) than that observed in the presence of DDM (63), but a majority of BR (∼90%) is present in a soluble form (**Figure 4**) and remains so over several months (5), whereas in DDM it tends to precipitate (63). BR expressed in the presence of NAPols and retinal folds to a greater extent (at least two-thirds) than in detergent, as shown by the appearance of the purple color characteristic of the holoprotein (**Figure 4**) (5). Preliminary data indicate that NAPol-assisted cell-free synthesis also applies to GPCRs, as exemplified by the leukotriene receptor BLT1 (E. Billon-Denis, F. Zito, J.-L. Banères, unpublished data).

NMR

Early NMR experiments were aimed at better characterizing MP/APol complexes and at exploring the resources and limitations of NMR to study their structure and dynamics. Three small, well-characterized β-barrel proteins from the outer membrane of either *E. coli* (tOmpA and OmpX) or *Klebsiella pneumoniae* (KpOmpA) were chosen as models. These three proteins, whose 3D structures are known, have been extensively studied by solution NMR, mostly in the presence of detergent. NMR studies of their complexes with

KpOmpA:
transmembrane domain of OmpA from *Klebsiella pneumoniae*

APols showed them to either retain or regain their native fold when associated to A8–35 (17, 99; M. Renault & A. Milon, personal communication), to SAPols (23), or to NAPols (5). The complexes appear slightly bigger than MP/detergent complexes, based on estimates of the overall correlation time, τ_c. For instance, the correlation time of A8–35-trapped OmpX (31 ns) (17) lies between that of OmpX in solutions of dihexanoylphosphatidylcholine (24 ns) (52) and that of OmpX inserted into small isotropic bicelles (35 ns) (53). The slightly larger size of MP/APol complexes compared to those formed with the most favorable detergents does not preclude obtaining high-resolution spectra nor achieving enough sensitivity to perform 3D experiments. Indeed, 3D ^{15}N-edited [^{1}H,^{1}H]-HSQC-NOESY spectra have been collected to explore dipolar interactions between KpOmpA and A8–35 (M. Renault & A. Milon, personal communication). Comparative 2D [^{15}N,^{1}H]-TROSY spectra of BR in detergent solution, in nanodiscs, and in A8–35 have been recently published (73).

A drawback of A8–35 for NMR studies is the need to work at pH >7 in order to avoid aggregation. A high pH renders the observation of exchangeable protons more difficult, if not impossible in some cases, when the chemical exchange becomes too fast (17, 99). This has been one of the incentives leading to the development of pH-insensitive APols, such as SAPols and NAPols, both of which make it possible to record NMR spectra under acidic conditions (5, 23). These new APols, however, suffer from other disadvantages: The purification of SAPols is demanding, which currently limits the amounts that can be conveniently produced (23), and NAPols, due to their complex chemical structure (**Figure 1c**), cannot be easily perdeuterated.

NMR spectroscopy has been used to examine the organization of MP/A8–35 complexes, yielding an increasingly detailed view of protein-polymer interactions. A general description of the distribution of the alkyl chains was first obtained, taking advantage of the variation of dipolar environment experienced by the amide protons of tOmpA upon trapping with either A8–35 or its deuterated homolog (99). Two further studies exploited intermolecular ^{1}H-^{1}H or ^{13}C-^{1}H dipole-to-dipole cross-relaxation phenomena. They yielded a qualitative detection of spatial proximities between ^{13}C atoms in the side chain of aromatic residues of OmpX and ^{1}H nuclei of A8–35 (18) and the identification of hydrophobic contacts between specific amide protons of KpOmpA and octyl and isopropyl chains of A8–35 (M. Renault & A. Milon, personal communication). The three studies showed that contacts with the alkyl chains of A8–35 are confined to the transmembrane, hydrophobic surface of the proteins. ^{1}H/^{2}H exchange measurements show that some of the amide protons of the membrane-spanning region of OmpX exchange much more readily than others, which likely reflects the dynamics of the barrel (17).

Application of solution NMR to GPCRs is rare, mainly due to the difficulty of producing sufficient amounts of functional receptors (10 and references therein). APol-assisted folding of BLT2 has made it possible to solve by NMR the structure of its natural ligand, LTB$_4$, in its receptor-bound state. Perdeuterated BLT2 receptor was overexpressed in *E. coli* as inclusion bodies and folded to its native state (24) in deuterated A8–35. The high-resolution structure of the BLT2-bound leukotriene was derived from transferred nuclear Overhauser effect signals (16). The ligand undergoes a drastic reorganization upon binding, switching from an elongated structure with many coexisting rotamers to a highly constrained conformation (**Figure 5**). This information is of great interest from a pharmacological point of view, because it opens new perspectives in BLT-targeted drug design (no antagonists of this LTB$_4$-mediated proinflammatory pathway are available today for clinical use). It also contributes to a detailed analysis of the molecular events that lead to receptor activation. Because the ability of APols to fold and stabilize MPs seems quite general, the same strategy can probably be transposed to many other systems.

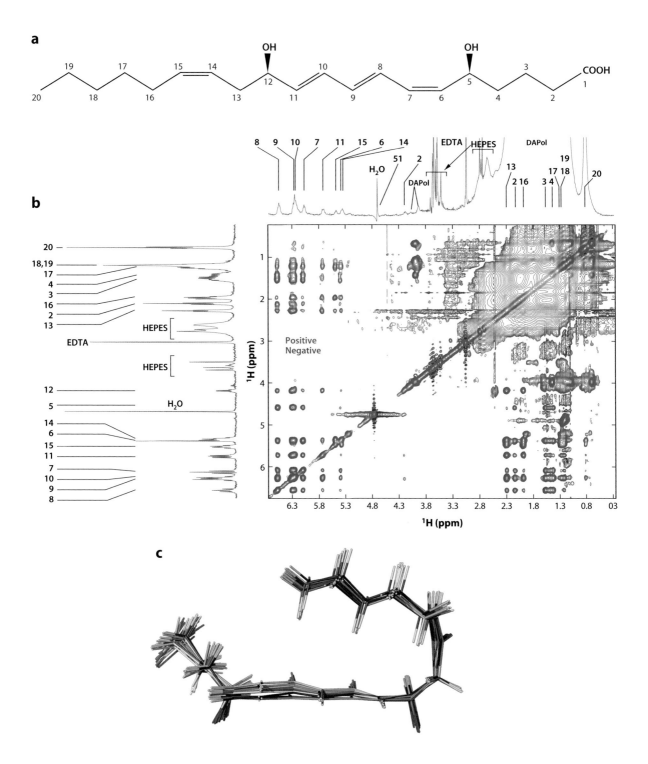

Immobilizing Membrane Proteins onto Solid Supports for Ligand-Binding Studies

The development of sensitive detection methods and the implementation of high-throughput technologies increasingly resort to the immobilization of functional proteins onto surfaces. On the one hand, protein arrays combined with optical detection provide an efficient mode of high-throughput screening, with a particular potential in diagnostics (95). On the other hand, surface-sensitive techniques such as surface plasmon resonance (97) and fluorescence microscopy provide information regarding the kinetics of interaction, in addition to the classical thermodynamic parameters (91).

Because of the poor stability of purified MPs in detergent solutions, many current studies are carried out in native (41, 55) or reconstituted lipid membranes (9, 78), which entails a series of technical constraints: nonspecific binding to other proteins and/or lipids, the difficulty of working out and standardizing protocols for MP reconstitution and for the immobilization of membrane fragments or vesicles, and the impossibility to access simultaneously the extracellular and intracellular domains of MPs.

Reversible or irreversible affinity tags are generally used to interface MPs with surfaces without disturbing the protein's function. Tags are typically added by fusing to the target MP a short peptide or a cargo-protein (38, 85). This approach is not applicable to proteins extracted from native tissues. In that case, the protein is chemically modified, usually on NH_2 or SH groups.

APols provide a novel, highly versatile way to immobilize MPs. Its advantages over the above approaches are many:

BAPol: biotinylated A8–35

- Trapping with functionalized APols does not alter the structure of the protein, and it leaves its N and C termini free for protein-ligand interactions or for other modifications.
- The protein can be trapped with a functionalized APol after its purification, during folding, or during cell-free synthesis.
- Although the MP/APol association is highly stable, it can be displaced by detergents or other APols. Affinity tags can thus be added or removed as desired.
- Because MPs are typically surrounded by at least five to six molecules of APols, several functional groups can be combined in a controlled stoichiometry, by the simple device of trapping the protein with a mix of appropriately functionalized APols.
- The MP benefits from the stabilization afforded by APols.
- The MP can (and, as a rule, must) be handled in surfactant-free solutions, which facilitates the handling of fluidics and detection using surface-sensitive techniques such as surface plasmon resonance.

It should be noted, however, that the location of the functional groups in the APol belt cannot be controlled and vary from one protein to the next, as well as over time, which, depending on the experiment, can be considered either a liability or an asset.

Whereas several APols carrying affinity tags have been or are being synthesized, biotinylated A8–35 (BAPol) is the best characterized to

Figure 5

Amphipol-assisted folding and stabilization of a G-protein-coupled receptor allow NMR determination of the structure of a receptor-bound ligand. The perdeuterated BLT2 receptor of proinflammatory leukotriene B_4 (LTB_4) was expressed as inclusion bodies in *Escherichia coli*, solubilized in SDS, and folded in deuterated A8–35 (DAPol). The structure of BLT2-bound LTB_4 was determined by analyzing transferred nuclear Overhauser effect signals. (*a*) Chemical structure of LTB_4. (*b*) Dipolar interactions in an LTB_4/[u-^{2}H,^{15}N]BLT2/DAPol sample, 2D NOESY spectrum (mixing time $\tau_m = 0.5$ s, $\nu_H = 600$ MHz, $25°$C). The corresponding 1D ^{1}H spectrum is shown above the 2D spectrum. The 1D spectrum of free LTB_4 in solution is displayed on the left side. Numbers refer to the protons annotated on the chemical structure of LTB_4 in panel *a*. (*c*) Three-dimensional structure of LTB_4 bound to BLT2 (hydrogen atoms are light gray, oxygen atoms are red; and carbon atoms are assigned a different color for each conformer). Adapted with permission from Reference 16. Copyright 2010 American Chemical Society.

date. As described above, trapping with APols as a rule does not interfere with ligand binding. The immobilization of BAPol-trapped MPs of various sources and sizes to streptavidin-coated chips or beads does not impair their function, e.g., the binding of a toxin and of a small acetylcholine analog to the nAChR, or the photocycle of BR, nor does it prevent the specific recognition of antibodies by tOmpA, BR, and cytochromes bc_1 and b_6f (21). The binding of antibodies and/or toxins to APol-trapped MPs has also been observed for OmpF (68) and MOMP (M.J. Cocco & L.M. de la Maza, unpublished data). APol-mediated MP immobilization thus appears as a promising system for diagnostic assays, as well as for functional investigations of ligand and protein interactions with MPs, with potential applications in drug screening. Prospects include the use of more sophisticated supports (electrodes, nanoparticles, and other nano-objects) and of other tagged APols. Histidine-tagged (F. Giusti & P. Kessler, unpublished data) and oligonucleotide-carrying APols (C. Le Bon & F. Giusti, unpublished data), for instance, would provide for the reversibility of the immobilization and, in the latter case, targeting, with a strong potential for the development of protein arrays. Tagged NAPols would reduce the probability of electrostatically driven nonspecific binding.

X-Ray Crystallography

The application of APols to MP crystallography would provide fascinating perspectives. This goal, however, may seem a priori hard to reach, because MP/APol complexes appear more polydisperse than MP/detergent complexes. In the case of charged APols, this difficulty is compounded by the need to achieve the right balance between minimizing the electrostatic repulsion between APol layers and preventing aggregation. Over the past few years, however, considerable progress has been made toward identifying factors that favor the formation of monodisperse preparations of MP/APol particles. Furthermore, repulsive interactions between MP/A8–35 complexes can be cancelled

or even turned to attractive interactions by appropriately modulating the ionic strength of the solutions (68).

Current attempts have focused on a model MP, mitochondrial cytochrome bc_1, which combines the advantages of being sturdy, colored, relatively easy to crystallize, and of featuring large extramembrane domains that provide anchoring points for the formation of well-ordered 3D contacts while keeping protein-bound surfactant layers away from each other (8). A8–35 was added to purified, detergent-solubilized cytochrome bc_1, and the detergent adsorbed onto polystyrene beads. Crystallization trials were carried out by vapor diffusion. Under the many crystallization conditions tested, complete elimination of the detergent precluded any crystals from forming. On the contrary, ternary cytochrome bc_1/A8–35/detergent complexes did crystallize, in the presence of concentrations of salt and polyethylene glycol higher than is required for crystallization from detergent solution (D. Charvolin, M. Picard, E.A. Berry & L.-S. Huang, unpublished data). The red color of the crystals indicated that they do contain cytochrome bc_1 (**Figure 6a**). The presence of APols in the crystals was established by growing them in the presence of FAPol$_{NBD}$ (**Figure 6b**). A few of these crystals have been tested on a synchrotron X-ray source and proved to diffract, with unit cell dimensions consistent with their being protein crystals. The current resolution of the patterns, however, is poor (>20 Å) (**Figure 6c**). It is not known whether this reflects an intrinsic property of the crystals, arises from problems with handling and/or freezing them, or is due simply to the limited number of trials that could be carried out. There are several possible reasons why cytochrome bc_1/A8–35/detergent complexes crystallize whereas bc_1/A8–35 ones do not. One of them is that the presence of detergent lowers the electrostatic charge density of the surfactant layer, and it may allow for easier charge redistribution. Another is that, as observed with tOmpA (100), ternary complexes may be more closely monodisperse than binary ones.

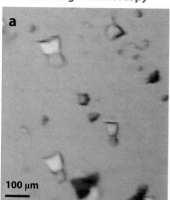
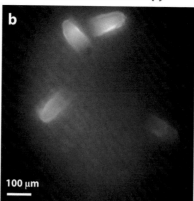
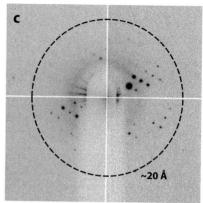

Figure 6

Crystals of ternary cytochrome bc_1/A8–35/detergent complexes. (*a*) Crystals obtained in the presence of A8–35, observed by visible light microscopy. (*b*) Crystals obtained in the presence of an NBD-labeled analog of A8–35, observed by fluorescence microscopy. (*c*) Diffraction pattern (collected at the ESRF facility in Grenoble) of one of the crystals shown in panel *b*. (From D. Charvolin, M. Picard, E.A. Berry and L.-S. Huang, unpublished data). Abbreviation: NBD, 7-nitrobenz-2-oxa-1,3-diazol-4-yl.

Although these experiments were interrupted, they show that crystallizing APol-trapped MPs is indeed feasible. The advent of NAPols opens new, exciting perspectives in this field, because their use would eliminate the electrostatic repulsion that is, presumably, the major handicap in using charged APols such as A8–35.

Electron Microscopy

APol-trapped MPs have been observed by scanning transmission electron microscopy (EM) (90), by transmission EM after negative staining (1, 28, 32, 67 and references therein) and by cryo-EM (1, 28). Because APols are mild surfactants, they may be particularly well suited to single-particle EM studies of fragile MP supercomplexes. Supercomplexes of the mitochondrial respiratory chain are stable and soluble in A8–35 after removal of the digitonin used for solubilization, and present NADH-dehydrogenase and cytochrome c oxidase activity after purification (1). Images of supercomplex B (consisting of one copy of complex I, two of complex III, and one of complex IV) adsorbed onto a continuous carbon support film were recorded under cryo-conditions

(**Figures 7*a,b***). The particles exhibited similar views as in digitonin (80). A 3D volume (**Figure 7*c***) was calculated by the random conical tilt method, revealing many details

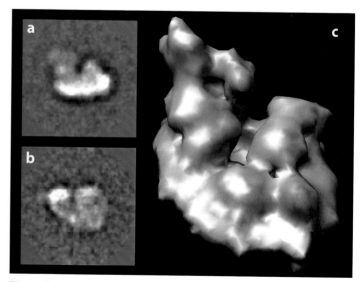

Figure 7

Cryo-electron microscopy (cryo-EM) reconstruction of mitochondrial supercomplex B ($I_1III_2IV_1$) trapped in APol A8–35. (*a,b*) Typical 2D projections from cryo-EM representing side (*a*) and top (*b*) views. (*c*) 3D reconstruction as seen slightly tilted from the matrix side. The red fringe indicates the likely distribution of A8–35. See Reference 1.

not previously visible. In keeping with NMR data, A8–35 seems to form a belt along the transmembrane region of the supercomplex (**Figure 7c**) (1).

Proteomics: 2D Gels and Mass Spectrometry

2D gel electrophoresis is a standard technique in proteomics. It is used, in particular, to study protein expression in cells. Single spots can be picked out from the gels and mass spectrometry performed to determine the mass of each protein and collect sequence information. In the first dimension, proteins are usually separated according to their pI by isoelectrofocusing in a pH gradient. APols that carry a net charge cannot be used for isoelectrofocusing, but NAPols can. The feasibility of this approach has been tested using NAPols whose solubility relies on the presence of either hydroxyl groups (70) or glucose moieties (71). In the second dimension, APols are expected to be displaced by SDS. [APols indeed do not affect SDS-PAGE patterns (59, 66, 90).] In brief, the results show that these two types of NAPols can indeed be used to run 2D gels (5; Y. Gohon, unpublished data) and the spots excised and subjected to mass spectrometry analysis (Y. Gohon, P. Bazzacco, E. Billon-Denis, C. Béchara, G. Bolbach & S. Sagan, unpublished data).

Matrix-assisted laser desorption ionization (MALDI)-time of flight (TOF) measurements have been performed on several APol-trapped MPs, namely OmpX (18), tOmpA, BR, and the cytochrome bc_1 and b_6f complexes (P. Bazzacco, C. Béchara, G. Bolbach & S. Sagan, unpublished data). Neither NAPols nor ionic APols are detected, probably due to their high polydispersity. No significant differences are observed between MP/APols and MP/detergent complexes. Thus, mass spectrometry could be used to check on the extent of isotopic labeling of A8–35-trapped OmpX (18). In preparations of A8–35-trapped and NAPol-trapped cytochrome b_6f, all but two of the eight subunits were detected, albeit with variable yields, as well as free heme and, for NAPol-trapped com-

plexes, some lipids (P. Bazzacco, C. Béchara, G. Bolbach & S. Sagan, unpublished data).

Altogether, these results indicate that A8–35 and NAPols are compatible with MALDI-TOF mass spectrometry analyses of both intact MPs and of their digests, and that NAPols are suitable to perform isoelectrofocusing and 2D gel electrophoresis. The stabilizing properties of APols ought to facilitate analyses of fragile MPs and MP complexes by both of these techniques.

Amphipols as Vectors for the Delivery of Membrane Proteins and Transmembrane Peptides

Upon injection of MP/A8–35 complexes into mice, no toxicity is observed. Antibodies are produced against MPs, not against APols (68). In cell cultures, A8–35 is not cytolytic, at least at the concentrations needed to deliver MPs to the plasma membrane (68). This has led to the examination of APols as vectors for immunization or other therapeutic purposes.

The transmembrane domain of bitopic receptors, which is made up of a single hydrophobic α-helix, plays an important role in their oligomerization (43, 56). Recent evidence indicates that peptides mimicking the transmembrane domain of neuropilin-1 (NRP1), a receptor controlling various biological effects ranging from cell migration to cell proliferation and cell death (76), are able to block its biological functions (77). In addition to helping to dissect signaling mechanisms, these peptides also present a therapeutic relevance in cancer. Preclinical studies have demonstrated that blocking NRP1 using transmembrane peptides in tumor cells or in the endothelial cells establishing the neovascular network supporting tumor growth largely reduces brain tumor expansion and dissemination both in vitro and in vivo (60). However, NRP1 transmembrane peptides are insoluble and prone to aggregation, which makes it difficult to ensure optimal distribution and efficacy in whole organisms. APols keep various hydrophobic peptides soluble (27, 30, 96), suggesting that they could be used as solubilizing and vectorizing agents. An analysis has been undertaken of the biodistribution of APol-trapped transmembrane peptides applied to COS cells (**Figure 8a**). Confocal microscopy showed that the complexes can reach and enter the plasma membrane within minutes. Both peptides and APols are endocytosed after a few hours, leading to a total clearance of the plasma membrane in four days. To examine whether APols could be potentially useful delivery agents in vivo, mice bearing subcutaneous tumors were injected with FAPol$_{NBD}$. Histological examination of tumor slices revealed strong staining of intratumor blood vessels (**Figure 8b**) (G. Crémel & D. Bagnard, unpublished data).

Another potentially promising therapeutic use of APols is the formulation of vaccines. Even though MPs from pathogenic organisms represent a priori privileged targets for the immune system, their production and handling pose formidable obstacles in the way of manufacturing affordable vaccines. The ability of APols to fold overexpressed MPs and to stabilize them may help to circumvent these difficulties. Recent data indicate that vaccines formulated with MOMP/A8–35 complexes protect mice against *C. trachomatis* infection better than vaccines containing MOMP/detergent complexes, to a level approaching that observed upon vaccination with the live organism (L.M. de la Maza, unpublished data). It remains to be established whether this improvement results from the strong stabilization of MOMP by A8–35, from a different presentation of the immunogen to cells of the immune system, or from a combination of both mechanisms. However, given how general MP stabilization by APols is, it is probably worth examining the possibility of extending this approach to MP-based vaccines against other infectious diseases, possibly also against some forms of cancer. The development of therapeutic uses of APols is still at a preliminary stage. However, the above data suggest that APols may provide an interesting alternative to the use of detergents, liposomes, and nanoparticles whenever organisms are to be injected with MPs or transmembrane fragments thereof.

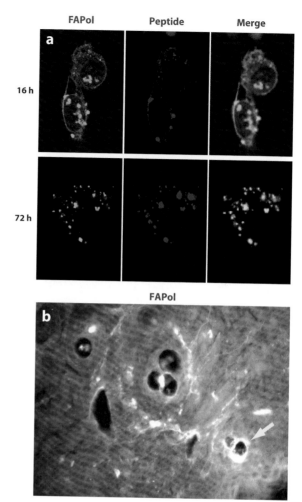

Figure 8

Using amphipols to deliver transmembrane peptides to cells and tissues. (*a*) Confocal microscopic analysis of the distribution of NBD-labeled A8–35 (FAPol$_{NBD}$; *green*) and a rhodamine-labeled neuropilin-1 transmembrane peptide (*red*) 16 and 72 h after application of peptide/FAPol$_{NBD}$ complexes to COS-7 cells. Note that the distribution of the two labels does not totally overlap, indicating at least partial segregation. After 72 h, the plasma membrane has been cleared of both labels. (*b*) Distribution of FAPol$_{NBD}$ in tumor slices. Mice bearing subcutaneous tumors (C6 glioma) were injected intraperitoneally with 100 μL of 10^{-7} M FAPol$_{NBD}$ 24 h before tissue collection. Vascular and perivascular staining demonstrate tissue penetration (*yellow arrow*). Unpublished data from G. Crémel & D. Bagnard. Abbreviation: NBD, 7-nitrobenz-2-oxa-1,3-diazol-4-yl.

CONCLUSION

Much is now known about what can and what cannot be advantageously done by substituting detergents with APols. Some properties of MP/APol complexes, e.g., the dynamics of

APol-trapped MPs, remain to be more fully explored. Many applications have been validated but not fully exploited, whereas the difficulties met by other applications, e.g., 3D crystallization, remain to be solved. Many more applications will appear and develop as novel APols become better characterized (nonionic APols in particular), as the variety of functions grafted onto APols increases, and as more laboratories familiarize themselves with their use. It is our hope that the work invested by so many people over so many years to turn APols into practical, reliable tools will provide the community of MP specialists with enough solid data and incentives so that their use becomes more widespread.

SUMMARY POINTS

1. APols are a special class of amphipathic polymers designed to keep integral MPs water soluble in the absence of detergents.

2. Various types of APols have been tested and validated. They have in common the ability to assemble in aqueous solutions in the form of small, globular, well-defined particles.

3. APols make MPs water soluble by adsorbing onto their hydrophobic transmembrane surface. The APol layer is thin and remains stable even at high dilutions.

4. MP-adsorbed APols can be readily exchanged for detergents, lipids, or other APols.

5. Most MPs are much more stable as MP/APol complexes than they are in detergent solutions. Several mechanisms contribute to this effect.

6. APols, as a rule, do not interfere with ligand binding or with MP function. They may, however, slow down some transconformations or interactions.

7. APols are an excellent medium in which to fold MPs to their functional state, and some APols can be used for MP cell-free synthesis.

8. APols lend themselves to many applications, including structural studies, proteomics, MP immobilization, and delivery.

DISCLOSURE STATEMENT

P. Bazzacco, D. Charvolin, F. Giusti, B. Pucci, C. Tribet, and J.-L. Popot are coauthors of patents on amphipols and their applications.

ACKNOWLEDGMENTS

This review is dedicated to the memory of two colleagues without whom this work would likely not have existed: Charles Tanford (1921–2009), whose role in introducing membrane biologists to the physical chemistry of surfactants can hardly be overstated, and Roland Audebert (1939–1997), colloid and polymer chemist, with whom C.T. and J.-L.P. launched this project in 1994. Particular thanks are due to C. Béchara, E.A. Berry, G. Bolbach, J. Broos, L.-S. Huang, P. Kessler, A. Milon, M. Opačić, F. Rappaport, M. Renault, S. Sagan, M. Tehei, F. Winnik, and G. Zaccaï for permission to cite ongoing work done in collaboration with them and/or for comments on the manuscript, and to J. Barra for her invaluable help for the preparation of the figures.

Affiliations for the coauthors of this manuscript are as follows:

P. Bazzacco, E. Billon-Denis, L.J. Catoire, D. Charvolin, F. Giusti, C. Le Bon, F. Zito, M. Zoonens: Institut de Biologie Physico-Chimique, CNRS/Université Paris-7 UMR 7099, F-75005 Paris, France

T. Althoff, W. Kühlbrandt: Max-Planck-Institut für Biophysik, D-60438 Frankfurt am Main, Germany

D. Bagnard: Université de Strasbourg, INSERM U682, F-67084 Strasbourg, France

J.-L. Banères: Institut des Biomolécules Max Mousseron, CNRS/Universités Montpellier I & II UMR 5247, F-34093 Montpellier, France

P. Champeil: Institut de Biologie et de Technologies de Saclay, CEA/SB2SM and CNRS/URA 2096, F-91191 Gif-sur-Yvette Cedex, France

M.J. Cocco: Department of Molecular Biology and Biochemistry, University of California, Irvine, California 92697-3900

G. Crémel: Centre de Neurochimie, INSERM U682, F-67084 Strasbourg, France

T. Dahmane: Institut de Biologie Physico-Chimique, CNRS/Université Paris-7 UMR 7099, F-75005 Paris, France; Present address: Biochemistry and Molecular Biophysics, Columbia University, New York, New York 10032

L.M. de la Maza: Department of Pathology and Laboratory Medicine Medical Sciences I, University of California, Irvine, California 92697-4800

C. Ebel, F. Gabel: Institut de Biologie Structurale, CEA/CNRS/UJF UMR5075, F-38027 Grenoble, France

Y. Gohon: Present address: Institut de Biologie Physico-Chimique, CNRS/Université Paris-7 UMR 7099, F-75005 Paris, France; Unité de Chimie Biologique, AgroParisTech/INRA, F-78850 Thiverval-Grignon, France

E. Goormaghtigh: Center for Structural Biology and Bioinformatics, Université Libre de Bruxelles, B-10050 Bruxelles, Belgium

E. Guittet, C. van Heijenoort: Institut de Chimie des Substances Naturelles, CNRS UPR 2301, F-91198 Gif-sur-Yvette, France

J.H. Kleinschmidt: Fachbereich Biologie, Universität Konstanz, D-78457 Konstanz, Germany

K.L. Martinez: Department of Neuroscience and Pharmacology & Nanoscience Center, University of Copenhagen, DK-2100 Copenhagen, Denmark

M. Picard: Institut de Biologie Physico-Chimique, CNRS/Université Paris-7 UMR 7099, F-75005 Paris, France; Present address: Laboratoire de Cristallographie et RMN Biologiques, CNRS/Université Paris Descartes UMR 8015, F-75006 Paris

B. Pucci: Laboratoire de Chimie Bioorganique et des Systèmes Moléculaires Vectoriels, Université d'Avignon et des Pays de Vaucluse, F-84000 Avignon, France

J.N. Sachs: Department of Biomedical Engineering, University of Minnesota, Minneapolis, Minnesota 55455

C. Tribet: Département de Chimie, CNRS/ENS/Université Paris-6 UMR 8640, Ecole Normale Supérieure, F-75005 Paris

F. Wien: DISCO, Synchrotron Soleil, F-91192 Gif-sur-Yvette, France

LITERATURE CITED

1. Althoff T. 2011. *Strukturelle Untersuchungen am Superkomplex $I_1III_2IV_1$ der Atmungskette mittels Kryoelektronenmikroskopie*, Dr. Phil. Nat., Johann Wolfgang Goethe-Universität, Fachbereich Biochemie, Chemie und Pharmazie, Frankfurt-am-Main, FRG. 248 pp.
2. Arcemisbéhère L, Sen T, Boudier L, Balestre M-N, Gaibelet G, et al. 2010. Leukotriene BLT2 receptor monomers activate the $G_{(i2)}$ GTP-binding protein more efficiently than dimers. *J. Biol. Chem.* 285:6337–47
3. Banères J-L, Martin A, Hullot P, Girard J-P, Rossi J-C, Parello J. 2003. Structure-based analysis of GPCR function. Conformational adaptation of both agonist and receptor upon leukotriene B_4 binding to recombinant BLT1. *J. Mol. Biol.* 329:801–14

4. Banères J-L, Mesnier D, Martin A, Joubert L, Dumuis A, Bockaert J. 2005. Molecular characterization of a purified 5-HT$_4$ receptor. A structural basis for drug efficacy. *J. Biol. Chem.* 280:20253–60

5. Bazzacco P. 2009. *Non-ionic amphipols: new tools for in vitro studies of membrane proteins. Validation and development of biochemical and biophysical applications.* Thèse de Doctorat, Univ. Paris-VII, 176 pp.

6. Bazzacco P, Sharma KS, Durand G, Giusti F, Ebel C, et al. 2009. Trapping and stabilization of integral membrane proteins by hydrophobically grafted glucose-based telomers. *Biomacromolecules* 10:3317–26

7. Berrier C, Park K-H, Abes S, Bibonne A, Betton J-M, Ghazi A. 2004. Cell-free synthesis of a functional ion channel in the absence of a membrane and in the presence of detergent. *Biochemistry* 43:12585–91

8. Berry EA, Guergova-Kuras M, Huang L-S, Crofts AR. 2000. Structure and function of cytochrome bc_1 complexes. *Annu. Rev. Biochem.* 69:1005–75

9. Bieri C, Ernst OP, Heyse S, Hofmann KP, Vogel H. 1999. Micropatterned immobilization of a G protein-coupled receptor and direct detection of G protein activation. *Nat. Biotechnol.* 17:1105–8

10. Bokoch MP, Zou Y, Rasmussen SG, Liu CW, Nygaard R, et al. 2010. Ligand-specific regulation of the extracellular surface of a G-protein-coupled receptor. *Nature* 463:108–12

11. Breyton C, Gabel F, Abla M, Pierre Y, Lebaupain F, et al. 2009. Micellar and biochemical properties of (hemi)fluorinated surfactants are controlled by the size of the polar head. *Biophys. J.* 97:1077–86

12. Breyton C, Pucci B, Popot J-L. 2010. Amphipols and fluorinated surfactants: two alternatives to detergents for studying membrane proteins in vitro. In *Heterologous Expression of Membrane Proteins: Methods and Protocols*, ed. I Mus-Veteau, pp. 219–45. Totowa, NJ: Humana

13. Breyton C, Tribet C, Olive J, Dubacq J-P, Popot J-L. 1997. Dimer to monomer conversion of the cytochrome b_6f complex: causes and consequences. *J. Biol. Chem.* 272:21892–900

14. Buchanan SK. 1999. Beta-barrel proteins from bacterial outer membranes: structure, function and refolding. *Curr. Opin. Struct. Biol.* 9:455–61

15. Cappuccio JA, Hinz AK, Kuhn EA, Fletcher JE, Arroyo ES, et al. 2009. Cell-free expression for nano-lipoprotein particles: building a high-throughput membrane protein solubility platform. *Methods Mol. Biol.* 498:273–96

16. Catoire LJ, Damian M, Giusti F, Martin A, van Heijenoort C, et al. 2010. Structure of a GPCR ligand in its receptor-bound state: leukotriene B$_4$ adopts a highly constrained conformation when associated to human BLT2. *J. Am. Chem. Soc.* 132:9049–57

17. Catoire LJ, Zoonens M, van Heijenoort C, Giusti F, Guittet E, Popot J-L. 2010. Solution NMR mapping of water-accessible residues in the transmembrane β-barrel of OmpX. *Eur. Biophys. J.* 39:623–30

18. Catoire LJ, Zoonens M, van Heijenoort C, Giusti F, Popot J-L, Guittet E. 2009. Inter- and intramolecular contacts in a membrane protein/surfactant complex observed by heteronuclear dipole-to-dipole cross-relaxation. *J. Magn. Res.* 197:91–95

19. Champeil P, Menguy T, Tribet C, Popot J-L, le Maire M. 2000. Interaction of amphipols with the sarcoplasmic reticulum Ca^{2+}-ATPase. *J. Biol. Chem.* 275:18623–37

20. Changeux J-P, Giraudat J, Heidmann T, Popot J-L, Sobel A. 1980. Functional properties of the acetyl-choline receptor protein. *Neurochem. Int.* 2:219–31

21. **Charvolin D, Perez J-B, Rouvière F, Giusti F, Bazzacco P, et al. 2009. The use of amphipols as universal molecular adapters to immobilize membrane proteins onto solid supports.** *Proc. Natl. Acad. Sci. USA* **106:405–10**

22. Corringer P-J, Baaden M, Bocquet N, Delarue M, Dufresne V, et al. 2010. Atomic structure and dynamics of pentameric ligand-gated ion channels: new insight from bacterial homologues. *J. Physiol.* 588:565–72

23. Dahmane T. 2007. *Protéines membranaires et amphipols: stabilisation, fonction, renaturation, et développement d'amphipols sulfonatés pour la RMN des solutions.* Doctorat d'Université. Univ. Paris-VII. 229 pp.

24. **Dahmane T, Damian M, Mary S, Popot J-L, Banères J-L. 2009. Amphipol-assisted in vitro folding of G protein-coupled receptors.** *Biochemistry* **48:6516–21**

25. Diab C, Tribet C, Gohon Y, Popot J-L, Winnik FM. 2007. Complexation of integral membrane proteins by phosphorylcholine-based amphipols. *Biochim. Biophys. Acta* 1768:2737–47

26. Diab C, Winnik FM, Tribet C. 2007. Enthalpy of interaction and binding isotherms of nonionic surfactants onto micellar amphiphilic polymers (amphipols). *Langmuir* 23:3025–35

27. Duarte AMS, Wolfs CJAM, Koehorsta RBM, Popot J-L, Hemminga MA. 2008. Solubilization of V-ATPase transmembrane peptides by amphipol A8–35. *J. Peptide Chem.* 14:389–93

21. Demonstrates how to use biotinylated APols to immobilize MPs onto chips or beads and study the binding of antibodies and pharmacological ligands.

24. Introduces APols as tools to fold GPCRs to their native 3D structure and stabilize them.

28. Flötenmeyer M, Weiss H, Tribet C, Popot J-L, Leonard K. 2007. The use of amphipathic polymers for cryo-electron microscopy of NADH:ubiquinone oxidoreductase (Complex I). *J. Microsc.* 227:229–35

29. Giuliani A, Jamme F, Rouam V, Wien F, Giorgetta JL, et al. 2009. DISCO: a low-energy multipurpose beamline at synchrotron SOLEIL. *J. Synchrotron Radiat.* 16:835–41

30. Gohon Y. 1996. *Etude des interactions entre un analogue du fragment transmembranaire de la glycophorine A et des polymères amphiphiles: les amphipols.* Diplôme d'Etudes Approfondies de Biophysique Moléculaire. Univ. Paris VI, Paris. 28 pp.

31. Gohon Y. 2002. *Etude structurale et fonctionnelle de deux protéines membranaires, la bactériorhodopsine et le récepteur nicotinique de l'acétylcholine, maintenues en solution aqueuse non détergente par des polymères amphiphiles.* Thèse de Doctorat d'Université. Univ. Paris-VI, Paris. 467 pp.

32. Gohon Y, Dahmane T, Ruigrok R, Schuck P, Charvolin D, et al. 2008. Bacteriorhodopsin/amphipol complexes: structural and functional properties. *Biophys. J.* 94:3523–37

33. Gohon Y, Giusti F, Prata C, Charvolin D, Timmins P, et al. 2006. Well-defined nanoparticles formed by hydrophobic assembly of a short and polydisperse random terpolymer, amphipol A8–35. *Langmuir* 22:1281–90

34. Gohon Y, Pavlov G, Timmins P, Tribet C, Popot J-L, Ebel C. 2004. Partial specific volume and solvent interactions of amphipol A8–35. *Anal. Biochem.* 334:318–34

35. Gohon Y, Popot J-L. 2003. Membrane protein-surfactant complexes. *Curr. Opin. Colloid Interface Sci.* 8:15–22

36. Gorzelle BM, Hoffman AK, Keyes MH, Gray DN, Ray DG, Sanders II CR. 2002. Amphipols can support the activity of a membrane enzyme. *J. Am. Chem. Soc.* 124:11594–95

37. Gorzelle BM, Nagy JK, Oxenoid K, Lonzer WL, Cafiso DS, Sanders CR. 1999. Reconstitutive refolding of diacylglycerol kinase, an integral membrane protein. *Biochemistry* 38:16373–82

38. Gronemeyer T, Godin G, Johnsson K. 2005. Adding value to fusion proteins through covalent labelling. *Curr. Opin. Struct. Biol.* 16:453–58

39. Hilf RJ, Dutzler R. 2009. Structure of a potentially open state of a proton-activated pentameric ligand-gated ion channel. *Nature* 457:115–18

40. Hirai T, Subramaniam S, Lanyi JK. 2009. Structural snapshots of conformational changes in a seven-helix membrane protein: lessons from bacteriorhodopsin. *Curr. Opin. Struct. Biol.* 19:433–39

41. Hong YL, Webb BL, Su H, Mozdy EJ, Fang Y, et al. 2005. Functional GPCR microarrays. *J. Am. Chem. Soc.* 127:15350–51

42. Huang K-S, Bayley H, Liao M-J, London E, Khorana HG. 1981. Refolding of an integral membrane protein. Denaturation, renaturation, and reconstitution of intact bacteriorhodopsin and two proteolytic fragments. *J. Biol. Chem.* 256:3802–9

43. Hubert P, Sawma P, Duneau J-P, Khao J, Henin J, et al. 2010. Single-spanning transmembrane domains in cell growth and cell-cell interactions: more than meets the eye? *Cell Adh. Migr.* 4:313–24

44. Junge F, Haberstock S, Roos C, Stefer S, Proverbio D, et al. 2010. Advances in cell-free protein synthesis for the functional and structural analysis of membrane proteins. *N. Biotechnol.* Epub ahead of print

45. Junge F, Luh LM, Proverbio D, Schäfer B, Abele R, et al. 2010. Modulation of G-protein coupled receptor sample quality by modified cell-free expression protocols: a case study of the human endothelin A receptor. *J. Struct. Biol.* 172:94–106

46. Katzen F, Peterson TC, Kudlicki W. 2009. Membrane protein expression: no cells required. *Trends Biotechnol.* 27:455–60

47. Kiefer H, Krieger J, Olszewski JD, von Heijne G, Prestwich GD, Breer H. 1996. Expression of an olfactory receptor in *Escherichia coli*: purification, reconstitution, and ligand binding. *Biochemistry* 35:16077–84

48. Kigawa T, Yabuki T, Yoshida Y, Tsutsui M, Ito Y, et al. 1999. Cell-free production and stable-isotope labeling of milligram quantities of proteins. *FEBS Lett.* 442:15–19

49. Klammt C, Schwarz D, Fendler K, Haase W, Dötsch V, Bernhard F. 2005. Evaluation of detergents for the soluble expression of alpha-helical and beta-barrel-type integral membrane proteins by a preparative scale individual cell-free expression system. *FEBS J.* 272:6024–38

50. Klammt C, Srivastava A, Eifler N, Junge F, Beyermann M, et al. 2007. Functional analysis of cell-free produced human endothelin B receptor reveals transmembrane segment 1 as an essential area for ET-1 binding and homodimer formation. *FEBS J.* 274:3259–69

32. Provides an extensive study of the structure and properties of BR/APol complexes.

33. Studies the solution properties of APol A8–35.

51. Ladavière C, Toustou M, Gulik-Krzywicki T, Tribet C. 2001. Slow reorganization of small phosphatidyl-choline vesicles upon adsorption of amphiphilic polymers. *J. Colloid Interface Sci.* 241:178–87

52. Lee D, Hilty C, Wider G, Wüthrich K. 2006. Effective rotational correlation times of proteins from NMR relaxation interference. *J. Magn. Reson.* 178:72–76

53. Lee D, Walter KF, Brückner AK, Hilty C, Becker S, Griesinger C. 2008. Bilayer in small bicelles revealed by lipid-protein interactions using NMR spectroscopy. *J. Am. Chem. Soc.* 130:13822–23

54. Martinez KL, Gohon Y, Corringer P-J, Tribet C, Mérola F, et al. 2002. Allosteric transitions of *Torpedo* acetylcholine receptor in lipids, detergent and amphipols: molecular interactions versus physical constraints. *FEBS Lett.* 528:251–56

55. Martinez KL, Meyer BH, Hovius R, Lundstrom K, Vogel H. 2003. Ligand binding to G protein-coupled receptors in tethered cell membranes. *Langmuir* 19:10925–29

56. Matthews EE, Zoonens M, Engelman DM. 2006. Dynamic helix interactions in transmembrane signaling. *Cell* 127:447–50

57. Merino JM, Møller JV, Gutiérrez-Merino C. 1994. Thermal unfolding of monomeric Ca^{2+}, Mg^{2+}-ATPase from sarcoplasmic reticulum of rabbit skeletal muscle. *FEBS Lett.* 343:155–59

58. Miller D, Charalambous K, Rotem D, Schuldiner S, Curnow P, Booth PJ. 2009. In vitro unfolding and refolding of the small multidrug transporter EmrE. *J. Mol. Biol.* 393:815–32

59. Nagy JK, Kuhn Hoffmann A, Keyes MH, Gray DN, Oxenoid K, Sanders CR. 2001. Use of amphipathic polymers to deliver a membrane protein to lipid bilayers. *FEBS Lett.* 501:115–20

60. Nasarre C, Roth M, Jacob L, Roth L, Koncina E, et al. 2010. Peptide-based interference of the trans-membrane domain of neuropilin-1 inhibits glioma growth in vivo. *Oncogene* 29:2381–92

61. Palmgren MG, Nissen P. 2011. P-type ATPases. *Annu. Rev. Biophys.* 40:243–66

62. Park K-H, Berrier C, Lebaupain F, Pucci B, Popot J-L, et al. 2007. Fluorinated and hemifluorinated surfactants as alternatives to detergents for membrane protein cell-free synthesis. *Biochem. J.* 403:183–87

63. Park K-H, Billon-Denis E, Dahmane T, Lebaupain F, Pucci B, et al. 2010. In the cauldron of cell-free synthesis of membrane proteins: playing with new surfactants. *New Biotechnol.* doi:10.1016/j.nbt.2010.08.008

64. Picard M, Dahmane T, Garrigos M, Gauron C, Giusti F, et al. 2006. Protective and inhibitory effects of various types of amphipols on the Ca^{2+}-ATPase from sarcoplasmic reticulum: a comparative study. *Biochemistry* 45:1861–69

65. Picard M, Duval-Terrié C, Dé E, Champeil P. 2004. Stabilization of membranes upon interaction of amphipathic polymers with membrane proteins. *Protein Sci.* 13:3056–58

66. Pocanschi CL, Dahmane T, Gohon Y, Rappaport F, Apell H-J, et al. 2006. Amphipathic polymers: tools to fold integral membrane proteins to their active form. *Biochemistry* 45:13954–61

67. Popot J-L. 2010. Amphipols, nanodiscs, and fluorinated surfactants: three non-conventional approaches to studying membrane proteins in aqueous solutions. *Annu. Rev. Biochem.* 79:737–75

68. Popot J-L, Berry EA, Charvolin D, Creuzenet C, Ebel C, et al. 2003. Amphipols: polymeric surfactants for membrane biology research. *Cell. Mol. Life Sci.* 60:1559–74

69. Popot J-L, Gerchman S-E, Engelman DM. 1987. Refolding of bacteriorhodopsin in lipid bilayers: a thermodynamically controlled two-stage process. *J. Mol. Biol.* 198:655–76

70. Prata C, Giusti F, Gohon Y, Pucci B, Popot J-L, Tribet C. 2001. Non-ionic amphiphilic polymers derived from *tris*(hydroxymethyl)-acrylamidomethane keep membrane proteins soluble and native in the absence of detergent. *Biopolymers* 56:77–84

71. Pucci B, Popot J-L, Sharma KS, Bazzacco P, Durand G, Giusti F. 2009. *Polymères comprenant une majorité de monomères amphiphiles destinés au piégeage et à la manipulation de protéines membranaires.* Patent FR 09 58072

72. Rajesh S, Knowles TJ, Overduin M. 2010. Production of membrane proteins without cells or detergents. *New Biotechnol.* doi:10.1016/j.nbt.2010.07.011

73. Raschle T, Hiller S, Etzkorn M, Wagner G. 2010. Nonmicellar systems for solution NMR spectroscopy of membrane proteins. *Curr. Opin. Struct. Biol.* 20:1–9

74. Reichardt G, Noll T, Packe I, Rotter P, Schmidt JS, Gudat W. 2001. Adaption of the BESSY I-3 m normal incidence monochromators to the BESSY II source. *Nucl. Instrum. Methods Phys.* 467:458–61

66. Demonstrates that APols can be used to refold denatured MPs.

67. Compares the applications of amphipols, nanodiscs, and fluorinated surfactants to MP studies.

75. Reinsberg D, Booth PJ, Jegerschöld C, Khoo BJ, Paulsen H. 2000. Folding, assembly, and stability of the major light-harvesting complex of higher plants, LHCII, in the presence of native lipids. *Biochemistry* 39:14305–13

76. Roth L, Koncina E, Satkauskas S, Crémel G, Aunis D, Bagnard D. 2009. The many faces of semaphorins: from development to pathology. *Cell. Mol. Life Sci.* 66:649–66

77. Roth L, Nasarre C, Dirrig-Grosch S, Aunis D, Crémel G, et al. 2008. Transmembrane domain interactions control biological functions of neuropilin-1. *Mol. Biol. Cell* 19:646–54

78. Salamon Z, Tollin G, Alves I, Hruby V. 2009. Plasmon resonance methods in membrane protein biology: applications to GPCR signaling. *Methods Enzymol.* 461:123–46

79. Sanders CR, Hoffmann AK, Gray DN, Keyes MH, Ellis CD. 2004. French swimwear for membrane proteins. *ChemBioChem.* 5:423–26

80. Schäfer E, Dencher NA, Vonck J, Parcej DN. 2007. Three-dimensional structure of the respiratory chain supercomplex $I_1 III_2 IV_1$ from bovine heart mitochondria. *Biochemistry* 46:12579–85

81. Schwarz D, Dötsch V, Bernhard F. 2008. Production of membrane protein using cell-free expression system. *Proteomics* 8:3933–46

82. Sebai S, Cribier S, Karimi A, Massotte D, Tribet T. 2010. Permeabilisation of lipid membranes and cells by a light-responsive copolymer. *Langmuir* 26:14135–41

83. Sharma KS, Durand G, Giusti F, Olivier B, Fabiano A-S, et al. 2008. Glucose-based amphiphilic telomers designed to keep membrane proteins soluble in aqueous solutions: synthesis and physicochemical characterization. *Langmuir* 24:13581–90

84. Spirin AS, Baranov VI, Ryabova LA, Ovodov SY, Alakhov YB. 1988. A continuous cell-free translation system capable of producing polypeptides in high yield. *Science* 242:1162–64

85. Terpe K. 2003. Overview of tag protein fusions: from molecular and biochemical fundamentals to commercial systems. *Appl. Microbiol. Biotechnol.* 60:523–33

86. Toyoshima C, Nakasako M, Nomura H, Ogawa H. 2000. Crystal structure of the calcium pump of sarcoplasmic reticulum at 2.6 Å resolution. *Nature* 405:647–55

87. **Tribet C, Audebert R, Popot J-L. 1996. Amphipols: polymers that keep membrane proteins soluble in aqueous solutions. *Proc. Natl. Acad. Sci. USA* 93:15047–50**

88. Tribet C, Audebert R, Popot J-L. 1997. Stabilisation of hydrophobic colloidal dispersions in water with amphiphilic polymers: application to integral membrane proteins. *Langmuir* 13:5570–76

89. **Tribet C, Diab C, Dahmane T, Zoonens M, Popot J-L, Winnik FM. 2009. Thermodynamic characterization of the exchange of detergents and amphipols at the surfaces of integral membrane proteins. *Langmuir* 25:12623–34**

90. Tribet C, Mills D, Haider M, Popot J-L. 1998. Scanning transmission electron microscopy study of the molecular mass of amphipol/cytochrome $b_6 f$ complexes. *Biochimie* 80:475–82

91. Tummino PJ, Copeland RA. 2008. Residence time of receptor-ligand complexes and its effect on biological function. *Biochemistry* 47:5481–92

92. Vial F, Cousin F, Bouteiller L, Tribet C. 2009. Rate of permeabilization of giant vesicles by amphiphilic polyacrylates compared to the adsorption of these polymers onto large vesicles and tethered lipid bilayers. *Langmuir* 25:7506–13

93. Vial F, Oukhaled AG, Auvray L, Tribet C. 2007. Long-living channels of well defined radius opened in lipid bilayers by polydisperse, hydrophobically-modified polyacrylic acids. *Soft Matter* 3:75–78

94. Vial F, Rabhi S, Tribet C. 2005. Association of octyl-modified poly(acrylic acid) onto unilamellar vesicles of lipids and kinetics of vesicle disruption. *Langmuir* 21:853–62

95. Weinrich D, Jonkheijm P, Niemeyer CM, Waldmann H. 2009. Applications of protein biochips in biomedical and biotechnological research. *Angew. Chem. Int. Ed.* 48:7744–51

96. Wolff N, Delepierre M. 1997. Conformation of the C-terminal secretion signal of the *Serratia marcescens* haem acquisition protein (HasA) in amphipols solution, a new class of surfactant. *J. Chim. Phys.* 95:437–42

97. Yu XB, Xu DK, Cheng Q. 2006. Label-free detection methods for protein microarrays. *Proteomics* 6:5493–503

87. Validates the amphipol concept by using A8–35 and three of its congeners varying in length and charge density.

89. Provides a thermodynamic investigation of the interactions of detergents and APols with the TM surface of MPs.

98. Zoonens M. 2004. *Caractérisation des complexes formés entre le domaine transmembranaire de la protéine OmpA et des polymères amphiphiles, les amphipols. Application à l'étude structurale des protéines membranaires par RMN à haute résolution.* Doctorat d'Université. Univ. Paris-VI. 233 pp.

99. **Zoonens M, Catoire LJ, Giusti F, Popot J-L. 2005. NMR study of a membrane protein in detergent-free aqueous solution.** *Proc. Natl. Acad. Sci. USA* **102:8893–98**

100. **Zoonens M, Giusti F, Zito F, Popot J-L. 2007. Dynamics of membrane protein/amphipol association studied by Förster resonance energy transfer. Implications for in vitro studies of amphipol-stabilized membrane proteins.** *Biochemistry* **46:10392–404**

101. Zubay G. 1973. In vitro synthesis of protein in microbial systems. *Annu. Rev. Genet.* 7:267–87

RELATED RESOURCES

http://www.ibpc.fr/popot/amphipol/

Cumulative Indexes

Contributing Authors, Volumes 36–40

Rivas G, 37:375–97
Ruda VM, 38:173–96

S

Sachs JN, 40:379–408
Sapra KT, 36:233–60
Scheraga HA, 40:1–39
Schulman BA, 36:131–50
Schulten K, 40:187–203
Schwille P, 36:151–69; 40:315–36
Selvin PR, 36:349–69
Semsey S, 39:43–59
Shell MS, 37:289–316
Shinzawa-Itoh K, 40:205–23
Shivashankar GV, 40:361–78
Singer RH, 38:173–96
Smith DE, 39:349–66
Smith SO, 39:309–28
Sneppen K, 39:43–59
Sokurenko E, 37:399–416
Soppina V, 40:267–88
Spector DL, 39:471–89
Stenmark H, 40:119–42
Stewart PL, 40:143–67
Strümpfer J, 40:187–203
Stuart D, 38:371–83
Suh B-C, 37:175–95
Sussman MR, 39:291–308
Swaney KF, 39:265–89
Swedlow JR, 38:327–46
Sweeney HL, 39:539–57
Sydow J, 37:337–52
Sykes MT, 38:197–215

Szobota S, 39:329–48
Szostak JW, 39:245–63

T

Tanaka S, 39:185–205
Terwilliger TC, 38:371–83
Thirumalai D, 39:159–83
Thomas DD, 38:347–69
Thomas WE, 37:399–416
Tijan R, 38:173–96
Toprak E, 36:349–69
Torres AJ, 37:265–88
Tribet C, 40:379–408

U

Udgaonkar JB, 37:489–510
Uversky VN, 37:215–46

V

van Heijenoort C, 40:379–408
Vannini A, 37:337–52
van Oijen AM, 39:429–48
van Oudenaarden A, 38:255–70
Verhey KJ, 40:267–88
Verkman AS, 37:247–63
Vogel V, 37:399–416
von Ballmoos C, 37:43–64
von Heijne G, 37:23–42
von Hippel PH, 36:79–105

W

Walz T, 38:89–105
Ward LD, 36:329–47
Weber PK, 38:53–74
Weikl TR, 37:289–316
White SH, 37:23–42
Wien F, 40:379–408
Williamson JR, 38:197–215
Wirtz D, 38:301–26
Woodside MT, 36:171–90
Woodson SA, 39:61–77
Wu M, 37:265–88

Y

Yang L-W, 39:23–42
Yang W, 39:367–85
Yao J, 38:173–96
Yeates TO, 39:185–205
Yokoyama S, 38:371–83
Yoshikawa S, 40:205–23

X

Xie XS, 37:417–44

Z

Zenklusen D, 38:173–96
Zhou H-X, 36:21–42; 37:375–97
Zito F, 40:379–408
Zocchi G, 38:75–88
Zoonens M, 40:379–408
Zuckerman DM, 40:41–62

subs
7/11